Earthquakes and Tsunamis in the Past

A Guide to Techniques in Historical Seismology

This handbook defines the discipline of historical seismology by comprehensively detailing the latest research methodologies for studying historical earthquakes and tsunamis. It describes the many historical sources that contain references to seismic phenomena, discusses the critical problems of interpreting such sources, and presents a summary of the various theories proposed (from ancient Greek to modern times) to explain the causes of earthquakes – indispensable factors for understanding historical earthquake descriptions.

The text presents numerous examples of interpretations and misinterpretations of historical earthquakes and tsunamis in order to illustrate the key techniques, with a chapter devoted to an explanation of the date and time systems used throughout history in Mediterranean Europe and the Near-East. The authors also tie historical seismology research to archaeological investigations and demonstrate how new scientific databases and catalogues can be compiled from information derived from the methodologies described.

This is an important new reference for scientists, engineers, historians and archaeologists on the methodologies for analysing earthquakes and tsunamis of the past. Illustrated with examples from a broad geographic region (including Europe, North Africa, the Middle East, central Asia and the Americas), the book provides a valuable foundation for understanding the Earth's seismic past and potential future seismic hazard.

EMANUELA GUIDOBONI, a historian by training, is a Senior Scientist and Head of the Historical Seismology and Volcanology Unit at the Instituto Nazionale di Geofisica e Vulcanologia, Bologna, Italy. From 1983 to 2007, she served as Director of Geofisica Ambiente (SGA), Bologna, a company specializing in the study of earthquakes and other historical environmental phenomena. Dr Guidoboni is a leading expert in the historical seismicity of the Mediterranean region, and the author of a number of important historical earthquake catalogues and more than 100 scientific publications.

JOHN E. EBEL is a Professor in the Department of Geology and Geophysics and Director of the Weston Observatory at Boston College, Massachusetts, where his research interests include theoretical, exploration and earthquake seismology. Professor Ebel was awarded the 2003 Jesuit Seismological Award and the 2004 Service to the Seismological Society of America Award.

Earthquakes and Tsunamis in the Past

A Guide to Techniques in Historical Seismology

EMANUELA GUIDOBONI

Istituto Nazionale di Geofisica e Vulcanologia, Bologna

JOHN E. EBEL

Boston College, Massachusetts

CAMBRIDGE
UNIVERSITY PRESS

University Printing House, Cambridge CB2 8BS, United Kingdom

One Liberty Plaza, 20th Floor, New York, NY 10006, USA

477 Williamstown Road, Port Melbourne, VIC 3207, Australia

314-321, 3rd Floor, Plot 3, Splendor Forum, Jasola District Centre, New Delhi - 110025, India

79 Anson Road, #06-04/06, Singapore 079906

Cambridge University Press is part of the University of Cambridge.

It furthers the University's mission by disseminating knowledge in the pursuit of education, learning and research at the highest international levels of excellence.

www.cambridge.org
Information on this title: www.cambridge.org/9781108462051

© E. Guidoboni and J. E. Ebel 2009

First published 2009
First paperback edition 2018

A catalogue record for this publication is available from the British Library

Library of Congress Cataloging in Publication data
Guidoboni, Emanuela.
Earthquake and tsunamis in the past : a guide on problems and methods of historical seismology / Emanuela Guidoboni and John E. Ebel.
    p. cm.
Includes bibliographical references and index.
ISBN 978-0-521-83795-8 (hardback)
1. Paleoseismology. 2. Earthquakes – History. 3. Tsunamis – History.
4. Earthquakes – History – Sources. 5. Tsunamis – History – Sources.
I. Ebel, J. E. (John E.) II. Title.
QE539.2.P34G85 2009
551.2209 – dc22   2008047549

ISBN 978-0-521-83795-8 Hardback
ISBN 978-1-108-46205-1 Paperback

Contents

Preface

In one sense, the person most responsible for this book is our colleague Jelle De Boer, Professor of Geology at Wesleyan University in Middletown, Connecticut. In October 2000 Emanuela Guidoboni made a trip to the United States and one of her stops was to visit Jelle at Wesleyan. Being a good host, Jelle wanted to introduce Emanuela to others in the New England region who were interested in historical seismology (at Yale University and Boston College). In the past, Jelle had worked with John Ebel at Boston College on the past earthquake activity at Moodus, Connecticut, and he was very familiar with John's work on the historical earthquake activity in northeastern North America. Thus, he arranged with John to have Emanuela visit Boston College and give a talk on her work in historical seismology. It was from this meeting that the collaboration of a historian of seismicity from Bologna, Italy, and a seismologist from Boston, Massachusetts began.

The idea for this book came from the mutual awareness that although in many countries of the world historical research into important earthquakes of the past has been performed and is still in progress, there was no handbook on how actually to carry out historical seismology research, with successful strategies and results highlighted and problems, pitfalls and mistakes specified. Such a guide could be a handy reference for professional researchers in many different countries, while students and amateur investigators who were interested in dealing with data on past seismicity could learn from such a text. A similar idea, but on a more limited scale, had been the topic of some work that Emanuela had conducted years before at the request of the International Atomic Energy Authority (particularly, Aybars Gürpinar), on behalf of whom she had studied the strong earthquakes of Armenia. Some sketchy 'guidelines' concerning how to carry out research in historical seismology were first drafted in Vienna in December 1994 by Emanuela and some colleagues (for the seismological part, by Agnés Levret and Claudio Margottini). Although some pages had already been written, for a number of years this early effort remained an idea to be expanded upon and brought to fruition. In the meantime, the methodological approach to historical seismology had become much better defined and accepted thanks to the research experiences of a number of investigators. Historical seismology was indeed emerging as a neo-discipline of its own.

Following her visit to Boston College, Emanuela approached John to work with her on her dormant idea of a handbook on historical seismology. John accepted her invitation to play an active role in this project. Because of their importance in the Mediterranean world, we decided to include historical tsunamis along with historical earthquakes as major topics in the book, and we defined historical seismology broadly enough to include archaeoseismology and the seismic effects on monuments. To keep the book at a manageable size and to maintain coherence in the presentation, we decided to omit some topics that are somewhat more peripheral to historical seismology. For example, historical volcanic eruptions and historical earthquakes associated with active volcanoes are important research topics that we have decided not to include directly in our presentation. Even so, many of the ideas that we present concerning research into historical earthquakes and tsunamis are quite pertinent for research into the historical traces of other natural hazard phenomena. Once we had converged on a scope and outline for our book, we convinced Cambridge University Press to be our publisher and set to work on the writing.

From the outset, it was our goal to write a book that would be a useful reference both to those seismologists and earthquake engineers who carry out research into historical earthquakes and to historians and archaeologists who want or need to know about past earthquakes and their consequences for the affected populations and their buildings. Hence we wrote the book as a kind of tutorial with these widely diverse audiences in mind. We have chosen to include many examples, both in pictures and in words, of the many details and subtleties that make accurate historical seismology research and the proper interpretation of seismological parameters from historical seismological sources such a challenge.

While efforts to compile information on historical earthquakes have been undertaken for several hundred years, in many ways historical seismology is still a comparatively young research discipline. It has only been in recent times that historians have brought to bear their full and significant interpretive tools on those historical sources that describe the earthquakes and tsunamis of the distant past. And for those seismologists who have been studying historical seismicity with an eye toward better defining the seismic hazard of different parts of the world, new analytical tools that give a modern understanding to past historical earthquakes have only been developed over the last decade or so. It is our sincere hope that this book will stimulate new research into historical earthquakes and will lead to the development of new seismological methods for interpreting the data that accrue from that new research.

Both of us owe our thanks to many people for their assistance in the production of this book: Alberto Comastri, for his assiduous, competent and invaluable support, Maria Giovanna Bianchi and Gabriele Tarabusi of SGA, for their help in preparing the figures and maps. Thanks also to Jean-Paul Poirier and Gianluca Valensise for their corrections and suggestions, and to Enzo Boschi, President of the Istituto Nazionale di Geofisica e Vulcanologia, who has supported the historical research into earthquakes in Italy with great foresight, and for his encouragement in writing this book. Dina Smith at Weston Observatory of Boston College provided a thorough proofreading of the book. Susan Francis at Cambridge University Press was of great help to us, and was extremely patient and encouraging when we were tardy meeting our deadlines. Finally, John wishes to thank his wife Martha, whose constant love and support during many evenings and weekends of writing and revising gave him the strength to carry on. Emanuela is grateful to her three wonderful grandchildren, Emmanuel, Luis and Lorenzo, who with their voices and games provided a pleasant background to this book over many a weekend.

The authors apologize for often having resorted to case studies they had themselves analysed or studies pertaining to research they themselves or their work group had performed, which have provided most of the discussion material.

The authors devised and discussed all of the chapters together and jointly reviewed them, commented upon them and at times added to them. However, as a result of their different scientific backgrounds, the drafting of the chapters was subdivided as follows:

> Chapters 1, 2: Emanuela Guidoboni and John E. Ebel
> Chapters 3, 5, 6, 7, 8, 9, 10 and 11: Emanuela Guidoboni
> Chapters 12 and 13: John E. Ebel
> Chapter 4: in Section 4.1 Emanuela wrote the part on the ancient
> world up to the eighteenth century, John the subsequent part;
> Emanuela wrote Section 4.2.

PART I DEFINING HISTORICAL SEISMOLOGY

1

What is historical seismology?

1.1 The interest in historical earthquakes and tsunamis

Mankind's interest in historical earthquakes and tsunamis arises from an innate human curiosity about history, both human history as well as that of the natural world. In highly concrete terms, these two strands of historical investigation and discovery are closely interrelated. The evolution of human history cannot be properly understood by society without appreciating the occurrences of the natural events of the past. At the same time, scientists seeking to achieve a better understanding of the natural world often rely heavily upon data from the historical records of past natural events when they put forward theories and models intended to predict the natural world's future evolution. Fundamentally, understanding human history means understanding natural science, whilst understanding natural science requires information contained in the documents recording human and natural history.

The interplay between developments in the civil societies of the past and of the forces and effects of the natural world is well known and has been extensively studied. This is especially true for climate and weather phenomena as well as the secondary effects caused by weather events (such as flooding, landslides, forest fires and plagues). The effects of droughts, storms and floods are well chronicled throughout recorded history, as they are phenomena that have affected the daily lives and even the existence of human populations throughout the world. The floods of the Nile, Tigris and Euphrates rivers have enabled great early civilizations to develop in the otherwise arid climates of North Africa and the Middle East. Throughout history the outcomes of many major wars were decided, apart from by the level of technology available to either side, by changes in the weather that favoured one army over the other. Epidemics and plagues, also well documented and widely studied, are natural biological menaces that have swept through human societies countless times in the past, causing much suffering and death. Invariably, these biological phenomena are also natural environmental responses to changes in the Earth's climactic, meteorological and hydrological systems.

One set of natural phenomena that on occasion have strongly affected human societies in the past but have been much less extensively studied by historians is that of earthquakes and tsunamis. Unlike weather events, which occur on a daily basis, earthquakes and tsunamis are much rarer occurrences, particularly major earthquakes and tsunamis that cause serious and widespread damage affecting human populations. Furthermore, earthquakes and tsunamis occur suddenly and without warning. In most parts of the world, many generations of people can live their lives without ever having experienced a major, damaging earthquake or tsunami, and many years may elapse between even mild events. Among people across the globe, there is much less understanding about the causes and effects of earthquakes and tsunamis than about much more frequent natural phenomena like storms and droughts.

Although thankfully rare, the impacts of some earthquakes and tsunamis on human societies have been great. In modern times, major earthquakes have had devastating consequences upon some major cities, such as Tokyo in 1923 and San Francisco in 1906, and some tsunamis have caused widespread damage and loss of life as did the 2004 tsunami in the Indian Ocean. Of course, the destruction of human structures and the disruption caused to human lives by recent major earthquakes and tsunamis have been documented in meticulous detail by modern scientific investigators. Also documented, although with less scientific precision and detail, have been the damage and destruction inflicted upon cities affected by strong earthquakes in earlier historic times, such as the 1755 earthquake and tsunami at Lisbon, Portugal, the 1356 earthquake at Basel, Switzerland and the 1117 earthquake at Verona in northern Italy. The same is true of large tsunamis that have caused large-scale destruction and many deaths along sea-coast areas. Thousands have been reported dead in some tsunamis, such as those in Japan in 1605 and in the Straits of Messina in 1783.

In general, the earlier in human history that a strong earthquake or tsunami took place, the less that is known about the event. This problem partly arises as a result of the inevitable deterioration and loss of historical records with time. However, it is also due to fact the ancient earthquake witnesses observed and interpreted the natural phenomena with the knowledge of their times. For seismic events in earlier historical times, descriptions of casualties, damage to structures, effects on water supplies, post-earthquake fires and so on, are very disparate in terms of quality and quantity. The chroniclers often did not know which effects to ascribe to the earthquake or tsunami. For example, the collapse of a building could easily be attributed to the sudden, violent shaking of the earth, but a writer may not have realized that the earthquake could also have caused groundwater springs to stop flowing. Compounding this problem throughout most of history was the difficulty in obtaining information (especially accurate and reliable information) from distant places that were also affected by the same event. In addition, elements that we might call a colouring of the descriptions of earthquakes and tsunamis arose from the mindsets of many authors who believed that these phenomena were supernatural events imparted for a special reason upon human society by some god or other supernatural cause. In these cases, the descriptions are not literal records of the natural event and its consequences but rather are evidence to support a religious argument. In such cases it is highly likely that the evidence cited to support the hypothesis has been carefully selected and perhaps even altered to make the case put forward by the author even stronger. As we describe in more detail later on in this book, all historical earthquake descriptions must be evaluated along with the context in which they are presented.

The need to better understand the occurrences and effects of earthquakes and tsunamis throughout historical times is one that affects both the scientists who investigate the causes and effects of earthquakes and the historians and social scientists who study the historical events of past societies. Scientists have learned that continued earthquake activity is a natural part of the Earth's plate-tectonic processes and that strong and damaging earthquakes and tsunamis are an unavoidable occurrence. The ever-growing human population on the Earth and its increasing concentration in urban centres make many societies more and more vulnerable to future earthquake-induced and tsunami-driven catastrophes. That is why research into earthquake causes, earthquake hazard and earthquake prediction involves many different crucially important disciplines in countries throughout the world. The search for earthquake causes, the identification of active earthquake faults and research into ways of predicting future earthquakes all rely on a database of earthquake information that is both spatially and temporally extensive. On any single fault, strong earthquakes have average repeat times of centuries to millennia or even longer. Seismically active countries may only experience a major earthquake catastrophe on average once every few decades. Thus, an ideal research database of earthquake information for a region would extend throughout historical time and as long ago into prehistoric times as can be reliably determined.

For the historian, a better knowledge of past earthquake and tsunami history in seismically active parts of the world is vital for understanding all of the forces that have shaped human civilization and determined its evolution. The historian might be rather surprised to see how certain societies and cultures have been deeply marked over time by their cohabitation with earthquakes. In historical research, there are numerous references to earthquakes and tsunamis, but the impacts of these events on the local human populations are often not adequately analyzed. Moreover, historians may not fully understand or appreciate the long-term impact that major earthquakes or tsunamis have had on some societies.

This book is addressed to two disparate and yet inevitably interlinked audiences: the scientist who needs to know about historical earthquakes and tsunamis and their effects, and the historian who wants to understand how earthquakes and tsunamis may have affected the societies that they study. For the scientist, this book is intended as a reference that describes the many complexities one faces when interpreting historical earthquake and tsunami reports as observations of past natural phenomena. It is crucial that the scientist is able to assess the accuracy and uncertainty of historical observations, since scientific analyses of the observations are used to relate the past events to the present and possible future earthquakes and tsunamis. It is important that a scientist has the most precise perception possible of the problems concerning the historical data used for the hazard estimations.

For the historian, this book provides a summary description of earthquake and tsunami phenomena, and it is a guide summarizing how these natural events have been described in the historical literature. It is primarily the historian's task to take hold of the historical reports and separate the facts from the interpretation and amplification, the wheat from the chaff. This demands an understanding of the natural phenomena as well as the historical and cultural context in which those phenomena are depicted. This book should be a useful reference for historians working closely with scientists as well as for those historians engaged in individual research in which earthquake or tsunami issues arise.

1.2   The historical approach to seismology

Modern seismology research is not confined simply to instrumental monitoring and physical–mathematical elaboration. Another vital part of the world of seismology research consists of the accumulation and interpretation of qualitative data, that is, descriptions of the effects of the earthquakes that occurred in recent times as well as in the distant past. *Historical seismology* is the branch of seismology that uses historical data in order to assess long-term seismic activity. It strives toward this end by seeking knowledge of earthquake effects that took place in the past, based on historical data. As a discipline, historical seismology stands at the recent intersection of historical research with the questions and issues specific to modern earthquake seismology. This means that the working methods and hermeneutic rules of historical seismology are derived from historical disciplines, while its aims and the questions it asks usually come from the science of seismology.

The convergence of these two disciplines, which are traditionally so far apart in terms of research instruments and aims, came about from the need to enlarge the temporal frame for which seismic activity is observed. For while it is true that instrumental data have been available for more than a century, only in the last 30 or 40 years have the instrumental data become sufficiently homogeneous and suitable for truly self-consistent analyses, and that is not a sufficient span of time for establishing the characteristics of seismic hazard and source areas that can only be delineated on the basis of long-term data. Especially in the case of earthquakes, a great many working groups and individual researchers recently have produced a strikingly large and valuable mass of historical earthquake data (not always easily accessible, however), which is helping to improve hazard and risk estimates in many parts of the world. A useful survey of these studies and of the different research levels and methods adopted is to be found in Albini et al. (2004).

The historical approach to seismology is a research field that has established its specific role in the last 30 years, even though it has its roots in work carried out by masterful scientific pioneers before the twentieth century who made many discoveries and encountered many obstacles in their efforts to accumulate information on historical earthquakes. During the past few decades research into historical seismology by scholars such as Nicholas Ambraseys, Pierre Gouin, S. J. (1917–2005) and Jean Vogt (1929–2005) has helped to greatly advance this neo-discipline, and their work is now enriching both historical and seismological studies. Historical seismology has proven to be most successful when one is dealing with an ancient written culture and a substantial number of available documents, but also it is of great importance even for regions where few documents have been preserved or have been available for the last two or three centuries. Even in the cases of earthquakes for which instrumental data are available, oral and written reports of non-scientific observations, the essence of historical seismology, usually are necessary to document otherwise unmeasured effects in inhabited areas. Indeed, instrumental surveillance invariably is of insufficient spatial density to document the complete scope and variations of the surface effects of earthquake shaking or tsunami inundation, and human observation is always required to accumulate some types of macroseismic data. Historical seismology is based on the qualitative data from human observers.

The current state of research into historical tsunamis remains at a very immature stage relative to that of historical earthquakes, partly because the tradition of tsunami studies is not directly linked to that of historical research and partly because the characteristics of the phenomenon concerned involve a number of special problems. Relatively few catalogues of historical tsunamis have been compiled, and those that exist suffer from many uncertainties that arise from the variable nature of tsunamis as well as from the ambiguities of historical records of tsunamis.

From an epistemological point of view, historical seismology might be defined as earthquake semiology in that it reconstructs the details of the effects of seismic events on the basis of signs or traces they have left behind. A number of other scientific disciplines, from medicine to psychology along with some sectors of geology, also adopt this evidential method of enquiry. The qualitative nature of many kinds of earthquake effects does not mean that the results obtained from their analysis are considered less scientific than direct instrumental observations. Indeed, the scientific validity of data does not depend so much their nature but rather on the methods used to analyse the data and on the interpretation of the results of the analysis. Repeatability of the decision-making processes and general coherence of the results with the levels of knowledge shared by a scientific community – in the case of historical seismology, on that of historians and seismologists – determines the value of historical seismological research.

1.3   Some key ideas in historical seismology

Historical seismology is a *multidisciplinary research* field in which the decision processes are *repeatable* and the analyses of earthquake and tsunami reports take due account of *regional history*, which means the human and historical context which colours the description of the seismic event. We summarize here some of the important details of each of these key ideas in historical seismology.

1.3.1   *Multidisciplinary research*

What historical seismology does in practice is to translate seismological questions into historiographical questions, find answers, and then translate the results back into seismological terms. However, this is no straightforward matter. Different cognitive models, such as those used in science or in the humanities, obviously have their own different cognitive paradigms. Seismologists have started to accept the fact that the human beings who have left written traces of earthquakes can be viewed as special 'seismographs'. They may be imperfect, subject to outside influence and sometimes very unreliable, but nevertheless they are irreplaceable. Furthermore, it must be accepted that the decoding rules for these particular traces, as well as their contextualization, fall within the specific research discipline of historians. On the other hand, historians need to begin querying historical sources from new perspectives and with new aims in mind. In other words, they should adopt a different attitude from that of traditional historiographical research by learning also to work out the answers to questions that come from scientific interests and needs.

Collecting and using historical data for seismological purposes calls for great care and a systematic approach, as well as transparent methods and a determination to check everything that is endemic to scientific research. Today it can be stated with confidence that historical seismology is no longer an ancillary discipline, a sort of 'sub-seismology' in which specific responsibilities and specializations were never made clear. The present profile of historical seismology is partly due to the importance it has attributed not only to a meticulous hermeneutics of the sources and the philological analysis of the texts, but also to the economic and social *contexts* of earthquakes. The set of these elements outlines the methodological heart of the historical seismology and makes possible a correct assessment of seismic effects in an inhabited environment.

1.3.2   *Repeatability of decision processes*

A second indispensable characteristic of historical seismology is the transparent nature of the various phases of analysis and thus of the results. The use of intensity scales (see Chapter 12) always involves a certain level of uncertainty and some subjective elements of interpretation. This is true for all historical reports, both those far removed in time and those closer to modern time. Although it is obviously easier to interpret the news from newspapers of a century ago than a text written on a medieval parchment, the conceptual problem remains the same: one needs to use an intermediary to know the effects on the inhabited world that cannot be seen directly. Thus, it is crucial that the assessment of seismic intensity, from which many other seismological parameters are calculated, should derive from a series of repeatable decisions, both in the descriptive elements of the intensity scale itself and in the reasons underlying choices of those descriptive elements. It is this repeatability of methodology that gives historical seismology its scientific value and makes it possible not only to carry out research and interpretation, but also to organize the various work-stages all the way to the final results.

In historical seismology, it is necessary to lower to a minimum the part of subjectivity inherent in the assignment of seismological parameters like seismic intensity. It is necessary to make available the basic historical data, the research and criteria used to accumulate those data and the analyses performed using the data. Also, extra-textual factors, typically historical as we shall see, can constrain the qualitative data and make them less vague and uncertain. In the last few years computational linguistic approaches (that is, automatic selection and classification) have also been tried in order to elaborate texts using a 'fuzzy' approach (Vannucci *et al.*, 1999b). However, this is only possible for events that are very well documented from a single type of source, such as newspapers or macroseismic questionnaires.

1.3.3   *Seismic effects in regional history*

A third key idea in historical seismology is that of situating seismic effects within their human and historical contexts. This makes it possible not only to differentiate more realistic effects from those that were exaggerated or did not take place, but also to explain information *gaps* by distinguishing two different reasons why some areas may be informationally mute regarding earthquake activity in the historic past (i.e. the silence of quiescent faults and the silence due either to a lack of research or to a lack of historical sources). Differentiating between these two possibilities requires a thorough knowledge of the history of the areas under examination on top of the already available documentary evidence.

In the past it was very often the case that information about historical earthquake effects was gathered and interpreted quite independently of the history of the region to which it refers. This prevented an assessment of the informational value of the historical reports in relation both to the total available documentary evidence and to the particular environment involved. Attention to common elements characterizing the context to which the historical sources of information belong can make a new contribution to historical seismology and, as mentioned above, also help seismological researchers to assess whether an earthquake catalogue is more or less complete with respect to given classes of events.

2

The importance of historical earthquake and tsunami data

2.1 The scientific understanding of earthquakes and tsunamis

It is important for both scientists and non-scientists investigating historical earthquakes and tsunamis to have a common understanding of these natural phenomena. For this reason, included here is a brief summary of what earthquakes and tsunamis are, what causes them, how often they occur and what their primary effects are.

2.1.1 Earthquakes

The term *earthquake* actually refers to a combination of two phenomena. An earthquake is a sudden fracturing or cracking of rock in the earth, where the rock crack (called a *fault*) releases vibrational waves (called *seismic waves*) that radiate in all directions away from the rock crack (Figure 2.1). All the rock in the Earth is under pressure, due to the natural gravitational pull holding the planet together, the movement of the tectonic plates over the surface of the Earth, local tectonic processes such as volcanism, and other effects such as variations in rock density and topography close to and at the Earth's surface. If the pressure in the rock in every direction is near the same value, the rock resists that pressure intact without breaking. However, if the pressure in the rock at some place in the Earth becomes much greater in one direction than in any other, then the rock can crack and slip along that crack. The slip takes place suddenly, occurring over just a few or several seconds even in moderately strong earthquakes. Through the orientation of the crack and the direction of the rock slippage on the crack, the Earth seeks to equilibrate the stress differences in the rock.

Globally, it is observed that faults of all orientations have experienced earthquakes, and all directions of rock slip have also been observed on faults. The

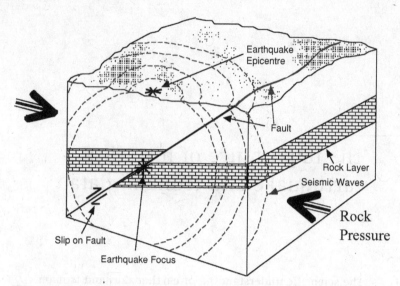

Figure 2.1 Illustration of the elements associated with an earthquake. Excess pressure in one direction in the rock causes it to crack and slip starting at the earthquake focus. The slip displaces formerly continuous rock layers along a crack that is called the fault, with the slip spreading away from the earthquake focus along the fault. The fault slip radiates seismic waves in all directions. The point on the surface directly above the earthquake focus is called the earthquake epicentre.

factor that controls the orientation of the fault on which an earthquake occurs and the direction of rock slip on the fault is the relative strength of the rock pressures in different directions in the rock around the fault. Geoscientists categorize earthquake faults orientations into three general categories, as shown in Figure 2.2. In a *strike-slip* earthquake, the fault is usually approximately vertical in orientation, and the rock slides past itself on either side of the fault. Strike-slip faults are categorized as *right-lateral* or *left-lateral* depending on whether the rock on the side of the fault opposite to the observer has moved to the right or left. *Thrust* faults dip at an angle into the earth and have earthquakes in which one block of rock is pushed up over another block of rock. This type of earthquake is commonly associated with the uplift of mountains. The third general earthquake type takes place on what is called a *normal* fault. In this case also, the fault dips at some angle into the earth. However, in a normal-faulting earthquake one block of rock slips down and away from the other block of rock, the opposite direction of slip that occurs in a thrust-faulting earthquake. In many places, normal faults are associated with the formation of major valleys. When faults emerge at the Earth's surface, *surface faulting* is observed from the earthquake, and the ground surface is permanently deformed by the earthquake fault movements. However, most of the Earth's earthquakes take place on cracks that

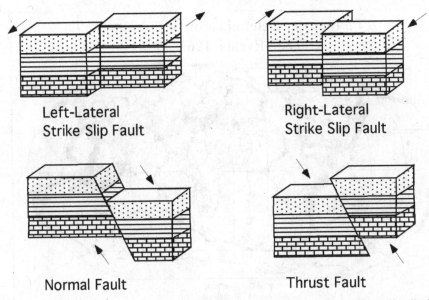

Left-Lateral
Strike Slip Fault

Right-Lateral
Strike Slip Fault

Normal Fault

Thrust Fault

Figure 2.2 Illustration of the major kinds of earthquake faults. The strike slip faults (top) move the rock horizontally on either side of the fault. Normal faults (bottom left) occur when the rock on one side of the fault slips downward relative to the other side of the fault. Thrust faults (bottom right) form when the rock on one side of the fault pushes up over the rock on the other side of the fault. The arrows in this figure show the direction of rock movement on either side of the fault.

do not come to the Earth's surface. For these events, it is only the seismic waves released by the earthquake that indicate the occurrence of an earthquake.

Earthquakes take place in many different parts of the Earth (Figure 2.3). Most of the world's earthquakes take place at the boundaries where two or more of the Earth's tectonic plates come into contact (Figure 2.4). However, a small percentage of the Earth's earthquakes also take place within the tectonic plates, sometimes well away from the plate boundaries. Most of the earthquakes take place within about 70 km of the Earth's surface, although some events at *subduction zones* extend to down to depths of about 650 km. Subduction zones are places where one tectonic plate is sliding deep into the Earth beneath another tectonic plate.

There are several different ways by which earthquakes can cause damage to manmade structures. The most common cause of damage comes from the shaking of the ground by the seismic waves released by the earthquake. Seismic waves shake the ground both vertically and horizontally. The strongest ground-shaking is almost always oriented horizontally. Since many manmade structures are less capable of withstanding strong horizontal ground movements compared to vertical motions, it is the horizontal shaking that invariably causes the

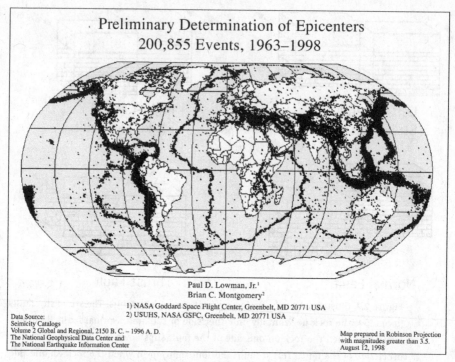

Figure 2.3 Globally recorded earthquakes from 1963 to 1998.

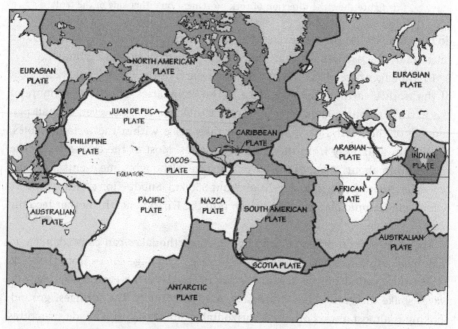

Figure 2.4 Locations of the major tectonic plates of the Earth. The heavy lines show the locations of the modern plate boundaries. Most of the Earth's earthquakes (Figure 2.3) occur at the boundaries between tectonic plates.

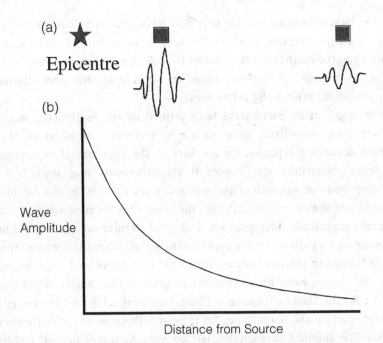

Figure 2.5 (a) Illustration of how the amplitudes of seismic waves attenuate as they propagate away from a seismic source. The amplitude of the seismic wave at the seismic station (square) that is farther from the earthquake epicentre (star) is smaller than that at the closer seismic station. (b) Graph showing the decay of seismic wave amplitude with distance from a seismic source.

greatest damage to susceptible structures. Another cause of damage from earthquakes is from surface earthquake faulting. Manmade structures, even those that are very well built, will sustain major damage if they happen to be built on a fault that experiences surface fault movement. While this is not a common occurrence, there are many cases where it has taken place. Earthquake ground-shaking can also cause other natural phenomena that can damage structures. The most important of these secondary effects are *rockslides*, *landslides* and *soil liquefaction*. Rockslides and landslides occur on steeper hillsides, where the ground-shaking can trigger the fall of rocks and soil. Liquefaction is the phenomenon where water-saturated layers a few metres below the Earth's surface effectively become liquid during strong earthquake ground-shaking. Buildings founded on layers that liquefy during earthquake shaking may sink, and the land overlying liquefied layers can slide downhill, even on the gentlest of slopes.

Like all waves, seismic waves decrease (*attenuate*) in intensity as they travel away from their source (Figure 2.5). This is observed for all earthquakes, although there are usually local variations to this generalization. For example, deposits of dozens of metres of soft soils can amplify some aspects of ground-shaking, leading to locally stronger shaking experienced on the soils than experienced at nearby rock sites. The rate at which the seismic wave intensity decreases with distance from the earthquake fault varies from region to region across the Earth. Globally, seismologists have documented the rates of seismic wave attenuation in many different seismically active areas.

The strength of an earthquake is quantified by the earthquake *magnitude*. There are many magnitude scales in use by seismologists, although the one that most accurately represents a measure of the true size of an earthquake is the *moment magnitude* (see Chapter 4). The earthquake magnitude is related to the amplitude or strength of the seismic waves radiated by the earthquake. All magnitude scales are logarithmic, meaning that an increase of 1 unit of earthquake magnitude corresponds to a factor of 10 increase in the amplitude of the seismic waves radiated by the event. In theory, all magnitude scales are open-ended, although in practice they are limited at both small- and large-magnitude values. The largest earthquake moment magnitude that has been observed to date is 9.5 for the 1960 earthquake in Chile. This is probably near the upper limit of possible earthquake magnitudes for the Earth (because of the finite strength of rocks). The smallest earthquakes felt by humans under normal conditions are around magnitude 2.0, although natural earthquakes as small as magnitude −2.0 have been recorded in special monitoring situations.

As a general rule, the higher the earthquake magnitude, the stronger the seismic waves radiated from the fault. Furthermore, the higher the magnitude of an earthquake, the longer the earthquake fault and the greater the amount of rock slip on the fault. Earthquake fault lengths range from hundreds of metres for magnitude 2 earthquakes to many hundreds of kilometres for magnitude 9 earthquakes. Earthquake fault slips range from a few millimetres for magnitude 2 earthquakes to more than 10 metres in magnitude 9 earthquakes. At all seismically active places on the Earth, earthquakes with smaller magnitudes occur more frequently than earthquakes of larger magnitudes. For each 1 unit increase in earthquake magnitude, there is a decrease by a factor of about 10 in the number of earthquakes that occur in a given time period. For example, the Earth annually averages about 100 000 earthquakes of magnitude 3.0 or greater but only about 14 earthquakes of magnitude 7.0 or greater.

2.1.2   Tsunamis

Tsunamis are another secondary phenomenon triggered by some earthquakes and by large submarine sediment slides. Tsunamis are caused by sudden vertical movements of the ocean floor. Because such undersea movements cause uplift or downdrop of the local ocean surface, the water in the ocean must readjust to the new configuration of the ocean bottom. The tsunami starts over the place where the vertical movement of the ocean floor occurred. It then spreads out away from this locality throughout the entire ocean basin. When the tsunami reaches land, it alternately rushes onto the land and draws away from the land over several minutes. The height of the tsunami wave at the shore is controlled by both the height of the tsunami wave in the deep ocean and by the configuration of the local coastline and nearshore bathymetry, which can focus or defocus the incoming water. Thus, *tsunami height* and *run-up* (the distance onshore traversed by the tsunami) can vary significantly along a coastline. Since tsunamis spread throughout the entire ocean, localities hundreds and sometimes even thousands of kilometres from the tsunami source can experience destructive tsunami run-ups if the tsunami source was especially strong.

Since tsunamis arise from vertical movements of the ocean floor, they can be caused by near-surface undersea thrust or normal earthquakes but not strike-slip events. The height of the tsunami is approximately related to the magnitude of the earthquake that causes the tsunami. While larger earthquakes are usually associated with larger tsunamis, sometimes earthquakes of only moderate magnitudes are associated with unusually large tsunamis. Such *tsunamigenic* earthquakes may be more efficient at moving the ocean floor than at generating seismic waves. Also, in some cases undersea earthquakes are thought to trigger submarine slides that enhance the tsunami caused by the earthquakes. Worldwide, tsunamis that are associated with damaging run-ups are almost invariably associated with earthquakes of magnitude 7 or more.

2.1.3   *Damage to manmade structures from earthquakes and tsunamis*

The proper discernment of historical evidence for damage from earthquakes and tsunamis is important for engineers wanting to design buildings to withstand earthquakes as well as for historians and archaeologists wanting to recognize accurately the effects of earthquakes and tsunamis in the records and at the sites they are studying. During earthquakes, most of the damage to manmade structures is caused by ground-shaking, although movement on the earthquake fault itself can have devastating consequences to any structure built across the fault. Earthquake shaking and fault movements can also cause secondary effects such as landslides and soil liquefaction that can cause significant damage. During tsunamis, it is the force of the rising and falling seawater against inundated structures and the undermining of the foundations of structures by the rising sea that are the causes of damage at coastal settlements.

As the seismic waves from an earthquake pass a point on the surface of the Earth, they cause ground-shaking in a complicated combination of rapidly changing horizontal and vertical motions. The strength of the ground motions increases to a time period of maximum shaking and then gradually decreases

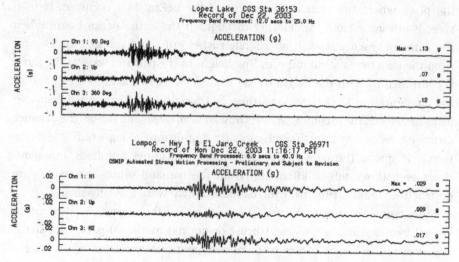

Figure 2.6 Examples of ground motion accelerograms from the 22 December 2003
magnitude 6.5 San Simeon, California earthquake. Each accelerograph station
recorded three components of motion: (from top to bottom) east–west, vertical and
north–south. The horizontal axis is time in seconds, while the vertical axis is
ground acceleration in g.

until it is no longer detectable (Figure 2.6). As a general rule, the greater the
earthquake magnitude, the greater the maximum ground motions caused by the
earthquake. Near an earthquake epicentre, the time for the earthquake shaking
to increase to its maximum value is only one to a handful of seconds, while this
time is much greater for sites far from the epicentre. The duration of the time
period of greatest ground-shaking can be anywhere from less than 1 second in
smaller-magnitude earthquakes to as much as 1 minute in the largest recorded
earthquakes. A number of factors affect this time duration, including the depth
of the earthquake faulting, the location of the site relative to the segment of
the fault that has moved during the earthquake and the near-surface geology
at the surface site. The time period for the earthquake shaking to decay to the
point where it is no longer perceptible is a direct function of the earthquake
magnitude. Earthquakes of small magnitudes (2 to 4) are perceptible for several
seconds to perhaps a dozen or so seconds, while earthquakes of major magnitude
(8 or more) are perceptible for up to several minutes.

In earthquake shaking, the horizontal ground-shaking is almost invariably
stronger than the vertical ground-shaking (typically by 30%–50%). This can pose
a major hazard because most manmade structures, especially those built in historical times, have relatively little resistance to horizontal ground movements.
Brittle structures, such as those made from stone, brick, adobe and unreinforced

Figure 2.7 Photo of the damage to a hospital at Tungurahua, Ecuador due to the magnitude 6.8 earthquake of 5 August 1949. X-shaped cracks radiate from the corners of many of the building's windows.

concrete, are susceptible to earthquake damage because they tend to crack when they experience strong horizontal ground-shaking. Typical damage in brittle structures subject to strong earthquake shaking include X-shaped cracks radiating from the corners of windows and doors (Figure 2.7), damage of the gable ends of buildings (Figure 2.8) or at the corner of a building (Figure 2.9), toppling of brick or stone chimneys (Figure 2.10) and the breaking of parapets (Figure 2.11). Massive stone structures, such as some large churches, sometimes withstand earthquake shaking better than smaller nearby brittle structures. The collapse of part or all of a brittle structure occurs when the cracking is so bad that the supporting or foundational elements of the structure lose their integrity. This can happen when the structure undergoes a very large horizontal ground movement after significant cracking has already started in the structure during earlier ground motion oscillations.

In contrast to brittle structures, structures made of wood with its walls, ceilings and roofs all firmly attached to each other can sway considerably in strong horizontal ground-shaking without sustaining any structural failure. However, if such structures also have brittle elements attached (such as a chimney or brick addition), the different natural periods of vibration of the wooden and brittle parts of the structure can cause damage where the two meet. This is often observed in the earthquakes of North America, where a wood-framed house will withstand earthquake shaking with relatively little damage,

Figure 2.8 Damage to the gable end of a masonry church in Watsonville, California due to the magnitude 7.1 Loma Prieta earthquake of 23 October 1989.

Figure 2.9 Damage to a masonry building in Turkey due to the magnitude 6.7 earthquake on 6 September 1975. Note the damage at the corner of the building.

Figure 2.10 Damage to a chimney at Newhall, California due to the ground-shaking from the magnitude 6.5 earthquake of 9 February 1971 at San Fernando, California. A section of the chimney collapsed, puncturing the roof of the house.

Figure 2.11 Damaged parapet in Seattle, Washington due to the magnitude 7.0 earthquake at Puget Sound on 13 April 1949. The parapet collapsed onto the fire escape below.

Figure 2.12 Collapse of a chimney in Seattle, Washington due to the magnitude 6.5 earthquake at Puget Sound on 29 April 1965. The damage to the house was caused by the collapse of the chimney.

whereas the chimney attached to the house will partially or totally collapse (e.g. Figure 2.12).

The amount of damage that strong earthquakes cause is also related to the surficial site conditions upon which the manmade structures are built. Structures that are built on thick layers of soft soil can experience significantly greater ground-shaking than nearby structures that are built on hard rock. There is an even greater potential for damage if a soil layer under the structures liquefies.

The strength of the ground-shaking experienced at a site due to an earthquake is described by seismologists and engineers in two different ways. One way is through instrumental measurements of the ground accelerations at the site. From such instrumental recordings, seismologists and engineers compute parameters called the *peak ground acceleration (PGA)* and the *spectral acceleration (SA)* at a number of different periods of vibration. PGA is simply the largest reading on the horizontal ground acceleration recordings, while SA at some period of ground oscillation is a measure of the highest level of ground acceleration experienced by a building with that natural period of vibration. Short buildings have a short natural period of vibration, while taller buildings have a longer natural period of vibration. For example, the natural periods of vibration of buildings of 1 storey and 10 storeys are approximately 0.1 second and 1 second, respectively. It is possible that the ground motions from an earthquake will preferentially cause damage to buildings only within a limited range of heights. This was dramatically demonstrated in 1985, when a magnitude 8.0 earthquake centred on the Pacific coast of Mexico caused strong ground-shaking with oscillations of about 2 seconds period at Mexico City, about 200 kilometres from the epicentre. In Mexico City, many buildings 20 to 40 storeys high collapsed or were severely damaged by the ground-shaking with this period of oscillation, while the large number of much shorter buildings in the city suffered no damage at all because they are not susceptible to such long-period ground motions. The location of Mexico City on the soft soils of the bed of a former lake helped contribute to the building damage due to this earthquake.

The second way in which seismologists and engineers describe the ground-shaking at a site in an earthquake is by the use of *earthquake intensity scales*, which are descriptive scales of ground-shaking strength based on felt and damage effects. Roman numerals are used to designate earthquake intensity values since these are not true quantitative measures of the earthquake shaking. There are a number of different earthquake intensity scales used throughout the world, which are discussed in detail in Chapter 12.

Maps of the spatial distribution of PGA, SA and earthquake intensities are commonly constructed after strong earthquakes. Maps depicting distributions of PGA and SA are called *shakemaps*, while those depicting the distribution of earthquake intensity are called *intensity* or *isoseismal maps*. On maps of the latter type, a contour that divides two regions of different earthquake intensity is called an *isoseismal*.

In general, the larger the earthquake, the greater the distance from the earthquake epicentre for which shakemaps and intensity maps must be constructed, and the greater the maximum values of ground acceleration and earthquake intensity at the earthquake epicentre. Seismic waves attenuate with distance at different rates in different parts of the world, which means that the area over which earthquakes of a given magnitude cause damage or are felt varies from one region to the next. Earthquakes above about magnitude 4.5–5.0 can cause damage near their epicentres to structures that are not earthquake-resistant, while the strongest ground-shaking in earthquakes above magnitude 6.0–6.5 may cause damage to structures even with some measure of earthquake resistance. As earthquake magnitude increases, so does the strength of the strongest ground-shaking, the potential for damage to structures and the area over which damage can occur. However, even in very large magnitude earthquakes, typically not every structure is damaged or destroyed in the area of strong ground-shaking. The type of structure, the building materials and methods used to construct a structure, and the kind and thickness of the soil or rock upon which a structure was built are all factors that affect the potential for damage to the structure when it experiences strong earthquake shaking.

Aftershocks are smaller earthquakes that follow a larger earthquake. All strong earthquakes are followed by aftershocks, which are most profuse immediately following the largest earthquake (called the *mainshock*) but whose rate of occurrence diminishes with time after the mainshock. Most aftershocks are felt within a few weeks of the mainshock, although some level of aftershock activity will typically be experienced for years after a mainshock. While the strongest aftershocks occur most frequently within hours or a few days of the mainshock, some occasionally stronger aftershocks can take place many months (or even later) after the mainshock. Aftershocks themselves can be almost as large as the mainshock, and large magnitude aftershocks can cause additional damage, especially to structures weakened by the mainshock. On average, the largest aftershock is about 1 magnitude unit less than the mainshock.

The damage from tsunamis comes from the inundation of land areas by the unusually high rise in the level of the sea as the tsunami comes onshore. The run-up of a tsunami at a shoreline takes only a few minutes until the maximum inundation is reached, after which the water withdraws to a point where the sea level has dropped to an unusually low level. There can be several such run-ups and withdrawal oscillations observed in a tsunami, with the largest run-up within the first two or three oscillations of the sea level and the run-up in each later oscillation being significantly smaller. An oscillation of the sea level during a cycle of sea run-up and withdrawal takes about 10–20 minutes to complete. The onset of a tsunami can begin with either a run-up or a withdrawal of the water. The latter has led to large human death tolls in some cases when people would go out to the water's edge to observe the unusual withdrawal of the sea following an earthquake, only to be caught unawares by the tsunami's subsequent inundation phase. The movement of the water towards the shore is so fast that people become trapped by the rising sea.

The force of the water movements, first onshore and then offshore, exert tremendous lateral pressures on any structures in the zone flooded by the tsunami. The greater the depth of the water, the greater the force on the structures affected by the tsunami. Most typically, light structures such as small beach dwellings are swept off their foundations by the force of the water movements (Figure 2.13). These structures are then carried by the water, sometimes being deposited onshore and sometimes being pulled offshore. Structures made of wood and other materials that are buoyant in water can be carried large distances (i.e. up to several kilometres offshore) if they are swept up by the tsunami (Figure 2.14). Damage to heavier structures, such as imposing stone buildings, can also occur if the height of the tsunami (and therefore the force of the water)

Figure 2.13 Photograph of the foundation slab of a building at the Yala Safari Resort on the coast of Sri Lanka. The building was swept away by the tsunami of the magnitude 9.0 Sumatra, Indonesia earthquake on 26 December 2004 (USGS).

Figure 2.14 Photograph of a floating house tied to a schooner in Little Burin Harbour, Newfoundland (Canada), following the 18 November 1929 magnitude 7.3 earthquake that was centred under the Grand Banks of the North Atlantic Ocean. The building was swept offshore by the tsunami, and it probably floated because a large amount of timber was stored under the house.

is great enough. Since it takes a great deal of force to move large stone blocks, these usually end up being deposited by the tsunami close to where they stood before the arrival of the water. The water movements in a large tsunami can also mobilize the sands and silts that are found along shores, scouring them in some places and depositing them in other places. Thus, another source of damage in large tsunamis comes from the undermining of structures by the erosion of the shoreline sands and soils upon which the structures are built.

2.2   Earthquake catalogues and their history

The search for the causes of earthquakes, the identification of active earthquake faults and work to find ways to predict future earthquakes all rely on a database of earthquake information that is both spatially and temporally extensive. The fundamental database for scientific studies of plate tectonics, seismic hazard and seismic risk is the *earthquake catalogue*. An earthquake catalogue is a listing of seismic events for some region of the Earth; each entry in the catalogue includes, if known, the earthquake date, time, epicentre location, magnitude determinations and/or other estimates of size, and other information as included by the earthquake compiler. The other information could comprise technical parametric information of scientific importance, such as the earthquake depth, the focal mechanism, the stress drop or the peak ground acceleration. On the other hand, the other information might involve parameters or descriptions of more general importance, such as the number of fatalities and injuries, the size of the area of damage or over which the earthquake was felt, or a listing of cities most affected by the earthquake. Although most modern earthquake catalogues have many common elements, there is no agreed-upon set of parameters that must be included in an earthquake catalogue. Furthermore, most catalogues do not include the primary data from which the catalogue entries were derived; earthquake seismograms cannot be conveniently incorporated into catalogues, and most catalogues of pre-instrumental earthquakes do not reproduce the primary historic reports of the earthquake activity. Today, the sector of seismology that deals with the formation of the data sets of the catalogues and the research into and interpretation of the historical data is one important facet of historical seismology.

Earthquake catalogues constitute an extraordinary corpus of worldwide information, and they are often the work of generations of researchers. Their quality and reliability can vary a great deal, however, so that close acquaintance with the sources of the catalogue information and analysis of those sources is essential if one is to draw benefits from the catalogue and to understand how to improve it. The very different learning traditions that have come together in

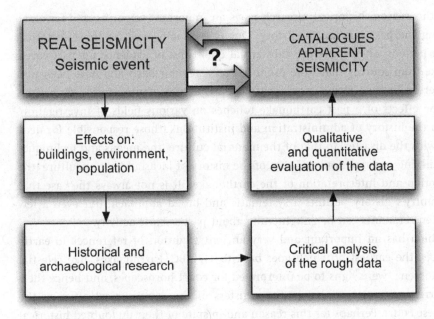

Figure 2.15 Relationship between real and apparent seismicity.

historical earthquake catalogues may be stratified and almost hidden within the numerical parameters, which lend apparent uniformity to very diverse data. Almost all present-day historical catalogues are in fact catalogues of catalogues. In order to assess the quality of any historical earthquake catalogue, therefore, it is necessary to go back to the descriptive data on which it is based.

Each country has a seismic history that can be used for the seismic hazard estimates, it generally has its tradition of historical studies in this sector, and has today drafted earthquake catalogues. One can get some idea of the breadth of these studies and their dissemination from the bibliographies of the various studies collected in Albini *et al.* (2004), which offers a near-global overview.

Many countries have also made their catalogues available online, mostly as synthetic parameters. The ongoing effort in some countries is to make transparent and critically ever more precise and accurate the data of the earthquake catalogues and to set up actual data sets with historically based data accessible on the Web.

2.2.1 The earliest catalogues

The aim in compiling historical earthquake catalogues is to bring the record of apparent seismicity (the seismicity which has been reconstructed by means of historical research) as close as possible to actual seismicity (Figure 2.15).

For this reason, historical research plays an important scientific role for elucidating the past earthquake history. This is especially true for those historical
time periods when there are substantial quantities of evidence for past earthquake from sources that were produced on a regular basis and have been well
preserved in such forms as archival documents and manuscripts. The analysis
of the effects of a past earthquake touches on various fields of investigation,
from the history of administration and institutions (those responsible for dealing with the disaster) to that of the material culture (the means and techniques
of building construction) and that of the history of ideas and social culture (the
reception and interpretation of the earthquake). It is not always the case that
a country's history permits a systematic and broad approach, but even a few
written remnants can be enlightening about past seismic activity.

China has an important and very ancient tradition of references to earthquakes, the earliest of which goes back to 1177 BC. Earthquakes, like celestial
phenomena, were signs to be interpreted for court horoscopes, and hence their
occurrences were recorded in the registers of the astronomical office of the
Chinese court. Perhaps for this reason and in spite of their undoubted historical
and cultural value, the data that have survived are rather slight in quantity and
quality and are now difficult to use for modern reinterpretation.

As far as the Mediterranean area is concerned, the earliest information about
real earthquakes (that is to say, those for which a location and time are provided,
as opposed to those which are in some way myth-related) goes back to about
760–750 BC. The earliest report concerns an earthquake that struck Jerusalem
and the valley of Hinnom. Of the few earthquakes mentioned in the Bible, this
is one for which there is other evidence in Hebrew and Greek sources. In other
cases, it is religious symbolism that predominates in Biblical references to earthquakes. Apart from their rarity and curiosity value, the earliest records are often
difficult to put to practical use since present-day historical seismology needs
to be able to establish from those records detailed seismic-effect scenarios in
order to make parameter estimates (such as seismic intensity, amount and types
of damage, possible epicentral location, etc.) from the observations. It is these
parameter estimates that add to the quality and quantity of the entries in historical earthquake catalogues. Unfortunately, this often cannot be done with these
early earthquake references.

The tradition of preserving the sort of historical records in Europe that are
most useful for historical earthquake research began in the Renaissance with the
imitation of classical texts. The first known catalogue of historical earthquakes
(De terraemotu libri tres) was written in 1457 by Giannozzo Manetti (1396–1459). It
is preserved in various manuscript codices recording about 70 seismic events
from antiquity to the fifteenth century together with the sources of those

earthquake reports. From the sixteenth century onwards, a learned fashion for compiling lists of past earthquakes started to take hold as part of a renewed interest in signs and all those phenomena that were generally considered to be omens. This resulted in the production of various earthquake lists, though they were obviously not meant for practical use. Amongst the most important were those of Conradus Lycosthenes, i.e. Conrad Wolffhart (*Prodigiorum ac ostentorum chronicon*) in 1557, Pirro Ligorio in 1570–71 (ed. 2005), Filippo da Secinara (*Trattato universale*) in 1652 and Marcello Bonito (*Terra tremante*) in 1691.

The great catalogue of Marcello Bonito (1691): an interweaving of information having diverse value

The famous earthquake catalogue of Marcello Bonito (1691), known to nearly all historical seismologists of the Western world, contains information on earthquakes that had occurred not only in Europe and in the Mediterranean area, but also in South America and in Japan (in the sixteenth century only). It starts its long series of seismic events around the year 1200 BC and extends it until AD 1690. In most cases the information, the result of unsystematic research, comes from texts that are not primary sources but rather are re-elaborations, made much later, of the narrated events. Their informational value is thus rather poor, and in many cases null. But Bonito was able to consult the archive sources of the thirteenth and fourteenth centuries that later were destroyed in the great fire of the State Archives of Naples in 1943. Fortunately he had transcribed them to the letter, and in this way they were preserved inside his text. Furthermore, Bonito was a keen observer of the earthquakes that took place in his time. He collected circumstantial information about these events, and he also made use of witnesses' letters. He is an important indirect source for these events. Lastly, he himself witnessed the earthquakes that occurred in Naples or were felt in Naples during his lifetime. For these events he is an authoritative, direct and independent source. Thus, a card indexing of Bonito's catalogue cannot assign this work a single classification (i.e. primary source, earthquake catalogue, etc.), but requires a different classification for each and every event listed in the catalogue (Figure 2.16 shows an example).

2.2.2   Seventeenth- to nineteenth-century catalogues

Bonito's catalogue, published in Naples, constitutes the ideal archetype for the early European seismic catalogues. It is a unique product of literary and historical erudition, with limitless geographical interests (it even lists earthquakes in Japan and Latin America). In the eighteenth century, especially after the 1755 earthquake in Lisbon, physicists and naturalists started to turn their attention to the study and theoretical interpretation of striking events in terms

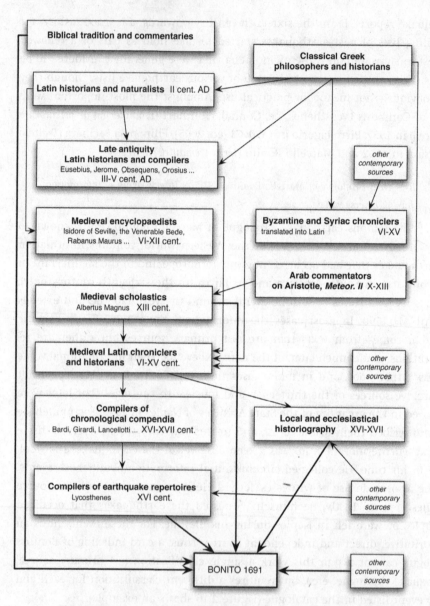

Figure 2.16 Scheme of the contributions to the compilation by Bonito, 1691
(from Guidoboni and Stucchi, 1993).

of their individuality, so that the compilation of earthquake catalogues became
a secondary issue. The main catalogues from this period were: Seyfart, *Algemeine Geschichte der Erdbeben* (1756); Bertrand, *Mémoires historiques et physiques sur
les tremblements de terre* (1757); Moreira de Mendonça, *Historia universal dos terremotos* (1758); Guéneau de Montbeillard, *Liste chronologique des éruptions de volcans,
des tremblements de terre, . . . jusqu'en 1760* (1761).

Only when a new volcanic theory spread in the beginning of the nineteenth century was the study of past earthquakes taken up again with renewed vigour. Within that theory, historical earthquake catalogues served to demonstrate the continuity of seismic centres (volcanoes) and to place emphasis on the correlation between earthquakes and volcanic eruptions, in a search for a pattern in seismic activity. It is from the early decades of the nineteenth century onwards, therefore, that historical earthquake catalogues began to be considered as genuine collections of empirical data, thus losing the guise of erudite oddities, which had generally been the characteristic of previous catalogues. Systematic research was carried out in Germany by Karl Ernst Adolf von Hoff (1771–1837), who compiled an enormous catalogue of earthquakes and volcanic eruptions, starting with the ancient Egyptians (*Chronik der Erdbeben und Vulcan-Ausbrüche*, 1840–41). His work was subsequently taken up and expanded in various ways by later European compilers. An authoritative school was founded in Great Britain by David Milne (1805–1890), followed by Robert Mallet (1810–1882) and then John Milne (1850–1913). With the assistance of his son John, Robert Mallet brought together the work of various authors in a huge new catalogue (*Catalogue of recorded earthquakes from 1606 BC to AD 1850*). Robert Mallet used the listed data to draw the first map of the seismic bands of the Mediterranean. In this map the historical seismicity largely follows (though he was obviously unaware of the fact) the great Afro-Eurasian plate boundary, where current seismic activity is concentrated. In his search for correlations between seismic, meteorological and lunar activity, the French scholar Alexis Perrey (1807–1882) showed great determination (but often a certain lack of accuracy) in gathering a new data crop concerning seismic events in France, Belgium, Holland, Switzerland and Italy (in a large output from 1841 to 1875) and also in Turkey and Syria (*Mémoire sur les tremblements de terre ressentis dans la péninsule Turco-Hellénique et en Syrie*, 1850), in the Danube basin (*Mémoire sur les tremblements de terre dans le bassin du Danube*, 1846), in the Rhine basin (*Mémoire sur les tremblements de terre dans le bassin du Rhin*, 1845–46) and worldwide from 1841 to 1875.

In the closing decades of the nineteenth century, the search for historical earthquakes saw scholars from all over the world join in. J. F. J. Schmidt (1825–1884) compiled a catalogue of historical earthquakes for the eastern Mediterranean area (1881). F. M. Montessus de Ballore (1851–1923) laid the foundations for an historical approach to seismology, including ethnology and folklore amongst his information sources in an attempt to cast light on the geographical distribution of earthquakes (*Temblores y erupciones volcánicas en Centro-América*, 1884; *Les tremblements de terre: géographie séismologique*, 1906; *Ethnographie sismique et volcanique*, 1923).

These collections of data at times also proved to be of value to original seismo-logical research. Thus, the Italian scientist Giuseppe Mercalli (1850–1914), who compiled the most famous intensity scale to classify earthquake effects, devoted many years of his working life to organizing historical seismic and volcanic data (*Vulcani e fenomeni vulcanici in Italia*, 1883).

2.2.3　Twentieth-century catalogues

Mercalli's research was continued and improved by Mario Baratta (1868–1935) and his network of correspondents, who worked intensively for a decade to produce the most innovative national catalogue of his generation (Baratta, 1901). It lists 1364 Italian seismic events from the first century AD to 1898 and provides more than 1500 bibliographical references. In the early decades of the twentieth century, research into historical earthquakes became more systematic. A valuable example of this new generation of research was the great work of Sieberg (1932b). He compiled an earthquake catalogue for Europe and the Mediterranean area, re-elaborating earlier catalogues and expanding them by means of new historical research and observations of earthquake effects performed in the field. There were also other seismologists who performed sim-ilar research for other areas: Wilson (1930) for the Persian region, Roux (1932) for Morocco and Morelli (1942) for Albania.

Even in Tsarist Russia (and subsequently in the former Soviet Union), there has been a prestigious research tradition in this field, ranging from historical work to the collection of data in the field on the effects of earthquakes. Those republics with high levels of seismicity – especially in the Caucasus – have established a substantial scholarly tradition in this field, which allowed them to produce the first maps providing seismotectonic information and a seismic classification for their country, works that were much imitated in Europe in the 1950s (see Kondorskaya and Shebalin, 1982). Japan, too, has been playing a leading role in this line of research. The first updating of historical records was published by Musha (1941–43, 1951).

In spite of differences in quality and chronological range, all of these cata-logues have in common the fact that they are works of descriptive mediation, in the sense that they are based solely on the language and cognitive background of a seismologist–compiler. The historical data themselves have nearly always remained distant and indirect.

2.2.4　The birth of parametric historical catalogues

By the mid twentieth century in Europe historical seismic data seemed to belong inescapably to the past, in spite of the fact that they had sur-vived changes in cultural fashions and interpretational theories. Research in earthquake seismology began concentrating almost entirely on instrumental data and theoretical parameters derived from those data. As research groups started up and grew at many different universities and government agencies, the instrumental data became increasingly dispersed among a wide variety of institutions, resulting in an infelicitous fragmentation of information and a great variety of parameters derived from the instrumental data. A decision reached by the scientific community concerning this problem, however, was indirectly responsible for renewed interest in historical data. An attempt to homogenize instrumental earthquake catalogue data and organize that field of study was launched at the Strasbourg meeting of the European Seismological Commission (ESC) in 1952.

At Strasbourg it was recommended that work should begin to organize and homogenize instrumental data into new catalogues. Between 1952 and 1964 criteria were established both for the form which these new earthquake catalogues should have and for the parameters to be adopted in relation to time of origin, location, intensity, magnitude and focal depth. It was hoped that this work would also be applied to the historical earthquake data that had been compiled by seismologists of earlier generations in view of the traditions of their own countries.

This renewed interest in descriptive data included *macroseismic* studies, that is to say observations of the non-instrumental effects of earthquakes made in the field. The aim was to enlarge the time-frame in relation to current seismic activity. Earthquake theories had changed, but basic historical data were recognized as retaining their basic information value. The historical information could be used to assess regional seismic characteristics (though still inaccurately) and to track down the active faults. That they could be applied to *seismotectonics* (earthquake information that shows the geological changes arising from local plate tectonics) was confirmed by the established Soviet and Japanese traditions. At a meeting in Utrecht in 1958, the ESC recognized the importance of Soviet and Japanese achievements, and it set up a special commission with headquarters in Moscow (under the presidency of V. V. Beloussov) to draw up the first seismotectonic map of Europe. Historical earthquake catalogues were among the data sources to be used. What was required, therefore, was a reassessment of descriptive historical earthquake catalogues, and this was carried out in historical catalogues using the same parameters established for instrumental earthquake data.

This innovative attempt to systematize the instrumental and historical data into new catalogues (often computer-based) bears the name of Vít Kárník (1926–1994). His introduction to *Seismicity of the European Area* (completed in 1964 but not published until 1968–71) is now part of the recent history of seismology. Between 1960 and 1980 a new generation of national earthquake catalogues spread throughout the world, and these were characterized by a blending of historical and instrumental data. This work also influenced the accumulation of tsunami catalogues, which prior to this time had not received the extensive attention accorded to earthquake catalogues. Some modern catalogues of tsunamis are those of Galanopoulos (1960), Ambraseys (1962), Papadopoulos and Chalkis (1984), Caputo and Faita (1984), Soloviev (1990), Soloviev *et al.* (2000) and Tinti *et al.* (2004).

The interpretative model for earthquake data was permanently changed as a result of this work, and the basic historical data became recognized as a useful factor for assessing regional seismic characteristics and for locating activity associated with *seismogenic* structures (geological features capable of being cracked in a large earthquake). This required a rereading of descriptive historical earthquake catalogues, which has to be carried out within the strictures adopted for instrumental data. The descriptive data in historical catalogues, which were the fruits of the literary and naturalist erudition of whole generations, are transformed into parametric data by means of a process that necessarily involves a compression of the historical information. This parametric transformation has often led to misunderstandings or misinterpretations that have now been consolidated within the catalogues currently in use. Thus, for these as well as other reasons, all kinds of errors became incorporated in the traditional descriptive historical catalogues, while the new parameterizations have produced new mistakes and further loss of information. That is the main reason why the historical catalogues need to be thoroughly analysed before putting them to use in detailed scientific analyses.

Two examples illustrate the difficulties faced in producing modern earthquake catalogues that include historical earthquake data.

Instrumental data as a model for descriptive historical data: a 'Bed of Procrustes'?

In the 1960s, the way national seismic catalogues (involving both historical and instrumental data) were organized reflected an unclear division between scientific research and the human sciences, with the result that the former were superimposed on the latter, ignoring the various specific qualities of human science research. Underlying this approach, at the same time, was the prejudice that seismic parameters based on an instrumental understanding of earthquakes were to constitute the standard descriptive model, even for earthquakes for which there were no instrumental data. This approach can be explained as a sort of *simplification*, arising from the need to classify earthquakes which had occurred at different times and in different places; but when such a rigid scheme was adopted by all earthquake catalogues, it created new problems.

Perhaps it is not over-fanciful to suggest that this process recalls the rigidity illustrated in the Greek myth of Procrustes. Procrustes was an innkeeper who lived on the road to Athens and offered a bed to travellers. But the fixed size of the available bed also governed the size of any ill-fated guest, for Procrustes either cut off or stretched the limbs of anyone using the bed until the length of bed and body were the same. Historical seismic data have similarly been fitted into parametric catalogues, thereby either causing much information to be lost or by failing to take into account the partial or doubtful nature of many indications of time and place.

A general lack of repeatability in decision-making processes

An important principle of scientific investigation is that it must be possible for the methods and data of one researcher to be repeated by following investigators. In historical seismology this principle is often very difficult to apply due to challenges in the availability and accessibility of existing historical materials to all researchers and to the influence of subjective decision-making in interpreting historical data and assessing seismic intensity. Even if two or more researchers use the same materials in their work, the resulting parameters they deduce from those data can differ as a consequence of the large degree of freedom they have when interpreting the descriptive information. For example, the evaluation of the extent of damage and the assessment of the relative degree of earthquake intensity from historic descriptions can be quite uncertain. On the other hand, in the past, earthquake intensity assessments for catalogue records were very often produced without reporting either the complete interpretative process developed by the researcher or all the data available to him. For these earthquake catalogues, it often becomes necessary for modern investigators to invest the time and resources to reproduce the entire process used originally to construct the catalogue before new catalogue entries can be added. The best way of solving this repeatability problem is to implement data bank/data base structures in which it is possible to reproduce all of the data and the interpretation steps from the available historical material to the catalogue records.

PART II   ISSUES CONCERNING
THE INTERPRETATION OF HISTORICAL
EARTHQUAKES AND TSUNAMI DATA

3

Written historical sources and their use

3.1   A definition of historical sources

Historians define an authentic and authoritative written testimony as being *strong* when it is placed in a chronological and geographical relationship of proximity to the events or the situations under examination. The term *testimony* clearly evokes the terminology of a court trial; the historian is, in the final analysis, a judge whose role is to get the true facts to emerge. This concept applies to the records of an earthquake as well, as there is the need to find out *when*, *where* and *how* a seismic event took place by following the winding, often curious and incident-strewn path of historical seismology.

It is important to remember that the meaning of the term *source* is very broad. The term refers not only to a single written text but also to an iconographic document (drawing, painting, photograph), an oral account or a handmade item produced by the material culture (an archaeological artefact). Obviously, each type of source requires a diversified critical approach and performs a different testimonial function, according to the time periods and areas involved, regarding the event under study. Historical seismology is, above all, based on written sources; however, other types of source data, as will be seen in subsequent chapters, can be encompassed within the set of reports used to get the most complete understanding possible of a seismic event of the past.

As historical seismology fully adopts the methods of historical research, the records used in this emerging field must abide with the criteria of reliability and authoritativeness established by a correct exegesis of the sources. Historical research is based on the gathering, analysis and interpretation of sources, which together represent a database from which interpretations, correlations and evaluations can be developed.

The historical sources used to learn about the earthquakes of the past belong to one of the two great memory banks of information about the past: institutions (public, private, ecclesiastical, etc., called *archive documents*) and private individuals (diaries, letters, chronicles, etc.). What are sources, and why is it important to recognize, differentiate and apply them appropriately to be able to study an earthquake? In relation to a given event, a source can be defined as the direct, contemporary and authoritative evidence of an event or as an indirect (or secondary) indication of an event. The latter generally refers to a source that is not close chronologically and/or geographically to the event being examined or that is unreliable or even partly or wholly spurious. Nevertheless, the criteria for making such an assessment are, in their essence, historical and textual. In historical earthquake research the criterion of contemporaneity cannot always be rigidly applied. Indeed, there are, as we shall see better later on, some particular historical contexts (such as for the ancient world or for poorly documented areas) where more complex critical approaches are required to define even the mere occurrence of an earthquake in a given area. Historical sources may be of different types, depending on who produced them. Differentiating between these types can be very complicated and difficult, but as a general rule-of-thumb it can be said that for the study of historical earthquakes, some large classification families may be identified, for which we shall provide details in the following sections.

The selection of the types of sources available for historic earthquakes largely depends on the period when an earthquake occurred and on the type of seismic effects to be investigated. For example, in the modern era, a destructive earthquake in areas with stable and well-organized administrations may have set in motion the workings the bureaucratic machinery responsible for making financial decisions for the revival of economic activity and living conditions. In such cases, the archival documentation has a good chance of being found and of making a detailed contribution to goals of the research. But if one wants to know only about the felt effects of an earthquake, one must resort mostly to acquiring the memories of the individuals who experienced the seismic event, subject to their personal capacity to describe and recollect their experiences. It is obvious that various types of sources should be used in the study of the destruction due to strong earthquakes. But outside the areas of great damage, often the only type of data available is that of the memorialistic sources, that is, the memory of the individuals themselves. Later we shall mention some strategies for maximizing the reliable use of such sources.

Another important category of sources for historical seismology is the documentation of observations of nature, starting from the most ancient observations of the natural philosophers and carried forward to the present time through modern treatises, academic research and the governmental seismological services. Broadly speaking, we label all of these as scientific sources in a very general historical sense. The peculiarities of this genre of sources calls for a thorough analysis, which we present in Chapter 4. Furthermore, in Chapter 5 we extend our discussion to still other important types of historical sources, such as cartography, drawings, frescoes, photographs and films. Especially from the mid nineteenth century onwards, these sources represent most valuable and irreplaceable sources of information for those earthquakes that they document. We also discuss unwritten sources, such as oral testimonies and the testimonies collected within the scope of ethnographic studies. While little used in the field of historical seismology for the Mediterranean region, unwritten sources are important in many other parts of the world.

3.2   Types of written historical sources

To be effective in evaluating the effects of earthquakes or tsunamis in a particular area over a long period, utilizing a variety of sources is both necessary and invaluable. Of course, time and place inevitably dictate the kind and the number of sources that are available. In the following sections we examine the various types of written historical sources available from throughout the long span of human history. The aim is not to compile an exhaustive list of all existing written sources, but rather to provide a general summary of this complex form of historical data, their information potential and their limitations. It is usually on the basis of these data that the historical seismologists can give the earthquakes of the past the most complete possible description of their scope and size. Our *excursus* starts with the most ancient written sources, which have very special characteristics, and eventually reaches the sources of the contemporaneous world.

3.2.1   *A glance at the most ancient historical sources*

Religion and the observation of nature were never clearly distinct in the ancient Mediterranean world, a region with many of the most ancient sources for historical seismology. The incorporation of great seismic events into religious tradition has lasted over three or four millennia, and traces of the effects of earthquakes on early human history can be found even today in linguistic relics, dialectal connotations associated with some place-names, the subject matter of myths, and knowledge gathered and transcribed a great time afterward the events took place. Divinities such as Poseidon or Zeus, mythological figures like the Giants, and even heroes and holy men were believed to be at the origin of seismic phenomena, following a *leitmotif* that foreshadowed the later legends of the Christians saints. In ancient Greece it used to be said that 'god shook', an image that also appears in the Christian era, while in the Byzantine chronicles the writers would speak of 'divine wrath' when alluding to an earthquake. In spite of modern reluctance to consider this set of beliefs and religious associations as an integral part of the observations of nature, they cannot be ignored altogether, and seismologists have already made use of some ancient religious texts to argue for strong earthquakes in ancient times. For the wealth of Greek and Roman texts, which constitute the literary sources of the ancient Mediterranean world, we highlight here some of their idiosyncratic features, as they have already been the subject of dedicated earthquake catalogues and academic analyses.

Myths concerning the divine nature of earthquakes and tsunamis: can they be used and how?

A book published in the 1970s opened up a window for Earth scientists, who until then had little knowledge of the geological origins of the myths (Vitaliano, 1973). Through an analysis of a large number of mythological characters from all over the world and accounts of their actions, the authors drew a sort of map of the great seismotectonic events that must have affected the very ancient world, revealing hidden traces and correlations that were perhaps distilled from a very ancient religious collective memory. Is the passing from myth to a rationalist interpretation critically acceptable and legitimate? In practice, can myths be used today for historical seismology without making arbitrary and questionable assumptions? These questions arise due to a perhaps excessive attraction of the layman towards these issues and their dissemination. They are also important from the standpoint of the earth sciences, which can lack rigour in terms of the methods used to interpret the ancient texts. This latter situation often exists due to the poor quality of the dialogue between seismologists and historical specialists who study the ancient world.

There is no cut-and-dried answer about whether or not these data can be used in historical seismology. Indeed, each case needs to be analysed individually, whilst remaining well aware of the fact that it consists of literary scenarios, replete with interpretative and critical pitfalls. Those scholars of antiquity who study myths do not favour the rationalist interpretation of the myths as a key for unlocking the meaning behind the myths themselves. For this reason, myths that refer to geological phenomena must be pondered with great caution, including those that refer to seismic phenomena. Very rarely is it legitimate to interpret ancient myths as though they were the literary representation of an event that had really occurred. We discuss here some early myths from the Mediterranean area with the aim of illustrating some of the difficulties

Figure 3.1 Poseidon (above, with trident), Oceanus and his sister and wife Tethys, in a mosaic from the archaeological site of Zeugma, Turkey (from www.pbs.org/wgbh/nova/zeugma/mosa_02.html).

interpreting these myths. We certainly do not explore the interpretation problems exhaustively, but we do note that interpreting the meaning of such myths is an area that already has its own rich, and from certain points of view, controversial literature (see below).

The main divinity invoked by the early Greeks to explain earthquakes and tsunamis was *Poseidon*, a very ill-natured, touchy god. Poseidon is a very ancient deity (there are Mycenean references dating back to the fifteenth century BC) recalled in the Homeric poems as the 'shaker of the earth' and lord of the depths of the sea (Figure 3.1). These associations with natural phenomena continued to distinguish Poseidon throughout later tradition. Besides the domination of the sea, Poseidon was thought to be endowed with control of underground waters and springs, which were considered to be in Poseidon's realm since they were believed to be in communication with the Earth's abysses. The marine abysses and the fearful energy that was propagated with tsunamis, a hazard with which even the ancients were familiar, were the elements over which Poseidon's power stretched.

Poseidon was the god of earthquakes, which overturn the depths of the Earth, shake mountains and uplift plains. In Homer's *Odyssey* (4.505–10) the god of the sea is described by his capacity to cleave open mountains and submerge coasts. Various mythical traditions attribute to Poseidon some earthquakes and tsunamis that actually occurred, and this belief persists throughout the classical age. In each of these cases the divinity was referred to as either the direct or the indirect cause of the event. For example, the Spartans believed that the cause of the earthquake in 464 BC (Thucydides 1.128.1) was a sacrilege committed against Poseidon. A similar opinion was handed down by the written tradition regarding the earthquake in the Gulf of Corinth which destroyed the cities of Helice and Bura in 373 BC. To confirm Poseidon's alleged culpability, Xenophon (*Hellenica* 4.7.4) recalls that when an earthquake began a paean (that is, a choral religious chant) would be intoned to Poseidon. This divinity represented the most violent form of energy with which ancient man came into direct contact (Burkert, 1985, p. 139).

In addition to earthquakes, mythological tradition ascribed a series of tsunamis of an indefinite and very remote date in the Aegean area to Poseidon. It is possible that the memory of real events contributed to consolidating such oral traditions since for time immemorial tsunamis and earthquakes had been part of the experience of the populations living on the islands and along the coasts of the Aegean Sea. According to historians of antiquity, however, it would be a mistake to interpret all of mythological traditions of this kind as a mere incorporation of seismotectonic phenomena that really occurred in historic or pre-historic times into the religious domain. Indeed, historians of the ancient world believe that the language of myths might have its own autonomous meaning, which must be respected and interpreted within the sphere of the magical and religious thinking of those times. Therefore, a deterministic ratio-nalization of myths is always at great odds with standard historical critical analyses.

Standard analyses of myths notwithstanding, some local myths are quite likely to have been connected to real seismotectonic phenomenon that actu-ally took place. For example, in the mythical past of the religious prehistory of the Argolides (a region to the northeast of the Peloponnesus), a primordial flooding of the sea is preserved, reportedly caused by Poseidon's ire. The famous tsunami of Argo is described by Pausanias (2.22.4) like this:

> Here [at Argo] there is the sanctuary of Poseidon Proclistius [bathing with the waves]. It is indeed said that Poseidon submerged most of the region, because Inacus and the other judges had decreed that the region belonged to Hera and not to him. So Hera got Poseidon to make the sea withdraw; the Argives then dedicated a sanctuary to Poseidon, as Proclistius, at the point where the waves had withdrawn.

If the remains of such a temple could be identified, this would give some mea-sure of the sea's inland penetration in the tsunami, although changes to the coastline must also be considered. In this case, the natural phenomena described are realistic, while the motivations to explain the causes of the tsunami are religious.

Even the epic tradition relating to the origins of Troy may possibly preserve a reference to a great tsunami in the form of the memory of a great wave (i.e., a sea monster). Pseudo-Apollodorus (2.5.9) wrote:

> It then happened that the city of Troy suffered as a result of the rage of Apollo and Poseidon [. . .]. For this reason Apollo sent a pestilence, and Poseidon a sea monster, brought on by the flooding, which swept away the people who were on the plain.

The myth can also be found rearticulated by the poet Ovid (*Metamorphoses* 11.199–215):

> [. . .] the lord of the sea directed all the waters towards the shores of the avid Troy, transforming the land into sea, depriving the farmers of the fruits of the land and covering the fields with waves.

It is interesting to point out that this legend was subject to a rationalist interpretation even during antiquity. Indeed, the second-century man of letters and philosopher Plutarch (*Moralia* 248 A–C) remarked:

> [. . .] a wave raised itself up and flooded the earth; it was a terrible sight to watch, the sea that followed it high up and covering the plain [. . .]. Some people, attenuating the fabulous part of this story, claim that Poseidon did not make the sea obedient with his maledictions, but that the most fertile part of the coastal plain was at a lower level than that of the sea.

Plutarch's account suggests that even in those times it did not escape the witnesses' notice of how the ground's morphology contributed to worsening the effects of a tsunami. Indeed, if the plain lay below sea level, the flooding could turn the fields into lagoons and make them unsuitable for agriculture for a very long time. In these cases such ancient data can be used, even if the dating cannot be inferred, not so much to hypothesize a datable event but in order to deduce an area's likelihood of having been flooded by a tsunami.

The Greek Poseidon was known to the Romans as their god Neptune. Neptune was nearly always portrayed as a sinewy figure, brandishing a great trident to symbolize his lordship of the seas. But Neptune was not the only divinity to be invoked by the Romans to protect them against earthquakes and tsunamis. Indeed, the ancient Latin texts display some uncertainty in recognizing to which precise god sacrifices should be dedicated in order to expiate a woeful prophesy, as the earthquake was thought to be (Aulus Gellius 2.28; 4.6). Among the gods invoked, there are references to the goddesses *Tellus* and *Bona Dea*, both representations of the Earth, as well as to *Ceres*, an underground divinity who was kidnapped and held prisoner by Pluto, the god of the underworld and fire. Even Mars, otherwise known as the god of war, received attention concerning earthquakes. A prodigy connected to earthquakes was linked to Mars through oscillations of the Spears of Mars, which were found in the Regia in Rome as well as in other cities (Dumézil, 1966). The spears were stuck into the earth, and their oscillations were believed to be an ominous sign related to the Earth's shaking. Aulus Gellius (4.6) recalls that in 99 BC the Roman Senate was informed that the Spears of Mars at the Regia had moved on their own. This led the Senators to suggest ceremonies dedicated to Mars, but they also decided to appeal to Jupiter and to other gods that the consul-in-charge proposed, thus appealing to the well-known pragmatism of Latin culture.

Early in the twentieth century, Lanciani (1918) hypothesized that the Spears of Mars in the Regia were predisposed to act as a rudimentary earthquake detector. It is true that oscillations of the spears could have been caused by far-off earthquakes, but it is hard to claim, as Lanciani did, that the correlation with distant earthquakes was evident to the people of Roman times. For the oscillation of such spears to be considered a prodigy, the Senators asked that they should move spontaneously (*sponte*), i.e. without anything or anyone intervening to disturb the objects. This requirement can lead to the conclusion that spontaneous oscillations might strongly favour correlation with a faraway earthquake, but it remains a fact that such a conclusion is too uncertain to be used in historical seismology (Guidoboni and Poirier, 2004). Even if a reported movement of the spears was indeed due to the passage of earthquake waves, this provides only a chronological clue about an ancient earthquake but no location for the event. Although the myths, legends and prodigies of the Mediterranean area are infused with undoubted appeal to the human imagination, their use in identifying specific earthquakes in early historical times contains more pitfalls, equivocations and chances of error than the elements of actual information that can be inferred.

In all parts of the world, the cases in which an element belonging to a myth or legend can really lead modern researchers to recognize a seismic effect are in actual fact rather few. More frequent are the cases in which elements corresponding to seismic events are inserted into a story by researchers in support of known geological data. Even with their great uncertainties, myths and legends still represent very appealing sources to examine for possible information about seismic events, above all for those areas that were poorly documented in very early times. An example of a modern re-examination of an early legend was put forward for Armenia by Karakhanian and Abgaryan (2004), who examine the Hittite myth of the slaying of the dragon and the myth of the missing god. Also, the legends of the war of Tigran and Azhdahak and that of the birth of Vahagn in the Armenian and Caucasian area represent some possible sources for identifying very ancient earthquakes. One needs to be reasonably cautious and critical regarding the desire to pursue such sources.

Myths and geology: study sector or hypotheses?

A field of interpretation that is as 'slippery' as ever and open to numerous objections is the matter of the legendary data reread and reinterpreted in the light of modern seismological and/or geological knowledge. Over the past few years, this theme seems to have exerted an almost magnetic appeal in a study sector that is perhaps unwittingly seeking its innermost soul, exploring the relationship between the world of legends and that of scientific knowledge. But often the results are most of all interesting for their curiosity value, rather than for their indisputable contribution as 'new' scientific data. As regards earthquakes, unlike the research based on written and documented sources, the results of this research are more often than not confirmation of the seismicity of areas or sites, whose seismic characteristics are already known from the written historical records. Still, it may be said that they provide a complementary or different perspective on such records.

An example of legendary data and palaeoseismology observations is that of the sanctuary of Monte Sant'Angelo (Figure 3.2) on the Gargano promontory (southern Italy), presented by Piccardi (2005). This sanctuary's crypt, originating in the early Middle Ages, preserves indications of a crack in the ground. A longstanding legend, connected to the apparition of the Archangel Michael at this location, suggests the occurrence of an important seismic event. The crypt and surrounding area have been explored from a geological standpoint, and a fault has been identified, upon which the monastery was later built. The main fault strand, which lies in the immediate vicinity of the sanctuary (Piccardi, 2005, pp. 118, 120), shows palaeoseismological evidence of recent earthquake movement. The estimate of the earthquake magnitude consistent with the palaeoseismological data indicates a maximum value of about $M = 6.7$. Obviously, such an indication is arbitrary, owing to the lack of elements compatible with the definition itself of magnitude. Perhaps the aim is just to indicate that there was a strong earthquake at this site. The cracks inside the sanctuary can be associated with movement of the main fault in the earthquake. The fact that the Gargano is a seismic zone is well known and its most important historical earthquakes (1627, 1646, 1893) have been studied ever since the beginning of the twentieth century, along with its active faults. It may be true that such an investigation is enthralling, but can the results be added to an earthquake catalogue?

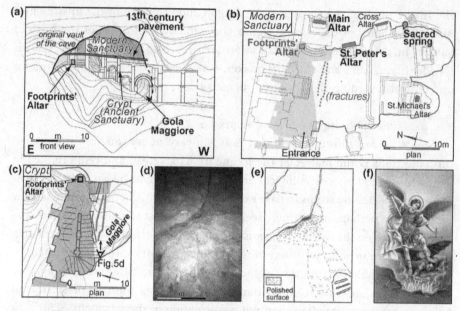

Figure 3.2 Evolution of the sanctuary at Monte Sant'Angelo (southern Italy). (a) Front view of the sanctuary vault, with the restoration dated 1273 highlighted. The ancient sanctuary (today the crypt), including the major gorge, is found below the newer (thirteenth-century) paving. (b) Plan of the present-day sanctuary. The main altars are located near the Footprints' Altar, which is at a lower level than the newer altars. (c) Plan of the crypt, with the morphology of the original access to the sacred vault. (d and e) Photograph and sketch of some openings along one of the fractures on the flank of the major gorge (see the fractures drawn in b and c). Three steps carved into the rock leading to the openings can be seen in the bottom right of the photograph and sketch. (f) Iconography of the apparition of Archangel Michael at Monte Sant'Angelo (from Piccardi, 2005, p. 122).

The publication of the book *Myths and Geology* (Piccardi and Masse, 2007), albeit by the authoritative Geological Society, does not dispel the doubts concerning such an approach, which is perhaps innately tied to opinion. Indeed, notwithstanding the enthusiasm or the skill of the individual authors, it cannot be seen how the barrier that the myths and the legends themselves pose our rationality can be overcome before they can ever be put in relation with definite scientific data. If geology undeniably adds a different point of view to the reading of myths and legends, it is nevertheless important to bear in mind that the specific disciplinary skills of historians of the ancient world, ethnographers and anthropologists should be seen as indispensable. In other words, multidisciplinary sensibility, as for historical seismology, should never be forsaken.

Ancient religious texts: how should they be used?
Bible earthquakes and seismologists' attraction to Sodom and Gomorrah

References to earthquakes and related phenomena in the Hebrew Bible undoubtedly spring from direct experiences of real events, but in most cases the experience is stripped of its historical dimension and is given a religious perspective. In other words, the earthquake becomes part of the theophany of Yahweh since a seismic event is one of the ways in which the God of Israel makes Himself manifest. According to Busi (1994), in order to understand the particular way in which this conception of earthquakes is applied in the Hebrew Bible, one must identify the cultural background of the authors who wrote those Bible passages. Unfortunately, in seismological tradition (Montessus de Ballore, 1915), even the most recent, there has been a tendency to interpret Biblical expressions literally as references to geological events (Bentor, 1989) rather than as theological statements.

In the ancient tradition of the Near East, earthquakes were seen as an expression of invisible forces – a manifestation of hidden power – and as such were integrated into the complex mesh of magical relationships that were believed to underlie the physical world. It is well known that, within a magical interpretation of things, a system of relationships exists between 'low' and 'high', the world of the senses and the one beyond the senses, the visible and invisible, heaven and earth, and earthquakes occupy their rightful position within this magical system. Indeed, earthquakes and other extraordinary natural phenomena have a particularly important part to play. It was the case for all of the ancient world that events such as earthquakes, which disrupt the regular behaviour of nature, are purveyors of a special message because they provide a higher and more intense level of communication between man and his physical and cultural surroundings. These beliefs can be found throughout the ancient Near East, and they underlie the Hebrew scriptures. Thus, in the pages of the Bible, the interpretation of events shifts from the symbolic system of magic to that of God's will and actions. In the case of earthquakes, the phenomenon is seen as a direct manifestation of Yahweh, showing His mood and His intervention in the worldly affairs of the chosen people. The shaking of the Earth is seen as announcing either the Lord's presence or His wrath.

The idea that earthquakes announce God's presence is described in certain passages from the Psalms, which provide a detailed account of the natural phenomena accompanying the presence of the *numen*. Thus, in Psalms 68.8–9 (EVV. 7–8), it is stated: 'O God, when thou didst go forth before thy people, marching across the wilderness, the earth trembled, the very heavens quaked before God.' Busi (1994) finds many other instances of this kind, and notes that the commonest term for earthquake in Hebrew is *ra'ash*. This term appears 47 times in the Old Testament. In 30 cases it appears as a verb and in 17 cases as a noun. The phonetically similar root *ragaz* is also used, though less frequently. There are 11 passages in the Bible where verb forms deriving from *ragaz* occur in reference to earthquakes, and one instance it is in a nominal form. This analysis definitely tends to restrict the number of passages in the Bible in which a reference to an earthquake can be recognized as an actual natural event, but this opinion on the interpretation of biblical references to earthquakes is not shared by everyone. For example, in Guidoboni *et al.* (1994) only the earthquake that struck Jerusalem and the valley of Hinnom in c.760–750 BC is accepted as being a real event based on the book of Amos, but other authors argue for other actual earthquakes as well. The c.760–750 BC earthquake caused damage to the temple in Jerusalem and is linked to the punishment inflicted by God on Uzziah, king of Judah, for his impious behaviour. This latter element provides the dating of this event: since it is known that when King Uzziah suffered from leprosy, Jotham succeeded him around 756 (or 759) BC. This provides a valuable piece of chronological evidence for dating the earthquake (Soggin, 1970, p. 120). King Uzziah caught leprosy in the sixth decade of the eighth century BC, and at the time of his illness there was an earthquake. These two unusual events were interpreted as being a divine punishment. Ben-Menahem (1979, p. 262) argues that the dating of this earthquake can be made even more accurate. In his opinion, the earthquake is very likely to have occurred 'at Yom Kippur, Monday, 10 Tishrei, 3003, that is October 07, 759 BC, during the daytime'.

According to excavations carried out by the Israeli archaeologist Y. Yadin in 1956, the Hazor site (in northern Galilee, 14 km north of Lake Tiberias) revealed traces of an upheaval of the earth that caused serious damage and an unexpected interruption to the construction of the settlement. Stratigraphic evidence suggests that the earthquake is likely to have occurred around 760 BC. Traces of the earthquake were found in the sixth stratum of the upper city (eighth century BC; see Yadin, 1961, p. 24, n. 73). Dever (1992) has attributed the signs of a sudden destruction found in a defensive wall at Gezer to the earthquake of c.760 BC. Other examples of archaeological studies in relation to this earthquake are Lemaire (1997), Herzog (2002), Knauf (2002) and Marco and Agnon (2002). For a critical evaluation of the archaeological evidence of this Biblical earthquake see Ambraseys (2005, in particular table 1).

Apart from the earthquake of the eighth century BC, the Bible also contains references to other seismic events. Harris and Beardow (1995) discuss four of these references: in Hazor in 759 BC as discussed above; in Sodom in 1900 BC, on the basis of Genesis 19; on Mount Sinai in 1500 BC, on the basis of Exodus 19; and in Jericho in c.1450 BC, on the basis of Joshua 4–6. The last of these earthquakes is supposed to have made the walls of Jericho collapse. However, more recent critical archaeology (Nur and Ron, 1997; Herzog, 1999; Finkelstein and Silberman, 2002; Ambraseys *et al.*, 2002) has raised the important counterpoint: that the cities of Cana'an did not have walls (a summary of the debate can be found in Guidoboni and Poirier, 2004, pp. 37–40).

In spite of doubts and conflicting interpretations, the thinking of geologists and engineers concerning the story of Sodom and Gomorrah in the Bible has converged towards a common viewpoint. The destruction of these two cities is described in the book of Genesis in rather picturesque, as well as enigmatic, terms. This destruction was the subject of interpretations in the 1940s within classical Bible exegesis, with a view to actually locating the two cities (Harland, 1942), for which ruins have never been found. Harris and Beardow (1995) used a geotechnical approach to interpret the destruction; they suggest that it 'may have been due to liquefaction-induced lateral spreading of deposits around the Dead Sea owing to an earthquake, with the remains of the cities ending up beneath the water'. Subsequently, Haigh and Madabhushi (2002) suggested a dynamic model to account for the destruction of Sodom and Gomorrah, as the result of 'alternate layers of liquefiable sand and silt'. But at least until the ruins are actually located, this event will remain shrouded in mystery.

Sibylline oracles and the 'prophesies' of actual earthquakes (end of the fourth century BC to fourth century AD)

In the classical tradition the term *Sibylline Oracles (Oracula Sibyllina)* refers to the collection of oracles attributed to the ancient Greek Sibyls. The term *oracle* refers to a prophetic response, given by a divinity or by a supernatural being even if not exactly divine, directly or by way of go-betweens. In actual fact, the work that is identified with *Sibylline Oracles* brings together oracles of a highly composite origin and comes under the category of Biblical apocryphs, which collect the oracles that actually came from pagan cults, to which additions and modifications were made, but also with substantial interpolations by Hellenist Jews and later by Christians. This work is chronologically set within a very long range of time, from some time between the end of the fourth century BC and the start of the third century BC to the fourth century AD.

Oracles cover a wide range of topics, from a reinterpretation of history in a religious light to prophesies concerning the end of Rome. According to Guidoboni and Poirier (2004) the interesting aspect of oracles for historical seismology is that they sometimes mention earthquakes and tsunamis that did indeed occur. The ancient authors, who at times made use of real seismic occurrences, cited them as a confirmation of the tragic nature of an event, whose exceptionality also lay in the fact that it had been envisioned beforehand. For example, Pausanias, a Greek author who lived in the second half of the second century AD, mentioned in his work on Greece the earthquake that in 227 BC had struck Rhodes, which made the famous Colossus guarding the port to fall over. This seismic event also hit the outlying regions of Carya and Lycia. According to Pausanias: '[The earthquake] damage to the cities of Carya and Lycia and above all the island of Rhodes was so great that the oracle of the Sibyl concerning Rhodes seemed to have come true' (Pausanias 2.7.1.). In the *Sibylline Oracles* (4.11) we can indeed read: 'Rhodes too will have its ultimate and greatest disaster.'

In oracle texts, an earthquake that had already occurred was transformed into a future event, which gave strength to the narrative and sustained a sort of sacred authority for posterity. As an example, the memory of the earthquake of Phrygia in 27 BC, which destroyed Laodicea and Tralles, is preserved in the *Sibylline Oracles* in the customary prophetic formula, in this case couched in the terms of a divine menace: 'Tralles, the neighbour of Ephesus – an earthquake shall destroy the well-built walls and the wealth of a troubled peoples; the earth shall spout up water boiling hot; the groaning earth shall swallow them down with a smell of brimstones' (3.459–62). And again: 'Woe unto you, oh delightful Tralles, and woe unto you, oh beautiful city of Laodicea, for you shall be destroyed and reduced to dust in an earthquake' (5.289–91). The evocation of a future tragedy could also assume the high tones of poetry, such as for the tsunami that hit Patara (near the present-day Kalkan, in Turkey) in AD 68: 'Fair Mira of Lycia, the earth shall shake and not remain firm; thou shalt fall headlong to the ground and pray to find another land of refuge, an an emigrant, when with thunderings and earthquakes the dark waters of the sea spread sand over Patara, for its godlessness' (*Sibylline Oracles* 4.109–123, 5.126).

The Patara tsunami is also recalled by Dio Cassius (63.26.5), who lived between the second and the third century AD and who drew upon other sources. In the 12th book of the *Sibylline Oracles*, datable to the third century AD, an earthquake in Phrygia is recalled in the formula of a prophesy *a posteriori*: 'Phrygia, too, with its rich in flocks, shall lament as a result of earthquakes. Alas! Laodicea, and alas! wretched Hierapolis, you were the first to be swallowed up by the yawning earth' (*Sibylline Oracles* 12.279–281). In this case, unfortunately, no other source is capable of adding information about this seismic event, which perhaps was accompanied by widespread landslides. Paradoxically, one can be sure that it really did happen because it was added to the Oracles.

Lives of Christian martyrs and saints: impossible or likely earthquakes?
(fourth to tenth centuries AD)

Some European earthquake catalogues use the life of a saint or an ancient martyr as a basis for their historical information on individual seismic

Figure 3.3 Sant'Agata, patron saint of the city of Catania, shown in a codex from the late tenth century (from Guidoboni and Marmo, 1989).

events. What kind of source is this? Does it contain reliable historical elements? According to Guidoboni and Marmo (1989), it is worthwhile reflecting on these kinds of sources, which concern periods, like those of late antiquity and the early Middle Ages, for which earthquake data are few and far between. The general narrative pattern depicted in a scene from the *Acts of the Martyrs* is that of the struggle between the Christian and the pagan, where the latter tries to make the former idolatrize his own idols. Paradoxically, the Christian God plays a secondary role in the episodes concerning the martyrs, performing the function that Propp (1971) argues comes under the sphere of action of the *Helper*. In the crucial moments of the clash between the hero and the antagonist, following an invocation by the Christian, God intervenes and then the earthquake makes its appearance as an instrument of divine intervention. It is precisely this narrative function that makes these stories unreliable as sources of information on historical earthquakes, even if the martyr story itself has an historical and chronological dimension. The earthquake as a literary *topos* is nearly always an event that had never actually occurred, unless it was an Act of God (the same goes for the *Acta Sanctae Agatae*, perhaps of the fifth century (see Figure 3.3), or the *Martyrium Sanctae Martinae*, of the seventh century). Narrative strategies of this kind can also be found in the Passion of the 10 000 martyrs (or the 1480, according to some versions) of Armenia (*Passio decem millium martyrum*), an application of the same storytelling devices on a vast scale. In this source, the earthquakes occur in the canonical hours and mark the rhythm of the passion of the 10 000 crucified soldiers. A far different narrative landscape presents the lives of the saints from the fifth century onwards. In this kind of source the earthquakes dictated by narrative demands appear to be less frequent. Often, especially in the biographies of the Oriental saints, the trend to insert the memory of an authentic earthquake, also attested to by other sources, into the lives of the saints becomes popular. In any case, the martyrs' texts must be evaluated with great caution in order not to make gross errors of in terms of historical accuracy resulting from a superficial use of an element of the storytelling narrative.

From the eighteenth century to the early twentieth century it was not infrequent for historians of European seismicity to use hagiographic sources for dating earthquakes, although later the seismic events in these sources were determined to be narrative fabrications, like the false earthquake of Tongres towards the end of the seventh century, of which Alexandre (1989) presented a critical study. There are also narratives of *Lives* that contain information on earthquakes and tsunamis that do belong to the historical genre. Generally speaking, the saint has the function of providing protection from the ongoing historical natural phenomenon, which wreaks havoc and makes fear run rife. One such example is in the *Life of Saint Hilarion*, written by Jerome, one of the most important intellectuals of his time (fourth century AD) and one of the Fathers of the Church. In this account Hilarion calms the sea on the beach of Epidaurus (present-day Cavtat, south of Dubrovnik, on the Dalmatian coast) and reassures the fear-stricken inhabitants with thaumaturgic actions. This story perhaps refers to a local effect of the violent tsunami in AD 365 that originated at Crete, of which there is some information for the other Mediterranean shores (Greece, Egypt). For the Dalmatian coast this cryptic reference is the only one known that might refer to this tsunami.

Other *Lives* that contain real historical data, such as those of Saint Severinus, Saint Theodosius and Saint Ignatius of Constantinople, and of Saint Simeon the Stylite, all have in common the fact of having been written a few years after the death of the protagonists. These contain references to earthquakes that had really occurred, and quite often the seismic events are attested to by other sources as well. For example, the earthquake of Antioch mentioned in the life of Saint Simeon the Stylite, in which the saint's father was killed, is that of May 526. It is also attested to by Malalas (419–20), Procopius (*Bella* 2.14.6), Theophanes (172–3), Evagrius (*Historia Ecclesiatica* 4.5) and other authors. Another *Life* that contains the only information so far available concerning a tenth-century earthquake in Calabria (Southern Italy) is that of Saint Nilus, written by the monk Bartolomeo da Rossano. The text states that while the saint was in the small village of Rossano, between the years 951 and 1004, there occurred an earthquake so strong as to reshape the landscape, while the inhabitants 'were no longer able to recognize the place'. This is an original way of paraphrasing the effects of this earthquake on the natural environment, as seen through the eyes of the inhabitants. There were probably landslides and changes in the local hydrological system, elements typical in this area, which have also been observed for modern earthquakes in the same region.

Ancient Greek and Latin literary sources: some clarifications

Although the quality of a seismic catalogue and the chances of it being used are not just linked to the length of the chronological period of the observations (the Chinese catalogue, for example, starts from 1177 BC but cannot be used for statistical purposes and for risk evaluations until after the thirteenth century), the continuity of information is nonetheless a very important factor. In the areas where a millennial written culture has existed, studies of the most ancient earthquakes and tsunamis are a relevant research sector for historical seismology. This research on antiquity is also fostered by a substantial production of studies, beginning in the nineteenth century. No other specialized historical sector has taken an interest in earthquakes and other calamities for such a long span of time as that of the scholars of classical antiquity, although one cannot speak of a fully fledged tradition of study. Two main issues emerge from this set of heterogeneous texts: on the one hand, an interest in single earthquakes or lists of events accruing from the written sources; on the other, the study of the cultural attitudes of the ancients with regard to seismic phenomena. The former topic has interested historians and archaeologists much more than seismologists (Nissen, 1883; Capelle, 1924; Momigliano, 1930; Downey, 1955; Hermann, 1962; Smidt, 1970; Croke, 1981; Ducat, 1984; Jacques and Bousquet, 1984); the latter has been developed by the historians of literature, philosophy and religion (Capelle, 1908; Chatelain, 1909; Cartledge, 1976; Dagron, 1981; MacBain, 1982; Henry, 1985; Autino, 1987; Waldherr, 1997; Olshausen and Sonnabend, 1998; Hine, 2002). The diversity between these two themes has led, only rarely, to examining the earthquakes and tsunamis within a historico-cultural perspective.

The ancient literary Greek and Latin texts are rich in information on seismo-tectonic phenomena. Earthquakes, tsunamis and volcanoes arouse great interest among the natural philosophers and the ancient authors. On the basis of the various seismic events that had occurred and were directly observed, a highly complex naturalistic thinking took shape that later underpinned the science of nature in the Western area (on these aspects and their development, see Chapter 4). Here we make only a few comments on the ancient literary sources and on the problems that they pose in historical seismology research.

The definition of a historical *source*, which we have given previously, is based on the criteria of geographical and chronological proximity to a described event and on the multiplicity and independence of the information concerning the event. For ancient sources these criteria become ever more difficult and at times almost impossible to fulfil. The ancient sources are indeed rare and have been passed down to us after surviving the great, and at times random, ravages of time. Over long time periods the losses are immense, and every fragment of information that has reached us is invaluable because of its rarity. It should not come as much of a surprise then if historical seismology uses texts that were written centuries after the events but which purport to give fundamental testimonies for ancient earthquakes. However, even in these cases, the philological criteria and those of textual criticism must be the guiding principles for a correct use of the texts. For the analysis of ancient texts the philological rule should be applied: *recentiores non deteriores* (the most recent texts are not the most corrupt). In other words, one piece of information is no less authoritative if testified to by sources produced even several centuries later. This element, which is not valid for the later periods with much richer documentary contexts, is motivated by the particular value of the written text in the ancient world and also by the particular ways that the texts themselves were transmitted.

As regards ancient historians and men of letters, it is important to take into account each author's working method. A good many ancient authors wrote their works by drawing upon codified oral traditions or by consulting other first-hand sources, such as the archives of Rome and of the *municipia*. Later testimonies such as these are sometimes the only sources that enable an earthquake to be dated and located. In other cases, earlier authors or even those contemporary to the event have left little mention, while later historians have picked up and perpetuated traditions of great value, often preserving details crucial to the knowledge of an earthquake. In order to better understand the informational discontinuity concerning the ancient world one should remember that a substantial part of the texts have reached us today in a fragmentary form, typically through more or less direct citations by other authors (consider the large and invaluable collection of fragments edited by F. Jacoby, *Die Fragmente der griechischen Historiker* for the Greek texts and that of H. Peter, *Historicorum Romanorum reliquiae*, for the Latin ones).

Something else that may help to add to the value of the relatively little information on ancient earthquakes is the almost sacred worth of the written text in the ancient world. Errors may have occurred during the copying of the manuscripts, but this is a problem common to the whole of ancient literature since, with very few exceptions, original texts dating back to the early Middle Ages no longer exist.

Substantial changes in the cultural and cognitive contexts of writing about seismic events over time mean that certain elements that are now considered important in the description of an earthquake were not at all thought so in the ancient world. The early authors' culture was essentially urban and the city represented a privileged watchtower. For this reason, minor seismic effects were remembered more easily than the disastrous effects hitting distant towns and villages only because the former were felt in a large city, such as Rome. Nevertheless, for the period of the Roman Republic (late sixth to early first century BC), various fragments of news concerning earthquakes occurring in different sites are known, coming from the lists of the so-called 'prodigies' of the various *municipia*. Such 'prodigies' concerned earthquakes, portentous births and generally whatever could be interpreted as an element unexplained by an idea of normality of nature (Traina, 1989, 1994). The lists preserved in the *municipia* did not, however, record all earthquakes systematically, because the earthquake itself was perceived as an inevitable natural phenomenon. For this reason, historical memory has recorded only those events that were considered in some way to interfere directly with human activity or could be charged with some particular significance, perhaps because the earthquake had occurred at the same time as an important social or political event.

Historiographic narrations are generally available for the whole of classical antiquity, allowing for a fairly accurate dating of seismic events. With the advent of Christianity, there was a mutation in the language of these sources. New values influenced the interpretation of natural calamities, which were seen as being a divine punishment. Moreover, Christian humanism did not initially impose the need to take into account new descriptive data, not as yet considered very relevant, such as the number of victims. The value of prodigy generally prevails in Christian historiographic narrations as compared with the value of creating a record of the effects on society of a seismic event. Information concerning earthquakes and tsunamis in this rich wealth of historiographic literature available today must be extracted from these sources keeping in mind the standpoint of their authors.

A large research effort has been made over the last few years to organize and reclassify the important data regarding the seismicity of the ancient Mediterranean world, which had been gathered by the rather confused tradition of past historical seismology research. These data are significant because they represent an island of memory concerning earthquakes from the distant past. While these data have been systematically selected and are now well analysed, the record of ancient earthquakes for this area still remains largely incomplete. In spite of these limitations, this temporally extensive record contains some extremely helpful information on those earthquakes that have lengthy return periods.

3.2.2   Sources written on stones: the inscriptions

By the term *inscription* we refer generically to a written text that is engraved or painted onto a durable material such as a stone wall or a tablet. Inscriptions chiselled or painted on these kinds of materials come not only from the classical period of Greek and Latin epigraphy but also from the very ancient cuneiform Assyrian–Babylonian tablets and, in much later eras, from the epigraphy of late antiquity, the Middle Ages and even more recent times. Inscriptions have been written in the many of the languages of the Mediterranean area and in the other parts of the world where they survive. We have chosen a rather arbitrary grouping of this chronologically extensive set of materials based on their functional use regarding reports of earthquakes and tsunamis. As a whole, this grouping does not correspond to a set of records that belongs to a single academic discipline, but rather these records are studied by a diverse array of historical and linguistic scholars. .

Although there is a wealth of inscription materials available that could be used in historical seismology research, it is our observation that inscriptions have not been used for scientific purposes for the dating and locating of earthquakes as frequently as one might have expected. Furthermore, there has been no shortage of epigraphers studying these sources who have been able to provide help on the subject of historical earthquakes and tsunamis. Epigraphers have pointed out inscriptions that have reinforced other literary testimonies concerning particular earthquakes, that contain a precise reference to a reconstruction resulting from previous seismic damage, or that, being part of funeral stele, mention people who had died as a result of earthquakes. The history of the use of inscriptions is by now a chapter in historical seismology. We recall that history here in a general way to give some idea how historical seismological knowledge correlated to epigraphic testimonies developed and how epigraphic studies have had repercussions for historical seismology.

Although there was no shortage of specific contributions (Lanciani, 1918, for Rome, Comparetti, 1914, for Cyrene–Libya) even at the beginning of the twentieth century, it is only in the second half of that century that epigraphic research started making contributions specifically addressed to earthquakes, and in some cases began to create a network of observations and exchanges with researchers interested in historical seismology. The publication by Thompson (1937) of a cuneiform thirteenth-century BC tablet, which recalls an earthquake at Nineveh, remains a historical contribution, but it is cited in Ambraseys and Melville (1982) for its information concerning an ancient earthquake. In another case, Camodeca (1971) made use of an inscription dated to 352–357 or to 356–357 AD (*CIL* 9.2338) concerning the restorations carried out in the Samnium (Central Southern Italy) by Fabius Maximus, the *rector provinciae*, to argue that the damage which was being fixed was a consequence of the earthquake of 346 AD. In this case, the epigrapher introduced the criteria of analogy by relating an inscription that has no reference to an earthquake to an earlier earthquake that was known to have occurred. Such an interpretation can be a rather hazardous one if it is extended uncritically. What is known is that a public figure (Fabius Maximus in this case) dealt with the restorations in a given area. On the basis of an inscription that explicitly reports an earthquake, a researcher asserts that other inscriptions for restorations to buildings from the same geographical and chronological period are an implicit attestation to the damage from the earlier seismic event.

Over the past two or so decades, the utilization of epigraphic sources for historical earthquake research has grown, as have historical studies of inscriptions. Robert (1978) published Greek inscriptions of Asia Minor and the islands of the southern Aegean Sea (Rhodes, Kos, etc.), from the end of the third century BC to the end of the fourth century AD. Burnand (1984) presented the first dossier of 11 epigraphic Latin sources for the Italian area in which there is mention of restoration actions carried out following earthquakes. In these cases the earthquakes are mentioned explicitly. Ambraseys (1988), in a study concerning earthquakes in the eastern Mediterranean area, among his other various sources also uses about 40 Greek and Latin inscriptions. Guidoboni (1989) increased the selection of the Latin inscriptions by Burnand, bringing the dossiers of the epigraphic earthquake testimonies for Italy to 16, by assessing all the epigraphic *corpora*. Also from this perusal, 415 inscriptions mentioning restorations, reconstructions and collapses also emerged, but unless the explicit cause of the damage is stated these alone cannot be used for the purposes of historical seismology.

There is one case of Latin African epigraphy that was the subject of a heated debate in the closing stages of the twentieth century. Di Vita (1964, 1980, 1982), Beschaouch (1975) and Rebuffat (1980) correlated a large number of inscriptions to the earthquake and tsunami on 21 July 365 AD in the central eastern Mediterranean. The inscriptions only mention restorations and reconstructions in areas that today are Algerian and Tunisian, with no mention of any earthquakes. Jacques and Bousquet (1984) and Lepelley (1984) have cast doubt from many standpoints on the validity of the hypothesis that the inscriptions refer to earthquake or tsunami damage. They confute it with rather sound arguments, which include a lexical analysis, the historical analysis of the building activities of some Roman consuls, and the use of written sources *ex silentio*. The clash between these two schools of thought has for several decades represented one of the most pointed debates on an ancient Mediterranean seismic event. Indeed, the consequences of the outcome of the debate are clear. To one group of scholars the damage caused by the event in AD 365 had a devastating impact across a huge area, including a large part of northern Africa; to the other group, there is no evidence to support this hypothesis and therefore the impact of the earthquake and tsunami is highly uncertain.

Hypotheses, clues and epigraphic riddles

What has been said above about inscriptions may provide some idea as to the value but also the difficulties in using these particular written sources. But what is the main difficulty in the use of these inscriptions within the specific discipline of historical seismology? First of all, one must remember that inscriptions are nearly always fragmented texts, often having been discovered fortuitously during archaeological excavations or during restorations of old structures. The ones currently preserved, which survived losses and destruction of all sorts, were transcribed and translated by generations of epigraphers and collected in specific *corpora* and specialized journals. Most inscriptions are mutilated and fragmented. The main problem concerning their use in historical seismology is their *interpolation*, where letters, words or even phrases are missing and must be inferred from *a priori* information. There are terms inserted in epigraphic transcriptions that are at times not even partially written in the original inscription but were deduced by the epigraphers on the grounds of analogies with other inscriptions or owing to congruence with the overall text and/or situation of reference. Missing letters, which at times might even occupy the space of several words, are inserted by the epigraphers to make sense of the inscription, much as one might complete a riddle. However, one can never be absolutely certain that the letters or words added to transcriptions, usually represented by square brackets, are the correct replacements for the missing information.

Integral inscriptions do not create this sort of problem, but present other problematic aspects, due to their intrinsic nature. We discuss as an example the best-known inscriptions, the Latin Italian inscriptions analysed by Guidoboni (1989). These 16 inscriptions, in which restorations or other interventions are reported following the effects of an explicitly mentioned earthquake, come from the period between the first century AD and the early years of the sixth century AD. That so few such inscriptions have been found is likely due to the paltry number of surviving texts. There is a general absence of compendiums on which to perform systematic research for the epigraphy of the Republican period and for post-antiquity (until the ninth to tenth centuries AD) epigraphy.

As part of his work, Burnand (1984) put forward some lexical analyses that sought to identify the party funding the restorations. In some instances, the funding party is identified in the construction, restoration and restructuring of public buildings with a public power or a person possessing with public functions operating in his own right. In other instances, the funding comes from a private figure who, for any of a variety of reasons, shoulders the cost of the restoration work. Burnand (1984) also evaluated the terms used to describe restorations with the aim of verifying whether there was a recurrent correspondence in the use of a specific term relating to collapses or to rebuilding work. His goal was to see if common terminology reflected the nature of the causes, whether earthquakes, wars or other events, underlying the collapses themselves that were being rebuilt.

Burnand (1984) notes that the terms used to indicate the destruction of a building are *collabi* for *collapse* (*conlapsum, conlapsam*, etc.), *dilabi* for *to be ruined* or *fallen* (*dilapsum*, etc.), *evertere* for *destroy* or *overwhelm* (*eversas*, etc.) and *prosterno* for *knock down* or *throw down* (*ruina prostravit*). There are also terms used to commemorate the reconstruction, such as *restituere* for *restore* or *reinstate* (*templum . . . restituit*, etc.) which can be accompanied by the expression *a solo* (*from the ground*) or *a fundamento, a fundamentis*, etc. (*from the foundations*). It is important to point out that these kinds of terms also recur in epigraphic texts that have no relationship to seismic events. Burnand (1984, p. 179) observes, from inscriptions definitely attributable to a restoration intervention following an earthquake, a special use of the verb *restituere*. However, this verb is normally used in most inscriptions bearing witness to restorations, whatever the cause of the damage may have been. Beschaouch (1975, p. 109), regarding African epigraphy, has interpreted the recurrence of the noun *ruina* (*ruin*) as a clue capable of substantiating the hypothesis of an earthquake. Lepelley (1981, p. 55; 1984, p. 487) refuted that hypothesis, providing a comparative table in which, from the examination of some inscriptions dating from the first half of the fourth century AD, the word *ruina* appears but was used in wholly different situations. Even the verb *collabor* (*collapse*) has the same variety of usages. Thus, one needs to proceed very warily in historical seismology research and be aware of the difficulties in using inscriptions that do not have the term *earthquake* clearly expressed as the cause of damage.

A stele of Nicomedia of the second century AD recalls two children and their childminder who died as the result of an earthquake

A violent earthquake struck Nicomedia in AD 120. The ancient city today is Izmit (Turkey) in a notoriously highly seismic zone. An inscription (now in the

Louvre) from the imperial period (erected by a man named Thraso in memory of his two small children and the man in charge of them, who were killed in the earthquake). It reads as follows: "Thraso, son of Diogenes, erected this stele for his two sons, Dexiphanes aged five and Thraso aged four, and for Hermes, aged twenty-five, who was raising them. In the ruins caused by the earthquake he embraced them like this." The inscription was published in *Corpus Inscriptionum Graecarum* (3293, 3352); see also Robert (1978).

> *A fragmented inscription bears witness to an earlier earthquake*
> *foreshadowing the damaging event at San Giuliano in*
> *central Italy in 2002?*

On 31 October 2002 a strong earthquake struck a limited area of the central southern Apennines in Italy. In a small village, San Giuliano di Puglia, it caused the collapse, amongst other things, of the roof of a school, killing 29 schoolchildren. It was a shocking event for the population as well as for the seismologists. The earthquake occurred during a period of administrative transition between old and new regulations regarding earthquake resistance in buildings. Although it was already established that the village was classified as a seismically active area owing to the proximity of some active faults, it was not known that an earlier damaging earthquake may taken place at San Giuliano in historic times. In actual fact there were some known indications of this possible earlier event, but its evidence was not properly examined until after the terrible 2002 earthquake.

Regarding this possible earlier earthquake, there was some possible evidence discovered in 1985 at Larino, a few kilometres away from San Giuliano, during restoration work that was being carried out on a private home. The restoration work uncovered some fragments of a Latin inscription (Figure 3.4) which celebrates a priestess named Gabbia, one of whose merits is that of having restored the sanctuary of Queen Juno in Larino after its collapse ([col]lapsam). Unfortunately, the inscription does not come from its original archaeological context. The stone had been reused as building material and, even before then, perhaps in the eighteenth century, had been reinscribed on its back. Based on the painstaking recent profile of the excavations carried out so far at Larino (De Felice, 1994), no sacred building that can be traced back to Juno has yet been found in the area. The editor of the newer inscription (Stelluti, 1997, p. 187) proposes a dating somewhere between the first and the second centuries AD. The portion of the text available does not enable scholars safely to associate the collapse of the *aedes Iunonis* to an earthquake. The only certain event is the collapse of the temple, which might simply have been due to the wear and tear of time. The phrase "collapsed due to its age" ([vetustate co]llapsam) – a formula often used – could

Figure 3.4 Reconstruction of the inscription found near Larino (southern Italy) (from Stelluti, 1997).

be equally compatible with the inscription fragment that remains today. Thus, until there is more accurate information about this inscription, it remains as a possible but questionable clue to a strong earthquake in an area that historically had not recorded any others before that of October 2002. Could there have been a predecessor earthquake?

The painted Greek Christian inscription of Cyrene: an earthquake in the second half of the fourth century AD

Until its abandonment in the seventh century following the Arab invasion, Cyrene (today close to Shahhat, in Libya) was a flourishing city in the historical region of Cyrenaica in northeastern Libya in northern Africa. This area is not known for being seismically active, and thus the report of a destructive earthquake here arouses particular attention. Evidence for the occurrence of this event has been preserved in two kinds of sources: a Greek Christian inscription painted on a wall and two letters written in AD 411 and 412 by Synesius (AD 370–413), the bishop of Ptolemais and Pentapolis.

The two letters of Synesius (*Epistulae* 42 and 66) are known because they have been the subject of critical historical analysis. The letters contain some terse references to the effects of an earthquake that had caused damage to the Pentapolis and had destroyed the village of Hydrax, bordering the Libyan desert. At Hydrax there was once a castle that served as a defence against invading barbarians (those were the years of the invasions of the Vandals in North Africa).

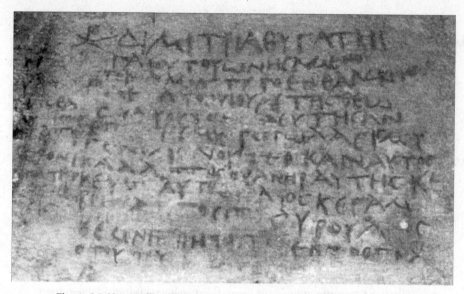

Figure 3.5 Necropolis of Cyrene, near Shahhat (Libya). This is a painted Christian Greek inscription from Demeter's Tomb recording the death of a mother and her son in an earthquake in the late fourth century (from Guidoboni *et al.*, 1994, p. 281).

Synesius had gone to Hydrax in his role as bishop, perhaps in 412, to discuss with some local clerics the matter of the use of those ruins. Synesius's references allow us to hypothesize that the earthquake had occurred several years before, although the date of the seismic event is not known.

The other source that bears witness to this earthquake is a Greek Christian inscription in the necropolis of Cyrene (Figure 3.5). This long inscription is painted with a brush in red on a rock wall in the northeastern necropolis. It recalls that a woman named Demeter had died with her son Theodulus at Myropola (perhaps a town near Cyrene) owing to an earthquake. This inscription independently confirms an earthquake in the Cyrenaica area, but is it the same earthquake as the one mentioned by Synesius? And in what year should the earthquake in the inscription be dated?

The interpretation of this inscription has a rather meandering history, which is briefly reviewed here as it may give some idea of how the accepted interpretation of an inscription can be the result of several attempts made at different times and by different researchers. The first mention of this inscription is found in a travelogue of a French artist, J.-R. Pacho, published in Paris in 1827–29. Without specifying his exact location, Pacho (1827–29) transcribed the inscription by hand, because there was no other way in those days to reproduce an image (Table LXIII, no. 9). But because the inscription was scarcely legible and the traveller did not understand Greek, he made many mistakes in his transcription. The first decodification of this text, made by Letronne in the same work (p. 395), is thus unreliable. The inscription was then reproduced in the fourth volume of the *Corpus Inscriptionum Graecarum*, under number 9136. In this work, Franz corrected some of the errors by Pacho but introduced others. For example, because of the length of the text Franz reckoned that the inscription consisted of two funeral inscriptions joined together. He postulated that together the inscription attests to the burial of two Christian martyrs, which occurred during the persecution of a proconsul who was hypothesized by Franz himself.

One day in 1911, just before the war between Italy and Turkey broke out, a German professor named F. Halbherr was part of an archaeological mission in Cyrenaica. He ventured alone into the semi-desert zone of Libya and spent three days visiting the ruins of Cyrene. He saw the inscription painted in red in the cave of the necropolis. Since he had a camera with him, he took some pictures of the inscription. Being uncertain on several points about the inscription text because the colour had faded, he had the photograph touched up by an archaeological draughtsman (designer), and then he handed it over to a renowned Italian expert in ancient Greek, Professor Comparetti. When at last Comparetti could examine the inscription, he realized that there were many mistakes in the earlier interpretations. Also, he believed that the inscription was in actual fact a single text. Besides highlighting the kinship relations correlated to the names in the inscription, Comparetti (1914) found that some letters in the marginal notes allowed for the identification of the symbol used to indicate the 15th year of the reign of the Emperor Theodosius or the year AD 394, albeit with some reasonable doubt. Many years later, other authors re-examined the inscription, casting some doubt over Comparetti's interpretation but without making significant variations to it. So, according to this story, the inscription of Cyrene is believed to bear witness to an earthquake in AD 394 that Synesius also referred to in his two letters.

At the current stage of research, it cannot be excluded that these destructive effects may be associated with the earthquake in Crete in AD 365, even if the inscription does not mention tsunami effects (see also Bacchielli, 1995).

The medieval inscriptions in the Mediterranean area

Epigraphy from the Latin, Greek, Arab and Armenian areas each tend to pose unique problems in the search for and analysis of earthquakes of the past. Since there is no single compendium available, research into these sources is a rather time-consuming process as it requires the consultation of a great number of erudite works, compendiums and works in the epigraphic literature. Furthermore, many inscriptions have never been the subject of study and are preserved inside monastic or ecclesiastical buildings, on bell towers, in mosques and minarets, and on the walls of ancient Armenian churches. In most cases they testify to the reconstruction of a great building, as well as its reconsecration if it is a religious structure. The custom of recalling with an inscription the reconstruction after an earthquake was a long-standing custom in the Mediterranean area until the last century. However, it is clear that for more recent time periods when many other authoritative testimonies are available, the value of this kind of source tends to be diminished.

The memory of an earthquake can be preserved in an inscription when it is mentioned as the cause of a reconstruction or a restoration. Although an inscription nearly always provides information for just one building, nonetheless it is an accurate piece of information concerning the damage at the locality where the building was located. This information may perhaps not be supported by other forms of documentation, and when this is true the earthquake inscription is uniquely important for historical seismology. New archaeological excavations or restorations of ancient or medieval buildings can lead to new epigraphic discoveries and to improve our knowledge of earthquakes in the Mediterranean area. The study of inscriptions is a pursuit that remains potentially rich in surprises for the historical seismology of the ancient and medieval period.

Byzantine inscriptions: a field for skilled and patient researchers

Research into Byzantine epigraphy is particularly complex owing to the lack of a single as well as a systematic *corpus*, as is available for the Greek inscriptions of the ancient period. Byzantine research work is therefore addressed to finding traces of the reconstruction of ecclesiastical buildings or of restorations of masonry sections of old structures. The nature of Byzantine inscriptions, however, is such that the reasons leading to restoration work described in the inscriptions are rarely given, and this must always make researchers proceed with the utmost caution when interpreting them.

Byzantine inscriptions are most easily understood if they are connected with other types of documentation and with the aims of those commissioning the restoration work. Indeed if those who commissioned the work were private individuals, the reconstruction donors usually tried to enhance their own image by emphasizing the novelty and difficulty of the reconstruction task rather than on the causes that had triggered it. The person who rebuilt or restored a building usually preferred to present himself to the public as the 'new founder' of the structure, and entrusted to the inscription the task of listing and praising the qualities and merits of his act. Moreover, from the middle Byzantine age and onwards the founders of private churches became increasingly independent of public initiatives, thus acquiring privileges or rights of lordship over the buildings constructed or refurbished (Herman, 1946).

From the standpoint of philological–textual analysis, the problems with interpreting inscriptions can be great owing to the fact that the verbs commonly used to denote the action of restoring, reconstructing or renovating tend to merge into and overlap with the normal meaning expressed by the verb 'construct' in the sense of a new construction. In Byzantine inscriptions the very same word means either 'built' or 'rebuilt'; as a consequence, the question of whether the building mentioned in an inscription was newly built or, on the contrary, was restored after a destructive event in many cases remains unresolved. Obviously, help with this problem may come from recovering as much architectural information as possible about the concerned buildings, but other problems may remain. In fact, in official inscriptions, those issued by political and religious authorities who took care of the construction, or reconstruction, of a certain building, usually the causes or events which led to such activities are ignored, while the solicitude, the *pietas* or the common good that moved the authorities to undertake these activities are stressed. This is the case of many Cretan texts, mostly from private individuals, reported in the fourth volume of the work by Gerola (1932), which can lend themselves to more than one interpretation.

In order to avoid the sometimes high level of uncertainty in attributing damage to a building as being the result of a seismic event, it is necessary to find architectural reports and news items on the history of the building. Alternatively, investigating the names of the donors of a reconstruction may facilitate comparisons of news reported in the historical and literary archival documentation (chronicles, political works, lives of saints, etc.), which are apt to offer a broader range of information about the building project. All of this is often possible by the painstaking perusal of bibliographic works, such as the annual bibliographic accounts of the *Byzantinische Zeitschrift* or the *Bulletin Épigraphique*, issued in the *Revue des Études grecques*, just to mention some of the most significant ones.

The *Corpus Inscriptionum Graecarum* (CIG) is a very dated but still helpful resource for Byzantine inscriptions. Ahrweiler (1973, p. 515) states that the study of Byzantine inscriptions remains sporadic and even neglected. An attempt to fill the knowledge gaps, especially the chronological ones, present in the CIG was announced but never completely carried out by Ševčenko (1973, p. 526) (see also Favreau, 1989). An excellent bibliographic summary concerning all of the Christian inscriptions in Greek edited up to 1991 (*Inscriptions of the Christian Empire*, ICE) is the one edited by J. M. Mansfield of the Department of Classics, Cornell University (*The Infimae Aetatis. A Textual Data Bank of Late Antique and Medieval Inscriptions*). It is always important to understand the ways in which the world of late antiquity and the Byzantine world perceived earthquakes and which answers the culture of the times tried to give concerning their occurrences. Some of these elements have been highlighted by Dagron (1981) and by Evangelatou-Notara (1993).

The Arab inscriptions: rare traces of seismic destructions

In Islamic culture references to earthquakes are present in the Koran in a verse (*sura*) called 'of the earthquake'. There the seismic event assumes a dual aura of exceptionality for the devastation that it causes and for its supernatural character, marking the advent of the Day of Judgement. Earthquakes are alluded to as analogues, and these can be found in different *ahadith* or sayings of the Prophet of Islam. The study of Arab epigraphic documentation for such a vast geographical area as the Arab–Islamic world presents a series of problems unique to this culture. Even so, these sources make a significant contribution to historical seismology thanks to the support of other non-epigraphic documentary sources. As regards the use of inscriptions for the location of seismic effects or at least the mention of earthquakes in the Mediterranean area, some specific research has been performed by Ciccarello (1996). This author has highlighted that the Arab literati, or Arabophones, utilized a quite consistent pattern of language in which neither the passing of centuries nor huge geographical distances are considered as great differences. In funeral tombstones (*stele*) the cause of death is never mentioned, except in the case of a martyr's death; most of an epigraphic text is reserved to Koranic and prophetic citations or to more worldly eulogies in honour of eminent personages. The inscriptions that mention the construction or restoration of a building only exceptionally contain any mention of the causes that have led to the ruin of the building to which it refers. At times a close linguistic analysis of the use of certain lexical items allows one to hypothesize that an earthquake might have damaged a structure. Ciccarello (1996) points out that some lexical markers allow one to hypothesize that the action of restoration attested to in an inscription may have been the consequence of damage caused by a seismic event, or at any rate a sudden destruction not due to anthropic causes. For example, the verb *saqata* excludes destruction by the hand of man (e.g. as a result of a battle), and the temporal conjunction *lamma* seems to indicate that a reconstruction took place immediately after the collapse of a building.

Thus, several hypotheses are possible for the numerous inscriptions attesting to reconstructions or restorations that took place in the years shortly after known earthquakes that are attested to by other written sources. But in the absence of explicit references to seismic events in the inscriptions, it is not always legitimate to construe a direct relationship between a damaging earthquake and an epigraphic documentation. Indeed, a number of different man-made causes, such as the frequent dynastic struggles, the attacks by armies and the sieges upon some centres by the Crusaders, and the rarer but no less destructive popular upheavals, could give rise to destruction to buildings that would make necessary far-reaching restoration measures. In order to differentiate from among the possible hypotheses of the cause of the destruction of a building, and whether the cause was anthropic and natural, it is essential to engage in interdisciplinary studies. This is necessary to maximize the possibility relating ambiguous inscriptions to earthquakes that are known from other sources. For the Arab world, the number of inscriptions that explicitly testify to a destructive seismic event, highlighted by Ciccarello's research, consists of five inscriptions out of a corpus of over 8000 inscriptions assessed. Thus, most Arabic inscriptions must be interpreted in light of other evidence.

Inscriptions in the Italian area

For the Italian area there is no *corpus* of inscriptions from the medieval or the modern era, as there is for the ancient period. Thus, most inscriptions hitherto known recall earthquakes that have either been identified directly for their effects on historical buildings (churches, bell towers, castles, mansions) or are known from works of scholarly historiography. There are some inscriptions that refer explicitly to an earthquake and thus preserve the precise date of the seismic event, and much more seldom they also contain references to the effects on a specific building. The language of inscriptions is generally terse with a minimum of information, although there is no shortage of truly literary compositions (for some examples, see the end of this section).

Numerous inscriptions refer to reconstruction or *ex-novo* construction works, above all of religious buildings, in which only the date of reconsecration is indicated. This is particularly true for the Middle Ages. The repairing of damage to the great religious buildings in medieval times could last for several decades. During the repairs the churches remained wholly or partly inaccessible for prayer, so that when the work ended, the sacred building underwent reconsecration. Thus, between the date of the foundation of a church and that of its reconsecration there could have occurred a strong earthquake that affected the church. If the start of the reconstruction work on a church is unknown, one may be strongly tempted to consider that the reconstruction phase is for repairs needed as the result of seismic damage. Although this hypothesis is rather enticing (even the great art historians of the twentieth century were attracted by it; e.g. Porter, 1915–17, and Arslan, 1939), caution is essential. Indeed, the association between large rebuilding projects and seismic damage is not so obvious or easy to show as one might wish. Although it is true that intense phases of rebuilding work and enlargements made to religious buildings had in the past been motivated by seismic destruction, it is nevertheless also true that intense construction activity could have been driven by economic and demographic factors alone.

As for the ancient inscriptions, many medieval inscriptions chose to celebrate the qualities of the promoters and the financial backers of the reconstruction work, rather than to recall the causes that had actually necessitated the work. In order to make the most accurate use of medieval inscriptions, it is worthwhile knowing the construction history of a building and its various construction phases. At the present time 14 Italian inscriptions are known that bear witness directly – that is with the explicit mention of the word 'earthquake' – to seismic events that had taken place between the twelfth and the fifteenth centuries, and these refer to six strong earthquakes. There are inscriptions that recall earthquakes for subsequent centuries, but these are less important reports of the local effects of an earthquake because in more recent centuries there other types of sources with greater information content that are also available to document the earthquake and its effects.

The two examples that follow are drawn from Guidoboni and Comastri (2005).

'Speaking' churches and chronological riddles in medieval inscriptions

In Padua in northern Italy an inscription, which is now lost but the text of which has been jealously handed down by the city's scholarly historiography, was in the cathedral on an architrave (or lintel) in the middle of the church. This inscription recalls the reconstruction work, under the direction of an architect named Macillo, that was completed in 1124. In accordance with medieval epigraphic convention, the inscription recounted the building's own collapse and reconstruction in the first person. The inscription, translated from the medieval Latin, reads:

> First the earthquake caused me to collapse completely, but Macillo raised me
> from the mud in beautiful form. In the year of our Lord 1124, in the 2nd
> indiction, Macillo built me up from the mud.

The earthquake that had caused damage to Padua and many other locations in northern Italy had taken place on 3 January 1117.

The date of this same earthquake is concealed almost as if it were a riddle within the text of another inscription, chiselled into the white marble of an architrave on the door of the main entrance to the church of the Abbey of Nonantola (near Modena). This long inscription (Figure 3.6), written using good Latin metre, records the collapse of the upper parts of the church building and the start of rebuilding work four years later:

> The top of the great church of the great Sylvester collapsed after the revolutions
> of the sun had registered one thousand one hundred and seventeen years since
> the birth of the Redeemer, and the rebuilding of the church was begun four
> years later.

Figure 3.6 Inscription from the church of San Silvestro at the Abbey of Nonantola (northern Italy). On the architrave of the entrance portal there is engraved a long inscription recalling the collapse of the elevated parts of the building resulting from the earthquake on 3 January 1117 (from Guidoboni and Comastri, 2005).

A story in golden painted lettering: an earthquake and plague in Venice (1348)

In Venice, in the lunette of an internal portal in the former Scuola Grande della Carità, there is a long text painted in gilded lettering, which was placed there in the years immediately following the earthquake on 25 January 1348. The text records both the earthquake itself and the serious outbreak of the plague that occurred a few months later. The effects of the earthquake are recorded in an ancient Venetian language, here in translation:

In the name of Almighty God and the Blessed Virgin Mary. In the year of the Incarnation of our Lord Jesus Christ 1347 [Venetian style; 1348 modern style] on January 25 on the feast of the Conversion of St Paul at about the hour of vespers there was a great earthquake in Venice and almost throughout the world and the tops of many bell-towers and houses and chimneys collapsed, as well as the church of San Basilio; and so great was the fear that almost everybody thought they were going to die, and the earth continued to shake for about 40 days; and after this a great plague began and people died of various diseases and from various causes; some spat blood from their mouths and some had swellings under their armpits and ears, while the skin of others became dark, and it seemed that these diseases passed from one person to another, that is to say from the sickly to the healthy. And people were so frightened that fathers did not want to be with their sons, nor sons with their fathers; and these deaths continued for about six months and it was often said that two thirds of the people of Venice had died.

Armenian and Georgian inscriptions: the walls of churches are like books of memories

In the Middle Ages in Armenia the production of inscriptions developed greatly. Even today there are well-preserved churches whose external and internal walls are covered with carved texts. These are essentially dedicatory inscriptions in churches and monasteries, along with funeral inscriptions that were often placed on the typical cross-shaped (*kachkar*) monuments (Papazian, 1991; Karakhanian and Abgaryan, 2004). The inscriptions represent a particularly useful testimony, in that they report accurate chronological data. The texts of the DHV (*Diwan Hay Vimagru yan, Diwan Hay Vimagrut'yan*) that make up the corpus of Armenian epigraphy are found only in the territories of the present-day Republic of Armenia and the autonomous region of the Karabak (a mountain region). In addition, there are some found in the city of Aní, today in Turkish territory, and there are Armenian inscriptions in the area of some of the ex-Soviet Union republics, particularly Georgia.

For the territories of Turkey, Azerbaijan and Iran, there are instead no systematic compilations of Armenian inscriptions, and their documentary use remains extremely difficult. There are inscriptions cited by antiquarian compendiums, local histories or monographic publications on individual monuments. Particularly frustrating is the presence of inscriptions reported within antiquarian works of the Venetian Mechitarist Father Levond Alishan, who lived at end of the nineteenth century. His monographs often report these sources on a regional basis with incomplete texts and without details that allow modern researchers to identify them accurately (Alishan, 1881, 1899). At the current state of understanding, only in a few cases do Armenian inscriptions indicate an earthquake exactly. During the research into the Armenian earthquakes only five such inscriptions have been identified (research by Igor Dorfmann and Giusto Traina cited in Guidoboni and Comastri, 2005). This is a very small data set for an area so often hit by strong earthquakes. It would be interesting to create a complete documentation of inscriptions for the region of Cilicia (present-day southern Turkey), where numerous inscriptions were made but unfortunately have never been systematically collected. An even more chaotic situation concerns the inscriptions in Georgian, of interest for the earthquakes of northern Armenia. Given the lack of more accurate and systematic data it is difficult to use the Armenian inscriptions (and, in general, the Caucasic inscriptions) directly in historical seismology research. There are some cases in which inscriptions attest to the

Figure 3.7 Coin of the Emperor Tiberius, who is thanked for the help given for reconstruction following an earthquake in Asia Minor. The coin (British Museum, K. 90535 D 36D) is dated between the years AD 22 and 23 (from Guidoboni *et al.*, 1994).

reconstruction of all or parts of monuments, but do not explicitly indicate the cause that necessitated the reconstruction. Thus, Armenian inscriptions may only be used if a very explicit chronological convergence appears with information from other sources.

Numismatic sources

A type of source akin to inscriptions comes from coins. Following particularly strong earthquakes in the ancient world, new coins were sometimes minted. These coins would celebrate the beneficial intervention of an Emperor following a destructive earthquake and served to spread an imperial image of generosity and splendour. One of the most famous coins of this type is the one minted after the earthquakes in the year 17 that damaged 12 important cities of Asia Minor. In this coin Emperor Tiberius is thanked for the aid given in the reconstruction work (see Figure 3.7). For the use of coins as a chronological indicator (*terminus post quem*) for dating collapses being studied in archaeological research, see Chapter 11.

3.2.3 Sources of individual memory: annals, chronicles, notulae and other memories

Human memory and sensitivity have made an extraordinary contribution to the knowledge of seismic events, enabling researchers to discover elements about the earthquakes of the past that could in no other way be known today. It is true, of course, that these sources are individual universes, immersed in their own time and permeated by their own cultures, and that they therefore act as extremely powerful *filters* between the earthquakes that had indeed occurred and our present chances of knowing them. In spite of such obvious drawbacks, these kinds of sources allow us today to obtain information on earthquakes that occurred many centuries ago. The texts associated with the memories of individuals contain elements that are indispensable for locating not only earthquake damage, from which an epicentral area can be traced, but even just the felt effects at localities where no damage was experienced. The latter element is very important in finding the extent of the strong ground-shaking from an earthquake throughout a territory, from which its epicentre and size can be estimated.

Annals

The annalistic genre constitutes one of the most resource-rich and powerful means of transmission of the memory of individual persons and their communities. This genre is characterized by the narration of facts *year by year* (*annum* in Latin) in a way that was nearly always schematic and chronological. Annals may concern kingdoms, cities or monasteries. Every civilization has had its own annalist traditional, whether of greater or lesser importance.

For a long period of time in the ancient Latin world, the Annals maintained the official and sacred character of the *Annales*, in which the *pontifices* (the priests dedicated to the pagan cults) recorded each year the most important city events. Latin historians often drew upon these annals of historical events and information, thus adding to their own works events recorded with some measure of officialdom. In the early and middle European Middle Ages, the *Annales* of the monasteries were records that narrated year by year the events that occurred and of which the monks were either the direct witnesses or recipients of oral information from others. The events were selected by their importance and by their significance to the view of the world at the time. Initially, the monastic Annals were created on the basis of annotations placed on tables indicating for each year the date of Easter; later they became a genre in their own right. The nature of such monastic texts was almost sacred. Not always are the authors of such texts known, because they are works whose contents have grown in time and they never were conceived of as the work of one author. Such valuable texts have been transferred and handed down obsequiously to subsequent generations, copied carefully or saved in their original form. Annals have preserved the memory of many historical events, amongst which are natural events such as earthquakes, comets, floods and so on, which would otherwise be totally forgotten today.

The ancient Latin Annals of Ravenna (sixth century AD): earthquakes recorded and represented

Ravenna achieved its utmost splendour starting in the fifth century AD, when Emperor Honorius (395–423) chose it as the capital of the Western Roman

Empire. During the time of foreign invasions into the Italian peninsula and the widespread disaggregation of the Roman administration, Ravenna offered important guarantees for defence, because it overlooked the sea and was protected from behind by a ring of canals and marshlands, which made it practically invincible. Even after the Western Roman Empire had fallen in AD 476, Ravenna maintained its central political position, becoming the capital of the Kingdom of Italy. Precisely for this reason the Annals continued to be written and jealously guarded, since they maintain the official chronicles of the city, as was customary in the ancient Roman cities. It was only thanks to the survival of these very ancient texts that we know that between 429 and 492 four earthquakes were felt in Ravenna, whose epicentres were probably located in the Romagna Appenines. The *Annales Ravennates* (edited by Bischoff and Koehler, 1939) is a sixth-century manuscript, which has reached us through an eleventh-century copy and today is preserved in the Capitular Library of Merseburg (Germany). Among the few other pieces of information it contains (in the typical style of the ancient annals), there is the mention of earthquakes dated 25 August 429; 15 April 443; 467 and 26 May 492. No more is known about these earthquakes, but it is the very fact of having been inserted into the Annals of the city that makes it legitimate to suppose that they were earthquakes that may have had a destructive impact, if not exactly in the city but perhaps in the outlying towns.

The news of an earthquake is accompanied on three occasions by an illustration that arouses some questions as to if the illustration could have been a fifth-century representation of the earthquake. The drawing (Figure 3.8) consists of two parts joined together. First, there is the bust of a person, a man or a woman, the left half of which is wrapped in a cloak. Second, from the cloak there departs the scaly body of a snake-shaped monster that ends with the head of a dragon. From its wide-open jaws, fire seems to emerge. This fire may be a breath, which according to the Aristotelian theory of the *pneuma* or the subterranean wind is the cause of an earthquake (see Chapter 4). The breath is directed towards the human figure facing it, which is holding his right arm up to the height of his shoulder, with the fingers of his hand straight. If this interpretation of these precious texts is right, then this is one of the most ancient depictions of an earthquake in the Western world. The strange dragon could be the simplification of an Echidna, a divinity of the underworld imagined by the ancient populaces as a half-nymph, half-serpent, closely related to the giant Typhoon.

A catalogue of medieval earthquakes based on the monastic Annals

The importance of the monastic Annals for learning about medieval earthquakes has been, so to speak, consecrated by a catalogue assembled using these types of sources alone. This catalogue is the one published by Alexandre

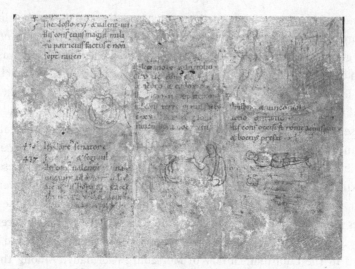

Figure 3.8 A fifth-century drawing that may depict an earthquake. In the centre of the page is the bust of a person, a man or a woman, wrapped in a cloak. Opposite the person is a snake-shaped monster with a dragon's head, with fire seeming to emerge from its mouth (Merseburg, Archiv und Bibliothek des Domstifts, ms. 202).

(1990). The area for which these kinds of sources make their greatest contribution is that of Western Europe. Earthquakes of Italy are partially attested to in this catalogue as well as some events from the eastern Mediterranean, although the latter are referenced quite indirectly. This catalogue is of high quality, being based on a critical reading of the various Annals written in the European monasteries. It systematically provides essential critiques that can be used for establishing the reliability of the annalist information. For example, it gives information about the place where the texts were written, their interdependence and their originality.

Almost complete for the period from the fourth to the thirteenth centuries for Western Europe, this catalogue is the result of a systematic perusal of a unique set of sources, and it bears witness to how the monastic culture assimilated earthquakes for almost a millennium. The earthquakes mentioned in these annals emerge from a unique perception of the real world, that is to say, the one pertaining to monastic religiosity and ethics. The representation of the sixth seal in the *Revelation* of John summarizes the profound religious meaning of the earthquake as a sign from another reality, the supernatural one (Figure 3.9). In spite of such a bias, an inherent charactersitic of the types of the sources themselves, the monastic annals preserve some elements of realism that allow modern researchers to locate the areas affected by the seismic events and to

Figure 3.9 A representation of the sixth seal in the *Revelation* of John preserved in the annals of a monastery, about 1300 (British Library, Add. 18633 29 39188).

learn about their effects. In this way, Alexandre (1990) has corrected the numerous errors made in the preceding tradition of European studies, from von Hoff (1840–41) and Perrey (1845) to Milne (1911) and Giessberger (1922), and has added new as well as priceless elements to the historical earthquake record.

The Mesoamerican handwritten codices

The handwritten codices of the native Mesoamericans have been analysed by historians of seismicity, who have shed light on these sources that were written before and just after the Spanish Conquest. Kovach (2004) presents a detailed contribution on Mesoamerican historical seismicity (in particular in Chapter 5, pp. 61–72). According to Kovach, the Mesoamerican codices can be divided into two large categories: ones written before the Spanish conquest (1519), or pre-Cortésian, and ones written subsequent to the conquest. These codices can be further subdivided between those of the Maya reign and those of the Mexico reigns.

Regarding codices of the Maya reign, only four are known dating from the time period prior to the Spanish conquest. These contain rituals and deal with time and divination. Amongst these, the *Dresden Codex*, dated to the first half of the thirteenth century of our era, is composed in the form of an almanac. It deals with matters of astronomy as well as many aspects of the daily life of the Maya. This codex also indirectly mentions earthquakes.

Eleven Mexican codices written before the conquest are known today, and they are pictographic and ideographic. Seven of these codices come from the Mixtec-speaking region situated in the valley of Oaxaca in southern Mexico. The Mexican codices written after the Conquest were drafted 'under Spanish patronage by native scribes' according to Kovach and often contain chronicles in the forms of annals (i.e. year-by-year chronicles).

Kovach (2004) cites as significant examples of Mesoamerican codices the *Codex Aubin* (c. 1576), the *Codex Telleriano-Remensis* (c. 1562), the *Codex Ríos* (1566–89), also known as *Codex Vaticanus A* or *Codex Vaticanus 3738* and the *Codex Mendoza* (1541), which deals with the life of the Aztecs at the beginning of the sixteenth century. It is interesting to note that this codex contains the history from 1325 to 1521 of the city of Tenochtitlán, which is the present-day Mexico City, and it also preserves valuable reports about earthquakes.

Kovach (2004, p. 72) lists 19 Mexican earthquakes occurring between 1455 and 1513, during the time period before the Spanish conquest. These earthquakes only have rather generic information about their locations (for example, the Valley of Mexico) and rather scanty descriptions of the shaking and damage effects. The texts share some similarity with the austere and bare language used in the annals of the European monasteries in the tenth to twelfth centuries, but are very far from the informational richness of the contemporary sources of the Mediterranean area (Latin, Greek and Arabic). For instance, the *Codex Telleriano-Remensis* contains some references to earthquakes. One of these is marked with seven dots and a symbol that stands for *Tecpatl* (= *7 Acatl*) (Kovach, 2004, Fig. 5.6) and reports additionally the date of the Christian era, in this case the year 1460. The text, written in Spanish and translated into English here, is the following:

> In the year 7 Flint and 1460 an earthquake occurred. It deserves to be said that since, according to their belief the world was again to be destroyed by earthquakes, they recorded in their paintings each year, the omens that occurred.

More personalized is the memory of a local scribe, contained in the same codex, of an earthquake that occurred in the year *6 Calli*, or 1537 in the Western calendar:

> This year of six houses [6 House] and 1537 the blacks tried to rebel, in the city of Mexico the instigators were hanged. The star was smoking and there was an

earthquake, the worst I have ever seen, even though I had seen many of them in these lands.

The *Codex Aubin* (1576) preserves the memory of an earthquake that occurred in the year 9 *Acatl*, which is the year 1475. The text is in the Nahuatl language, which translated into English reads something like this:

Here [the Earth] shook much. Many hills were crumbled and all the houses were crushed. (Kovach, 2004, p. 69)

Chronicles and diaries

Chronicles are a kind of source that are very similar to annals, but in the middle and early Middle Ages (in particular from the twelfth to the fifteenth century) in the more intensely urbanized areas, they generally took on a less universal character than annals. Chronicles are usually connected to short-term events within the area of influence of a city. These important fragments of memory and personal perception are produced by notaries, merchants, men of law, educated citizens and intellectuals belonging to some religious orders. These texts are nearly always extremely complex linguistically, semantically and culturally; they cannot therefore be used in a naive way by merely extracting the information that apparently serves to describe a seismic event. On the contrary, interpreting these texts requires a thorough knowledge of the cultural context and means of expression in which they were written.

In the case of chronicles, it is important to know where the author was when he did the actual writing. This is necessary to discern whether the author reports a personal and direct testimony in the text, or rather records news that he had received after it had been passed on by word of mouth, perhaps several times. Indeed, a writer could often gain access to letters, dispatches or other diplomatic documentation and use them in his text. This is an important reason why a re-examination of chronicle texts already known to the seismological tradition, if based on revised critical tools, can make a fresh contribution to the interpretation of seismic effects.

Chronicles containing other sources: letters of Florentine merchants on the 1348 earthquake in a work by Giovanni Villani

Giovanni Villani is the author of one of the most important fourteenth-century chronicles. He was the son of merchants and was a merchant himself, and hence he moved in the circles of the great companies of Florentine bankers such as the Peruzzi and the Bardi. He made frequent journeys to France and

Flanders in the early fourteenth century, and was later in Naples and Siena. He carried on intense political activity. He died in Florence during the outbreak of plague in 1348.

Villani began the 12 books of his chronicles about 1306. The earlier part of the work, up to Book 6, is a compilation of earlier texts and information. In these earlier books he uses Biblical and Roman material, as well as myths about the origins of Florence. From the sixth book to the twelfth, he deals with more recent events, finally reaching those of which he had personal experience. For the most recent events, he used first-hand official documents. The chronicle is of considerable documentary importance for the history of Florence in the thirteenth and fourteenth centuries, because of the wealth and reliability of the information it provides. Villani records important information which he had obtained, he tells us, from some merchants who had been travelling at the time of the earthquake between the villages of Sacile and Carinthian to the north of Villach. The long passage in his chronicle that deals with the earthquake must therefore be considered a very important piece of evidence for establishing the effects of the seismic events of 1348:

> In that year, on the night of Friday January 25, there were numerous very severe earthquakes in Italy: in the cities of Pisa, Bologna, Padua, and especially in Venice, where a great many of its numerous beautiful chimneys collapsed; and church towers and other buildings in these cities split open, some of them collapsing. And these things brought damage and plague to these areas, as you will find if you read on.
>
> The most dangerous shocks were felt that night in Friuli and in the city of Aquileia, as well as in parts of Germany. They were of such a kind and so destructive that they would seem incredible if described in spoken or written words; but to provide a true and accurate account in this treatise, we will set out a copy of a letter which certain reliable Florentine merchants sent from there, and we will explain the tenor of their letter, written as it was and dated at Udine in the month of February 1347 [1348, Incarnation style].
>
> You will have heard of the various dangerous earthquakes that occurred in these countries, causing very severe damage. In the year of our Lord 1348 according to the church calendar, but still 1347 in our Annunciation calendar, on Wednesday January 25, the Feast of the Conversion of St Paul, at eight and a quarter hours towards vespers, which is the fifth hour of the night, there was a tremendous earthquake lasting for more than two hours, unlike any known to living memory.

First of all, in Sancille [Sacile] the gate towards Friuli collapsed completely. Part of the patriarch's palace in Udine collapsed, together with other houses. The castle of Santo Daniello in Frioli [San Daniele del Friuli] collapsed, and a number of men and women were killed. Two towers of the castle of Ragogna collapsed and slid down to the river called the Tagliamento, and a number of people were killed.

More than half the houses in Gelmona [Gemona del Friuli] collapsed in ruins, and the tower of the principal church cracked and broke open, and the stone figure of St Christopher split from top to bottom. Spurred by these miraculous events and by fear, the moneylenders of the place repented and announced that everyone to whom they had lent money at interest should go to them to get it back; and they continued giving it back for more than a week.

Half the town bell-tower at Vencione [Venzone] cracked, and a number of houses collapsed.

The castle of Tornezzo [Tolmezzo] and those of Dorestagno [Arnoldstein] and Destrafitto [Stassfried] almost completely collapsed in ruins, and many people were killed there.

The castle of Lemborgo [Wasserleonburg], which is in the mountains, was shaken. As it collapsed, it was moved ten miles from its original position by the earthquake, and was completely destroyed.

A very large mountain, near the road leading to Lake Dorestagno [Arnoldstein], split open and half of it completely collapsed, blocking the road in question.

And the two castles of Ragni [Rain] and Vedrone [Federaun], with more than fifty villages, situated by the contado of Gorizia, on either side of the river Gieglia [Gail], were reduced to ruins and buried under two mountains, and almost all the inhabitants perished.

In the town of Villach in Friuli, all the houses collapsed, except for that of a virtuous and just man, full of Christian charity. And in the surrounding area, more than 60 castles and villages above the river Atri [Afritz] were similarly reduced to ruins and buried under two mountains, and the valley where the river ran for more than 10 miles was filled in; and the monastery of Orestano [Arnoldstein] was reduced to ruins and submerged, and many people died there. And since this river had lost its exit and usual course, a large new lake formed on top. Many strange things happened in the town of Villach. A cross-shaped crack appeared in the main square, and out of the crack there came first blood and then a great quantity of water. And in the town church of San Jacopo 500 men who had taken refuge there were found dead, and there were many other victims in the town, amounting to more than three quarters of the

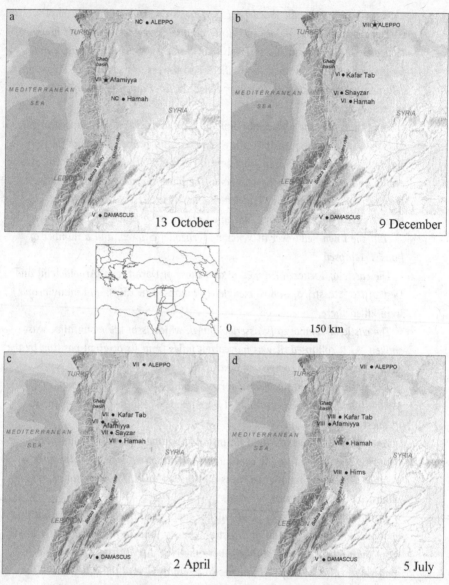

Figure 3.10 Area affected by the destructive earthquakes of September 1156 to May 1159, reconstructed on the basis of the contemporary Arab sources (from Guidoboni and Comastri, 2005).

inhabitants; but by a miracle, Latins, people from other parts and the poor escaped with their lives. In Carnia, more than 25 000 people were killed in the earthquake; and all the churches in Carnia collapsed, and the houses and the monastery of Osgalche [Ossiach] and Verchir [Feldkirchen] were all reduced to ruins.

*In Bavaria, in the town of Trasborgo [Oberdrauburg], as well as at Paluzia
[Paluzza], La Muda [Mauthen] and La Croce [Monte Croce Carnico] beyond the
mountains, most houses collapsed, and many people were killed.*

*And the reader should note that all the above earthquake damage and
danger is a great sign and judgement from God. Its great cause lies in God
and in the prediction provided by those miracles and signs which Jesus Christ,
as he preached the gospel, gave to his disciples as due to appear at the end of
time. (Giovanni Villani, ed. Porta, 1991, vol. 3, pp. 562–566.)*

*Medieval diaries from the Islamic area: how the twelfth-century seismic
crisis was told*

Between late September 1156 and late May 1159, a long and destructive seismic sequence struck an area comprising present-day northwest Syria,
northern Lebanon and the region of Antioch (Antakya, in Turkey) (Figure 3.10).
The duration and destructiveness of this long seismic sequence, as well as the
breadth of the affected area, have contributed to the event being remembered
and passed down to subsequent centuries. These earthquakes entered the cultural and erudite European tradition, and therefore the early earthquake catalogues. The seismic sequence is first recorded in Italian texts dating back as
far as the end of the seventeenth century (Bonito, 1691). Since then, nearly all
historical seismic catalogues of the eastern Mediterranean, both descriptive and
parametric, record great earthquakes in the Middle East in the mid twelfth
century.

Retracing the seismological tradition of the last two centuries, one does
indeed encounter substantial fluctuations in the chronological parameters of
this sequence (the date varies between 1152 and 1160), which in some cases has
generated earthquakes that in actual fact never happened (false earthquakes).
But perhaps the most important factor is that this seismic sequence has for
a long time been mistakenly interpreted, sometimes as a single earthquake or
sometimes with at most two or three separate events. The shocks are never
related to one another and are ascribed enormous uncertainties regarding the
locations of their epicentres. As a consequence, an understanding of the real
nature of these important earthquakes has been lacking. It has not been appreciated that this was a very long series (it lasted more than two and a half years)
of seismic events, all originating across a well-defined area.

The most detailed account of this series of earthquakes is provided by Ibn al-
Qalanisi, a historian from Damascus who died in 1160. He was an eyewitness to
the long seismic sequence, and devotes many pages to an account of the earthquakes, including numerous details on the effects of individual earthquakes.
The keenness of the observations and the accuracy of the story are surprising

if compared with the Latin sources of the same period. These elements in al-Qalanisi's descriptions are probably due to a cultural propensity for accuracy, and to the author's personal characteristics.

On the basis of information supplied by Ibn al-Qalanisi, who was in Damascus at the time of the shocks and received news about them, it is possible to identify at least the more important shocks in the sequence between 13 October 1156 and 16 April 1158. The earthquakes in 1156–1159 took place in a particularly difficult period for the Latin and Frankish states. Edessa had fallen in 1144–1146, a fact that later constituted the justification for the Second Crusade in 1147–1149. From 1151 to 1157 the attack upon Nur al-Din was incessant (and ended in 1160). In the years 1156–1157 there were attempts by the Christian army to reconquer some fortresses, including that of Shayzar, precisely because it had been hit by the earthquake.

> During the night of 9 Sha'ban in the year 551 [27 September 1156, corresponding to 27 Elul], there was a tremendous earthquake at the second hour, which caused the earth to shake three or four times; then, by the will of Him who had brought it about, the Most High, the Omnipotent, it grew calm again. During the night of Wednesday 22 Sha'ban [10 October 1156], there was another earthquake followed by two more of similar intensity during the day and at night. Then there were three more weaker shocks, making a total of six. During the night of Saturday 25 of that month [13 October 1156], another earthquake filled hearts with fear from dawn and then throughout the day, until, by the will of Him who had brought them about, the Most High, the Omnipotent, the earthquakes subsided. News came from Aleppo and Hamat of disasters in many parts [of those provinces], and the destruction of a tower at Afamiyya, caused by these earthquakes sent by God. About forty shocks were counted, but only God knows the truth of the matter. Certainly, nothing of the like had ever been seen in previous years. On Wednesday 29 Sha'ban [17 October 1156], there was another earthquake, after the one already mentioned, at the end of the day, and then yet another during the night. On Monday [from this point on the days of the week do not match the days of the month] 1 Ramadan [18 October 1156], there was a terrible earthquake followed by a second and a third. On Tuesday 3 Ramadan [20 October 1156], there were three earthquakes, the first of which occurred at the beginning of the day and was dreadful, whereas the second and third were of lesser intensity. There were more shocks in the afternoon, and another dreadful one woke people up in the middle of the night. Glory be to God, who devises such formidable trials. At the ninth hour of the night on Friday 15 Ramadan [1 November 1156], there was another terrible earthquake, stronger than the previous ones; in the morning,

there was a less powerful shock, followed by a second and a third at the beginning and the end of the Saturday night. On the following Monday, there was another dreadful earthquake. It was followed by another violent earthquake during the first third of the night of Friday 23 Ramadan [9 November 1156]. At midday on Sunday 2 Shawwal [18 November 1156], an even more violent earthquake than the previous ones spread panic and terror amongst the populace. On Thursday 6 Shawwal [22 November 1156], there was a shock at the time of midday prayer. On Monday 16 of the same month [2 December 1156], there was a dreadful shock at the same hour. The following Tuesday there was another similar shock, followed by a second, weaker one, and then by a third and a fourth. During the night of Sunday 23 Shawwal [9 December 1156], there was a tremendous earthquake that disturbed men's spirits. Other earthquakes followed – too frequent to number. God spared Damascus and its districts from this terror, displaying his mercy to the inhabitants, all praise and thanks be to Him. But news from Aleppo, not to mention Shayzar, spoke of many houses destroyed [at Aleppo], falling in on their inhabitants and so causing many deaths. The inhabitants of Kafar Tab fled from their town in panic. The same thing happened at Hamat; we heard nothing from the other Syrian provinces as to what happened after the earthquake. (Ibn al-Qalanisi, *Dhayl ta'rikh Dimashq,* pp. 334–6.)

The contributions of the notulae *and* colophons

Before the time when books were mechanically printed, the writing system used for manuscripts was a matter of individual choice. Bearing in mind that the medium used for manuscripts (above all parchment) was expensive and not abundant, it was quite normal that it should be used as efficiently as possible. For this reason, it was not unusual for the owners of various kinds of texts to note within them, amongst other things, the memory of an earthquake that they had themselves felt. These observations were written inside another text, in the blank spaces within or at the foot of a page, or on the cover page. These are called *notulae* (from the Latin, which means 'small annotations'), but in actual fact they could also be very long.

The *host* text, so to speak, for an earthquake *notula* could have been of any of various kinds and was mostly related to some custom followed by the writer of the *notulae*. *Notulae* could have been entered into a liturgical book, a juridical codex, a monastic annal, a chronicle or a book of prayers. The publication of such short texts started in the early twentieth century, with the works of Lampros (1910, 1922). Numerous other *notulae* were subsequently published by Spyridon and Eustratiades (1925), Darrouzès (1950, 1951, 1953, 1956), Schreiner

(1975–79, vol. II), Evangelatou-Notara (1982), Turyn (1980) and Constantinides and Browning (1993). These kinds of notes have some distinctive elements that differentiate them from other notations. For example, the important *notulae* on the earthquakes that occurred in the Middle Ages in the Byzantine area are preserved in some liturgical books of local churches (see the very useful editions of Lampros, 1910 and Evangelatou-Notara, 1982). For these reports, usually only the date of the earthquake is indicated, sometimes accompanied by an adjective that describes the event: strong, very strong, etc. At times, together with the date (year, month and day) there is also some reference made to the time of the event. This element is often complementary to other sources that were written in different places, which can help gain an improved understanding of the territory affected by the earthquake.

Monks tended to note in summary form the memory of an earthquake in the spaces left empty in liturgical calendars, Easter tables or obituaries from the monastery. In these cases the earthquake references are even barer than those in the annals. The memory of those far-off earthquakes thus survives in shreds of terse testimony. In other cases prayer books are the works that preserve scraps of earthquake history, likely because the trust–fear–earthquake association that has been very enduring over time can lead to such memories being set down in writing.

Some Hebrew prayer books also contain similar kinds of notations, some of which have been accurately transcribed in order to be used for the purposes of historical seismology (see a list of Hebrew *notulae* concerning earthquakes in Busi, 1995). In many cases, the persons who left these records showed a marked tendency for precision in noting the times and effects of the seismic events; this can be inferred from the many commentaries left by the authors and is highlighted by modern research into historical seismology.

Notes made by a scribe or the owner of a manuscript sometimes prove to be of great interest for the realism and vivacity of their style. Here, for example are the notes (*notulae*) written at various moments by Galaction Madarakis regarding the seismic sequence of October–November 1343, which originated in the Marmara Sea and which caused damage to Constantinople:

> While I was writing this book and had reached this point, there was an earthquake in Constantinople, in the year 6852 [1343], on 14 October, in the twelfth indiction.

> While I was writing this codex, and as I reached this point, there was a great earthquake in Constantinople, in the year 6852 [1343], in the twelfth indiction, on 18 October. The earthquake lasted for twelve days. Galaction.

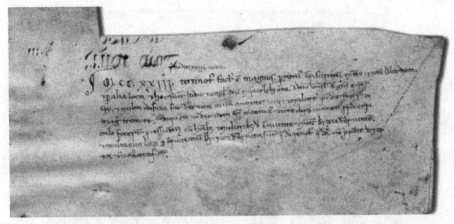

Figure 3.11 *Notula* in the inside cover of a thirteenth-century Italian codex, in which an anonymous scribe recalls the earthquake of 25 December 1222 (from Guidoboni and Comastri, 2005).

> *On the eve of the Feast of the Presentation [of the Virgin Mary in the Temple, i.e. 20 November 1343], as I was writing and had reached this point, a great earthquake occurred.* (Oxford ms. Bodl. Barocc. 197, in Turyn, 1980, pp. 108–112.)

Still more striking for the modern reader are the words written by an anonymous owner of a manuscript on the occasion of the strong earthquake which struck Constantinople in 1296:

> *On June 1, in the ninth indiction [1 September 1295 – 31 August 1296], on the sixth day [Friday], before cockcrow, there was a great earthquake and many walls of the city collapsed and churches were split open. It lasted for eight days. In the year 6804.*

> *And again on 13 of the same month, on the fourth day [Wednesday], there was a tremor at the seventh hour of the day. And a(gain)* . . . (Now Athos Vatop. 290 and Paris. Suppl. gr. 682, in Richard, 1955, pp. 332–333.)

At that point where the note ends, the writer must have been taken by surprise by another earthquake, as he did not finish the note. A contemporary hand has added '*A great fright!*' Expressions such as this last addition are very difficult, if not impossible, to find in the work of a medieval historian (see Figures 3.11, 3.12 and 3.13).

An important case of *notulae* and earthquakes concerns the seismic history of the island of Cyprus. For many of the earthquakes of this area the *notulae* represent the only sources that bear witness to their occurrences, particularly for

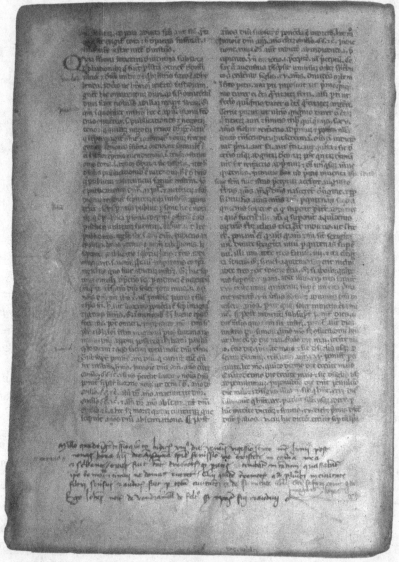

Figure 3.12 A folio of a fifteenth-century codex of law in which an Italian notary public noted his own personal testimony of the earthquake dated 26 June 1405 (Biblioteca Nazionale Marciana, Venice, LT. V, 61 [2515], fol. 52v.). Translated from medieval Latin, it reads: 'In the year 1405, thirteenth indiction, Friday 26th June, after the ninth hour, at the eighteenth hour of the day, in inclement weather, while I was in my room intent on my writing, I noticed such a great earthquake that the walls of the building shook greatly and swayed to the rhythm of the earthquake, so much so that I feared the house might collapse; such an earthquake was noticed by a great many people in the whole of the city of Feltre and from the summit of the church of Santo Stefano a large slab became detached. Written by my hand, Giovanni Vendramello, notary public, of Feltre, witness to and participating in the event.'

Figure 3.13 Codex no. 75, in Armenian, from the monastery of the Holy Trinity of Chalchis. The *notula* in the right-hand margin recalls an earthquake which occurred on 28 January 1292 at Sudak (Crimea) (by kind concession of the Ecumenical Patriarchy of Constantinople, Istanbul).

those in the second half of the fourteenth century and for the fifteenth century (1350: Delehaye, 1907, p. 289; 1392: Darrouzès, 1951, p. 42 and Constantinides and Browning, 1993, p. 77; 1395: Darrouzès, 1953, p. 89; 1397: Darrouzès, 1951, p. 43, and Constantinides and Browning, 1993, p. 77); 1479: Darrouzès, 1953, p. 89; 1497: Darrouzès, 1956, p. 57).

Some *notulae* are more than mere random notations of an earthquake that was experienced locally. Sometimes *notulae* also preserve an indication of the seismic effects in places very far away from the epicentral area, thus providing some information of the spatial extent over which the earthquake was experienced. For example, a Florentine merchant happened to be in Munich in Bavaria on 25 January 1348, when a strong earthquake took place that strongly hit Carinthia (Austria), Slovenia and the northeastern part of Italy ($I_0 = $ IX–X; $M = $ 7.1). From this information, it appears that the merchant was some 240 km from the epicentre when he felt the shock: '1348 – 4 *florins which I paid in Munich in front of the city while an earthquake was happening*' (in Stolz, 1957, p. 35).

Particularly long commentaries, so long at times as to represent small chronicles, are really colophons, i.e. texts added on in the final part of any of several different kinds of manuscripts. Especially useful for research into earthquakes are the colophons of Armenian manuscripts. Many of these were published at the beginning of the twentieth century (see below).

A comparison of the different reports of independent chronicles can often allow researchers to identify the common elements as well as the dissimilarities in the individual memories and perceptions of the same shocks. Analytical comparisons of different texts are not merely literary exercises but in fact can often be useful attempts to use the human reminiscences as though they were very special seismographic recordings. In some cases, individual memory sources can be correlated with public authority sources in such a way as to identify complementary or corroboratory elements concerning the occurrences of effects of an earthquake.

The invaluable colophons of Armenian manuscripts
(thirteenth–eighteenth centuries)

By the term colophon we usually mean the final part of a manuscript, which bears the name of the author (or the amanuensis), along with the place and date of writing. Armenian authors are noteworthy for their tendency to compile long colophons, called *y'atakaran* (meaning *memorials*) in accordance with their tradition of learning. In practice these are *notulae*, which can either be short or long and more or less accurate, concerning events held to be important for the author, who is nearly always anonymous. Current editions, though very rich in material, are not quite complete. Yovsep'ean (1951), Xac'ikyan (1955) and Hakobyan (1951–56) published *notulae* and minor chronicles along with their extensive philological and critical examination. Thanks to these editions there exists today some information about seismotectonic phenomena that occurred in a very wide area that comprises present-day Armenia and eastern Turkey (the area of historical Armenia).

The data concerning individual seismic events are often sparse, like the colophon that we report here that discusses an earthquake of Marmet that occurred between 20 February 1117 and 19 February 1118:

> *In that year [567 = 1117/8], there were terrible omens from heaven and*
> *earth. The blue skies turned blood red, the stars were seen to be in a state of*
> *turbulence, and there were earthquakes in a [. . .]* (Yovsep'ean, 1951,
> pp. 329–30.)

In some cases the text can be quite complex. In 1441, a strong earthquake, which may have been related to an eruption of the Nimrud volcano, shook the locality of Van (in present-day Turkey). The colophon that recalls this event can be found in a menology (i.e. a liturgical or martyrological calendar) written by a scribe called Vardan:

> *In this year [890 = 1441] a great sign appeared, for the mountain called*
> *Mamrut, which lies between Xlat' and Bales, suddenly began to rumble making*
> *a sound like heavy thunder from the clouds. Children cried and moaned to*
> *their parents, for they saw the ice crack open over an area as broad as a city*
> *and other fragments; and as the flames arose from this cleft, they were*
> *shrouded in dense, whirling smoke of a stench so evil that the children*
> *breathing it became ill. Stones glowed in the terrible flames, and boulders of*
> *enormous size were hurled aloft with peals of thunder. And the city of Xlat'*
> *was gripped with terror because of this shock. In other provinces, too, men*
> *clearly saw it all.* (Xac'ikyan, 1955, pp. 515–516, no. 581.)

Another colophon, which further illustrates the kind of information that can be contained in this type of record, is one that describes the earthquake dated 23 April 1457, which hit the city of Eznkay (present-day Erzincan, in eastern Turkey]:

> *And then, in those days, a terrible earthquake occurred at Eznkay, destroying*
> *the city and reducing all the walls to ruins, and at that time, 12 000 people*
> *perished in the earthquake, believers and unbelievers, young and old, and pious*
> *priests; and we cannot measure the sorrow, affliction, tears and sighs caused*
> *by their death. We were the eyewitnesses of a tremendous punishment, when the*
> *Creator, the only God, visited calamity and despair on his servants. It happened*
> *in the year 906 [27 November 1456 – 26 November 1457], on April 23, the first*
> *Saturday after Easter, at the first hour of the day* (Hakobyan, 1956, p. 215,
> no. 26.)

Letters and epistolaries

Loose single letters and/or epistolaries, i.e. collections of letters, constitute a self-standing kind of source when they are written by witnesses of earthquakes and tsunamis or by persons interested in these phenomena. These sources might be defined as 'an earthquake memoir addressed to someone in particular'. Precisely because of the first-hand nature of the observations that are contained in letters, these eyewitness reports are of substantial interest for historical seismology. The effects of an earthquake as well as of an entire seismic sequence are often described in minute detail. Epistolaries are something different from letter collections in that the latter also contain the replies.

Collections of letters of earthquake reports by scientists make up a particular genre of information about earthquakes, to which much attention has been dedicated in the past few years. Such material can make available valuable information about individual earthquakes along with reflections about the nature of earthquakes. Some reports may contain descriptive aspects of earthquakes felt in places where the author had happened by pure chance to be at the time of the seismic event (see the following example on Darwin and the Chilean earthquake in 1835). In Europe a project is currently being carried out especially dedicated to the scientific letters and in particular the correspondence of interest for the Earth sciences (see the section in Chapter 4 on the scientific sources). It is not unusual for such letters to contain important information both for historical seismology and for the history of seismology and scientific instrumentation.

From the diary of an ambassador travelling to the court of Samarkand (1403)

In the fourteenth century, Ruy Gonzalez de Clavijo, who was a Castilian aristocrat from Madrid, became chamberlain at the court of King Henry III of Castile and Leon. He was placed in charge of an embassy to the court of Tamerlane in Samarkand (in present-day Uzbekistan), for which he left Cadiz (Spain) on 21 May 1403. He kept a diary during the long journey to and from Samarkand, and once he was back in his own country in 1406, he wrote a complete and detailed narrative of his extraordinary experience. He visited Mytilene on the island of Lesbos (in the Aegean Sea, opposite the coast of Turkey, north of the Gulf of Smyrna) in October 1403. There he learned from the inhabitants about the strong earthquake that had struck the island many years before, on 6 August 1384. Although indirect, this source is unique evidence of that strong earthquake. One may be struck by the randomness and fortuitous nature of the surviving record about this event, but this is not unusual when one seeks the basic data of historical seismology from remote centuries. The report from Ruy Gonzalez reads:

> We heard an extraordinary story about the present lord of the island, according to which, about twenty years ago, while he and his father, mother and two brothers were sleeping in a castle building, the island was shaken by

an earthquake. The building collapsed and everyone was killed, except for John,
who was protected by his cradle. Amazingly, he was found the next day, safe
and sound, in a vineyard below the castle, at the foot of a very high crag. (Ruy
Gonzalez de Clavijo, The Spanish embassy to Samarkand 1403–1406,
ed. 1971.)

Letters with information nested like Russian dolls in the Diaries of Marin Sanudo
on the earthquake of Constantinople in 1509

Marin Sanudo (1466–1536) was an indefatigable Venetian scholar and
collector of printed books and manuscripts, and in addition to minor writings he
was responsible for a substantial historical work entitled *Le vite dei Dogi*. This treatise begins at the earliest times and comes up to Doge Agostino Barbarigo (1494).
The work underwent successive revisions and additions, at least until 1533. His
monumental work is the *Diarii* (in 58 volumes) which he regularly updated from
1496 to 1533, and which has proven to be of great importance because Sanudo
was personally involved in many of the principal events in the Venice of his day.
In the *Diarii* various letters are transcribed, which are invaluable sources for
some earthquakes. One of these letters was written by Niccolò Giustinian from
Constantinople on 15 and 25 September 1509, and was addressed to his brothers
Alvise and Piero Giustinian, who lived in Venice, where the letter arrived on
17 October 1509. Sanudo diligently writes:

Some letters have arrived from the sea, with the ship that set sail on September
25, from Costantinople, and arrived here in three days from Zante assisted by
the scirocco. It brought letters to messers Alvise and Pietro Giustinian, sons of
Marco di Niccolò their brother, relative to a great earthquake that had occurred
at Constantinople and Pera, as I will say in the chapter of the letters copied
hereafter, as it was read in the college (dei Savii). [. . .]

I do not think I can linger too much over the present as I am still almost out
of my mind, owing to the extreme fear that we experienced on the 10th of the
month at 4 o'clock at night owing to an earthquake so strong as no others ever
experienced in the memory of man, neither here not elsewhere, nor like the one
of Candia and Rhodes that had happened many years ago, that was said to
have been very strong. According to those who had felt them both, this one was
incomparably stronger, and this has been clearly demonstrated, considering the
fact that between here, Pera and Costantinople, it is said that over 1500 houses
collapsed, more than 4000 people died and more than 10 000 people were
injured. We Venetians, praised be the Lord, all survived without injury and
damage. From a house inhabited by Florentines and that had completely
collapsed, of the 10 people who were there 7 miraculously survived, while 3

instead were injured and died, among the latter there was also the wealthy gentleman Antonio Miniato. God forgive the sins and give rest to his soul. Most of the walls of Constantinople, both the ones overlooking the sea and those towards the countryside, collapsed down to their foundations. The minaret [of the mosque] [. . .] is in danger of collapse, and so are most of the mosques. To conclude, neither here nor at Pera or Constantinopole were any houses unscathed, above all mine, which nevertheless, miraculously has not collapsed. By an even greater miracle, in the whole stretch of coast both at Pera and at Constantinople the ground facing the sea has opened as far as an abyss, something really beyond belief. And every day and every night the shocks still do not cease, so that all of the population is camping out in the gardens. And another extreme thing is this, that the earthquake did not only happen here, but at the same time also at Bursa, Gallipoli and Adrianopoli, and it is believed in all of this village, and in all of these places it was very strong. As a result of that, the chiefs meet in great anguish, because this earthquake seems to them to be a bad sign for their state. I have heard news of the recovery of Padua and the capture of the earl of Mantua, etc. In Pera, on September 15, 1509. (Marin Sanudo, *I Diarii*, vol. IX, cols. 260–262.)

In the same vol. IX, cols. 563–565, under the title *De terremotu magno in urbe Bisantii*, correspondence dated 12 October 1509 by Mihnea Voyvode of Wallachia (Romania), is transcribed. He in turn had received the news from his son who happened to be in Costantinople/Istanbul. This is the case of a letter that contains another letter that at times was inserted in the text.

Travellers and their accounts

In every age people have travelled to and described new countries and landscapes. When travellers witnessed earthquakes or came to know about them through the accounts of others, they produced a particular kind of literature that is a wonderful source of information for historical seismology. This kind of text has existed since the remotest centuries and encompasses a great variety of works. There is now a vast amount of literature about such sources, and almost every year new editions of this literature are published. Every country now has its collection of sources about travellers. Of particular value to historical seismology are the accounts of lands that were sparsely inhabited, were a long way from ordinary means of communication, or which suffer from a dearth of written sources for one reason or another.

As in the case of all sources produced by a single human hand and not by an institution, accounts of travels, though always valuable and to be sought after with all diligence, may appreciably distort the reality that they are describing.

This can happen for a variety of reasons attributable to the author's personal motives (see also Chapter 10). The fact that travellers' writings, which one can use to learn about earthquakes and tsunamis, may distort the truth should not discourage their use in historical seismology, but rather one needs to be cautious and particular about the veracity ascribed to the literal details of such accounts. Even very authoritative travellers may produce a distorted description or recollection, as is seen below in the case of Goethe. Exaggerations and distortions are usually less frequent in the writings of naturalist travellers than of less informed visitors (see below the examples of Pallas for the Crimea in the eighteenth century, and Darwin in Chile).

Many travellers' writings concern regions that were far from towns, sparsely populated and lacking in communications – places where the production of written sources is likely to have been rare or non-existent. Information about seismic effects from this kind of source will always be a matter of chance, and nearly always impossible to check because of a lack of other documentation. On the other hand, the uniqueness of such information is irreplaceable. For these reasons vigilance is essential in using such sources to document past earthquakes and their effects.

A twelfth-century Spanish Jewish traveller in the Mediterranean

A Spanish rabbi, Benjamin of Tudela, wrote a diary (*Massa'oth*) of his wanderings between 1160 and 1173 from Spain across the Tyrrhenian coast of Italy to the eastern Mediterranean lands. It is a fundamental source for knowledge of the Jewish communities in the Mediterranean area, as well as of the political situation in the Holy Land. In two passages he refers to violent earthquakes that struck Tripoli and Hamah, in Syria, causing the deaths of countless Jews. Unfortunately, Benjamin's work presents no explicit chronological reference, except for the mention of the year 4933 of the Jewish calendar, corresponding to 1173, inserted by the editor of his work, in order to indicate the time of his return to Spain. On the grounds of the implicit chronological references it is possible to date Benjamin's journey to between approximately the first half of the 1160s and 1173.

In the part referring to Tripoli in Syria, Benjamin writes that the earthquake had occurred 'some years ago', while the one affecting Hamah also was 'some years ago'. Perhaps these two references are to the same earthquake. It is known from other sources that the Syrian region, as well as Tripoli and Hamah, were hit by a seismic sequence in September–December 1156. Like Prawer (1988, pp. 193–4), we believe it to be more likely that Benjamin was referring to the earthquake of 29 June 1170, which also seriously affected Tripoli and Hamah. The two passages are as follows:

*Some years ago there was an earthquake at Tripoli, in which many Jews and
Gentiles lost their lives, because houses and walls collapsed on top of them. At
the time, the whole of 'Eres Isra'el was laid waste, and more than twenty
thousand people died there.*

*Hamah, that is to say Hamath, is one day's journey [from Hims]; it stands on
the banks of the river Jabboq, at the foot of Mt Lebanon. Some years ago, there
was a great earthquake in the city and in a single day twenty-five thousand
people were killed, of whom about two hundred were Jews, but seventy survived.
(The Itinerary of Benjamin of Tudela, ed. 1907.)*

Did travellers always tell the truth? The case of Goethe at Messina in 1787

In 1786 and 1787, J. W. Goethe went on his famous journey to Italy, following the tradition of the Grand Tour which intellectuals and artists had been
undertaking ever since the sixteenth century and which became very fashionable in the eighteenth and nineteenth centuries. When Goethe left Karlsbad
on 3 September 1786, three years had gone by since the great seismic crisis of February–March 1783, when five dreadful earthquakes reduced Calabria
to ruins (Boschi *et al.*, 1995, 2000; Guidoboni *et al.*, 2007; 5 February:
I_0 = XI, M = 7.0; 6 February: I_0 = VIII–IX, M = 6.2; 7 February: I_0 = X–IX,
M = 6.6; 1 March: I_0 = IX, M = 5.9; 28 March: I_0 = XI, M = 7.0). These
earthquakes acquired great fame throughout Europe. The region which had
been struck was Magna Graecia – that land of ancient and illustrious memory,
which the aesthetic yearnings of European culture liked to see as a lost *terra
felix*. The earthquake of 5 February also struck Messina, situated at the north
end of the kilometre-and-a-half-wide straits that separate Sicily from Calabria.
But Goethe did not go to Calabria; he left Naples on 8 May 1787 and travelled
around the coast of Sicily, collecting minerals and observing the countryside
between Taormina and Messina. It was evening when he reached Messina, and
together with his companion, the painter Kniep, he took lodgings at the only
inn he found open. Here is what he has to say:

*The immediate effect of this decision was to give us a frightening impression of
a city in ruins, since we rode for a quarter of an hour past ruin after ruin until
we came to an inn – the only one in the whole quarter to have been rebuilt. All
that could be seen from the upper windows was a desert of jagged ruins.
Beyond the boundary of the farmstead, neither man nor beast could be seen,
and the silence of the night was tremendous. (Goethe, Italianische Reise, 1816.)*

This image of dark despair and destruction is in contrast with other contemporary reports, including the authoritative remarks of Sir William Hamilton, who
visited Messina himself:

Many houses are still standing, and some little damaged, even in the lower part
of Messina; but in the upper and more elevated situations, the earthquakes seem
to have had scarcely any effect, as I particularly remarked. (Hamilton, 1783.)

Another illustrious witness was Michele Sarconi, a member of the Naples
Accademia delle Scienze. He wrote:

All we claim to show is that when the earthquake reached there [Messina], it
was so fragmented and weakened that only its last vestiges were felt. (Sarconi,
1784.)

Do these accounts taken together mean that what Goethe wrote was not what
he actually saw? The contrasting damage assessments have led some scholars
to make a careful analysis of Goethe's evidence and to come up with various
explanations for it (Puzzolo Sigillo, 1949; Di Carlo, 1955–56; and especially Placanica, 1985b, referred to below). Although Goethe kept a travel diary, his *Reise*,
the version that is known today was actually written in 1813, some 26 years
after his journey. Goethe burned his original notebooks, but there is no reason
to think that it was his intention to distort the truth and destroy the evidence.
So why is his description markedly different from those of other witnesses?
According to Placanica (1985b), who has made a penetrating study of this aspect
of Goethe's work, there are various explanations. First, since Goethe had not
seen the destruction in Calabria nor perhaps the effects of any earthquake in
Germany, he may have been much impressed by even minor damage in Messina.
Second, the naturalistic aspects of the event were not of particular interest to
Goethe, although he was a keen connoisseur of minerals and of the geological
aspects of the countryside. It seems likely that, in recalling Messina in this way,
Goethe was primarily displaying his poetic sensibility, perhaps seeking an effect
of contrast with the love of life that he subsequently observed in the people of
the city.

A German naturalist in the realm of Catherine the Great of Russia:
earthquakes in Tauris and the emergence of islands in the Sea of Azov

Peter Simon Pallas (Berlin, 1741–1811) is well known in the international
world of science as one of the most important botanists and naturalists of the
eighteenth century. He was a member of the St Petersburg Academy of Sciences,
and spent the years 1768–1774 travelling in the Ural Mountains and Western
Siberia. In the years 1793–1799 he travelled the length and breadth of the Crimea
(ancient Tauris) and the Taman Peninsula, and he wrote a detailed description of
the regions he visited. He did not confine himself to noting plants, environmental characteristics and temperatures. In the isolated villages where he stayed he
also recorded accounts of natural disasters that had occurred some years earlier.

Furthermore, he experienced some natural events himself. His accounts are usually precise and detailed, and the importance of his evidence did not escape the notice of the first Russians to catalogue earthquakes in the Crimea (Smirnov, 1931). The care with which he collected information from eyewitnesses whom he considered reliable means that we today are aware of events which might otherwise be unknown due to a lack of other written sources.

It is from Pallas, for example, that a large landslide is known, preceded by two earthquakes on 11 and 18 February 1786, which seriously affected the south coast of the Crimea (Kutschuk-Koe) (Pallas, 1802–03, vol. I, pp. 148–51 and vol. II, p. 13). In this case his description can be compared with that of Dubois de Montpéreux (1839–43), a French traveller. The landslide swept away a village with its houses and mills, and caused the inhabitants, a Tartar people whose culture was principally oral, to flee in terror. In addition to mentioning other local earthquakes, Pallas records that shocks originating in the Carpathians (e.g. that of 27 November 1793) were also felt in the Crimea. But perhaps his most interesting description is that of the emergence of an island in the Sea of Azov on 5 September 1799:

> . . . at sunrise, opposite Temruk, at a distance of 150 French toises from the shore [c. 300 metres] a subterranean noise was heard, accompanied by a thunderous roar. Those watching were already in a state of suspense, and now their attention was drawn to the sight of an island rising from the bottom of the sea, which must have been about 5 or 6 toises deep, after an explosion similar to that of a cannon shot. The island seemed to rise up and threw out mud and stones until an eruption of fire and smoke covered the whole spot . . .
> (Pallas, 1802–03, II, pp. 316–17.)

The event lasted for about two hours. The sea was so rough that day that no one dared take a boat out to the island, which appeared to rise about 3 metres above the waves of the sea and was entirely black, being formed from an emission of mud. According to measurements provided by Pallas, the island was more than 140 metres long and about 100 metres wide. One year later, it could no longer be seen. On 10 May 1814, in the same area, almost opposite the island of Tyrambe, another island emerged. In this case, the evidence comes from the French traveller Dubois de Montpéreux (1839–43).

Charles Darwin and the earthquake of Valparaíso (Chile) in 1835

After three years of journey aboard HMS *Beagle*, Charles Darwin found himself in March 1835 becalmed in the seas off Valparaíso, Chile. He took the opportunity to write letters to friends and relatives, including this letter to his sister Caroline (10–13 March 1835).

*We now are becalmed some leagues off Valparaiso & instead of growling any
longer at our ill fortune, I will begin this letter to you. [. . .] The voyage has
been grievously too long; we shall hardly know each other again; independent of
these consequences, I continue to suffer so much from sea-sickness, that nothing,
not even geology itself can make up for the misery & vexation of spirit. [. . .]*

*We are now on our road from Concepciòn. – The papers will have told you
about the great Earthquake of the 20th of February. – I suppose it certainly is
the worst ever experienced in Chili. – It is no use attempting to describe the
ruins – it is the most awful spectacle I ever beheld. – The town of Concepcion is
now nothing more than piles & lines of bricks, tiles & timbers – it is absolutely
true there is not one house left habitable; some little hovels built of sticks &
reeds in the outskirts of the town have not been shaken down & these now are
hired by the richest people. The force of the shock must have been immense, the
ground is traversed by rents, the solid rocks are shivered, solid buttresses 6–10
feet thick are broken into fragments like so much biscuit. – How fortunate it
happened at the time of day when many are out of their houses & all active: if
the town had been over thrown in the night, very few would have escaped to
tell the tale. We were at Valdivia at the time the shock there was considered
very violent, but did no damage owing to the houses being built of wood. – I
am very glad we happened to call at Concepcion so shortly afterwards: it is one
of the three most interesting spectacles I have beheld since leaving England – A
Fuegian savage. – Tropical Vegetation – & the ruins of Concepcion – It is
indeed most wonderful to witness such desolation produced in three minutes of
time.* (from Burkhardt and Smith (1985), *The correspondence of Charles
Darwin*, vol. 1, pp. 433–4)

Antiquarian historiography, books of prodigies and local histories

Antiquarian historiography began in Europe in the fifteenth to sixteenth centuries. Its aims and methods in earlier periods are very different from
those of historiography today, which is based on a more rigorous and accurate
mentality than in earlier times. Any information about earthquakes found in
earlier historiographical works should be subjected to a critical examination.
In many local historiographical traditions, works of this kind serve as a substitute for more direct and authoritative sources that are lost or missing. This
situation is not without its risks, for only after a genuine lack of sources has
been verified should antiquarian historiography be used as a reference record.
In any case, the historical record based on texts of this kind should be given
a suitable reliability classification before they can be used by the historical
researcher.

Within the historiographical tradition, the transfer of information between texts in different languages has often led to mistakes that have later become consolidated in currently used earthquake catalogues. Indeed, different historiographical texts may derive from different sources and thus have incorporated different aspects of the same earthquake. In some cases surviving historiographical sources may have misunderstood earlier texts that cannot be checked today as they have been lost. Since such texts often underlie many nineteenth-century earthquake catalogues, an examination of antiquarian historiography is an essential stage in the revision of historical catalogues of seismic and tsunami events. In special circumstances, works of antiquarian historiography may have the same value as direct or indirect sources. This is the case when the historiographical records contain complete transcriptions of official documents or private testimony that are no longer available, or when the author of the text himself offers direct evidence of a seismic event that he has witnessed.

A tradition of texts that had been very popular in the ancient world was that of the *prodigies*. From the sixteenth century onward, thanks to the advent of the printing press, some works on prodigies spread with great success. The term *prodigy* defines a variety of natural phenomena, whose common denominator is that of being statistically rare or infrequent. In this sense, then, earthquakes were part of the prodigies. The work that became most famous in this field was the *Prodigiorum ac ostentorum chronicon*, by Conradus Lycosthenes, i.e. Conrad Wolffhart, published in Basel in 1557 (Figure 3.14). The author compiled this work in Latin, systematically examining ancient sources, lives of saints, universal chronicles (the ones available in editions) and many works and printed reports contemporary to him. The earthquakes are listed together with floods, snowfalls and exceptional hailstorms, comets and other astronomical phenomena, and also with other more fanciful events such as monstrous births (of animals and humans), speaking animals or newborns, babies having several heads, and other oddities. These events are listed from Biblical times until 1557. The popularity of this work was due, apart from the topic, to its illustrations. They are present in every page (an example is in Figure 3.14b) and are repeated with some variations, depending on the type of event described, using small icons that indicate the category of the event listed.

A few decades after Lycosthenes put out his *Prodigiorum*, many other works of this kind were published in Europe in the local languages. In this way, news of events that had occurred in the past was disseminated. Unfortunately, the information about the past events was extrapolated from sources that were no longer cited, often containing errors in dating or other misunderstandings. The heterogeneous set of information from the prodigies was then channelled into many minor local histories. These histories absorbed the data about earthquakes

(a)   (b)

Figure 3.14 (a) Frontispiece and (b) a page from the book *Prodigiorum ac ostentorum chronicon* by Conradus Lycosthenes, published in Basel in 1557. This work was a best-seller in its day.

from the prodigies nearly always in an acritical manner. Since the data concerning seismic events were mostly wrong or inaccurate, these questionable data influenced the authors of the earliest earthquake catalogues. The mistakes then propagated to later catalogues.

There are also local stories with information about seismic events that can be close to the reality that modern researchers seek (Figure 3.15). However, the relationship of these local stories with the primary sources is inevitably quite variable. Often the authors of local histories of centuries past have nearly always expressed particular points of view and aims, such as demonstrating the noble origins of a town or of a great family, or focusing on the political history of their city or village. Even if such works contain references to earthquakes that had occurred, it is necessary to exercise a critical sense and go back to the original sources to evaluate the accuracy of the occurrence and details of the seismic events. For many earthquakes and tsunamis it should not be necessary to resort to this kind of source to know about the effects of the earthquakes. However, local histories, when they are authoritative and well documented, may

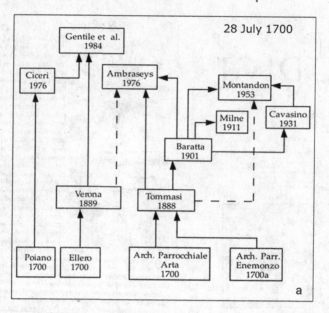

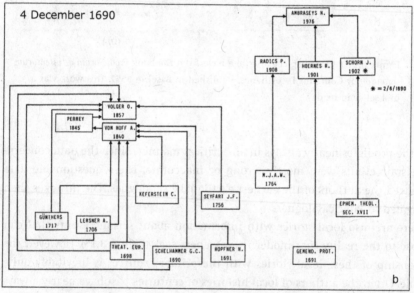

Figure 3.15 Two examples from among many of the analyses that historical seismologists make in order to highlight the dependencies among earthquake catalogues, local historiography and sources. These derived patterns show the often indirect and distorted manner in which primary sources were assimilated into the subsequent historiographic tradition (from Barbano *et al.*, 1994 and Cergol and Slejko, 1991).

be used to retrieve some chronological indicators of earthquakes not contained in the seismic catalogues currently in use. Thus, local histories can supply some indications for new research paths. Furthermore, they can be quite useful and authoritative when they contain direct transcriptions of known sources, such as inscriptions that have been lost.

3.2.4   Sources of institutional and administrative memory: the archives

The archive sources that are most useful for learning about the effects of a destructive earthquake are those of the public institutions, especially the departments that deal with administrative and financial activities. Other sources relevant to historical seismology that might be contained in archives are correspondence between different hierarchies, between figures from the same offices or ministries, between central and outlying administrations, between central headquarters of political power and local branches, etc. One characteristic of archive sources is that they refer directly to the structure of the authority producing the document. It must always be kept in mind that such reports are not inseparable from their bureaucratic contexts (this is true even for very recent archives). Indeed, the current state of archives, from the medieval to the most recent ones, is the result of selections of materials that have been saved and of losses of materials which have occurred both indirectly, by rejecting or reordering of archive records, and directly, by destruction or loss over the course of time. The characteristics of archival records, external to the production of the documents themselves, are controlled by elements of and variations in bureaucratic practices, particularly in premodern administrations. Bureaucratic practices that help produce archive sources include reports between individual people in power; variations or superpositions of offices and responsibilities; and extensions, proxies and arrogation of procedures. These records all are invariably linked to each other and serve to fulfil the specific needs of the authorities producing the material. This means that the archive context, seen as the image of the authority (or office) producing the documentation, is almost impossible to reconstruct in its entirety, not only because of material losses but also due to local variations in the requirements of production of the surviving records.

Knowledge of extra-textual elements, historical features in the broad sense, is of particular importance for the use of public archival material in historical seismology research. These extra-textual elements concern not only the way the individual institutions operated (i.e. the administrative history), but also the different types of relationships among institutions and authorities. These relationships among institutions and authorities fall into two general categories. On the one hand, there are those relationships between institutions and the territories affected by the earthquake, in other words the interactions between the central powers and the peripheral powers in the affected region. On the other hand there are those relationships among the different institutional powers and figures (they can be public, ecclesiastic or private individuals and institutions).

In dealing with an archive for research purposes, it is important to recognize that an archive, even one that has been developed in an organized fashion over the long term, may have been managed differently in different time periods and for different administrative areas. This inevitably is reflected in the complexity of the various offices and functions whose records are preserved in the archive. Archived reports sometimes differ among themselves in content and style for the same territory affected by an event like an earthquake because each report is derived from a different political and administrative area. Dealing with the complexity of different administrative hierarchies and authorities producing documentation for the same geographic area obviously calls for a researcher to pay careful attention to the administrative context of each report.

Besides early and recent damage and devastation to archival administrative records, other problems encountered when searching for data concerning past earthquakes and tsunamis include: (1) reordering of material generally during the eighteenth and nineteenth centuries, and (2) the practice of rejecting material. As regards the reordering of archives, a direct consequence was obviously to alter the original order of the records, which if it had been preserved would today make it easier to reconstruct the context and history of the administrative papers. A further consequence of reordering is that it is now difficult or impossible to locate many documents cited in historiographic tradition.

Archives of public administrations and institutions: kinds of documents potentially useful for learning about the effects of earthquakes and tsunamis

Every archive is in actual fact a set of archives. Each archive can be considered to be the organized and selective memory of an administrative, political or financial function of the past. The more that is known about the nature and evolution of the offices that produced the documents in an archive, the easier it is to carry out research using that archive. When planning to search an archive, it is useful to have some idea about which documents may contain information concerning damage from the earthquake that is being researched.

By taking a chronological approach, that is beginning from the date of an earthquake until a dozen or so years afterwards, one may maximize the informational potential of an archive about that earthquake (with obvious variations depending on the affected area and the time period of the seismic event). In the *chancelleries* and the *secretariats* of the central powers the letters of the governors and other representatives of the central administration are kept. These were written in different places within a state or a kingdom and were sent to the central authority in order to inform the people there about what had happened. Such letters are always very interesting for the immediacy of the data that they contain, for the chronological elements that they contain on the succession of earthquake shocks (when there were multiple mainshocks or strong aftershocks) and for the emotional aspects of coping with the effects of the event. Regarding this last point, however, one must be careful since the account may tend to overemphasize what had indeed happened. A strong earthquake is also a memorable happening in the lives of the survivors. The response of the central powers to the early letters from a region affected by a strong earthquake were usually requests for more information, particularly if the areas affected included the military installations, harbours and other sensitive sites, or if the earthquake had compromised the roadways (due to the fall of rocks onto roads, damaged bridges, etc.).

Until the sixteenth century the central administrations in Europe were not interested in the private building heritage. For this reason reports of earthquake damage can appear today, judging from the maps of the central administration, to be more limited and less severe than it actually was. Therefore, to get a full picture of damage caused by a seismic event, it is necessary to add from other sources what had been left out of these types of documents. The official standing of the writers does not signify that the information about damage was exhaustive.

Prior to the twentieth century, if an administration already had a modern structure and outlook, a rudimentary emergency machine might have been put in motion following a destructive earthquake or tsunami. The most common responses were the dispatch of soldiers to oversee public order and of materials for building basic shelters. But before the twentieth century the lack of mechanical means for search and rescue and of effective first-aid techniques and medicine made actions for the removal of the rubble and the rescuing of trapped survivors very ineffective. There is, therefore, an almost total lack of indications of that phase of natural disaster response that today is called emergency management in archival documents until well into the nineteenth century.

Even without an official emergency management capacity, administrative centres often dealt with requests for aid and reconstruction, and indications of these requests can be contained in archives. In general, in the space of a few weeks or months the pleas from the various local communities that had been hit by a seismic event started to arrive at the central government offices. From these documents it is often true that some seismic scenarios emerge that tend to be more serious that they actually were in truth. This is understandable, because the aim of such requests was to appeal for material aid, and exaggeration or embellishment often helped one's cause. The requests range from pleas for mere charity from the lord of the area to requests for public funding measures. The kinds of requests that were made vary greatly, depending on the historical period and the geographical area of the earthquake or tsunami.

Depending on the administrative complexity of the centre of power that received such pleas, the response at times could be financial help, and at other times no help at all. The issuing of checks for financial assistance following a seismic event were official actions, at times detailed and at other times hasty, that were related to the overall amount of damage. If the damage was substantial, other economic measures also might been undertaken. Such measures almost always took the form of tax break for local administrations. Since local administrations were nearly always in debt, an exemption for one or more years from some tax or levy (which might have been numerous and heavy) was a help to the local administration. The documents that bear witness to such fiscal exemptions may contain an indirect reference to the gravity of the effects of the earthquake. A greater number of years of tax exemption may indicate a greater amount of damage. However, notwithstanding the comparisons with other sources, pleas for assistance following a seismic event cannot be considered to be objective, as political elements generally influence these requests, such as a desire for political protection by local authorities, the presence locally of influential individuals who promote the requests, etc. From the correspondence, at times dramatic, between the central and outlying administrations, one can infer important elements about the aftermath of a strong earthquake or tsunami, such as food crises, the outbreak of epidemics, the demand for manpower to shore up or demolish buildings. One can also learn about other related problems, including disruptions of social disorder, looting and the fleeing of the survivors, all elements that can help researchers to evaluate the impact of the earthquake.

Lists of damaged private buildings and other structures are rare before the eighteenth century, and these generally seem to be found only for relatively more complex administrations. Generally beginning with the early eighteenth century, the public administrations specialized in functions that were more sensitive to the functional requirements of the inhabited areas, and this is also evident in the organization of the archives and in the types of technical documentation available.

An ancient fiscal practice of seismic damage assessment in a document from 1273 (southern Italy)

An ancient document of the Angevin Chancery, dated 18 December 1273, attests to an administrative practice of the government for controlling earthquake damage. The document is important because it is one of the most ancient testimonies of administrative practices of this kind in the Western world. On the grounds of an earthquake damage assessment, the amount of the tax exemption was then calculated. The document belonged to King Charles I of Naples and is addressed to his representative (justiciar) in the region to the south of Naples, the Basilicata region. In this letter the king orders the gathering of detailed information about damage caused by an earthquake that had struck the city of Potenza a short time before. For lack of sufficient resources, most inhabitants were unable to undertake both rebuilding and paying their taxes, and so they sought temporary tax exemption from Charles of Anjou. Potenza and its surrounding area were going through a period of social and economic difficulties. For the few years before the earthquake the city had fought a war with Charles of Anjou, after which the city walls had been knocked down.

The text of the document has survived by way of a transcription by Bonito (1691), because the original document was destroyed in the great fire of the State Archives of Naples in 1943 during the Second World War. A copy of this document is presented in Filangieri's edition (1958, doc. 151, pp. 56–7):

A letter to the justiciar of Basilicata, etc. A petition has been made to us as Lord by our loyal people of Potenza to the effect that the city has been devastated in its buildings and all their furnishings by an earthquake which raged there terribly in recent days, and so the men of this city are equally devastated because they are completely unable to maintain themselves, to rebuild their houses, repair their furnishings and cope with the burden of taxation, for many people have abandoned the city and continue to do so, [and so a petition has been made] asking our majesty to deign to grant them immunity for a fixed period, so that by means of such immunity men will be encouraged to return to the city and reside there and restore it to its previous state. In this regard, therefore, we wish to take steps both to avoid harming our curia and to provide succour to the petitioners, and so, trusting in your loyalty, we order you to go in person to the city in question, to examine carefully and ascertain diligently what wealth these people had before the earthquake, in what way and to what extent they have been harmed, and to what they have been reduced by so powerful an earthquake; whether they are able to sustain contingent taxation without difficulty, and to what extent; or, if you think that for the time being they cannot contribute any taxes, for how long they need immunity both for their own succour and to maintain the curia; and when you have established this, be sure to provide us with a faithful, clear and precise report in your letters, and take care not to be negligent in carrying out these duties and not to report to our court things that are not true. Issued at Corato, on 18 December of the second indiction. (Bonito, 1691, pp. 523–4.)

Unfortunately, the report made by Charles I's justiciar as to the amount of damage caused by the earthquake and any intervention by the public authority in aid of the stricken communities has not survived. This relevant document from the Angevin Chancery, to be found in the register of documents for the year 1274, uses the indiction cycle system, and is dated to 18 December of the second indiction. Bonito held that this chancery used the Greek or Constantinopolitan indiction system, which began the numbering of cycles from 1 September, so that 18 December of the second indiction fell in the year 1273. The fact that the document was put into the register for 1274 was the result of a later reorganization of the documents in the archive, in the early sixteenth century.

An earthquake remembered on the painted cover of a financial ledger: Siena 1467

The *Biccherna* was one of the main financial magistratures of Siena, active from the twelfth century until 1786. At the State Archives of Siena over a thousand pieces are preserved, between ledgers and envelopes of this magistrature. Many of these ledgers had a painted wooden tablet as the cover (from around the mid thirteenth century). Today just over a hundred are preserved, but it is known that many have been lost, looted at various times in the past (Morandi, 1964).

One of these ledgers has a beautiful tablet as its cover, commissioned by the Biccherna to commemorate the occurrence of an earthquake that had struck Siena and had caused some damage. The image is subdivided into three sections. At the top there are some protecting divinities (Mary and some female saints); in the middle the city is depicted, above which there is the writing that reads *ALTENPO DETREMUOTI* ('At the time of the earthquakes'). The pink walls of Siena are surrounded by tents and wooden shacks, of which, as is attested to in the chronicles, the frightened and homeless citizens lived for several months. These precarious dwellings are painted in the foreground, almost as though to underline the pivotal nature of this memory. In the lower part of the picture there are the coats of arms and the names of those in charge of the magistrature of the Biccherna from January 1466 to December 1467. The earthquake is the one on 3 September 1467, of moderate intensity (I_0 = VI–VII; M = 4.9), which caused widespread damage in Siena and caused much panic. This tablet attests to the importance that this event had in the social and urban history of the city, so much as to be set down in a financial register (Figure 3.16).

Archives: technical damage documentation

The potential value of archival sources is enormous, in particular when these are the expression of an efficient public administration. The assessments

Figure 3.16 The cover of a ledger from the *Biccherna* (State Archive, Siena), a financial magistrature of Siena (Italy), painted by Francesco di Giorgio Martini. The cover of the ledger commemorates the earthquake of 3 September 1467.

of damage made by the masons, architects and engineers of the day are irreplaceable documentation of the effects of past earthquakes, although these assessments are not easy to use. Like all technical documentation, they should be evaluated within the scope of the knowledge of the times (to understand the technical terminology that they contain) and within the aims of the local administrative practices, of which they are an integral part.

Technical reports may contain either real descriptions of the damage to the individual buildings, or they may contain just an estimate of the financial loss due to the damage. In the former case, the information is very useful for many research projects, especially when it gives some indications as to the vulnerability of the buildings before the earthquake. In the latter case, the financial loss estimates are much more limited in use in discerning what kinds of damage were experienced. Historical seismology studies based on expert reports have developed important methods for the evaluation of seismic effects on lesser civilian building and for interpreting the technical or idiomatic terms used to describe these effects (see, in particular, Ferrari, 1988; Albini *et al.*, 1991).

Evaluations of the damage suffered by settled areas in earthquakes became the bases upon which government decisions were made, decisions that were carried out by the local administrations. The evaluations tried to balance in some way the opposing interests of the public powers who wanted to give as little money as possible and private individuals wanted to get the most they could in government assistance. No matter what the central administration decided, from a financial point of view the estimate of damage following a seismic event always became the subject of bargaining between well-to-do private individuals and the public administration, as well as between local and central administrations. That explains, from the point of view of administrative history, the various assurances requested by all governments for surveys of seismic damage. It also explains the sometimes sharp differences in the descriptions of earthquake effects between the damage assessed by the local communities and that assessed by the central governments through their own reconnaissance experts. The latter's evaluations are always judged to be more reliable than the declarations of private individuals. Such divergences in public and private viewpoints often paved the way for endless legal proceedings that sometimes were resolved by the intervention of local or central figures of power, but sometimes they ended up as a stalemate, especially if wars, revolts or other accidents disturbed these negotiations. All of this happened because the citizens were allowed something that did not belong to them by right. In the past (as even for today, for that matter) no legal principle or text explicitly calls for an entitlement to be paid to rebuild one's own building assets following an unforeseen seismic event. Payments for post-earthquake reconstruction have always been decided on an ad-hoc basis, and for this reason the surveys performed after a destructive earthquake were often the basic documentation for deciding what reconstruction measures were to be undertaken and supported.

Post-earthquake government surveys of damage are the technical documents that are today of significant importance for historical seismology (see an example in Figure 3.17). These documents allow one to know, nearly always to an incomplete degree, the damage from an earthquake. Indeed, rarely do they concern all of the damage for an inhabited area, and more often than not they focus on only one or a few categories of buildings, such as churches, fortresses or mosques. When a modern investigator has the chance to work with such documentation, the researcher learns, one could almost say 'from the inside', about the vision that societies of the day had of building vulnerability to seismic shaking.

Even projects of reconstruction sometimes set off very lengthy legal proceedings owing to the government's need to obtain money and to keep under control the requests that came from the different social classes affected by the earthquake or tsunami. The documentation of these proceedings and the bureaucratic

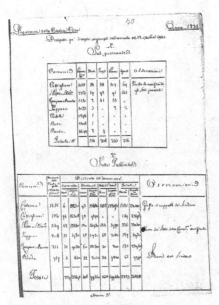

Figure 3.17 One of a number of examples of technical archival documentation that attests to a cursory classification of the damage affecting the building stock. In this case, the Bourbon engineers documented the damage in Calabria caused by the 12 October 1835 earthquake (State Archive, Naples, *Ministero dell'Interno*, fascio 319, fasc. 1, 1835–1836).

practices that developed from these legal actions can provide an important contribution for evaluating the impact of an earthquake on a town. In this sense they can be usefully employed in historical seismology.

The cases of fruitful and specific use of archival documentation for the analysis of a individual earthquakes are by now very numerous in the world's seismological literature. To cite some of these can give an idea of the many possibilities of new information about earthquakes that becomes available with this type of archival research.

The examinations of the Grand Duchy of Tuscany of the damage due to the earthquake on 11 April 1688

An example of documentary sources is preserved at the State Archives of Florence, Italy. As far as earthquakes are concerned, the sources in the Florence archive often provide very extensive and thorough information on the official response to an earthquake, starting with the occurrence of the event in an inhabited area of the Grand Duchy and finishing with the complete reconstruction of the damaged sites.

Figure 3.18 Damage levels for a town from an engineer's report of the earthquake of 11 April 1688 in the Grand Duchy of Tuscany, Italy (State Archive, Florence, *Fabbriche Granducali*, 1928, no. 65).

On 11 April 1688 a strong earthquake struck a mountainous area of the Grand Duchy of Tuscany, an ancient Italian state which then also included a part of the present-day region of Romagna. The Grand Duchy was characterized by a highly efficient public administration. A few days after the news of this disastrous event, a ducal engineer was ordered to check out the state of the military structures of the towns which were part of the Grand Duchy. The engineer scrupulously reported on the damage caused by the earthquake (Figure 3.18), and he supplemented his report with some drawings that represent the effects of the earthquake both on the military buildings and on civil houses. His accurate official examination was the basis for the process of reconstruction, which was substantially supported by public funds. Technical examinations with beautiful as well as detailed drawings of the damage are not unusual (see Figure 3.19).

 A clash of power and a case of double examination of the damage to a city: Rimini 1786

On 25 December 1786 a strong earthquake widely damaged the city of Rimini as well as many villages along the Romagna (Adriatic) coast and nearby

Figure 3.19 Technical examinations by an engineer of the Kingdom of Naples of the damage to a watch tower in Calabria, caused by the earthquakes occurring in February–March 1783.

inland areas of Italy. Today this is one of the most popular tourist areas in Italy, comprising numerous resorts. At the time of the earthquake this area was a *Legazione* of the Church State, at the head of which was the powerful Cardinal Colonna. He was entrusted with institutional powers due to his position, in addition to commanding his personnel. Immediately after the earthquake, the Cardinal summoned the most renowned architect and expert of the region, Camillo Morigia, and gave him the task of performing a detailed analysis of the earthquake damage, including surveying every building in the city of Rimini and as well as in the outlying villages. Camillo Morigia, a scrupulous professional, with a profound knowledge of local buildings and their construction immediately got down to work in faithful obedience to the Cardinal's orders. At the same time, Pope Pius VI, informed of the earthquake and without consulting Cardinal Colonna, decided to have his own examination of the earthquake damage undertaken, on the grounds of which financial measures for reconstruction and their allocation were to be decided. He summoned one of his protégés, an architect who was little more than 20 years old. This skilled and very enterprising young person was Giuseppe Valadier, who in the subsequent decades was to become a dominant promoter of Roman public works and the architect of neoclassical Rome.

Figure 3.20 Two independent technical examinations by the Italian architects Camillo Morigia and Giuseppe Valadier of the damage in Rimini (northern Italy) of the earthquake on 25 December 1786. The arrows indicate the descriptions of the same buildings in the two handwritten texts (State Archive, Rimini).

Morigia and Valadier, both first-rate observers, ended up together in the field to carry out their investigations. Neither of the two experts alone could turn out their assigned task, and the two people who had commissioned the work, Cardinal Colonna and the Pope, ignored each other, each wishing to promote his own power and prestige. The expert examinations were very expensive, but each drew from its own separate source of funds, albeit from the same state. This allowed both investigators to go ahead with their job. The two experts worked long and hard, and the result was of outstanding quality. Perhaps the competition between the two contributed to the high quality and detail of their surveys. It is probably thanks to this administrative competition that today there are two complete and independent examinations of the same damage for this earthquake.

The two examinations, written in bulky registers, are today preserved at the State Archives of Rimini (see Figure 3.20). They have been compared, building by building, after reconstruction of the two different reference systems that were used. Morigia referenced his work to the *isole*, that is blocks defined by the roads, while Valadier used the names of the roads to situate the structures that he described. Both of these systems of reference have today disappeared or are no longer in use because of heavy bombing of the city of Rimini during the

Second World War. The pinpointing of the damage in Rimini has required careful research concerning its urban history. From the examination of many damage descriptions, located not only in the urban area but also in small hamlets or isolated houses in the outlying territory, it has been possible to formulate a very accurate estimate of the earthquake intensity pattern. Besides descriptions of the earthquake damage, both of the architects also observed the quality of the seismic response of the buildings, shedding light upon relevant characteristics of the state of the building prior to the earthquake. Among the causes that helped to worsen the damage, Camillo Morigia identified structurally weak elements of the local construction. Above all, poorly repaired earthquake damage from an event that had hit this same area in 1672 led to new damage in the 1786 event. Indeed, from his examinations he noted that 'little attention and skill had been given' to fixing the cracks, which had been filled with lime or gypsum and hidden beneath the plastering, thereby leaving the cracked walls repaired. Furthermore, Morigia had found in the buildings many walls supported by temporary metal chains or by wooden props, which the century before had been used to avoid the collapse of the crooked walls. The walls were in a very bad condition because their wooded rafters were only resting precariously upon the load-bearing walls; they could thus be detached from the walls with even the slightest horizontal shaking, causing the roof to collapse. Unfortunately, the temporary repair measures after the 1672 event (see effects in Figure 3.21) had never been replaced with subsequent proper works of consolidation.

Giuseppe Valadier, the young pupil of Pius VI, also stressed the lack of craftsmanship in repairing the damage from 1672. He concluded that for the new earthquake in 1786 not everyone had performed the work necessary after the earlier seismic event to ensure the stability of their homes. In the end, Morigia's analysis was reckoned to be the more reliable and accurate of the two regarding the damage estimate in cash terms. On the other hand, the analysis made by the young Valadier was reckoned to be more suitable for the allocation of the contributions and the exemptions that the Church State would choose to grant. Valadier did not stop at merely surveying the earthquake damage. He also took an interest in the theory then in fashion concerning the origin of earthquakes, that of *electricism* (see for a discussion of this theory, see Chapter 4). The young Valadier proposed the use of *torri para-terremoto* (anti-earthquake towers) that, much akin to upturned lightening conductors, were supposed to shield Rimini from the effects of future shocks.

Archives: diplomatic correspondence

The current state archives of European countries nearly always preserve the archives produced by the premodern states. A substantial section of those

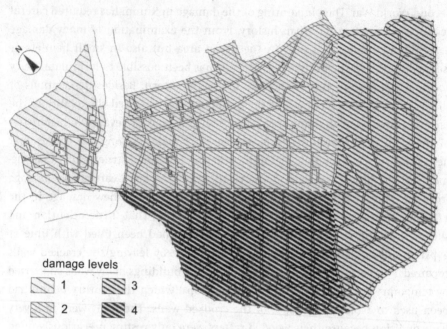

Figure 3.21 Map of the city of Rimini showing the different levels of seismic effects caused by the earthquake in 1672, gleaned from technical descriptions of the damage. The greater density of the lines corresponds to greater levels of seriousness of the damage (from Guidoboni and Ferrari, 1986).

archives is made up of diplomatic dispatches, letters or short notices written by ambassadors or by other official representatives who were residing at other courts. The purpose of such letters, addressed to their lords or sovereigns, was essentially to inform their home authorities what was happening abroad. In the past as well as today, information is the basis of every political decision, especially decisions that affect the relations between realms and nations.

The wealth of such documentation, available starting from the mid fifteenth century, is quite obvious. The news that reached the ruling courts from areas hit by strong earthquakes was often received by ambassadors or other diplomatic agents, who in turn would divulge them to the other courts that were represented. For the worst-hit areas, the diplomatic dispatches became a testimony regarding not only the economic and political difficulties caused by the earthquake, but also the potential loss of international prestige and standing of the state that suffered the disaster. The ambassadors communicated with great attention to detail the effects of an earthquake both to inform the central authorities about elements that might influence foreign policy and to apply for monetary aid if the buildings where they themselves were living needed to be repaired.

Numerous studies in historical seismology have fruitfully used diplomatic correspondence, and specific research projects have been carried out to expand this data base (see, in particular, for the State Archives of Venice the work of Daltri and Albini, 1991 and Albini, 1993). Communications of this kind can be found in many archives that preserve diplomatic correspondence with other states. This kind of correspondence, which unites officialdom with information that was written immediately after an earthquake or tsunami, are of great value for historical seismology researcher and, fortunately, are not rare.

A dispatch from Naples to the Senate of Venice on the earthquake of 5 June 1688 in southern Italy

On 5 June 1688, at about 15:30 hours universal time, a violent earthquake hit an area of the Kingdom of Naples and the Church State (I_0 = XI; M = 7.0) (see Boschi *et al.*, 2000; Guidoboni *et al.*, 2007), causing the deaths of some 10 000 people. Dozens of villages were extensively damaged, so much so that they needed to be rebuilt almost from scratch. Naples also experienced destructive effects from the shock. Numerous churches and other major buildings were cracked, and houses became uninhabitable. In Naples there lived, amongst others, a diplomat of the Venetian Republic, Antonio Maria Vincenti, whose dispatches to the Venetian Senate are interesting sources of information about the effects of this earthquake. He wrote his first letter three days after the earthquake, that is on 8 June, providing a chronicle of what had happened on the previous days.

Serenissimo Dominio

Saturday 5th of this month at the twenty-first hour [Italian time = 16:30 local time = 15:30 universal time] in Naples, for the duration of a Credo a horrible and fierce earthquake was felt, from which only by divine miracle, it may be said, the whole was not destroyed.

The main churches, and those that were the marvellous ornaments of the city, are mostly dilapidated, and all the houses and mansions have been considerably affected by the damage. It was repeated however with only one shock on Sunday at the ninth hour, and on Monday at the eleventh hour. On this day of Tuesday (praise be to God) nothing has been felt.

In the lands and the territories nearby even more horrible and fatal events occurred, the memory of which will remain alive in this Kingdom for many years, and the fear of which for a very long time.

[. . .] the very ancient building of the Serenissima Republic is still standing through God's mercy, but is so unstable and ruined by visible cracking, that I and my wife and children, owing to the evident danger, have moved into just

one room below the terrace that overlooks the garden, and I did not take the room above. My family lives in the garden underneath blankets made of curtains and mats.

As all the architects were very busy, I have with difficulty and promises made, been able to see the damage and the imminent needs [here follows a list of urgent works required to hold up the building] *[. . .]*
Naples June 8, 1688.
Your Illustrious Excellency, your devoted servant,
Antonio Maria Vincenti (Venice State Archives, *Senato*, Dispacci, Napoli, filza 98.)

The same correspondent added on another sheet:

All the people, closing their houses and shops in a great hurry, and leaving many of them open, ran out into the squares, and the largest places of the city, and outside, not believing they would find safety in the churches, being soon disseminated, indeed all too clearly visible, the damage done to many of the same.

The great dome of the church fell and the house of the Jesuit Fathers with a further three lateral domes fell from the same building, with the death of 4 fathers, who were saying confession, nor is the number of the other people buried before known, dead under a mound of stones, that has filled the whole of the church, and even shaken the façade in which great cracks can be seen, with the bell-tower being about to collapse. Said church was the most noble ornament of Naples, and its most marvellous, and the damage will be of vast sums for the paintings, the gildings, the stuccoes and the very precious stones that formed it. [Other descriptions follow of damaged churches.]

There is no mansion, nor house that has not suffered a serious damage with cracks, and highly apparent gaps in the walls, and that threaten to fall into ruin, with many collapses as well. [There follows information on religious rites.]

Most of the nobility has left Naples, and is under the shelters and the tents. An endless number of ladies and other people are in the carriages in the larger squares, spending the night there, after sending home the horses, and the terror, the confusions, the groans, and the fear of repetitions make for a horrible spectacle of sadness for every social class. [More detailed information follows for the days of 6, 7 and 8 June.]

Two great archives of interest for the study of the earthquakes

Archivo General of Simancas (Spain). About 10 km north of Valladolid lies the small town of Simancas, with its great castle built in the fifteenth century. A Royal Order dated 5 May 1545 transformed that fortress into the greatest container of documents of the Crown of Spain, then the most powerful empire of the Western world. Since then the Archivo General has guaranteed the preservation of and made available for consultation three million documents, which represent the archival product of the intense political, administrative and military activity of the Spanish Empire across all its dominions of the time. The preserved documentation covers the period between the end of the fifteenth century and the end of the eighteenth century. The geographical area comprises Portugal, Flanders and Italy directly, all being Spanish dominions; there is also correspondence with the rest of Western Europe among the documents. The enormous importance of the wealth of documents is evident for historical seismology. Indeed, if a destructive earthquake occurred in an area of the Spanish dominions, the local governors would promptly inform the central government of it, and complex bureaucratic procedures and lengthy correspondences would begin because each decision about post-earthquake actions had to be communicated and approved. The vertical structure of the imperial Spanish government had cast-iron rules controlling the work of the various viceroys and governors, as well as of the agents and the representatives who resided in very distant countries. This form of vertical control also called for an incredible number of spies and correspondents of various kinds among the civilians, the clerics and the nobles. There were efficient control networks for the central powers. One can get a fairly good idea of how this bureaucracy operated by directly consulting the innumerable papers in this archive. It was this particular centralized political structure that produced sources, priceless today, concerning natural events that occurred in countries located even at the frontiers of the empire. Destructive earthquakes were important events; they could make military defence systems (city walls, fortresses, watchtowers, etc.) and communications infrastructure (bridges, roads, etc.) either useless or inadequate. The impact of strong earthquakes on the big cities of the empire could harm the image itself of Spanish imperial might.

Dispatch riders carrying the latest news would travel on horseback with special wooden boxes (even today they are the containers of abundant correspondence) and in a few days they would cover large distances. The two violent and extensive earthquakes that hit eastern Sicily on 9 and 11 January 1693 are a case in point, in which the documentation of Simancas has been a fundamental source of information. About one week after the second earthquake, the viceroy of Sicily, the Duke of Uzeda, informed the imperial Spanish secretariat about what had happened, reporting on the physical damage and on the number of dead. This correspondence implies an extraordinary organizational capacity. As a matter of fact, with the first letter on 20 January, the viceroy advised on the intervention task force that would begin a complex and very expensive reconstruction programme. The documentation of Simancas analysed in order to evaluate the effects of these two strong earthquakes comprises 850 handwritten sheets, between letters of the viceroy to the Council of State, letters of reply, and letters and reports by other people in power with decision-making roles (De la Torre, 1995). This is a set of data having an extraordinary interest for historical seismology.

The Archivo General de Indias *of Seville (Spain).*   Emperor Charles III in 1785 ordered the preservation, in a palace of Seville of which the construction had been ordered by Philip II between 1584 and 1598, of all the deeds and the correspondence that were part of the archives of the conquest and the colonisation of Latin America. These materials comprise over 35 000 dossiers, whose historical value is incalculable for the history of the American countries and therefore also for the knowledge of their historical seismicity. This archive also preserves the documentation produced by the religious orders that were leading the missions in the various Latin American countries. Important manuscripts are found in these archives concerning earthquakes, amongst which was one over 350 pages long that is a report on the devastating earthquake of Guatemala in 1716. The Franciscan monks wrote down a list of the damage, village by village, in order to claim subsidies and aid (Figure 3.22).

The General Archives of the Indias have the rare feature of being 'Continental Archives'. In other words, the archives provide information on nearly the entire American continent, including Brazil and the United States. These archives also contain information for the countries of the South Pacific and for other territories in Asia. There is also preserved a great collection of charts and maps that cover the period from the sixteenth to the nineteenth century. At the present time, the *Archivo de Indias* is in the process of being digitized, and by the end of the project 15 million pages will be accessible for consultation online. Once completed, their contents, which have until now been reserved to researchers who could only examine the documentation at the archive itself, will be easily accessible to scholars everywhere.

3.2.5   Ecclesiastic, monastic and capitulars archives

The territorial organization of the Catholic Church as a religious institution and as a State

Besides its own political and administrative structure as a state, from its very beginning the territorial organization of the Roman Catholic Church was based throughout the territory of the practice of Catholicism on a network of territorial structures that survives today as dioceses and parishes. However, it was

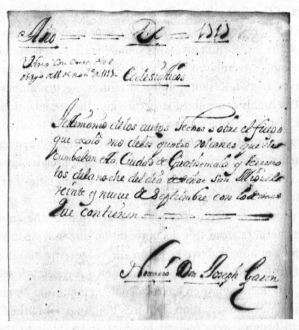

Figure 3.22 Manuscript report from Franciscan monks on the damage caused by the earthquake that hit Guatemala in 1716 (General Archive of Indias, Seville, *Guatemala*, 305).

only after the application of the Counter-Reformation promulgated by the Council of Trent in 1563 that the structure of the Catholic Church in dioceses and parishes was formally established in its specific administrative and bureaucratic forms that still exist today. In a sense this structure formed an administration of the souls and the goods of the Church. The production of documents to satisfy this administration became systematic and widespread towards the end of the sixteenth century.

The sources produced by the Catholic Church as an institution across a territory can be studied within the scope of historical seismology both for their documentary homogeneity and for their extraordinary territorial coverage. Bishopric and parish archives contain registers, visits and papers produced by parish priests and bishops in their official roles as ecclesiastical administrators. They supervised a considerable building heritage, the survival of which was a necessary condition for the work being performed in those buildings. As part of their normal record-keeping, they maintained registers of deaths, which can be useful for determining numbers of fatalities due to seismic events. Often their records also include considerations about the event that caused the deaths and the damage to church property.

The headquarters of the Catholic Church always was and is still today in the city of Rome. Its affairs are recorded in the *Secret Vatican Archive*, a great and extraordinary depository of documents, correspondence and reports between the Pope and his complex secretariat. The various interlocutors whose correspondence is preserved include bishops, political figures, agents, ambassadors, princes and kings. The highly centralized structure and the organization of the Church throughout historical time today preserves, in a single location, a collection of a very large number of records of ecclesiastic territorial management and of religious leadership and control. Letters of bishops, prelates, governors and representatives of the Holy See, which were written from cities not under papal control or influence, are valuable correspondence as well.

The Vatican sources provide a contribution of great value for the knowledge of earthquake effects because from this documentation one may deduce many elements of the impact of a destructive event on the ecclesiastic buildings and other Church properties (hospitals, houses, buildings, etc.). A systematic selection and examination of this documentation has also enabled important elements to emerge regarding the state of preservation of church buildings. From the Council of Trent onwards, all bishops were obliged to write and submit to the Pope reports on the state of their dioceses. These reports (*Sacra Congregatio Concilii – Visitae ad limina*) were due at intervals of once a year to once every few years. Not infrequently from these reports one can infer the conditions of the buildings prior to a major earthquake. This information is important since it can be used to determine whether a church was new, restored, in poor repair, or even had been abandoned.

The letters of the Vatican archives are a treasure-trove of details on the effects of past earthquakes, and often they also provide a chronology of the various seismic shocks that the writers experienced. A knowledge of the places and dates of important seismic events makes it possible to find in the various sections of the papal secretariat the correspondence that might contain mention of the earthquakes and their effects. Since the Catholic administrative network crosses the boundaries of individual states, this documentation offers a high degree of homogeneity in the point of view and the style of information about natural events for a very wide geographical area.

A great ecclesiastical archive of interest for the study of the earthquakes in non-European countries: the archive of the Congregatio De Propaganda Fide *at Rome*

The Catholic Church, besides its organization as a religious institution, also has had a territorial structure in place in various areas of the world for the dissemination of the Catholic faith. This intense apostolic activity, which was particularly energetic in the seventeenth and eighteenth centuries, has produced a vast body of documentation that today is of great interest for historical seismology research in many non-European countries. The policy of expansion of Catholicism and the ecclesiastic jurisdictional structures, which the Catholic Church has continually pursued in all the parts of the world where it could send its emissaries, led to the formation of a complex communications network. Through these channels copious documentation flowed towards Rome, following the timescales and the methods of the development of the organization of the network itself. The foundation of the congregation *De Propaganda Fide* by Gregory XV (6 January 1622) constitutes a fundamental development of the interests of Rome in the world beyond Europe, which more than a century after the discovery of America and the inauguration of the great seafaring routes, was becoming increasingly well known. Previously, in Rome there was no lack of news from countries even very far afield, but this news had reached Roman Catholic officials mainly as second-hand reports, mainly thanks to the information of the Nuncios (the Pope's representatives) or through the accounts of clerics belonging to the Catholic religious orders. Through time, the founding of new religious congregations that were involved in missionary work abroad made the exchange of knowledge between distant places and Rome a continuously increasing and regular event, something that had previously been occasional and indirect.

Indeed, from 1622 there occurred the development of an increasingly sophisticated process of centralization both of the missionary jurisdictions themselves and of the collection of information from those jurisdictions. These were two interdependent elements that were necessary to plan for the founding of new missions and to optimize the working of the existing missions. Whilst before 1622 the clerics who went to the various places for missionary work were nearly all members of religious orders who reported to the superiors of their congregations, after this time the missionaries depended on and were required to report their activities to the congregation *De Propaganda Fide* of the central Church itself.

Beginning with its foundation, the congregation *De Propaganda Fide* drafted bureaucratic criteria for the management of the material it received so that the material could be used as the basis upon which central decisions could be taken. Furthermore, this Vatican congregation organized the material it received into an archive that was subdivided into various series that corresponded to the decision-making activities of the congregation itself. This bureaucratic organization has allowed scholars to connect the various documents and make possible the reconstruction of the history of single missionary provinces. As part of its work, the *Propaganda* had to know the geographical and political situation of each area where missionary work was ongoing or planned. Moreover, the management of the missions also required knowledge of the funding for the subsistence of the clerics and for the construction and the maintenance of the convents, churches and other buildings, all questions of great importance about which this congregation wanted to be duly informed. If a destructive earthquake hit both the people and the Catholic-owned buildings of a foreign province where missionaries were at work, then the *Propaganda* was informed of this.

In order to undertake today systematic research in the archive of the *Propaganda* it is necessary to understand how the material of the archive is organized into the various series and how the bureaucratic and decision-making activity of this institution was performed. The archive makes possible the development of two research paths: one internal to the Roman documentation (through the Acts, the Registers and the Scriptures, mostly edited) and one that is external, that is, in the archives of the single regular monastic orders that had sent missionaries to distant or remote territories.

The strong earthquakes on 1 November 1755 (Lisbon) also destroy the church of Meknès (Morocco)

The documentation and archival letters of the congregation *De Propaganda Fide* (Rome) indicates that the church of Meknès (Morocco) was an important structure with three naves, which could contain 600 faithful. The hospital was not a small structure, containing 100 beds. On 1 November 1755 a strong earthquake caused terrible damage to the buildings of Meknès. The church and the hospital were seriously damaged and became inaccessible. On the same day Lisbon was totally destroyed and numerous towns in Spain and Portugal were damaged by this earthquake. Two different faults may have been activated, and in Morocco the damage was substantial. The information from the Franciscans on Meknès, like the tiny stones of a mosaic, adds to the collections of testimonies concerning this most destructive earthquake.

The monastic complex of Meknès was rebuilt in 1756, but the repairs were only enough to help the buildings remain standing. The damage from the earthquake in 1755 was so profound that in 1767, owing to remaining weaknesses in the structure, the church fell to the ground, dragging some adjoining structures down with it. The documentation states that the foundations of the destroyed church needed to be reconstructed, after which there was built a new structure that apparently was smaller than the previous one.

Besides its role as a international religious institution and thus having a presence in many countries of the world, the Church of Rome has also been a political state. In a large area of present-day Italy, the Church has had direct administrative control, and church leaders were also governmental leaders for this area. The two tiers of management, which is as a religious institution and a political institution, often overlapped, creating for us today complex archival situations. The Church State had comprised, at the time of its greatest expansion, almost the entirety of five present-day Italian regions (Molise, Lazio, Umbria, Marche, Romagna and part of Emilia), along with some individual cities that were enclaves inside other kingdoms.

The Church State comprised an area that suffered a significant number of strong earthquakes. The damage caused by these earthquakes was an economic and political problem that had a considerable impact upon the secular governmental operations. Starting from about the mid seventeenth century, the ecclesiastic administration specialized its bureaucracy for coping with earthquake disasters by setting up offices to intervene and to control post-earthquake reconstructions. One of these administrative entities was the *Sacra Congregazione del Buon Governo*. Documentation contained in the archives of the Pontifical Chamber, today preserved at the State Archives in Rome, is one of the most important documentary data sources for historical seismology. Known for its homogeneity and continuity, since the mid seventeenth century, this archives has enabled dozens of researchers to learn in great detail about the effects of many Italian earthquakes and their related reconstruction phases (e.g. Figures 3.23, 3.24 and 3.25).

The archives of religious orders' congregations and of the Capitoli *of the churches*

The archives of the great religious orders of the Catholic Church are different from the ecclesiastic ones described above. This type of documentation, belonging to monasteries, convents, orders, religious societies and congregations, contains much specific information about seismic effects on the building heritage belonging to the religious orders. A vast amount of information is preserved in these archives concerning the history of monument complexes and buildings (the majority of which still exist), and it represents a detailed and continuous record since the seventeenth century. Following destructive earthquakes, there were restoration efforts, sales of goods, and works of renovation that were inventoried and described in great detail. The quality of the information in this documentation is quite good, and the data are interesting also for other historical reasons, such as the history of construction types and styles and of techniques of restoration (Figure 3.26).

Various religious orders have made specific contributions to the study of earthquakes, some of the most important orders being the Jesuits, the Barnabites, the Benedictines and the Piarists. Some religious orders founded geophysical observatories, described earthquakes that they had witnessed, formulated

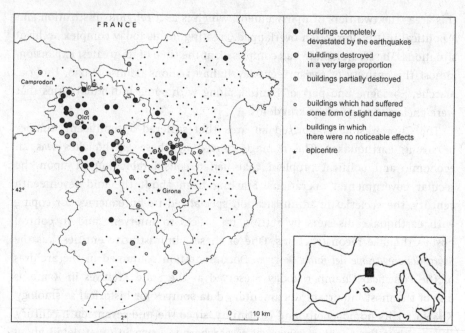

Figure 3.23 Map constructed with the data contained in a Pastoral Visit in 1432 of the dioceses of Gerona (Spain). The ecclesiastical buildings are located with a code concerning their damage as documented following the earthquakes in 1427 and 1428 (from Riera Melis *et al.*, 1993).

theories about seismic phenomena and produced seismic instruments. For example, as noted by Vogt (2005), the Canadian Jesuit Father Pierre Gouin has produced some real masterpieces of historical seismology research. The earthquake catalogue of Ethiopia (Gouin, 1979), the country where Gouin lived and studied for several decades, and the earthquake catalogue of Québec (Gouin, 2001) are both high-quality compendia of information on historical earthquakes. Gouin's versatile linguistic skills and his tenacity in pursing information from native sources make his earthquake catalogues of immense value. Apart from the printed works of the best-known authors, there also are letters and correspondences in the archives of the religious order. These documents are of particular interest for the earth sciences and for the knowledge of past earthquakes. The archives of the great religious orders are thus another resource where historical seismology research may uncover important new information about earthquakes and tsunamis in historical times.

 Earth sciences and the Society of Jesus

Ever since its origins in 1540, members of the Society of Jesus have been renowned for their special dedication to the study of the natural sciences. Hence,

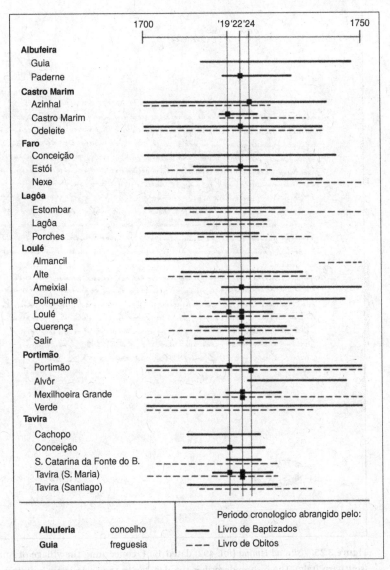

Figure 3.24 An example of the use of parish books in historical seismology. A summary of the data gleaned from the parish registers consulted from 1700 to 1750 by De Noronha Wagner (1993). The squares indicate the dates of earthquakes as noted in christenings (Livro de Baptizados) or records of deaths (Livro de Obitos). Place-names in bold are municipalities, those in light type are civil parishes.

it is not unusual to find references to earthquakes and their causes in the many books written by Jesuits. When seismology eventually developed as a science, the Jesuits were one group who made important contributions. Among the first Jesuits to document earthquake catastrophes was José de Acosta (1539–1600), a Spaniard who spent several years in South America. In his book *Historia Natural*

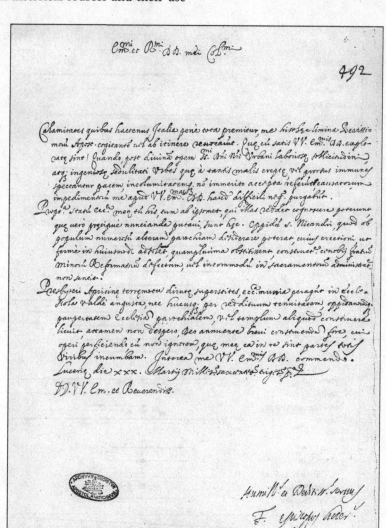

Figure 3.25 *Visita ad Limina* (fol. 492) dated 1631, concerning the village of Apricena (southern Italy). The bishop describes the damage to the ecclesiastic buildings caused by the earthquake on 30 July 1627 (Secret Vatican Archive, *Sacra Congregatio Concilii*, Relationes dioecesium, 465A, *Lucerina I*).

y Moral de las Indias (1590), Acosta described the effects caused by earthquakes that had occurred in Ecuador, Peru and Chile, and he even hypothesized about the causes that might have produced the seismic events. From his observations, he decided that there was no relationship between earthquakes and volcanoes. Also, he investigated the size and extension of the tsunamis related to earthquakes (Udías, 1999). Athanasius Kircher (1601–1680), a professor at the Collegio

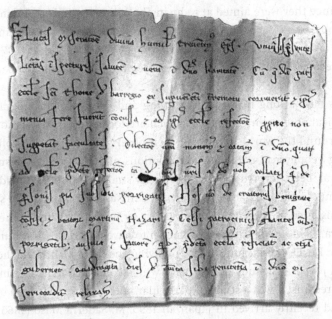

Figure 3.26 Parchment dated c.1260 describing a paid-for indulgence granted by the bishop of Trivento (now in the province of Campobasso, southern Italy) to raise funds for the restoration of a church damaged in the earthquake of 19 February 1258 (Archive of the Abbey of Montecassino, parchment no. 386) (from Guidoboni and Comastri, 2005).

Romano, was another leading Jesuit who investigated the causes of earthquakes. His ideas are presented in his book *Mundus subterraneus* (1664). He suggested the existence of three systems of conducts in the Earth's interior through which fire, water and air circulated (for de Acosta and Kircher see also Chapter 4). Earthquakes are related to the flow of these elements. The Society of Jesus was suppressed by the Pope from 1773 to 1814. During this time there was a hiatus in the Jesuits' scientific activities. Nonetheless, after the Restoration, the Jesuits regained their dedication to academic investigation, including to the natural sciences.

During the past century or so, the Jesuits have founded many seismic stations around the world. Udías (2003) points out that the Jesuits' long tradition in astronomical observatories led naturally to the establishment of geophysical and meteorological observatories by the end of the nineteenth century. Due to the increasing importance of geophysics (especially the study of weather, earthquakes and volcanoes) as a scientific discipline, Jesuit ideals caused this order of priest not only to study these natural phenomena but also to search for ways to mitigate their effects and to forecast their occurrence. These activities carried a social value since they were aimed at reducing both human and economic losses from natural disasters. It is worth recalling here that the first hurricane forecast made on a sound scientific basis was that made by Benet Viñes (1837–1893), the Jesuit director of the Colegio de Belen Observatory at Havana, Cuba, on 11 September 1875.

The Jesuits have a long history of installing and operating seismographs in their geophysical observatories, most of which are located at their missions and universities around the world. Because of their observatories' focus on collecting and analysing geophysical data, the contributions of the Jesuits to seismology have, in general, been devoted more to seismicity studies and related fields than to developments in theoretical seismology. As early as 1867, some early seismoscopes were installed at the Manila Observatory by Juan Ricart under the direction of F. Faura (1840–1897), who is also known to have issued the first hurricane forecast in the Far East, on 7 June 1879. Following the large Manila earthquakes in 1880, Faura prepared a study (Faura, 1880) which combined macroseismic and instrumental records. It was a study of particular note, as acknowledged by John Milne who had recently arrived in Japan. In 1895, M. Saderra Masó (1865–1939) published *La Seismología en Filipinas* (Saderra Masó, 1895), an impressive catalogue of earthquake data that was comparative study on the seismicity of the Philippine Islands. By 1902, J. Algué (1856–1930), as director of the Philippine Weather Bureau, began to deploy and take charge of the first network of seismic stations in the Philippines (Batlló, 2002).

The Jesuits were the first to install seismographs in many countries in the Far East, in Central America and in South America. Suffice it to mention a few of the Jesuits who started or maintained earthquake monitoring sites at locations throughout the world: G. Heredia (1859–1926), who began operating seismographic instrumentation in 1877 at Puebla, Mexico; L. Froc (1859–1932), who installed seismographs at Zi-Ka-Wei, near Shanghai, China, in 1904; S. Sarasola (1871–1947) who started earthquake monitoring in 1923, at Bogotá, Colombia, later followed by J. E. Ramirez (1904–1983), founder of the Instituto Geofísico de los Andes Colombianos in 1941; M. Gutierrez-Lanza (1865–1943), who began recording earthquakes in 1907, at Havana, Cuba; E. Colin (1852–1923), who initiated seismographic instrumentation in 1899, in Tannannarive, Madagascar; and E. F. Pigot (1858–1929), who started monitoring seismic activity in 1909, at Riverview College, Australia. Special mention should be made of the San Calixto Observatory in La Paz, Bolivia: P. Descotes (1877–1964), who was commissioned to begin geophysical observations in South America after the second General Assembly of the International Seismological Association (Manchester, 1911), advised the Jesuits to install a seismic station in the central part of South America. This led to the founding of the San Calixto Observatory. This observatory was praised by Gutenberg and Richter (1949) as the most important seismic station in the southern hemisphere.

In Europe, the first seismoscope installed by the Jesuits was set up in Italy, at the Tuscolano Meteorological Observatory (Rome) by G. Egidi (1835–1897). Also, in Europe in 1902 the Jesuits founded the important seismic station of Cartuja, in southern Spain, where its director, M. Sánchez Navarro-Neumann (1867–1941), developed new seismic instrument designs, later of great importance for the development of seismic stations in South America (Batlló, 2003).

Finally, it is worth mentioning the contribution of the Jesuits to seismology in the United States. There, F. L. Odenbach (1857–1933) installed seismoscopes at John Carrol University, Cleveland, Ohio, in 1900. In 1909, he created the Jesuit Seismological Service, a network comprising sixteen stations distributed at cooperating Jesuit Colleges spread out across the United States and Canada (Udías and Stauder, 1996). This service was continued by J. B. Macelwane (1883–1956) under the name of the Jesuit Seismological Association (JSA). The JSA was responsible for the publication of the first unified seismic bulletin for the USA. Macelwane also made important contributions to theoretical seismology, through his book *Introduction to Theoretical Seismology* (Macelwane, 1936), which was the first seismology textbook published in the United States. In 1925 he founded the Department of Geophysics at the Saint Louis University, one of the first in North America to offer Ph.D. degrees in geophysics. Finally, W. Stauder (1922–2002), remembered for his contributions to focal mechanism studies (Stauder, 1962), can be considered to be the last representative of this important group of North American Jesuit seismologists.

Since the 1960s, Jesuit involvement in seismology has declined, and almost all the observatories founded by the Society of Jesus have been closed down or transferred to universities or public agencies. It can be said that the Jesuits' contribution to seismology has been important in many countries, especially those in the Far East and South America. While their contributions to theoretical developments in seismology are not significant, their contributions in the fields of seismicity and seismotectonics were quite prodigious.[1]

Private archives

In a number of cases the administration of the holdings of some private individuals, families and organizations has evolved into formal structures in the not-too-remote past. This development, which varies for different time periods and geographical areas, has created a very special administrative resource which can be exploited for information on historical earthquakes. In Europe,

---

[1] This item was written by Josep Batlló, whom we thank warmly.

the presence of feudal states, owned by great noble families and apportioned from lands belonging to the state, brings with it a specific problem for historical seismological research. There are two reasons for this. First, the types of documentation and the situations represented in those documents vary greatly according to the administrative style, the importance and the seniority of the feudatory families. Second, due to relationships formed between feudal lords and the state, following a calamity private individual often submitted their own requests and petitions for aid in competition with those requests from the local communities. A large number of state archives contain family archives, whose inventories are often found to be incomplete. Cataloguing of the private archival information was often hasty and simplified. The owners of private materials often withheld some of the papers when a bequest to an archive was made. This means that the family and other private materials that are available today in public archives are often incomplete. There are also notable exceptions to this, which can lead to positive contributions to research into the seismic effects of historical earthquakes.

To reconstruct the effects of a past earthquake by using data from many different points of view, it may be useful to explore one or more private archives. Since historic banks have had a tradition of lending money for reconstruction work, their records may contain information useful for learning about earthquake damage. For the Kingdom of Naples, for example, the *Banco di Napoli* has been a great lending institution ever since the Middle Ages. The documents preserved in their archives, some already examined by Bonito (1691), contain information of financial transactions between private citizens or between institutions that stem from applications for money for post-earthquake reconstruction (Lattuada, 2002). In order to support the financial requests, very often reports of examinations of the religious or noble buildings that had been damaged were enclosed. Furthermore, accurate descriptions of the effects suffered by single large buildings, above all churches, in earthquakes may be preserved in the archives produced as part of the construction of the building. Many churches were under construction for very long time periods (some for several centuries), so plans and other documentation, such as the building materials used and their costs, related to the construction needed to be preserved. For this reason, small archives have generally survived within the churches themselves, and these are separate from the institutional ecclesiastical archives.

The notaries

Starting in the early Middle Ages in Europe, the notary was the man of public trust, that is, the guarantor of the truthfulness of a deed or similar document. For various reasons, numerous notary deeds may preserve indications of the occurrence of a seismic event, the destruction resulting from a seismic event, or even just the great fear generated by an earthquake. Many notarial documents, including donations, wills, concessions and sales, have been used in studies of historical seismology because they contain a reference to or information about an earthquake. Of course, these earthquake-related entries have been deliberately sought in order to improve the understanding of the effects of an earthquake, especially for those seismic events where few details are otherwise known due to the lack of public sources or memoirs.

Much could be said about the documentation collected in the resources of notaries' archives, present in vast quantities in nearly all European state archives with generally systematic property organization since about the fourteenth century. Many notarial documents contain valuable information about the seismic damage caused to the private and monastic buildings. The importance of these sources is due to their reliability, since they are direct testimonies of damage or restoration, and to their extraordinary coverage from across a territory. However, it is not realistically possible to make systematic use all of these kinds of sources because of their practically limitless quantity. Due to the presence of various notaries even in small towns, covering the same sites, full use of these records would require a very time-consuming selection process. In Europe notary archives have mostly been used in historical seismology research when other types of sources are missing as the result of destruction or are lacking due to particular administrative situations (such as the existence of fiefs which results in the lack of some historical documents). Research into notary archives has produced good results in cases for filling information gaps concerning some earthquakes (Figure 3.27).

A slow notary or an unknown earthquake in September 1348 in central Italy?

This case shows how it is not enough to bring to light a piece of documentary evidence, even if it is authoritatively supported, as is the case of notary deeds. It is also necessary to comprehend the documentary evidence within the historical and cultural context of the area being examined. Shortly before 13 September 1348 (that is to say, about a year before the great earthquake of September 1349 – see below), an earthquake caused damage to the castle at Subiaco in the province of Rome. This building, also known as the 'Palace of the Abbots', was occupied by the abbots of the local Benedictine monastery, one of the oldest in Italy. The only known report of the earthquake is a notarial document, drawn up on 13 September 1348 by notary Paolo di Cervara. What is of interest to historical seismology research is the date of the document itself, for it declares that the deed was drawn up in the garden of the church of San

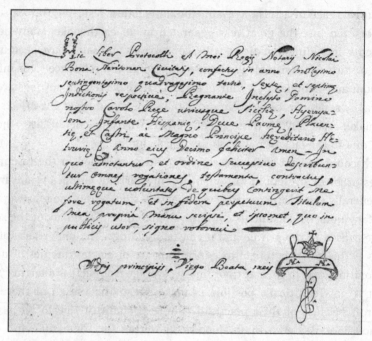

Figure 3.27 Example of a notary's deed. In this document the damage caused by the earthquake on 20 February 1743 at Nardò (southern Italy) is mentioned (Lecce, Notary's Archive, vol. 66/16, Book of the protocol of the notary public Nicola Bona).

Francesco because Abbot Pietro had taken refuge there, as the castle had been damaged by an earthquake:

> In the name of Our Lord, amen. In the year of Our Lord 1348, in the seventh year of the pontificate of Clement VI, in the first indiction, on the thirteenth day of the month of September. [. . .] Drawn up in the garden of S. Francesco di Subiaco of the order of Friars Minor, where the said abbot had taken up personal residence because of the very great earthquake which seriously damaged the castle of Subiaco. (Biblioteca del Monumento Nazionale di Santa Scolastica.)

The abbots' residence at the Benedictine monastery complex was separate from the monastic buildings themselves. The latter consisted of the monastery of Santa Scolastica (founded in the sixth century) and the monastery of the Sacro Speco of San Benedetto (founded around the eleventh century). The church of San Francesco was also just outside the inhabited area of Subiaco. About a year later, this area, like much of central Italy, was struck by the great earthquake of September 1349, for which there are a great many reliable sources (Boschi *et al.*, 1995). There are legitimate reasons for casting doubt on the accuracy of the above deed, as has been done by Molin *et al.* (2000). However, from a palaeographic point of view, the date given in the deed is unquestionable, since all its elements fit together perfectly; 13 September 1348 was indeed in the seventh year of the pontificate of Clement VI (May 1348 – May 1349), and in the first indiction (25 December 1347 – 24 December 1348). The indiction calculation used in this document is based on what is known as the Roman or pontifical system, according to which the indiction begins on 25 December (or 1 January). A careful examination of the parchments dating to those years in the Subiaco archives shows that the Roman indiction system was used by all the notaries working in that area.

The only chronological element that appears to clash in the deed described above is the reference to Abbot Pietro. He was in fact not officially appointed head of the Subiaco abbey by Clement VI until 30 March 1349 (ASVat., *Registra Vaticana*, 187, fol. 125r). But there is an explanation for this apparent discrepancy. When his predecessor Giovanni died of the plague in 1348, Pietro was immediately (between 31 May and 13 September 1348) designated abbot by the monks of Santa Scolastica. It was only later (in March 1349) that this appointment was confirmed by the Pope. This explanation is confirmed by a study of Benedictine customs in the thirteenth and fourteenth centuries regarding the appointment of abbots at Subiaco (Cignitti and Caronti, 1956), as well as of clashes with the Papal See and the fact that there are other notarial deeds that mention Abbot Pietro before 1349.

Based on the analysis described here, Guidoboni and Comastri (2005) think that there are no intrinsic or extrinsic factors to justify doubts about the dating of the document. But it must also be asked whether an authentic notarial deed may contain, for reasons unknown to us today, a falsehood. Could the date, for example, have been tampered with, and the deed backdated by one year? Or could the document be referring to the earthquake of 1349? Such a view is not considered tenable, for if the notary wrote the document in 1349 and wished to backdate it, why did he include such specific references to the great earthquake? The seismic crisis of 1349 was very famous, and so no one at that time and in that region could have had any doubt about dating it to that year. These two considerations lead us to set aside the hypothesis, put forward by Molin *et al.* (2000), that the date of the notarial deed in question is simply wrong, and hence the reference to an earthquake in that year is to be rejected. By its very nature, a notarial deed cannot be rejected as being false or true without adequate proof. The only hypothesis that can support this interpretation of the document is that, for unknown reasons, the notary let a year pass by between the writing of the document (1348) and its official completion, thus explaining how the collapse of the abbots' building was included in the text. But there is no line of textual criticism which supports that hypothesis.

Molin *et al.* (2000) prefer to accept what the *Chronicon Sublacense* has to say, having compared the two texts and decided that only one of them can be right. The *Chronicon* is the only text close to the time of the earthquake which mentions effects at Subiaco; it was written between 1370 and 1377, i.e. about 30 years after the seismic event took place. However, the chronological inaccuracies of this chronicle are well known to historical criticism. The anonymous chronicler backdates the famous plague of 1348 by a year, in spite of the fact that there are a great many wills drawn up in 1348 which bear witness to the plague at Subiaco. He also places the death of Abbot Giovanni in 1347, though the abbot was in fact still alive in May 1348. Furthermore, the chronicler goes on to state that Abbot Pietro died in 1349 (shortly after the great earthquake) when in fact he was still active in September 1350 (BMN Subiaco, arca VIII, parchment 81; Egidi, 1904, p. 215).

Set out here is the text of the *Chronicon*:

> In the year of Our Lord 1349, the abbot after Giovanni was Pietro of Perugia, in whose time an earthquake almost completely destroyed the monastery and castle of Subiaco and other fortified abbey buildings. The abbot himself was in his room in the castle when the destruction occurred, and he was so frightened that he lived for only a short while after [the earthquake].

It is reasonable to suggest that when the anonymous compiler of the *Chronicon* was writing almost 30 years after the great earthquake of 1349, he absorbed into it the report of the 1348 earthquake at Subiaco. Perhaps the report from 1348 and memories of the earthquake in 1349 became superimposed in his memory. One building that both sources mention is the abbots' castle. According to the notarial document it was badly damaged in 1348, whereas the *Chronicon* states that it completely collapsed in 1349. Even if there were two different earthquakes, the two reports about the castle do not appear to be completely contradictory. For example, they allow for an ongoing deterioration in the castle's condition due to damage from the first earthquake (assuming that the castle had not yet been completely restored), which then led to collapse of the castle in the second earthquake.

3.2.6   Newspapers from the sixteenth to the eighteenth century

This group of sources comprises widely circulated printed reports that gradually developed into a common form that are categorized today as newspapers. An event of fundamental importance occurred in 1453, when the first mechanical process for reproducing written texts was invented. Within a few decades, the circulation of information via the dissemination of multitudinous printed copies created a new urban culture based on the use of this more widely available information. This development gave a boost to the acquisition of literacy amongst an ever-growing segment of the population. The spread of printing increased the knowledge of earthquakes and the human curiosity about them, causing earthquakes to lose, in a way, their strictly religious interpretation which had monopolized the thinking about earthquakes from very early times. It is clear that the spread of news about the occurrences of seismic events promoted a process of rationalization by providing details of damage and other strong seismic effects, thereby exerting a powerful influence on theoretical views about their origin.

One of the characteristics that developed during the growth of Europe's renowned tradition of popular printed gazettes and newspapers was a substantial news circulation network, promoted by a widespread urban culture. As a whole, this type of source has made a substantial contribution to historical seismology, not only by outlining the effects of earthquakes and tsunamis in inhabited areas, but also by describing the social conditions in the period of the immediate post-earthquake emergency. These sources have also thrown light on the attitudes of local culture toward earthquakes. From the sixteenth to the seventeenth century, printed matter in the form of notices and reports, often preceded by terms such as truthful and original (*veridica, distinta*), was aimed at a poorly educated public that evidently felt a new need for information (Figures 3.28, 3.29 and 3.30).

Although newspaper texts were influenced by the particular literary tastes and religious atmosphere of the time, the descriptions they provide do not follow a strict pattern. They usually took the form of compendia or letters, with their popularity being the result of their direct communication of the events of interest, which satisfied the strong thirst for information. These texts are a sign of the increased literacy that spread through the middle classes in the seventeenth and eighteenth centuries, and they also display the influence that empirical observations of natural phenomena brought to bear on the choice of what information to print and circulate. Like today, the choice of printers about what to print was aimed at selling as many copies as possible. Even in the early days of printing it was realized that spectacular events sell more copies. For example, extraordinary events such as visions and prodigies with religious overtones were a favourite subject in early newspapers. There was no strong differentiation between the natural and the spiritual aspects of an event that was reported. These elements were superimposed upon one another, coexisting within a single cultural framework. One of the direct results of this type of thinking was the insertion of the descriptions of earthquakes and their effects into religious and

Figure 3.28 Frontispiece of an ancient printed report (*Copia de una lettra*) on the earthquake in Sicily on 17 December 1542 (Biblioteca Nazionale Marciana, Venice, *Miscellanea*, 2230/29).

social contexts in ways that today seem completely alien. Many descriptions of seismic effects appear alongside reports of miraculous signs, prodigies and religious interpretations of events, giving the impression to the unprepared reader that one is simply reading fanciful tales that include earthquakes. In fact, the complexity of the mental attitudes involved means that appropriate critical tools are required if the evidence about real earthquakes that these texts contain is to be thoroughly examined and properly interpreted to discern factual from literary or religious elements.

Taken as a whole the descriptive material provided by published notices and reports cannot be used in isolation. Rather, it must be related to the information contained in other types of source. Some of these popular newspaper-style journals are at times embellished with stereotyped drawings of the disaster concerned (such as ruined houses or whirlpools in the sea) or on rare occasions with maps of effects. These texts were much used and appreciated by the nineteenth-century seismological tradition, but nowadays they are not always easily found in public libraries because many have found their way into the private collectors' market.

Figure 3.29 Frontispiece of a report–newspaper, printed in 1570 and circulated in the German-speaking areas, on the earthquake at Ferrara in November of the same year (the left image) and of lightning in Florence (the right image). In this case the association of these two unrelated events in the paper later generated in the literature the false news of an earthquake in Florence in 1570 (Hans Moser, Nuremberg, from Strauss, 1975).

After the appearance of gazettes in the seventeenth century, deriving and interpreting information from these kinds of journals becomes even more difficult, but in a different way from that of earlier times. During the eighteenth century, some journals succeeded in achieving a pre-eminent status as purveyors of information. Since there were no press agencies at the time, minor journals took their news from their more prestigious brethren and then often elaborated on what they had taken, for no clear reason but in such a way as to supply apparent confirmation of the news. Such sources have made an important contribution to the knowledge of local seismic effects, but there are numerous risks involved if they are used uncritically, as has been shown by a number of scholars who have studied this material in detail (see the interesting work by Quenet, 2005 with its complex use of the gazettes for the earthquakes of the seventeenth and eighteenth centuries felt in France).

There are two principal limitations to the use of these sources. One relates to the way in which they are consulted, and the other concerns the actual content

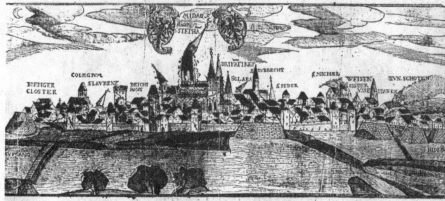

Figure 3.30 Vienna damaged by the earthquake of 15 September 1590 in a proto-newspaper of the time printed in Augsburg in the same year (from Gutdeutsch *et al.*, 1987).

of the news item. As far as the former is concerned, it is important to note that any assessment of such sources must be systematic and chronologically broad. It is sometime found, for example, that a mistake which entered the seismological tradition was based on an initial false or seriously inaccurate report in a gazette. A subsequent denial or correction may have appeared later in that or another publication, but this was never discovered by the original earthquake catalogue compiler. Thus, publicly retracted events or reports can remain taken as fact in earthquake catalogues (Figure 3.31).

As far as the actual content of news items is concerned, limitations lie both in the time taken for news to be disseminated and in the ways journals relied on one another. As pointed out above, news gazettes and journals had the habit of republishing news items with appreciable embellishment. Because of this, it is always important for researchers to seek out the primary source responsible for publishing and disseminating the news item by identifying the original news report as well as its later variants (Musson, 1986). It was only after the mid nineteenth century, that is to say, after the invention and spread of telegraphy, that coordinated press agencies began to operate in many countries. Critical use of press agency dispatches can be extremely valuable today for research purposes, because they allow time and resources to be saved for research. Historical

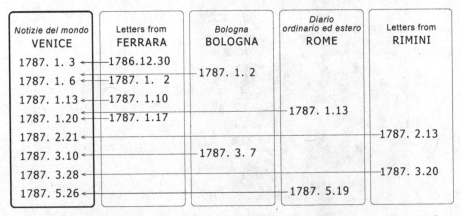

Figure 3.31 Relationships of the times that several gazettes reported the news of the earthquake on 25 December 1786 at Rimini (northern Italy). Some newspapers copied the news from the others and therefore are not independent sources (from Guidoboni and Ferrari, 1986).

seismological research using these reports is also considerably facilitated by the fact that some of the most important press agencies have set up digital archives of their historical material.

3.2.7 Newspapers from the nineteenth and twentieth centuries

The newspapers of the last two centuries constitute the main source of information for many earthquake catalogues in use. Therefore, it is important to analyse carefully the quality of the information in newspaper reports and how the collection of the news came about. Beginning in the mid nineteenth century newspapers became quite specialized, but they also contained services and correspondences concerning earthquakes (main shocks and aftershocks: see, for example, Seeber and Armbruster, 1987) and tsunamis that occurred in areas very far away from the place of publication of the newspaper (Figure 3.32). The press also became enriched with visual images. Drawings, sketches and etchings, often taken from photographs, were published in large numbers, contributing to the creation of a new cultural image of the earthquakes and the tsunamis that had been taking place (Figure 3.33). In many countries there were newspapers that specialised in illustrations. Earthquakes, together with great floods, avalanches, landslides, hurricanes, shipwrecks and wars, became topics that attracted the public and encouraged the sale of the newspapers themselves. Some newspapers became famous, like the *New York Times*, the *Illustrated London News*, *L'Illustration* (Paris) or *Le petit parisien* and *L'Illustrazione italiana*, etc.

Figure 3.32 *L'Illustration, Journal Universel*, issue of 30 January 1858, no. 779, vol. 31 p. 69, with some images of the earthquake on 16 December 1857 which devastated a hundred or so villages in the Kingdom of Naples.

Figure 3.33 A detail from the preceding figure, highlighting the quality of the images that were published in the illustrated newspapers of the mid nineteenth century. This depiction shows damage to the church of Paterno (Potenza southern Italy).

Figure 3.34 The earthquake wave at St Thomas striking the Royal Mail steamship *La Plata*, from the *Illustrated London News*, 28 December 1867 (36 × 25 cm).

The news of a tsunami off the coast of the island of St Thomas in the Caribbean in 1867

On 28 December 1867, the *Illustrated London News* presented a thought-provoking picture of a ship, the Royal Mail *La Plata*, which crashed into a great wave that seemed almost a wall of water (Fig. 3.34). The illustrator had not forgotten to draw the crew as human figures that seemed to wander around in terror on the ship's deck or huddled in a lifeboat. The caption called the phenomenon 'an earthquake wave'. The news refered to calamitous events that had happened a few months before, between October and November of that year. 'Misfortunes never come singly' is the newspaper's measured comment. On 29 October the area had been hit by a violent hurricane, which caused heavy damage to the port, sank two merchant ships and damaged many buildings. Perhaps 1000 people perished. The population was just recovering from this calamity, when on 18 November at 3 o'clock in the afternoon, a strong earthquake hit the same area:

> In a moment the houses shook from doorstep to roof; they cracked and groaned; the earth heaved and danced again, and much of the damage which the hurricane had commenced was now completed. The whole population, some

> 15 000 to 20 000 souls, rushed into the streets, shrieking, praying, and
> imploring the Divine mercy. All the night one shock succeeded another in rapid
> succession. The shocks continued for several days.

Twenty minutes afterwards a tsunami of alarming proportions hit St Thomas
and the surrounding islands:

> But this was not all. About twenty minutes after the great shock on the 18th, a
> confused cry was heard from the sailors and workmen on board the shipping in
> the harbour, while numbers of them were seen pouring, like strings of ants
> down a wall, over the sides of the ships and pulling for life to the shore.
>
> In a few seconds the cause of the panic was apparent. A monstrous breaker,
> or rather a sea wall, variously estimated at from 30 ft to 60 ft [about from 9 to
> 18 metres] high, was seen racing in towards the harbour, at a pace of fifty
> miles an hour [80 km per hour]. It seemed as if the town at least if not the
> island, would be swept away.

The twentieth-century press has contributed a great deal to the knowledge of
seismic effects, not merely through the spread and penetration of news services
but also by sensitizing public opinion to disasters. It must be kept in mind, however, that, as far as historical seismology is concerned, the use of these sources
must be subject to a critical awareness of their political and social context. Since
strong earthquakes could have an indirect influence on social control and public
order, news of damage has been subject to press censorship in some political
situations. In such cases, the extent to which censorship has been brought into
play obviously needs to be assessed, and other independent sources must be
relied on as far as possible.

In most recent times, the press has adopted new attitudes toward reporting
earthquake effects, which seems to be leading to an abandonment of many
descriptive elements in favour of greater attention to social and emotional
responses to the earthquake. Journalists in general seem also to be experiencing
some difficulty in establishing a clear role for themselves in the communication
process concerning earthquake effects, for their written descriptive pieces are
placed almost in competition with other specific summaries and depictions of
the event and its effects. Indeed, increasingly available to everyone are pictures
and scientific reports, especially now with the availability of information on the
internet that is processed by research centres and specialized agencies.

Newspaper accounts of a large earthquake are reports of a devastating event,
with the result that the language used is sometimes far removed from that
of research. Statements made in newspaper accounts are often scientifically
inappropriate or incomplete and may convey inexact information about

magnitude scales, levels of ground-shaking, etc. Thus, one can conclude that while the news media offer an increased speed of communication, breadth and diversity of reporting, range of locations from which reports come, etc., this is offset by a segregation of different aspects of the knowledge about an earthquake and its effects. This could be a problem if the many data available today on the internet or on digital media are not available in the future. One can imagine an historical seismologist in the year 2500 who is looking for data on the earthquakes at the start of the twenty-first century. If he or she only found today's newspapers on paper or microfilm, he or she would certainly have difficulty reconstructing an image of the seismic event that is close to the reality.

There have also been in the past particular periods of time in some places where the press was subjected to government censorship. In these places the news of a destructive earthquake may have been subjected to censure because alarming news of destruction or reports of inefficiency in the system of aid provision could trigger problems of public law and order. For these reasons such news has often been filtered by dedicated governmental offices.

When historical seismological research concerns particularly difficult periods from the point of view of archival sources and is conducted in the absence of the consolidated traditions of other sources of information, newspapers represent an important sector for their continuity of information and for their territorial diffusion. From the parsing and the comparisons between the various pieces of information that has been performed in a wide-ranging and critical manner for many newspaper reports earthquakes, the data obtained from many newspaper reports have been shown to be quite reliable.

Jack London and the 1906 San Francisco earthquake

At the time of the 1906 San Francisco earthquake, Jack London was living on his farm in Glen Ellen, California, about 40 miles (65 km) from San Francisco. London was 30 years old and had published his most popular work, *The Call of the Wild*, three years earlier. Soon after the earthquake London was asked by *Collier's Magazine* to write a report from San Francisco. London arrived as the fires that followed the shock were consuming the city.

> *Not in history has a modern imperial city been so completely destroyed. San Francisco is gone. Nothing remains of it but memories and a fringe of dwelling-houses on its outskirts. Its industrial section is wiped out. Its business section is wiped out. Its social and residential section is wiped out. The factories and warehouses, the great stores and newspaper buildings, the hotels and the palaces of the nabobs, are all gone. [. . .]*

Within an hour after the earthquake shock the smoke of San Francisco's burning was a lurid tower visible a hundred miles away. And for three days and nights this lurid tower swayed in the sky, reddening the sun, darkening the day, and filling the land with smoke. [. . .]

There was no opposing the flames. There was no organization, no communication. All the cunning adjustments of a twentieth century city had been smashed by the earthquake. The streets were humped into ridges and depressions, and piled with the debris of fallen walls. [. . .]

Dynamite was lavishly used, and many of San Francisco's proudest structures were crumbled by man himself into ruins, but there was no withstanding the onrush of the flames. Time and again successful stands were made by the fire-fighters, and every time the flames flanked around on either side or came up from the rear, and turned to defeat the hard-won victory.

On Thursday morning at a quarter past five, just twenty-four hours after the earthquake, I sat on the steps of a small residence on Nob Hill. [. . .] I went inside with the owner of the house on the steps of which I sat. He was cool and cheerful and hospitable. 'Yesterday morning,' he said, 'I was worth six hundred thousand dollars. This morning this house is all I have left. It will go in fifteen minutes.' He pointed to a large cabinet. 'That is my wife's collection of china. This rug upon which we stand is a present. It cost fifteen hundred dollars. Try that piano. Listen to its tone. There are few like it. There are no horses. The flames will be here in fifteen minutes.'

Outside the old Mark Hopkins residence a palace was just catching fire. The troops were falling back and driving the refugees before them. From every side came the roaring of flames, the crashing of walls, and the detonations of dynamite. (From the Museum of San Francisco website, www.sfmuseum.org/hist5/jlondon.html)

4

Types of scientific sources: historical interpretations of earthquakes (an *excursus* from the ancient world up to the twentieth century)

Historical scientific sources provide a valuable, albeit little-used, wealth of data about the earthquakes of the past. In general, these texts have been examined only since the second half of the nineteenth century, which is when scientific research and written communications in the natural sciences had become focused into the distinct disciplines that are still recognized today. It is worthwhile remembering that from a historical perspective, the inquiry into nature had very remote origins. From the ancient Hellenic and Roman worlds up to the defining of modern sciences, it was natural philosophers who conducted observations and articulated theories about the natural world.

This chapter presents a concise examination of a very large set of sources featuring a great variety of texts. The common denominator of these sources is not so much a record of individual earthquakes and tsunamis themselves (as found in historical sources) but rather a set of thoughts and observations about the causes and effects of these natural events. As many of the theoretical observations were based on directly observed phenomena, one can find in these texts some important references to actual earthquakes and tsunamis of which we would otherwise have no knowledge today. This short *excursus* proceeds from the ideas of the ancient Greek philosophers up to the theories that have been the forerunners of modern seismology.

4.1 Theories and treatises of the past

4.1.1 *From the ancient world to the early Middle Ages*

If today something is known about the seismicity of the Mediterranean that can be used to identify dangerous, active faults, then a debt is owed to the ancient philosophers. Their reasoning and observations have reached us today because their texts were venerated, transcribed and passed down; these were recognized as cultural roots and important references and have been preserved for over 2500 years. As has been observed from the beginning of history, in the ancient Mediterranean region beliefs deriving from religion and from the observation of nature were never distinctly separate. Naturalist interpretations of earthquakes were not considered as aspects of a rational thinking to be separated from their religious interpretations. On the contrary, religious thinking was a fundamental element of the whole of the Mediterranean culture. What is presented here is a systematization useful for chronicling the history of the theories concerning earthquakes.

Interest in earthquakes and tsunamis was constant in the Mediterranean world because the recurrence of these phenomena is a natural characteristic of the entire Mediterranean area. A complete review of theories about these phenomena would mean going over the entire history of ancient science and philosophy (Lloyd, 1970, 1973, 1991; Russo, 2004). It is possible to summarize this topic here only by reducing it to general terms and leaving a full consideration of the detailed issues to more specific literature. Contrary to common belief, the ancient world was characterized by a plurality of interpretations of seismic phenomena.

The earliest surviving theories about earthquakes and tsunamis are known only indirectly through the works of Aristotle (384–322 BC), Pliny the Elder (AD 23–79) and Seneca (4 BC – AD 65). The most ancient known theory of the causes of earthquakes is attributed to Thales of Miletus (c.624–c.545 BC), who was indicated by Aristotle as the initiator of philosophical inquiry into the principles of nature. In the view of Thales, the cause of earthquakes coincided with the material cause of all things, i.e. the water (or 'wet element') upon which the Earth floated like a great ship. This theory was supported by the observation that an earthquake would make new springs flow, a phenomenon explained by its similarity to what happens when a ship is rocked by the waves of the sea (Seneca, *Naturales quaestiones*, 6.6.2).

Unlike Thales, Anaximenes (586–528 BC) identified air as the element from which seismic events originate. Rarefaction and condensation of the air give rise to all the elements (fire, wind, clouds, water, etc.). This explanation of all things was given in terms of the *efficient cause*: while air or water are the prime elements of things, the objects of the world, once constituted, have their own autonomous existence. To illustrate the generation of earthquakes, Anaximenes used a curious analogy in which the Earth was compared to an old house, whose lower parts can suddenly give way, making the upper parts become unstable or even fall suddenly (Seneca, *Naturales quaestiones*, 6.10.2). Thus, the processes of the Earth's ageing as well as the drying up or excess of water were responsible for earthquakes.

Aristotle (and Seneca a few centuries later) seems to disagree about Anaxagoras (499–428 BC). Aristotle attributes to Anaxagoras the thesis according to which earthquakes occur when the ether, which has a natural propensity to rise, is trapped in underground cavities and cannot escape because the pores of the Earth are closed by rain. Aristotle discards this explanation as being an oversimplification, given that Anaxagoras presupposed a flat Earth floating on ether. Furthermore, Anaxagoras's floating ether theory, according to Aristotle (*Meteorologica*, 2.7.365a), does not explain the correlations between earthquakes, the seasons and the places where earthquakes took place. As Marmo (1989) observed, Seneca presents Anaxagoras's theory as an attempt to explain in a unified way both celestial and subterranean phenomena by resorting to the action of fire (*Naturales quaestiones*, 6.9.1). Seneca also recalls other philosophers who, like Anaxagoras, thought that internal combustion of the Earth provokes collapses and thus clefts in the Earth's crust (*Naturales quaestiones*, 6.9.3). Probably connected to the theory of underground collapses and to that of underground waters is a later opinion that traces the prime cause of earthquakes to the action of 'underground vapours' which are released by underground waters. An analogy that illustrates this theory, used in medical treatises, is the metaphor of water inside a small container bubbling over when heated over a fire. According to some commentators, Anaxagoras was the first philosopher to conceive that matter was infinitely divisible into particles of differing qualities (what Aristotle later called *omeomerias*). This atomist conception of matter was subsequently developed in a systematic way by Democritus of Abdera (460 – c.370 BC). He conceived of the real world, from a physical point of view, as *discontinuous*, being made up of indivisible atoms, but he admitted that from a logical and mathematical point of view, matter could be thought of as infinitely divisible, i.e. as a *continuum*. Democritus thus began a debate on the continuity or discontinuity of nature that lasted until the twentieth century. The various forms of this debate modified within the changing scenarios with time of human knowledge had repercussions for theories about the origin of earthquakes, as is seen below. For Democritus, the cause of earthquakes lay in the relationship between the amount of water that flowed within the Earth and the capacity of empty spaces to contain this water. The pressure exerted by the mass of water suddenly erupts in an outflow, thus producing earthquakes (and giving rise to new springs). Democritus's explanation of earthquakes may seem rather superficial when compared with the complexity of his conception of nature. But the chief merit of the ideas attributed to Democritus is that he tried to construct a materialistic and mechanistic system capable of accounting for the whole of reality, without having recourse to extra-natural forces. His theory enjoyed great popularity and was put forward on a number of occasions across the centuries.

Different theories, which are presented here in a unified way albeit with different shades of meaning, resorted to the 'force of the air', i.e. *pneuma*, to explain seismic phenomena. The *pneuma* was not considered to be air *per se*, but rather the force of the air. This concept was revived and systematized, as is seen below, by Aristotle. The first to present this thesis seems to have been the philosopher Diogenes of Apollonia (fifth century BC), a pupil of Anaxagoras. Aristotle associates Diogenes with Anaximenes. Diogenes was a supporter of the air doctrine as a principle cause of natural things (Aristotle, *Metaphysica*, 1.3.984a). He is thought to have separated himself from his master's ideas by having attributed the responsibility of earthquake shocks to air. Air enters the bowels of the Earth through the pores that naturally open on its surface and through the voids that derive from the erosive action of rivers and tides. When the pores are closed again, however, the air finds its return path barred and becomes agitated. Unable to continue its natural movement, it veers upwards and shakes the Earth (Seneca, *Naturales quaestiones*, 6.15.1).

Even if Diogenes of Apollonia (or Seneca, on his behalf) does not clarify the nature of the *spiritus* that manages to shake the Earth with its force, it seems clear that he is referring to the wind. It was Archelaus (fifth–fourth century BC), presumed to be the master of Socrates and a pupil of Anaxagoras, who indicated that the cause of earthquakes lay precisely in the winds (Seneca, *Naturales quaestiones*, 6.12.1–2). Aristotle (384–322 BC) believed that there were common elements shared by the winds and earthquakes. In his *Meteorologica*, the discussion of earthquakes follows that of the winds precisely because their causes were of the same kind (*Meteorologica*, 2.7.365a). The Aristotelian systematization of the theory of the winds as the origin of earthquakes represents a seminal development in the history of the theories of seismic events, as it remained practically unchallenged for over nineteen centuries.

According to Aristotle (*Meteorologica*, 2.4), there are two types of emissions from the Earth: a humid one, called *vapour*, and a dry one (2.4.359b). The latter is the origin and natural substance of the wind (2.4.360a), and therefore it is more than merely the movement of air. Under the name of *pneuma*, a dry steam constitutes the common trait of earthquakes and winds. When the Earth is heated up by the Sun and by its internal fire, it produces a huge amount of *pneuma*, both internal and external. When the *pneuma* leaves the Earth, it gives rise to the winds; when it heads towards the inside of the Earth, it is amassed and causes earthquakes (2.8.365b). According to Aristotle, the relationship between dry and humid emissions in the Earth also explains the climatic conditions in which earthquakes usually seem to occur. In the *De mundo*, which some scholars attribute to Aristotle (Reale and Bos, 1995), seven types of earthquake are mentioned: horizontal, heaving, sinking, splitting, thrusting, oscillating and roaring (*De mundo*, 4.396a).

In Aristotle's works there are several references to earthquakes, the emergence of islands, and volcanic eruptions known in his time. Such records are invaluable for us today, for they are actual empirical examples that he used to support and illustrate his theory. For example, he wrote:

> *Evidence of all this is available from observations in many places. For there*
> *have been earthquakes in some places which only ceased when the wind, which*
> *was their motive force, burst through the earth into the air like a hurricane, as*
> *recently happened at Heracleia in Pontus, for example, and before that at the*
> *island of Hiera.* (Meteorologica, 2.8.367.)

In listing the various types of earthquakes, Aristotle refers to tremors which come up from below.

> *Whenever this type of earthquake does occur, large quantities of stones come to*
> *the surface, like the chaff in a winnowing sieve. It was this kind of earthquake*
> *that devastated the countryside surrounding Sipylus, the area known as the*
> *Phlegraean Plain, and the districts of Liguria.* (Meteorologica, 2.8.368.)

Theories on the earthquakes of ancient China: an extraordinary point of contact

We may wonder whether the ancient Mediterranean world and that of late antiquity had elaborated more backward or more advanced theories concerning earthquakes than the Chinese world, which we often tend to believe has always preceded us in terms of theoretical and technological complexity. According to Needham (1959), who offers a pivotal and by now a classical point of view on the sciences in China, as concerns earthquakes, the ancient Chinese did not express theories that differed substantially from those circulating in the same period in the Mediterranean area. Around the eighth century BC – the period when the information on the earthquake occurring in the Mediterranean area started – the Chinese explanations resorted to *yin* and *yang*. The former represents what is cold, dark, damp, lifeless, while the latter stands for what is warm, light, dry and moving (identified in philosophy, as the universal female principle and the male one, respectively). *Yin* can be translated approximately as the side of the hill in shadow and *yang* as the sunny side of the hill: the relationship of one in relation to the varying of the other responds to the awareness of the endless nuances of reality. It is a more complex vision, but not unlike the states of cold and damp/warm matter that generates the *pneuma* (see above). According to Needham, the Aristotelian principle of *pneuma* is surprisingly similar to the Chinese idea of the imprisonment of the *chhi* in the bowels of the Earth.

In China, it was also believed that earthquakes could be predicted with the help of astrology. The seismic events that occurred were associated by soothsayers and court astrologists to actual horoscopes on the life of the reigning emperors. In China, as in Europe, the pneumatic theory survived for a long time and was extraordinarily popular for almost two thousand years, becoming the most enduring and widespread theory that has ever been developed.

Theophrastus (373/370–287 BC) (he was given this name, meaning 'divine speaker', by Aristotle) and his pupil Strato of Lampsacus (328–278/270 BC) expanded on some parts of the Aristotelian theory on earthquakes. Theophrastus developed scientific and antimetaphysical aspects of the first phase of Aristotelianism. Typical of his approach is the investigation of real phenomena, the systematic collection of data and the elaboration of a flexible and open logic to explain his observations. As far as his interpretation of earthquakes is concerned, he points to the *pneuma* as their cause, but he also includes underground collapses as arising from the same cause. Strato of Lampsacus, on the other hand, introduces the dynamics of the relationships between warm and cold *pneuma*, whose continuous alternations in the bowels of the Earth (reminiscent of a sort of thermal convection) account for the *pneuma's* action on the Earth's surface (Seneca, *Naturales quaestiones*, 6.15.2–6). Strato accentuates the importance of empirical observation and experimental research in the Aristotelian theory. In this perspective, he adds new types of earthquakes, subdivided on the basis of the phenomena that accompany them (emissions of wind, rocks or mud, underground noises, etc.), or the type and number of shocks or vibrations ('strike earthquakes' if they cause a single shock, 'vibrating events' if they cause oscillations in opposite directions).

Besides those examined above, there are also early theories that one could call *pluralist*. Democritus put the action of the winds and of the waters contained in underground cavities on the same level. The dynamics are the same for both kinds of fluids (e.g. compression, search for a way out, impetus), even if at times it is the wind that thrusts the water forth (Seneca, *Naturales quaestiones*, 6.20.1–4).

According to Epicurus (341–270 BC), all the causes mentioned (that is, water, earth, fire and air), along with others, can bring about an earthquake. He disapproved of those people who pointed to just a single cause of earthquakes (Seneca, *Naturales quaestiones*, 6.20.5). The reason for this stance lies in the fact that, for Epicurus, the causes of earthquakes were 'obscure in themselves', that is, they could be known only by inference or conjecture based on 'signs'. One can understand Epicurus's statements as an invitation to build explanatory models by means of the systematic application of analogy, especially analogies that appeal to the human senses (Marmo, 1989). For Epicurus, the *plurality of explanations* for phenomena like earthquakes, which go beyond the possibilities of direct human observation, contributes to preserving the unperturbed peace of the soul, believed to be an ideal condition for a man who is wise.

Chrysippus (281/277–208/204 BC) believed that the cosmos was a living, rational and animate being. His general theoretical framework can be traced back to the earlier idea of the *pneuma*, and his theory was subsequently written down by Posidonius (135–51/50 BC). Chrysippus resorts to an analogy with the workings of the human body in order to postulate a new theoretical interpretation for seismic activity. Posidonius, like Aristotle, suggests a classification of earthquakes, but into four types: 'oscillating, catastrophic, vortex-like, and jolting' (Seneca, *Naturales quaestiones*, 6.21.2; Diogenes Laertius, 7.154).

Roman culture was mostly heir to the classical Greek world, and harked back to ancient thinking with almost no changes at all. Lucretius (98–54 BC) with the *De rerum natura*, Seneca with the *Naturales quaestiones*, and Pliny the Elder with the *Naturalis historia* were the most important cultural go-betweens by means of which the classical theories on earthquakes have reached us today. They did not articulate new theories but merely presented the traditional thinking that had preceded them, often inserting descriptions of earthquakes or volcanic eruptions that had occurred during their own times. Seneca's treatise, for example, was written immediately after the earthquake on 5 February 62 AD which struck Pompeii, Herculaneum and other cities in Campania (southern Italy), 17 years before the famous Vesuvius eruption in AD 79. The works of these Latin authors were very popular and well known, and were quoted throughout the Latin Middle Ages. Pliny, in particular, represents the *summa* of the whole of ancient encyclopaedic knowledge.

4.1.2   Late antiquity and the Middle Ages

The Christian culture of the ancient and medieval period did not mark a break with the ancient world but rather limited itself primarily to grafting its new values onto those earlier theories passed on from classical tradition. Christian culture incorporated the concept of an earthquake into its religious viewpoint and emphasized an earthquake as a supernatural sign. Religious thinking was already highly developed in the classical world, and as Christianity became the dominant culture in Europe and elsewhere, it tended to be radicalized to the point where it discouraged the rationalist observation of natural phenomena. A symptom of such radicalization in the West may be found in the *Liber de haeresibus* by Philastrius, bishop of Brescia (who died between AD 391 and 397). In this work Philastrius regarded as heretics those people who believed an earthquake to be a phenomenon having a natural origin.

Figure 4.1 Isidore of Seville, encyclopaedist of the sixth century AD in a twelfth-century miniature (Bibliothèque Municipale d'Amiens, *Vitae Sanctorum*, fol. 246v).

A great Christian heir to ancient wisdom was Isidore of Seville (560–636), one of the first and most famous encyclopaedists of the Christian Middle Ages (Figure 4.1). In his *De natura rerum liber*, Isidore observes the opposition between the knowledge of this world, which provides a causal and rational explanation for seismic and other natural phenomena, and that of the sacred scriptures, which interpret earthly events as signs of a Divine Will that is otherwise inscrutable. Isidore seems to want to consider the opposed theories of collapses within the Earth and the movement of underground waters as corollaries to the theories of enclosed winds (46.2, p. 76). To Isidore, the image of the wind also lends itself to portraying the spirit of God acting in the Earth, since it was believed that God would come to judge the world at the end of time.

In the early Middle Ages there were not only theoreticians but also experimenters. The experimenters were constructors, mathematicians or physicists who applied existing knowledge rather than elaborating new ideas, as the case described below demonstrates. The activity of these practitioners is not reported in the texts of the natural philosophers, so today one learns of their existence almost by pure chance. A few years before the birth of Isidore of Seville, perhaps in AD 557, Anthemius of Tralles died in Constantinople. He was one of the builders of the great church of Saint Sophia and was considered to be an expert on problems of physics and meteorology (Traina, 1989, pp. 186–90). According to the Byzantine historian Agathias (*Historiarum libri quinque*, 5.7), Anthemius tried to simulate an experimental earthquake in his house by applying Aristotle's theory. Constantinople had been hit by strong earthquakes in 554 and 557 that had damaged the great church of Saint Sophia, and apparently Anthemius wanted to observe in a laboratory setting, so to speak, the shaking of a building. He built a boiler with some spouts that terminated beneath the floor of his neighbour's dwelling (the rector Zeno, a detractor of his). Anthemius then created a steaming vapour so violent as to create a sort of shaking, which made the unfortunate neighbour flee panic-stricken. The experiments of Anthemius followed the Alexandrine tradition of steam machines (like the *aeolipile* built by Hero, perhaps in the first century AD). It is well worth pointing out here that the thinking that underpins this story highlights an experimental approach to understanding natural phenomena, a thinking that tends to become popular only in later eras.

About one century later, the Venerable Bede (c.672–735) in his *De natura rerum liber* reiterates the theory of the winds based on the scheme of Isidore. He adds an original note by drawing a parallel between the production of earthquakes and that of thunder and tsunamis.

A religious vision of the world prevailed increasingly over a naturalist one in the medieval centuries, and for a long time religious thinking dominated the interpretation of natural phenomena. Contemporaneous to this Christian thinking is Islamic thinking (as expressed in the Koran), which began in the seventh century AD. The image of the earthquake in the Koran occurs just once in the title and the first verse of *sura* XCIX, 1, not as a premonitory sign of the end of the world but as a representation of its end. The nature/culture bipolarity that later became codified in the Latin West does not seem detectable in Islamic religious thinking. The Islamic religious vision is substantially anthropocentric and the concept of law of nature does not exist. Rather the natural elements follow a custom in which there is a creation that is continually being formed and changed (Khalidi, 2005). This was a religious conception, but there was also genuine naturalistic thought.

The great Islamic cities, such as Baghdad, were active centres for the study of Greek philosophy and science. The greatest Arab philosopher and scientist in the first millennium, and one of the greatest in the whole of the Middle Ages, was al-Farabi (872–950). He was of Turkish origin, but trained in the Baghdad school. The theories he worked out blended naturalism and religiosity, and their complex systematization had a profound influence on later thinkers. Islamic philosophies of the natural world did not invent new interpretative theories of earthquakes, but added new empirical observations to corroborate existing theories. The study of the origins of earthquakes was thus based on the earlier, thinking of the ancient Greek world, and especially on Aristotelian theory since his texts were known to Arab philosophers, and the Arabs were the first to write commentaries on these works from antiquity.

Another great thinker of the medieval Arab world was Abu Ali al-Husayn Ibn Sina, who was known in the West as Avicenna (980–1037). He was born in Persia in the period of the Arab conquest, and became a physician and philosopher. His medical texts were used in Western universities up to the seventeenth century. His thought is a mixture of Aristotelian and oriental theories, very much like those of al-Farabi, for his system of rational thought also included religious views. According to Avicenna, volcanic and seismic activity was considered evidence of the existence of winds imprisoned within the bowels of the earth. Fundamental amongst the works of Avicenna is his translation of Aristotle's *Meteorologica*, with his own added commentary. It was widely disseminated in the Latin Middle Ages. In Avicenna's work, the nucleus of ancient Aristotelian thought was described anew and commented upon for contemporary readers, and for several centuries following its production it represented standard medieval geological thinking. Avicenna's treatise *De mineralibus* became the basis for subsequent treatises on nature.

In the Islamic culture of that time there were no original developments regarding explanations of the causes of earthquakes, but only a more thorough analysis of the older ideas of Aristotle. This may strike one today as being rather strange, because it is well known that Arab culture was very innovative in such scientific disciplines as medicine, astronomy and mathematics.

Muhammad ibn Ahmad Muhammad ibn Rushd was known in the West as Averroes (1126–1198). He was a Spanish philosopher and physician whose famous *Commentaries* on Aristotle led to his being known as *The Commentator* in the Middle Ages. He was opposed to Avicenna's thinking, which regarded the origin of the world as being from God, and he postulated the eternity of matter. In the years when Averroes was writing, a work by the philosopher al-Ghazali (1058–1111) enjoyed a considerable following. This work was titled *The incoherence of the philosophers* (later known as *The destruction of the philosophers*). In it, al-Ghazali attributed absolute control of reality and nature to God, opposed Aristotle, and pointed to fideism and mysticism as the only way to arrive at the truth. Averroes was opposed to such thinking, and systematically refuted al-Ghazali in his *The destruction of destruction* (or *The incoherence of the incoherence*), which has remained famous for the originality of the principles enunciated. Averroes declared that the world had a necessary and rational order (hence the existence of natural laws) and described God as the principle which guaranteed that order. Science (that is, the knowledge of nature) was the way to rejoin God, the ultimate truth.

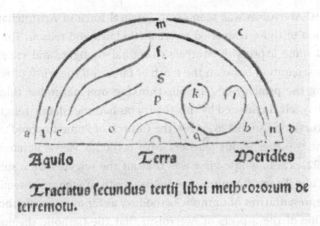

Aquilo   Terra   Meridies

Tractatus secundus tertij libzi metheozozum de
terremotu.

Figure 4.2 Beginning of the treatise *De terraemotu* in the third book entitled
Metheororum from the work *Phisicorum* by Albertus Magnus (1260), here in a rare
Venetian edition dated 1494.

Averroes tended in particular to affirm the autonomy of rational research in relation to faith. These ideas caused him to be found guilty of heterodoxy in 1195 by
the Islamic religious courts at Marrakesh (Morocco), where he was teaching. He
fled but was later imprisoned, although he may have been freed a few months
before his death.

Albertus Magnus (Albert the Great, 1193–1280), was deeply influenced by the
Arab philosophers, and especially by Avicenna, whom he greatly admired. He
was a German Dominican who taught theology at the University of Paris (where
Thomas Aquinas was one of his pupils). He was an important and original
thinker, who held that philosophy and the sciences had their own autonomous
value. He also proposed to expound and comment, in Latin, on the corpus of
Aristotelian works and those of the Arab commentators, but he never completed this project. He set out a scientific method based on experiment and
direct observation. His treatise *De mineralibus et rebus metallicis* (c.1260) became
the point of reference for all Western seismological thinking until the dawn-
ing of the sixteenth century. Albertus asserted that the natural sciences (which
he called 'profane') should be developed using methods distinct from those
proper to theology. He also argued in favour of research inspired solely by
reason (Figure 4.2).

Arab scientific thought continued to influence the Western Latin world into
the later Middle Ages. While the ideas of Averroes were execrated by exponents
of the Islamic religion, they found followers in the early decades of the thirteenth century at Naples and at the court of Frederick II in Sicily. From there,
the theories of the Averroists spread to European universities, and took root in

particular at Paris. Averroism was seen as an integral form of Aristotelianism, in direct opposition to those who tried to reconcile faith and reason. The Averroists radicalized some aspects of Averroes's original thought, and supported the existence of a separation between the truth of faith and the truth of reason, thus undermining the primacy of religious thinking over naturalist thinking. Christian orthodoxy felt threatened by this theory (as had the Islamic religion in the previous century), and so in 1267–1268 the Church of Rome imposed on the University of Paris two prestigious teachers with orthodox views: Bonaventura da Bagnoregio (1221–1274), whose view was that all the sciences were subordinate to theology, and Thomas Aquinas (see below), both of whom subsequently became official representatives of Catholic orthodoxy as far as scientific thought was concerned. One of the aspects of Averroism that the Catholic theologians most feared was that the idea of a dual truth (science separate from theology) should take hold. Averroism was later (1270 and 1277) officially condemned as heresy by the bishop of Paris, Étienne Templier.

A brief digression may perhaps convey an idea of the impact on contemporary culture of the thinking of natural philosophers. The philosophical dispute about the relationship between religious and rational truth and concerning how the investigation of nature should be carried out had a direct impact on the future development of interpretative models of earthquakes and on the nature of the Earth in general. The idea of a clash between revealed truth and the truth of nature met solid philosophical opposition from Thomas Aquinas (1225–1274), a Dominican philosopher and theologian. As regards the interpretation of earthquakes, he set a hierarchy of values that did not wholly resolve the conflict between religion and the rationalist interpretation of natural phenomena. Aquinas put God as the primary cause of earthquakes and matter as a secondary cause.

4.1.3   *The Italian Renaissance: from Giannozzo Manetti to Leonardo da Vinci (fifteenth century)*

Belief in the existence of two types of earthquakes, one natural and one supernatural, had been an undercurrent in Western medieval culture but became explicit in the seismological thinking of the Italian Renaissance. For those studying the earthquakes of the past there is a very important manuscript codex: *De terraemotu libri tres* by the famous Florentine humanist Giannozzo Manetti (1396–1459), a politician and man of letters. This outstanding work has not had a critical edition (only a quick translation into Italian by Scopelliti, 1983), yet it has been the subject of study several times. It has reached us today in eight codices, four of which are preserved in the Vatican Library (Figure 4.3), one in Naples (Biblioteca Nazionale Vittorio Emanuele III), one in Paris

Figure 4.3 Title page of the treatise *De terraemotu libri tres* by the Florentine humanist Giannozzo Manetti (1396–1459) (Biblioteca Apostolica Vaticana, *Palatini Latini*, 1077).

(Bibliothèque Nationale de France), one in Madrid (Library of San Lorenzo de l'Escorial) and one in New York (Kraus Library). The treatise was written in 1457, immediately after a large earthquake in the Kingdom of Naples that Manetti himself witnessed, since at that time he was working at the court of Naples as secretary to King Alfonso of Aragon. Some years earlier, Manetti had written his *De dignitate et excellentia hominis*, a work entirely devoted to an earthly exaltation of human virtue which completely burst the bonds of the medieval world.

The treatise *De terraemotu* is divided into three parts. In the first book Manetti sets out an Aristotelian theory on the origins of earthquakes. It is interesting to note that he analyses the problem of dual truth. Manetti denies the doctrinal opposition between faith and reason, but accepts that both divine will and determinism of matter are the causes of earthquakes. He compares two opinions: the theologians' view that earthquakes are 'not natural phenomena' (Manetti claims this opinion is 'insignificant and inconsistent'), and the opposite view, held by natural philosophers, that earthquakes occurred within the order of nature. Both opinions were for Manetti 'true up to a point and neither of them was wholly false'. Manetti's was a carefully argued and complex vision, which harked back to the diverse ideas of the theoretical theology and natural philosophy of the past.

In the second part of his treatise, Manetti set out the first catalogue of historic earthquakes in the West, and for this reason he can be considered a forgotten father of historical seismology. His learned exposition cites the historical sources for 100 earthquakes in the Mediterranean area from ancient times up to the Middle Ages. The third part of Manetti's treatise is entirely devoted to a description of the effects of the earthquake of 5 December 1456. Manetti probably collected the reports of the various governors who came to the court of Naples, and so was able to outline, location by location (he cites 153), the entire territorial picture of the effects. For the historical seismologist Manetti's treatise is a sort of *summa* of the knowledge of an era, a text that ought to be read by anyone dealing with historical earthquakes. *De terraemotu* is an extraordinary work, but it was unknown to his contemporaries and even in later times until the closing decades of the twentieth century. It was ignored after Manetti was condemned by the court of the Inquisition in 1584 and consequently made no contribution to culture in the centuries following its composition.

In spite of the many doctrinal obstacles, fresh attention began to be paid to contemporary earthquakes by philosophers and men of learning beginning with the end of the Renaissance, especially in Italy. Indeed, from that time onwards an earthquake became a sort of laboratory; observations that were made served to confirm or undermine more or less widely accepted theories.

The 1456 earthquake, which had captured the attention of Manetti, must also have deeply affected other contemporary natural philosophers. Indeed, a treatise probably written in the summer of 1457, the *Tractatus de cometa atque terraemotu* (ed. 1990), outlines another interpretation of the causes of the earthquake. As pointed out in the title of the treatise, the author, Matteo dell'Aquila (c.1410–1475), drew a correlation between the passing of a comet in June 1456 (identified by modern astronomers as Halley's Comet), the earthquake in December and the passage of a subsequent comet in the summer of 1457 (Yeomans, 1991). Thus, astronomical observations were brought to bear on the question of the causes of earthquakes.

Leonardo da Vinci (1452–1519) was another Renaissance thinker who dealt with earthquakes and tsunamis, but unlike Manetti he did not feel the need to mediate with theology. Leonardo's solely naturalistic track foreshadows a new outlook for research and a sensitivity oriented exclusively to the observation of nature. But his interpretation of earthquakes still remained couched in the terms of classical philosophical language. As highlighted by Di Teodoro and Barbi (1983, pp. 28–38), in Leonardo the causes of earthquakes are various: water, earth, air and fire. However, it is possible to trace his observations and thoughts back to just two theories, the ones most widely adopted and commented on from antiquity to the Renaissance, namely those of Aristotle and Epicurus.

Probably encouraged by his experiments on firearms and landmines, Leonardo hypothesized the existence of subterranean fires (not an original theory, as has been seen) and their complex relationship with the water mass of the Earth (*Codex Leicester*, formerly *Codex Hammer*, fol. 34v). This thought is associated with an anthropomorphic conception of the earth, which Leonardo borrowed

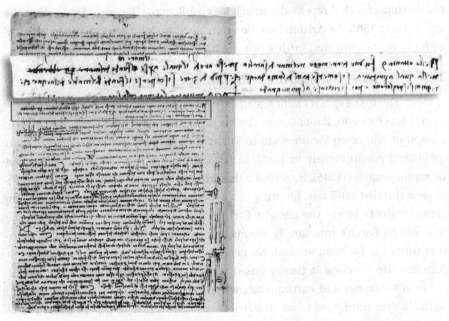

Figure 4.4 *Codex Leicester* (formerly *Codex Hammer*) fol. 10v by Leonardo da Vinci. The note (in the foreground) in Leonardo's characteristic mirror-image writing records effects at Antalya of the tsunami associated with the earthquake of 3 May 1481.

from ancient wisdom. In the *Atlantic Codex* (at fol. 289v) one finds a change in his thinking. Here Leonardo draws on the theory that subterranean collapses generate earthquakes. What theory was actually supported by Leonardo is hard to say. It may be that he supported a pluralist stance, in the sense that he believed that several different causes were possible. Leonardo also developed some ideas about how to build structures in order to defend oneself from earthquakes (see Chapter 11).

Leonardo has left us today with a record of a tsunami in 1481 at Adalia on the southern coast of Turkey, between the islands of Rhodes and Cyprus. Leonardo's annotation is in the *Codex Leicester* (= *Codex Hammer*) at fol. 10v (Figure 4.4):

> In eighty 9 an earthquake occurred in the sea of Adalia near Rhodes, as a result of which the seabed opened up, and such a quantity of water poured into it that for over three hours the floor of the sea was left uncovered by reason of the sea which was lost into it, after which it closed again to its former level.

Another humanist, Filippo Beroaldo senior (1453–1505), famous in his day for his broad culture and the sheer range of his interests, dedicated a short treatise to the earthquake, *De terraemotus ac pestilentia* (1505), written on the occasion of the earthquakes that repeatedly struck his city, Bologna, between December 1504 and January 1505. An Aristotelian, he interpreted the earthquake as a 'precursor' sign of other calamities, in this case plague epidemics.

4.1.4   *The start of the modern era: the sixteenth century*

The first explicit challenge to Aristotle's thinking on the causes of earthquakes was by Georg Bauer (known by his Latinized name of *Agricola*, 1494–1555), a pupil of Francesco Vicomercato in Turin. In Agricola's treatises *De re metallica* (published posthumously in 1556), in *De natura eorum quae effluunt ex terra* and *De natura fossilium* (1546), he writes a dry critique of Aristotle based on an exclusively naturalist rationale. Ignoring any existing philosophical and religious concerns, Agricola severs the ancient Earth–Sun bond as the prime origin of the heat of the Earth's interior. By means of a simple empirical observation Agricola turns the problem around. No heat descends, he states; suffice it to watch furnaces. He develops a theory based on the inner fires of the Earth as the cause of volcanoes and earthquakes, without resorting to the heat of the Sun. Agricola does not refer to fires in a strict sense but rather to inflammable conditions. The cause of combustion in the Earth lies in certain chemical substances (i.e. bitumen, nitre and sulphur). The detonating fuse, so to speak, may be the phenomena of compression and overheating of the external air. Several times Agricola resorts to the examples of mines and of the relationship between the mass of gunpowder and the size of the combustion chamber to explain seismic phenomena. He traces the first correlation between hot zones (geothermal areas) of the Earth and the frequency of earthquakes. In spite of the striking novelty of Agricola's treatise, only 20 years later his opinions were considered to be just an oddity.

His extraordinary technological culture (he was a specialist in the extraction of metals) and his attention to empirical data make of him a key person in the observation of nature. Agricola did not formulate new earthquake theories, but his critical thinking in regard to Aristotelian theory, the primacy of the empirical datum, his pragmatism, together with his great capacity for observation and correlation of natural phenomena (e.g. thermalism, the Earth's internal heat, the propagation of seismic effects) definitely marked the beginning of the modern era. But his contemporaries found it difficult to understand the bearing of his innovative contribution, and so his works only spread slowly during the sixteenth century.

In the mid sixteenth century a new contribution in support of naturalistic thought was published in Rome in 1565 by Bernardino Telesio (1509–1588), an Italian Benedictine monk and philosopher. Previous philosophers had been careful to develop their thinking along the lines of the Aristotelian tradition, but Telesio began his work *On nature in accordance with its own principles (De*

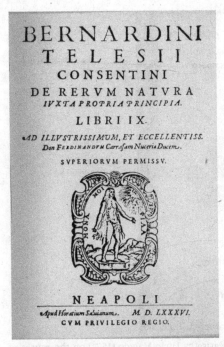

Figure 4.5 Title page of the treatise *De rerum natura* (1586) by Bernardino Telesio, which contains a part entitled (in translation) 'On things that are in the air and on earthquakes.'

rerum natura iuxta propria principia), declaring that what he had written had been worked out by him. His thesis was that nature should be investigated in accordance with specific principles and independently of metaphysics, because its rules are intrinsic. This view was considered revolutionary at the time. It had been put forward a number of times in earlier centuries, but had always encountered objections and obstacles to its acceptance.

In his tract *On things that are in the air and on earthquakes* (*De his quae in aere fiunt et de terraemotibus*) (Figure 4.5), Telesio re-examined Aristotelian theory, but modified the wind theory only in part. He held that winds were caused not by dry but by damp exhalations, caused by the heat of the Sun both within the Earth and at its surface, and especially on the sea. More important than his theories were his methods and the principles of intellectual freedom that underlay them. Telesio, too, was condemned by the court of the Inquisition, and his works were placed on the Index and prohibited. The crisis created by Telesio's criticism of Aristotle (not so much Aristotle's original thinking, but rather his works as interpreted by medieval Christian thinkers), was not brought to a halt by condemnations of Telesio and his ideas. There was developing a deep cultural crisis, and the search for new explanations for earthly phenomena such as earthquakes and volcanic eruptions was opening up new lines of enquiry.

Figure 4.6 Title page of the treatise by Lucio Maggio (1571) on the causes of earthquakes.

During the last few decades of the sixteenth century at the famous Este court in Ferrara, a school of new ideas about earthquakes started to take shape. The stimulus was provided by the strong earthquake of November 1570. This seismic event had hit not just the town of Ferrara but also numerous other villages, as well as affecting the image of ducal power itself, represented by the Duke Alfonso II d'Este. Indeed, the 1570 earthquake had been believed to be supernatural by the natural philosophers of the time because it had occurred in a plain that was not believed to be subject to earthquakes, in that it was indeed a lowland plain. The actual data clashed with those of the tradition, hence with what was reckoned to be *valid*. According to the Pope, Sixtus V, the 'prodigious feat' (i.e. the earthquake) was a sign of God against the duke because he was harbouring Jews. In that context the naturalistic interpretation of the earthquake could have been a card in the duke's favour (an example of how politics could influence the scientific research of the day). It is in that context that a rationalizing tradition of the earthquake developed in contemporary treatises. This earthquake became an extraordinary laboratory of ideas and thoughts that were disseminated in treatises and dialogues. The literary form itself of such texts privileged the comparison between different ideas, the capacity for demonstration, the casting of doubt and unresolved problems. The treatises of Iacomo Antonio Buoni (1571), Lucio Maggio (1571) (Figure 4.6), Gregorio Zuccolo (1571), Alessandro Sardi (1586),

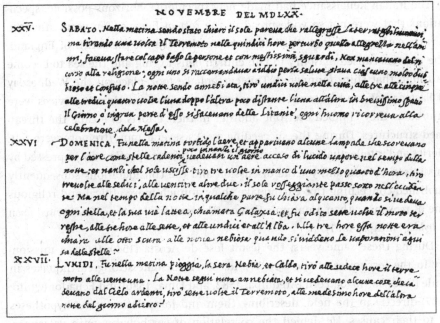

Figure 4.7 Detail of a manuscript page by Pirro Ligorio, in which he lists and describes with extraordinary precision and rigour the numerous shocks he felt in Ferrara between November 1570 and January 1571 (State Archive, Turin, *Codici ligoriani*, 19–30bis, vol. 28 (Ja.II.15), fol. 75v, Pirro Ligorio, *Libro di diversi terremoti*).

Annibale Romei (1587) and Pirro Ligorio (1570–71, ed. 2005) marked an era of critical reappraisal of the Aristotelian theory.

From the point of view of historical seismology, these treatises have left us today with a large amount of descriptive data on the earthquake of 17 November 1570 and on other earlier events. In particular, the treatise by Pirro Ligorio (1513–1583), a famous architect at the papal court in Rome, historian, archaeologist, the figure of an artist of elevated stature, who subsequently moved to the court of Ferrara, contains explicit reference to foreshocks of the earthquake on 17 November, which started about fifteen days before the main shock. Such detailed descriptive elements appear to indicate that a new sensitivity towards natural phenomena was taking the place of the certainties contained in older thinking (Figure 4.7). The explicit awareness of 'not knowing' and being faced with events, such as that earthquake, still unknown, pushed Ligorio to write an outstanding treatise (Ligorio, ed. 2005): on the one hand, he wanted to show, as did other contemporaries of his, that in history other earthquakes had occurred in the plains, and that the earthquake was a phenomenon that was mostly natural, even if he did not exclude 'miraculous' earthquakes. Also, Pirro Ligorio is the author of a project for an earthquake-resistant house (see Chapter 11).

The Italian Renaissance had made new speculative horizons possible, speculations that were not so widespread in the rest of Europe. Ten years after the Ferrara earthquake, on 6 April 1580, an earthquake hit the south of England. The tremors caused masonry to fall, church bells to ring and the sea to become so rough that ships foundered. This earthquake took place on the Wednesday of Easter week, and in London many people who were at church services were injured as they climbed over one another in their efforts to escape the threatened structures. During the succeeding weeks several more tremors were felt. Clerics and intellectuals at the court of Elizabeth I were profoundly impressed by this seismic event, and yet the discussion and the considerations subsequently published concerning this earthquake show a much more ethical and religious line of reasoning than a naturalistic one (*The wonderfull worke of God shewed upon a chylde*: Phillips, 1581).

During these same years, the Jesuit José de Acosta (1540–1600), a missionary in the territories of the New World, observed some strong earthquakes in Chile (Valdivia, 1575), Peru (Arequipa, 1582, and Lima, 1586) and Ecuador (Quito, 1587) directly in the field, describing them and formulating some hypotheses as to their causes. He denied the correlation of earthquakes with volcanoes. Furthermore, he observed tsunamis associated with some earthquakes and the heights of the tsunami waves that lashed the coasts. He noticed that the earthquakes on the coasts of Chile and Peru propagated in a direction from south to north (Udías, 1999). His work, *Historia Natural y Moral de las Indias*, published in 1590 (translated into English in 1604), was little known to his European contemporaries. Even so, de Acosta is considered to be a pioneer of geophysics (Udías, 1986). Thanks to his curiosity about natural events and his knowledge on the subject, he has been called the 'Pliny of the New World' (Stoetzer, 1986).

The sixteenth century drew to a close with the production of new rationalizing treatises, heralding a period of scientific unrest that had started during the Renaissance. One of the most striking elements of these treatises was an awareness that there was no shared theory of natural phenomena. This lack of an explanatory theory led the treatise authors to describe contemporary natural events as meticulously as possible.

4.1.5   *From the seventeenth century to the mid eighteenth century*

The seventeenth century was a time of great importance in the history of the interpretation of earthquakes because a mechanistic interpretation of natural events, from the origin of the Solar System to the structure of the Earth, was taking hold. In spite of the great novelty of this new kind of thinking, which marked a fundamental turning point for ideas on terrestrial phenomena, the seventeenth century was also a time when there was an attempt to harmonize the religious and providential vision with mechanical philosophy. The sciences were not yet separated from philosophy, and the theories of earthquakes were still a part of the philosophers' cosmogonies.

During the seventeenth century, hypotheses on the causes of earthquakes were situated firmly within the great scientific and philosophical issues that characterized the thinking of the time. Some of the important issues investigated were the origin and age of the Earth, the occurrence of the great Biblical flood and the interpretation of fossils. New geological observations, in particular in the fields of sedimentology and mineralogy, highlighted elements that were disturbing for the religious mindset of the day. Religious thinking started to be separated from the sciences beginning in the mid eighteenth century.

It was with René Descartes (1596–1650) that an important new phase in the thought about natural phenomena began. In the *Principia philosophiae* (1644) Descartes argued in favour of a mechanistic interpretation of the Earth. According to this argument, each natural phenomenon, from the origin of the Earth to its extinction, could be explained as the effect of the movement of the elements that constituted all matter. A model of the full history of the Earth from a luminous star to a cold planet provided for the presence of a central fire along with smaller or minor fires. Descartes explained such phenomena as the discontinuities of the Earth's crust, landslides and internal cracks as a result of the process of cooling. In paragraphs 76, 77 and 78 of the *Principia*, Descartes lays down his theory about the origin of earthquakes. The fumes that come from inside the Earth could combine in various ways to form sulphur, bitumen, petroleum and clay. When such substances catch fire, the explosion generates an earthquake on the surface, the violence of which varies according to the quantity of inflammable fumes and the extent of the Earth's discontinuity. Within Descartes' new and modern system of thought, the *continuity/discontinuity* dichotomy, which had first been seen in the atomist philosophy of Democritus, put in a new appearance. However, in the geological thinking of the day this dichotomy was not connected with the atomist interpretation of matter. Rather, it was connected to a great chronological scale of natural events. This dichotomy had repercussions in the geological bias of the times, which stemmed from philosophical or religious concepts (as will be seen below), and it has had some influence even in modern geological thought.

The German Jesuit Athanasius Kircher (1601–1680) was a younger contemporary of Descartes, whose works he had probably read. Kircher was a witness to the strong earthquake that struck Calabria in March 1638. He was at sea at the time of the earthquake and felt the first great shocks from his boat. He later wrote

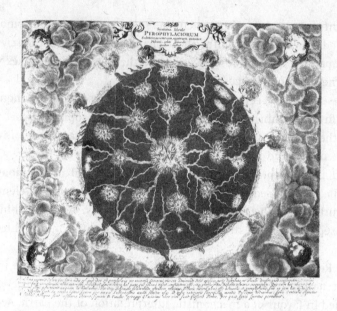

Figure 4.8 Illustration of the theoretical model of Athanasius Kircher of a great fire at the centre of the Earth that was connected to smaller fires and surface volcanoes (*Mundus subterraneus*, 1664–65).

that he had seen the sea 'bubbling and boiling'. In actual fact these may have been gas emissions from the sea floor released by the strong earthquake shaking, but Kircher believed the phenomenon to be a sign of the heat that was being released by nearby volcanoes (Etna, Stromboli and Vulcano). Kircher reported what he had seen in his treatise *Mundus subterraneus* (only published later, in 1664–65). In his opinion, the connection between the earthquake and the volcanic activity confirmed the thesis of an igneous origin of earthquakes. Kircher believed in the existence of a great fire at the centre of the Earth, connected to smaller hotbeds through a complex mesh of underground canals and galleries (Figure 4.8). He published a representation of this model, which became very well known. With the *Systema ideale pyrophilaciorum subterraneorum* as he himself called it, he attempted to explain the origin of earthquakes and volcanoes within a single theoretical model, using the pre-existing theories on inflammable matter. There were a number of scholars of the seventeenth century who followed Kircher, such as José Saragoza y Vilanova (1627–1679), professor at the Imperial College of Madrid.

The new conceptions of nature, above all those theorized by Descartes, were based on matter and movement, and they were governed only by natural laws. Yet Descartes' theory was an interpretation of nature that had great difficulty in prevailing. Indeed, many seventeenth-century geologists accepted Descartes' hypothesis of the stratified structure of the Earth but decidedly rejected his naturalist philosophy on how natural effects occur. Among those who still sought agreement with the Bible were Thomas Burnet (1681–89), John Ray (1691), John Woodward (1695) and William Whiston (1696). During the years of the publication of the famous 'sacred theory' of the Earth by Burnet, Martin Lister (c.1638–1712) had published in the Philosophical Transactions of the Royal Society (Lister, 1684, pp. 512–15) a thesis on the origin of earthquakes as the effect of the decomposition of pyrites. This chemical thesis was very successfully received. For example, it offered a number of ideas to Woodward, who believed that sulphur and nitre were the causes of volcanic eruptions. According to Woodward, the underground fire acted as the regulator of the entire flow of terrestrial waters. Owing to some blockage, that flow could become perturbed, creating an anomalous pressure in the waters contained in the abyss inside the Earth. The blockage could press upon and crack the surface of the Earth, thus causing an earthquake. According to Woodward, this interpretation could also explain the frequently observed disequilibriums in underground and surface waters during earthquakes.

For the creationists who were the supporters of Burnet's sacred theory, earthquakes were not primary dynamic elements to be understood in a general model. Excluding Whiston, this school of thought did not put forward an original contribution to the interpretative theories of earthquakes. For his part, Whiston imagined the origin of the universal flood to lie in the sinking of a comet inside the Earth, after which he imagined the end of the Earth in a great earthquake, again caused by a comet.

The theories on the causes of earthquake were various and expressed different cultures and visions of the world. Alongside the nascent sciences, there were also popular disciplines such as astrology that were deemed to be as much a science as astronomy. In 1672, a brochure was published in Bologna in which the author, Flaminio Mezzavacca, presented the astrological coincidences with the occurrence of some recent earthquakes (Figure 4.9).

Alongside the fanciful rather than scientific conjectures, there was no shortage of much more concrete approaches. The physicist Francesco Travagini, after having felt in Venice the 1667 earthquake at Ragusa (today Dubrovnik, Croatia), published a small original work (Travagini, 1669) on the propagation of seismic motion. He wondered how the water of the canals of Venice had become so agitated at the time of that earthquake, although Venice was so distant from the apparent centre of the earthquake effects. To try to understand this problem, Travagini made use of the movements of suspended masses,

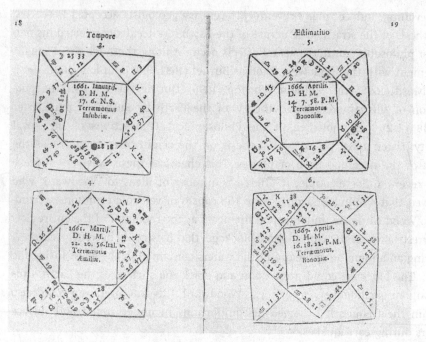

Figure 4.9 Flaminio Mezzavacca's connections between earthquakes occurring and the conjunction of the planets (Mezzavacca, 1672).

since at this time there was still no theory or understanding of seismic waves (Figure 4.10).

Towards the end of the seventeenth century and at the start of the eighteenth, the prevalent theory of the cause of earthquakes claimed that they resulted from the explosions of inflammable substances together with the existence of a central fire in the Earth. Lazzaro Moro (1687–1764) and Antonio Vallisneri (1661–1730) both supported the theory of the continuity to explain the causes of earthquakes, as opposed to Descartes' thought that discontinuities in the materials in the Earth led to earthquakes. Moro harked back to the theory of fire to explain neotectonic phenomena, and considered an effective test of this theory to be the sudden appearance of a new island in the Aegean on 23 March 1707 in the vicinity of Santorini. Using Newton's axiom according to which 'natural effects of the same kind can only have one and the same cause', Moro stated that all islands and mountains were of igneous and volcanic origin.

The fire theory was adopted with a few variations by many other physicists and naturalists, often within a conceptual framework in which God was the engine behind every phenomenon. Along with those who reflected on the recent theories and experiments, there was no lack of esteemed men of culture who

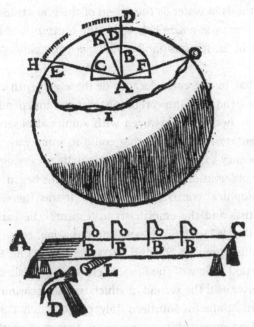

Figure 4.10 Francesco Travagini's record of his experiments concerning the propagation of seismic motion (Travagini, 1669).

expressed a wisdom still tied to the thinking of the ancient classicists, in particular Aristotle and Seneca. One such person was Giorgio Baglivi (1668–1707), who described in a very appreciable way the great earthquake that hit central Italy in 1703. Being an anatomist, Baglivi was influenced by an anthropomorphic theory of the Earth. He made use of medical analogies to explain the seemingly endless sequence of shocks of that seismically active period. His work was very popular, testifying to the fact that there was no generally shared earthquake theory but rather that various older interpretations coexisted. Although his theories have almost no technical value, his minute and careful description of the long seismic sequence in 1703 remains a source of great importance for historical seismology.

Descartes' theory, a point of reference for the rationalists, was returned to in the cosmogony of Gottfried Wilhelm Leibniz (1646–1716), a German mathematician and philosopher, whose intention was to reconcile the mechanistic and creationist hypotheses (his *Protogaea sive de prima face telluris* was published posthumously in 1749 but had already been completed in 1690). As for his ideas on geology, Leibniz was the first to observe that some rocks were endogenous and others exogenous, and he had some novel ideas about changes in the Earth before the appearance of man. He accepted the idea of the central fire, but excluded the hypothesis of water as the origin of the great transformations in the Earth's crust. Leibniz was a firm believer in the continuity of nature, a view which he summed up in the famous phrase '*Natura non facit saltus*' ('Nature does not make leaps').

It is interesting to see that in the early decades of the eighteenth century in Europe the theoretical earthquake interpretation, not yet independent as a discipline, slowly started to become interwoven with empirical observations and the building of rudimentary instruments that could in some way 'record' seismic movement. The tendency to turn some of the actually occurring earthquakes into a laboratory for observations and measurements thus began, on the margins of the great philosophical constructs on Nature. It was therefore up to the minors, the pragmatists and the empiricists to 'capture' the early seismic traces. It was Nicola Cirillo (1671–1735) a Neapolitan physician, botanist and naturalist who pioneered this approach (Baldini, 1981; Baratta, 1896; Ferrari, 1992). In 1727 he was appointed Fellow of the Royal Society. He published three memoirs in *Philosophical Transactions*, the second of which on the occasion of the earthquake that hit northern Apulia (in Southern Italy) on 20 March 1731. This strong earthquake provided an opportunity to apply the Cartesian method of which Cirillo was an enthusiastic proponent. In the course of 1731, Cirillo, with the help of two collaborators, observed the oscillation of two pendulums, one palm long (c.26 cm), applied to a graduated semicircle which he had installed (Cirillo, 1733–34, 1747).

The frequent agitations and the more substantial oscillations observed in the pendulum, convinced Cirillo that he had found the experimental proof that would demonstrate how the laws that govern the propagation of every type of movement also held good for seismic movements. The axioms of geometric physics were thus fully confirmed and could successfully refute the arguments put forward by Giambattista Vico. The latter was a friend of Cirillo's but was very critical of the attitudes of the Cartesian school. The necessity to demonstrate the universal validity of the geometric method persuaded Cirillo to set up an embryonic network of instrumental observation on the occasion of the occurrence of a seismic phenomenon. The dates relating to the observations were then correlated with a description of the damage effects, thus freeing the use of instruments from the narrow perspective of a naturalist's mere curiosity (Ferrari, 1992).

It was a century after Descartes' *Principia* that the compromise between Genesis and the natural sciences came to an end, but not without some ambiguity. The permanent separation began with the first work by Georges-Louis Leclerc de Buffon (1707–1788), *Histoire et théorie de la Terre* (1749). In this work Buffon argued in favour of a hypothesis that earthquakes are caused by an explosion of inflammable materials, due to the central terrestrial fire. Subsequently (in 1780), he redefined his thinking about the great dispute among *neptunists*, *volcanists* and *plutonists*, and came to favour water as an important factor in the transformation of earth materials and as the cause of earthquakes.

A new theoretical phase in the understanding of natural phenomena started after the mid eighteenth century and was triggered by experiments concerning the Earth's electricity and magnetism. The discovery that atmospheric electricity was identical to that obtained with Leyden's jar (Winkler, 1745; Franklin, 1751) paved the way for new interpretative hypotheses about the generation of earthquakes. A reanalysis of the causes of earthquakes was also stimulated by a close sequence of important seismic events. In 1749 Franklin and Hall reprinted for their publishing house a partial English translation (*A true and particular relation* . . .) of a more extensive anonymous report on the Peruvian earthquake of Callao and Lima in 1746 (Anonymous, 1749). Then there were the two earthquakes in England in 1749 and in 1750 that shocked the English, as it had been the widespread opinion that the island was virtually aseismic (Melville, 1985). In 1750 William Stukeley (1687–1765) published in London *The philosophy of earthquakes, natural and religious*, in which he expressed the opinion that earthquakes derived from an electrical discharge between the Earth and the atmosphere. Stukeley also believed that seismic motion derived from vibratory interferences, similar to 'a string of an instrument or of a glass when you rub its rim with your finger'. Stukeley speculated that from a vibratory movement a substantial oscillatory mass movement could derive, capable of leading to considerable mechanical effects. To explain the explosive nature of seismic shocks he used Lister's seismo-chemical theory (1684).

Bina was positioned in that broad group of scientists and naturalists from the early eighteenth century who, impressed by the potential of the electrical forces that were then being discovered, attributed a predominant role in the explanation of all the natural phenomena to the latter, giving rise to a sort of electricist determinism. Bina's first scientific essay, *Electricorum effectuum explicatio*, published in Padua in 1751, mirrors a similar vision of nature, which was consolidated in the same year with the publication of the book *Ragionamento sopra la cagione de' terremoti* (Reasoning on the cause of the earthquakes) (Bina, 1751a, 1751b). Thanks to this work, Bina became a convinced proponent of the theory according to which the earthquakes were caused by disequilibria in the underground electrical fluid triggered by the explosion of gas fuels in the underground caves. Among the first in Italy and the world, Bina was concerned to demonstrate his own theoretical arguments through *measurement* of the seismic phenomenon. For that purpose, he designed a vertical seismoscopic pendulum

(Terrenzi, 1887) which, by means of a needle lodged in the mass, was supposed to record the ground movements on the sand contained in a wooden box, floating in a water container (Ferrari, 1992).

The theory of electricity in general was also held by important scientists who studied electricity independently of earthquakes. Some of them did not resist the temptation of extending their scientific ideas to the seismic domain. Giambattista (name in religion of Francesco Ludovico) Beccaria (1716–1781), the author of the first rigorous European treatise on electricity, had been corresponding with Benjamin Franklin and was an authority in the field of experimental electricity (Heilbron, 1979). In his *Dell'elettricismo artificiale e naturale libri due* (1753) he expressed a significantly different opinion from that of Stukeley (see also Beccaria, 1776). Beccaria reckoned that the inside of the Earth was permeated with 'electrical vapour', a sort of atmospheric source of electricity. Like Benjamin Franklin, Beccaria thought that such electricity was seeking its own state of equilibrium. If an 'imbalance' as he called it occurred in the electricity of the atmosphere, then lightning occurred. If the 'imbalance' occurred underground, then there was an earthquake. Electricity was considered by Beccaria to be like an 'elastic vapour', endowed with very great expansive energy and great speed of transmission. Another supporter of the electrical theory as the earthquake cause was the Spaniard Benito Jerónimo Feijóo y Montenegro (1676–1764). His treatise (*Nuevo Systhema sobre la causa physica de los terremotos explicado por los phenomenos electricos*, 1756) bears witness to the diffusion and the following that the electrical theory enjoyed in the Europe of the day.

Besides the supporters of the Earth's central fire (boiler) theory and the electricians (with all their variants), today one can recognize from that same time period a third theory, based on the hypothesis that compressed air, above all hot air, could be the cause of earthquakes. In 1703 Guillaume Amontons (1663–1705) had presented a dissertation on the causes of earthquakes based on the strong pressure and temperature of air trapped underground (Amontons, 1703). This idea of the push of air was in some way connected to contemporary experiments on compressed air rifles. Fifty years later, Stephen Hales (1677–1761), basing his work on recent experiments concerning the velocity and elasticity of air, argued that air pressure was capable of producing a 'push' that then caused the earthquake (Hales, 1749–50). The coexistence of different interpretations, experiments and theories on the origin of earthquakes characterized all of the eighteenth century.

One of the most important summaries of eighteenth-century knowledge, the great *Encyclopédie* by Denis Diderot and Jean-Baptiste D'Alembert, presented an interpretative theory of earthquakes (see entry 'Tremblemens de terre', by Holbach, vol. 16, 1765) involving an eclectic vision that included references to fire, air and water in joint action. They cited an experiment by Nicolas Lémery (1645–1715), which reproduced 'seismic explosions' by mixing together sulphur and iron filings. The authors also reported critiques of the work of the chemist M. Rouelle, who argued that the experiment was not valid because in nature iron was not found in the pure state. An English encyclopaedia translated into Italian in those years (*Ciclopedia*, 1754, vol. 8, p. 461) proposed a version of the fire theory, with the addition of a curious paragraph entitled 'Artificial earthquakes'. Here the reader was given a recipe for preparing an earthquake, making the explosive paste himself, burying it, and then enjoying the spectacle (reminiscent of Anthemius of Tralles).

4.1.6   *From the Lisbon earthquake (1755) to the late eighteenth century*

The Lisbon earthquake of 1755 was a major seismic event that destroyed that great city along with dozens of other towns and villages in Portugal, Spain and Morocco (Oldroyd *et al.*, 2007). About 30 years later, in 1783, Calabria (a southern Italian region) was struck by a succession of five disastrous earthquakes over a period of a few weeks. Within ten years or so of the latter events, European seismological thinking underwent an extraordinary theoretical acceleration, in both scientific and philosophical terms. A chronology of the publications from this period highlights the intensification of this theoretical research.

Soon after the great Lisbon earthquake in 1755, which so perturbed the European culture of those years, the scientific debate broke new ground (Poirier, 2005). While still very young, the German philosopher Immanuel Kant (1724–1804) wrote three essays dedicated to an interpretation of earthquakes (1756a, 1756b, 1756c). These essays were entitled *Von der Ursachen der Erderschütterungen*, *Geschichte und Naturbeschreibung* and *Fortgesetzte Betrachtung* (Placanica, 1984). Kant was a disciple of Buffon's geological ideas as well as a Newtonian and a supporter of the Earth's central fire (boiler) theory. He did not devise a new interpretative theory, but he pointed out some pivotal ideas that consolidated rationalist thinking on the origin of earthquakes. He proposed: (1) that seismic effects can be propagated over great distances; (2) that there is a cause triggering the shocks; (3) that great seismic sequences tend to recur over time.

As regards the first point, Kant argued in favour of the existence of underground canals and tunnels that ran parallel to mountain chains and the flow of the great rivers (in Italy, for Kant, the most efficacious propagation was north–south, while in Portugal it was east–west). As regards the second point (the cause triggering the shocks) Kant referred back to the chemical explanation of earthquakes, probably influenced by the experimental chemistry of his day. He thought that a seismic 'explosive mixture' was made up of two parts vitriol oil, eight parts water and two parts iron. Kant based his thinking on the mechanical hydrodynamic experiments of Carré and asserted 'that even water can react like a solid body when suddenly compressed. Thus, it can transmit shocks almost without softening them.' As regards the third aspect (the multiplicity of shocks), Kant argued that earthquakes were due to hot compressed air in underground cavities beneath the high mountains. The intervals and recurrence of the shocks were determined by a 'chemical respiratory process' that also involved the release of hot air from volcanic vents. Kant considered electricity to be a marginal curiosity and declared that too little was still known about magnetism to advance any theories.

In 1757, Elie Bertrand (1713–1797), a Swiss geologist and a priest in Berne, published a book entitled *Mémoires historiques et physiques sur les tremblemens de terre*. Along with a historical catalogue of the earthquakes felt in Switzerland from the sixth century AD to his own day, Bertrand set out his interpretative theory of earthquake causes (which can be traced back to Buffon and Stukeley), but he included his doubts and other remarks about his ideas. Bertrand was convinced that many things about earthquakes still remained to be explained. Above all, not unlike Travagini a century earlier, he was surprised by the great velocity of the propagation of seismic effects and the vastness of the area affected by a strong earthquake. Indeed, he had observed the effects (seiches) of the Lisbon earthquake in 1755 on Lake Neuchâtel, near Berne, in Switzerland. Bertrand's doubts and his minute observations were of importance for John Michell (1724–1793), as Michell himself observed. Michell's *Conjectures concerning the cause and observations upon the phenomena of earthquakes*, published in the *Philosophical Transactions* (Michell, 1761), contains about 60 pages that left their mark on the thinking of the people of the time. He highlighted the connection between seismic and volcanic phenomena, surmising that their origin might be a common one. His theory became known as 'volcanism'. Michell carefully studied the direction, force and breadth of ground motion, along with the propagation time of ground vibrations, applying wave theory for the first time. Using such methods he located the source of the Lisbon earthquake at 10–15 leagues (50–75 km) out to sea between Lisbon and Oporto. He wanted to demonstrate that the fundamental motor of the seismic vibrations, as well as of the explosive force of the volcanoes, was water vapour produced by contact with incandescent terrestrial masses. According to Davison (1927, pp. 18–24) the modernity of his approach is beyond dispute. However, his contemporaries had great difficulty in understanding him since most still sought links to a metaphysical cause and used a deductive line of thinking.

From this time on, the glorious fire theory went into decline. Alongside volcanism, the 'post-Lisbon' scientific disputes highlighted an expansion of electrical theory, upon which numerous experiments had been carried out. The famous

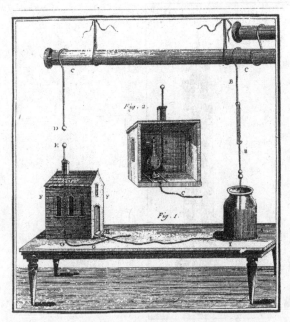

Figure 4.11 Design of the demonstrative table planned by Pierre Bertholon to demonstrate the electrical origin of earthquakes. The device triggered an electrical discharge, which caused the paper house to burn out (from Bertholon, 1787).

chemist Joseph Priestley (1733–1804), who was also an electrician, had performed as a young man various laboratory experiments to corroborate the electrical interpretation of earthquakes (Priestley, 1775). Another supporter of the electrical theory of seismic events was Tiberius Cavallo (1749–1809), an authoritative member of the Royal Society (his work *A Complete Treatise on Electricity in Theory and Practice, with Original Experiments* was published in London in 1777). But probably the most strenuous, almost fanatical, supporter of the electrical origin of earthquakes was Pierre Bertholon (1741–1800), a priest of the Congregation of Saint Lazare. In his *De l'électricité des météores* (Bertholon, 1787) he argued in favour of this interpretation using some experiments that aroused great curiosity because it was thought that they reproduced on a small scale the real dynamics of earthquakes. On a flat electric condenser Bertholon constructed some cardboard houses leaning on a paper and cardboard hillock, and then he released the electrical charge, which provoked jolting (Figure 4.11). He also designed 'para-earthquakes' as a means of defence against shocks. This remedy consisted of placing some long iron rods with diverging tips deep into the earthquake-prone ground. The lower tips had the task of attracting the 'electrical fire', while the upper parts were supposed to disperse it. It was, in practice, a kind of upside-down lightning conductor. This idea was subsequently elaborated

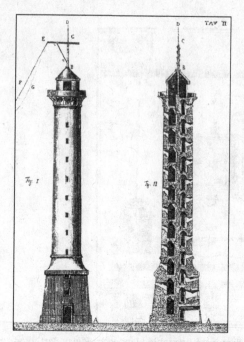

Figure 4.12 'Para-earthquake' tower, designed by Giuseppe Valadier, and drawn by Pietro Santi (in Comastri, 1986). The tower, designed after the 1786 earthquake at Rimini (Italy), has a square base from which a cylindrical structure rises. On the wooden antenna BC an iron rod CD is fixed to a screw at point C; from C a wooden board CE, called *falcone*, stretches horizontally, supported by an arm known as *saettone*; CEF represents the path of the *filo scaricatore* that was supposed to channel the dangerous electric vapour (believed to be the root cause of the seismic shocks) into the sea.

by Giuseppe Valadier, then a young and as yet unknown architect in the service of the Pope. His 'para-earthquake' tower project (Figure 4.12) was inserted in a scientific leaflet of Giuseppe Vannucci (1787), right in the middle of a lively academic discussion on the causes of the Rimini earthquakes of 1786 (Comastri, 1986).

Although the idea of the lightning rod was in some way at the root of this curious application, the inventor of the lightning rod did not wholly agree with that. Indeed, according to Dean (1989), Benjamin Franklin (1706–1790) himself, to whom all the electricians of his time referred, made no specific contribution on this subject and stayed out of the direct debate among the electricians. Even Luigi Galvani (1737–1798) and Alessandro Volta (1745–1827) were supporters, although not outspoken, of the electrical interpretation of earthquakes.

Some doubts of Luigi Galvani concerning the electrical
interpretation of earthquakes

Luigi Galvani had already become an established physiologist of the electrical phenomena on animal bodies, when in 1779–80 there was a long and intense seismic sequence in Bologna (Italy), the town where he was living. It was an important earthquake series because, besides causing widespread damage, it paralysed the city's social and commercial life, leading to an economic depression (Boschi and Guidoboni, 2003). As always, actual earthquakes were a laboratory for new research and analysis. What did Galvani think of electricity, the theory most in fashion in those days, concerning the origin of earthquakes? In his *De viribus electricitatis in motu musculari*, published in 1791, there is but a fleeting mention of earthquakes:

> ... as such muscular contractions, which, as we have said, occur during atmospheric storms, offer new and undoubted evidence of the electricity of the atmosphere and its influence on the animal economy, it may, even if not without difficulty, lead to the discovery not so much of the causes of an earthquake as its effects on such an economy; and for this reason it will be useful, in our opinion, to repeat the same experiments during earthquakes. (Galvani, 1791, p. 415.)

It is hard to believe that a scholar of such phenomena had ignored the problem almost completely, for electrical theories had been disseminated through dozens of pamphlets and scientific articles in his day and had drawn the best-known people of his town into a fierce debate. Indeed, an examination of the preparatory notebooks for his famous work (cited above) reveals a consideration that Galvani later removed from the published text. Galvani mentioned earthquakes, avoiding the identification of a specific cause but bearing in mind a possible link between seismic activity and electricity. The following is some text translated from his notebooks (originally written in Latin):

> 18. The effects that occur in the animal economy and above all on the nerves when an earthquake occurs, are similar to those obtained by the fluids of artificial or natural electricity, and seem to offer evidence of electrical action in the earthquake itself, and akin to the whole of the bodies and in the soles of the feet, in the places of the earth. From such presuppositions it seems to be clearly apparent that dangers exist in those regions in which in the course of time earthquakes have produced shocks [. . .].
>
> 19. Certainly, the phenomena and effects that are in relation to the animal economy during an earthquake are easily explained by means of the electrical

> *fluid.* (Archivio dell'Accademia delle Scienze di Bologna, cart. II,
> plico IX, fasc. D, fols. 6–7.)

Galvani was a very serious-minded scientist. Having insufficient evidence for a correlation of earthquakes with electricity (because experimental evidence was lacking), he relegated what had seemed to him to be a certainty to the status of a vague hypothesis.

The Lisbon 1755 earthquake had stimulated a very vast and lively debate in Europe (Poirier, 2005). In the *Philosophical Transactions of the Royal Society*, 22 contributions were published. The great earthquakes in 1783 at Calabria reignited this debate with new observations and polemics. For years, in a singular spurt of scientific production, more or less well-known authors compared data and opinions, famous theories, and more modest interpretations. From this enormous outflow of ideas there emerged three schools of thought, which together represented the evolution of the theory of earthquake interpretation in late-seventeenth-century Europe. Indeed, the proponents of each set of ideas could be distinguished in the following way:

(1)   the supporters of the Earth's central fire (boiler) theory, a minority at this time, with points of contact with the volcanists;
(2)   the pure volcanists and those with points of contact with the electricians;
(3)   the pure electricians and those with points in common with the supporters of the Earth's central fire (boiler) theory and the volcanists.

According to Placanica (1985a, pp. 75–6), the person among these who should in particular be remembered for the originality of his ideas was Benvenuto Aquila (1784). Aquila hypothesized that in a seismic shock, triggered by a central fire inside the Earth, there was a huge thrust of rocky masses against one another that continued until their state of equilibrium was upset.

This hypothesis, which to us appears to be quite innovative in regard to its historical context, was formulated in clear-cut opposition to the electricians, who were still fighting for the interpretative supremacy. That record was often supported by instruments, as in the case of the seismoscope devised by the academic from Calabria, Nicola Zupo (1784).

The most important supporters of the volcanism theory were William Hamilton (1730–1803) and Déodat de Dolomieu (1750–1801). Their contribution is also important for historical seismologists. Hamilton was above all interested in volcanoes, which were his true passion. While he was living in Naples as a diplomat, some earthquakes struck Calabria in 1783, as noted earlier. He wrote about them in the *Philosophical Transactions* in 1783, providing his theoretical interpretation, and in diplomatic correspondence he added various items of news. Hamilton's 14 letters, still preserved at the Foreign Office in London, were written between 18 February 1783 and 29 March 1785. Hamilton was particularly struck by shoals of dead fish that had come from very deep seas some days before the first shock on 28 February 1783 (this is also described by other sources). There were probably gas emissions from the sea bed, and the dead fish on the surface were the consequence of this. But Hamilton, as others of that time, attributed the death of the fish to a change of temperature instead of gas emissions in the sea. Dolomieu, an indefatigable and very keen observer of landscapes and rocks, developed the correlation between seismic effects and the nature of the ground. This line of investigation later became a very fertile field of observation intended to comprehend the effects of earthquakes on buildings.

Hamilton and the Calabria tsunami on 6 February 1783

While Sir William Hamilton was living in Naples as a diplomat, earthquakes struck Calabria in 1783. He was impressed by the strong tsunami on 6 February, which hit the shores of Calabria and Sicily and was caused by a major landslide. In an 'Account' published in the *Philosophical Transactions* he described the tsunami as follows:

> On the 17th of May I left Messina, where I had been kindly and hospitably treated, and proceeded in my Speronara along the Sicilian coast to the point of the entrance of the Faro, where I went ashore, and found a priest who had been there the night between the 5th and 6th of February, when the great wave passed over that point, carried off boats, and above twenty-four unhappy people, tearing up trees, and leaving some hundred weight of fish it had brought with it on the dry land. He told me, he had been himself covered with the wave, and with difficulty saved his life. He at first said the water was hot; but as I was curious to come at the truth of this fact, which would have concluded much, I asked him if he was very sure of it? and being pressed, it came to be no more than the water having been as warm as it usually is in summer. He said, the wave rose to a great height, and came on with noise, and such rapidity that it was impossible to escape. The tower on the point was half destroyed, and a poor priest that was in it lost his life. From hence I crossed over to Scilla. Having met with my friend the Padre Minasi, a Dominican friar, a worthy man and an able naturalist, who is a native of Scilla, and is actually employed by the Academy of Naples to give a description of the phenomena that have attended the earthquake in these parts, with his assistance on the spot, I perfectly understood the nature of the formidable wave that was said to have been boiling hot, and had certainly proved fatal to the baron of the country, the

Prince of Scilla, who was swept off the shore into the sea by this wave, with 2473 of his unfortunate subjects. The following is fact. The Prince of Scilla having remarked, that during the first horrid shock (which happened about noon the 5th of February) part of a rock near Scilla had been detached into the sea, and fearing that the rock of Scilla, on which his castle and town is situated, might also be detached, thought it safer to prepare boats, and retire to a little port or beach surrounded by rocks at the foot of the rock. The second shock of the earthquake, after midnight, detached a whole mountain (much higher than that of Scilla, and partly calcareous and partly cretaceous), situated between the Torre del Cavallo and the rock of Scilla. This having fallen with violence into the sea (at that time perfectly calm) raised the fatal wave, which I have above described to have broken upon the neck of land, called the Punta del Faro, in the island of Sicily, with such fury, which returning with great noise and celerity directly upon the beach, where the prince and the unfortunate inhabitants of Scilla had taken refuge, either dashed them with their boats and richest effects against the rocks, or whirled them into the sea; those who had escaped the first and greatest wave were carried off by a second and third, which were less considerable, and immediately followed the first. I spoke to several men, women, and children here, who have been cruelly maimed, and some of whom had been carried into the sea by this unforeseen accident. (Hamilton, 1783, pp. 201–3.)

The electricians did not just write theories. Often their texts are rich in descriptive information on the effects of the earthquakes that they had observed. For example, Giovanni Vivenzio, a physician at the court of Naples, translated Bertholon's memoir on the electrical earthquakes (in the first part of his *Istoria e teoria de' tremuoti*) (Vivenzio, 1783, 1788) and left a very detailed description of the effects and chronology of thousands of earthquake shocks in Calabria in 1783. To corroborate the electrical theory, in 1785 Vivenzio had the experiment with the models of houses first performed by Bertholon repeated before the Queen of Naples. The 'great and marvellous' machine devised by Vivenzio was supposed to have convinced everyone of the correctness of the electrical interpretation. Vivenzio also designed an earthquake-resistant house built on a wooden trellis (wood was renowned for being a poor conductor of the electrical fluid). It is curious to note that one of these structures, the prototype of which is known of by the name of *casa baraccata*, completely withstood the disastrous Messina earthquake of 1908 (see Chapter 11).

Among the other better-known electricians of those years was Andrea Gallo (1732–1814), who left a thorough description of the earthquake shocks in Calabria in 1783 (Gallo, 1784). Cristofano Sarti, a professor at the University of Pisa, compared the motion of houses shaken by the earthquakes to that of a pendulum. He also wrote several observations on earthquake-resistant construction (Sarti, 1783).

At the end of the eighteenth century, from the standpoint of theory, the crux of the general debate rested with the preservation of the *continuity* principle in nature. Because this principle was a kind of obsession of natural philosophy, it was necessary to show that a continuous cause (such as fire or water) produced a discontinuous effect (like an earthquake). But electricity seemed to adapt itself nicely to the thesis of a discontinuous cause, and for this reason it was very popular in those times. The thinking of natural philosophers and physicists was immobilized in this dispute which, as we have seen, had very deep roots. The *continuity/discontinuity* polemic only began to die down in contemporary research in physics. For example, theories about light show that a single phenomenon can be analysed as being both continuous (waves) and discontinuous (particles) at the same time.

From an epistemological standpoint, one can observe that the logical procedures followed by the thinkers of this time were mainly deductive, based on axioms and analogies. As compared with contemporary scientific disciplines, above all chemistry but also the geology of the last few decades of the eighteenth century, seismological thinking appeared at the end of the seventeenth century still to be interwoven with metaphysical issues, and it remained under the spell of cosmogonist perspectives because it could not develop its own experimental side.

However, the experimental motor, so to speak, had been triggered. The different theoretical positions that nearly always set the scholars one against the other, from different disciplinary angles, gave rise to lively debates and endless discussions, the true soul of research.

4.1.7 The early nineteenth century

While the electrical interpretation of earthquakes continued until the second half of the nineteenth century, though vigorously confuted by Timoteo Bertelli as late as 1887, new mathematical developments were setting the stage for the modern understanding of seismic waves. In 1821 and 1822 Claude-Louis-Marie Navier (1775–1836) and Augustin Louis Cauchy (1789–1857) developed the theoretical foundations of elasticity theory, which describes how solid materials such as rock can behave like three-dimensional springs. In 1830, Simeon Denis Poisson (1781–1840) showed that seismic waves will traverse as P and S waves through the interior of a solid body, and this was later confirmed by Sir George Gabriel Stokes (1819–1903). Later, John William Strutt, more commonly known as Lord Rayleigh (1842–1919), articulated the theory of an elastic wave that propagates along the outside surface of a solid object. These waves have come to be known as Rayleigh waves. Later still, in 1911, Augustus Love (1863–1940) discovered different elastic surface waves that have come to be known as Love waves. While primarily theoretical in nature, these developments in the understanding of seismic waves formed the basis from which the modern theory of earthquake sources was eventually developed.

4.1.8   *From the second half of the nineteenth century to the first half of the twentieth century: the development of the modern theory of earthquakes*

The investigation of a strong earthquake in Basilicata (southern Italy) on 16 December 1857 led Robert Mallet (1810–1881), an Irish engineer who studied earthquakes, to propose some new ideas about earthquakes and how they occur. In his two-volume report of field observations of the effects of this earthquake (Mallet, 1862), Mallet used his data to argue that the earthquake waves that caused the destruction must have been generated by an explosive source beneath an extinct volcanic edifice called Monte Vulture. Mallet reasoned that the earthquake was caused by an explosive source of volcanic origin, and he used a map of the directions in which building contents and building damage were thrown to deduce the location of the source of the seismic energy. While this work was based on an assumption that is now known to be false (that earthquakes are caused by explosive underground sources), it did herald a new form of seismological research in which the theory and observations of seismic waves can be used to learn more about the sources of earthquakes.

The development of seismographic instrumentation capable of recording the vibrations from distant earthquakes was another important milestone in the evolution of the modern view of earthquakes. Simple devices that were sensitive in some way to the ground vibrations of earthquakes had been around for centuries. However, it was not until the second half of the nineteenth century that the first seismographic instruments capable of recording the ground motions from distant earthquakes were invented. Filippo Cecchi (1822–1887), a Piarist father, built a seismographic instrument in 1875, but it was not sensitive enough for recording any but the strongest earthquakes (Cecchi, 1876). However, when working in Japan in 1880, John Milne (1850–1913) and his colleagues James Alfred Ewing (1855–1935) and Thomas Gray (1850–1908) constructed what is considered the first practical seismograph (Ewing, 1880, 1881; Gray, 1883; Milne, 1897, 1898, 1901). During the next two decades many other researchers invented new seismographic instruments or made improvements to those already in existence. By 1900 the data gathered by seismological services throughout the world, combined with the seismographic detection of earthquakes, enabled Milne and Fernand de Montessus de Ballore (1851–1923) to construct maps of worldwide earthquake epicentres.

The connection between earthquakes and faults began to be made through field investigations of strong earthquakes at the end of the nineteenth century. Milne had argued that the seismic waves that were experienced in an earthquake in Japan in 1891 might have been due to fault slip. In 1897 a major earthquake struck the Assam hills of eastern India. Richard Dixon Oldham (1858–1936), the head of the Geological Survey of India, undertook a detailed field investigation of this earthquake (Oldham, 1899). His report documents what were subsequently recognized as two faults, the Chedran fault and the Samin fault, with slip up to 11 m in this earthquake. Less than a decade later, the 1906 earthquake on the San Andreas Fault at San Francisco in California again revealed a strong connection between fault slip and seismic waves. Based on the field data of ground deformations that were noted after the 1906 earthquake, Harry Fielding Reid (1859–1944) developed what is known as the elastic rebound theory (Reid, 1911). In this theory, the rock formations on either side of a fault try to slide past each other in opposite directions, but friction prevents rock slip along the fault itself. The rock along the fault gradually deforms until so much stress builds up that the friction is overcome and the rock along the fault slips suddenly and quickly. This slip releases the elastic waves that are felt and instrumentally recorded as the ground-shaking. The slip seems to start on the fault at a single point that is now called the earthquake focus. Reid's elastic rebound theory remains today a foundation in the understanding of how earthquakes take place. Subsequent work by seismologists has led to methods to locate an earthquake, compute its size, and use seismic waves to estimate the fault orientation and slip direction.

Another discovery worth mentioning here is that of deep-focus earthquakes. In 1922 Herbert Hall Turner (1861–1930) provided evidence that the foci of some earthquakes must be deep (hundreds of kilometres) below the surface of the Earth and not merely at shallow depths of a few kilometres to a few tens of kilometres (Turner, 1922). In 1928 Kiyoo Wadati (1902–1995) in Japan confirmed Turner's ideas (Wadati, 1928). From this time on, the depth of an earthquake became an important source parameter in earthquake catalogues.

As the recording of earthquakes using seismographic instruments matured into a widely practised scientific endeavour, scientists refined their ideas about what happens on a fault when an earthquake takes place. In 1935 Charles Richter (1900–1985), using an idea put forth by Kiyoo Wadati in 1931, proposed the first widely used instrumental earthquake magnitude scale, now called M_L, which was derived for earthquakes in Southern California. Beno Gutenberg (1889–1960) teamed with Richter in 1942 to propose the first magnitude scales for global earthquakes, those scales being the body-wave magnitude scale m_b and the surface-wave magnitude scale M_s. While many subsequent magnitude scales have been developed, all are calibrated with and similar to these original magnitude scales. In 1954 Gutenberg and Richter published the second edition of an important book, called *Seismicity of the Earth and Associated Phenomena*, which was a compilation of maps and other information about worldwide earthquakes. This book showed that most of the Earth's earthquakes took place along confined belts, an observation that was used a decade later as part of the evidence for plate tectonics.

From the middle of the twentieth century onwards, there have been a number of elaborations and extensions of the elastic rebound theory that came from Reid's studies of the 1906 San Francisco earthquake. Work by a number of scientists led to the theory that the seismic energy radiated by earthquakes has the same spatial pattern as if that energy had been radiated by a set of forces called a double couple without net moment. This discovery has allowed seismologists to use earthquake seismograms to deduced the fault orientation, slip direction, amount of slip, and the fault size for earthquakes that are instrumentally recorded, even if the fault for that earthquake cannot be observed directly at the Earth's surface. Keiiti Aki (1930–2005) showed in 1966 that earthquake observations can be used to determine the seismic moment of an earthquake. Unlike earthquake magnitude, seismic moment is a fundamental measure of the size of an earthquake, and it forms the basis of the modern moment-magnitude scale M for earthquakes, now considered the best measure of the size of a seismic event. Much laboratory and theoretical work that began in the middle of the twentieth century has shown that earthquakes result from frictional slip on faults to release pressures that have built up asymmetrically in the rock. In the 1960s the development of the theory of plate tectonics provided the explanation for the cause of the rock pressures that cause earthquakes.

4.2   Scientific studies and services

The importance of seismological studies and services is often underestimated by historians. But this appears to be a gross error, since these sources as a whole reflect different types of knowledge and provide cognitive pictures of the past that are of considerable help towards an understanding of descriptions of the same phenomena provided by other sources. The next two subsections describe the main scientific sources of seismological data that have been created since the mid nineteenth century: scientific reports and field surveys, which have been produced since the middle of the eighteenth century, and macroseismic questionnaires, of which production started at the end of the nineteenth century.

4.2.1   Scientific reports and field surveys in the eighteenth and nineteenth centuries

Ever since seismology became a separate discipline from meteorology, it has produced sources of outstanding interest for an understanding of seismic events. In the eighteenth century and the first half of the nineteenth century, field surveys after a strong earthquake were regarded as valuable scientific evidence which governments were very keen to have. On the basis of such surveys, theories and interpretations were discussed, administrative decisions were taken, and seismic classification criteria or standards for choosing new building sites and constructing buildings were established. The sensitivity to scientific and social factors of the various authors involved led to a fresh evaluation of building characteristics, the relationship between surface geology and seismic effects, and the social conditions of earthquake victims. Field research began with the Lima earthquake of 1746, was developed and extended with the Lisbon earthquake of 1755, and became an important observational instrument during the Calabrian earthquakes of 1783.

An important contribution to the investigation of the Calabrian earthquakes was provided by the scientists of the *Reale Accademia delle Scienze e delle Belle Lettere* (Royal Academy of Sciences and Belles-Lettres of Naples), coordinated by Michele Sarconi. The novelty of their approach consisted of meticulously detecting and documenting earthquake-induced damage to buildings (150 towns were visited) and to the ground surface, presenting their data in 70 iconographic plates (Sarconi, 1784; Schiantarelli and Stile, 1784). The most important contributions of this organization lay precisely with the important wealth of field data, which today remain a subject of analysis by historical seismology researchers. The information that these scientific reports provide today is irreplaceable, and it complements in many ways the information provided by traditional historiographic sources. Such reports usually take the form of travels to the place concerned, and are almost always embellished with drawings, sketches, maps and photographs, thereby bringing together a body of material that is invaluable today.

Another important early contribution in the history of scientific reports on seismic events was the field survey carried out after the earthquake in December 1857 in the Kingdom of Naples by Robert Mallet (1862), who honed a method for describing his observations of earthquake effects. This method was applied on various occasions by Giulio Andrea Pirona and Torquato Taramelli for the earthquake of 29 June 1873 in Belluno (northeastern Italy) (Pirona and Taramelli, 1873), by T. Taramelli and Giuseppe Mercalli for the earthquakes of Andalusia (Spain) in October 1884 (Taramelli and Mercalli, 1886) and again by the latter two authors for the earthquake of western Liguria (northern Italy) and southern

France on 23 February 1887 (Taramelli and Mercalli, 1888). It was precisely as a result of such direct and detailed observations of earthquake effects that in 1897 Giuseppe Mercalli published his scale of macroseismic intensity (see Chapter 12). Mercalli's scale, composed of ten degrees of increasing macroseismic intensity, later became, with subsequent additions and elaborations, one of the most widely used in the world. Many other scales of macroseismic intensity that rated the different effects of earthquakes and their shaking were elaborated beginning in 1811 by J. Brooks. Ferrari and Guidoboni (2000) present a list of the macroseismic intensity scales and their predecessors that have been applied during the past several centuries.

There were usually two aspects to scientific surveys that were carried out to investigate the effects of natural disasters like earthquakes and tsunamis. On the one hand, the researchers made observations about seismic effects on buildings, the natural environment and people, while on the other, they provided a series of interpretive reflections of possible causes to explain their observations, in which the writer brought together empirical data and theory. The merging of observations and theory was ideally done in a way where the empirical data were not regarded from a previously established point of view. Epistemology argues, it is true, that every observation is limited by already being an interpretation, but only rarely has that limitation prevented these works from providing information of value to modern researchers.

Perhaps the most serious drawback to many reports of seismic surveys for this time period lies in the fact that the area observed was only part of the entire region subject to damage or notable ground-shaking. Much of the damage area usually was not described, either because travel difficulties prevented all villages from being reached or because the scientists restricted their attention to the most seriously damaged centres. The fact that writers had a predilection for certain particular aspects of seismic effects (such as rotational movements in vertical structures, cracks in the ground, the behaviour of buildings, etc.) has meant that an important collection of case histories of these effects has been preserved for a large number of seismic events.

4.2.2   Macroseismic questionnaires and cards: their history and use

On 9 November 1852 an earthquake hit north Wales and was also distinctly felt in Ireland. Robert Mallet (see above) wrote to *The Times* of London and asked them to publish a sort of questionnaire on the effects of that earthquake. Unfortunately, *The Times* refused to do so. That refusal was a pity, as it would have been the first attempt to print and disseminate a macroseismic questionnaire using newspapers. Some 20 years later there was another attempt to disseminate a macroseismic questionnaire. This was successfully done by Alessandro

Serpieri (1823–1885), and was intended to collect information on the effects of the earthquakes on 12 March 1873 (I_0 = VIII; M = 6.0, Marche – central Italy) and on 17 March 1875 (I_0 = VIII; M = 5.8, southeastern Romagna – northern Italy) (Serpieri, 1873, 1876).

From the closing decades of the nineteenth century onwards, many countries began to set up national seismological services. These seismological services developed from earlier meteorological services involving the collection of meteorological data. The use of macroseismic investigations to obtain information about earthquake effects was initiated by the seismological services in a number of countries. This was generally done by means of preprinted questionnaires in the form of postcards. These questionnaires were compiled at observatories, meteorological stations, town halls and telegraph offices, and sent postpaid to an office whose scientific staff compiled seismic bulletins from them. Early bulletins contained a transcription of the descriptive content of the questionnaires, with additional information from newspapers and other sources. As time went by, the space occupied by these data in the bulletins increasingly gave way to instrumental data, following the increasing realization of the importance of instrumental readings that was gradually becoming established in the seismological community as technological advances were made in instrumental seismology. Descriptive data of local effects of earthquakes started to be assembled in conjunction with earthquake parameters measured using the increasingly widespread instrumentation that was being deployed worldwide.

In a little less than 100 years, macroseismic questionnaires have undergone a number of modifications in order to adapt them to the various revised forms of intensity scales and to new interpretative models in seismology. As a direct documentary source, they are generally speaking of great scientific value because of the part played by their compilers, their eyewitness nature, the immediacy of the evidence they provide, their statistical value and the fact that evidence is standardized through the use of a list of precise questions. This last factor has its disadvantages as well as its advantages, in that the information provided is limited by the questions asked (Figure 4.13).

4.2.3   *Letters and correspondence between scientists*

Among the historical materials produced by scholars in general, and by the Earth sciences in particular, scientific letters play a specific role. This correspondence helps both in the reconstruction of the scientific pathways leading to the design of the seismic instruments of the time, and in the analysis of the different phases of development of the science of earthquakes. A close mutual relationship between meteorological, astronomical and seismic observation methods can also be seen in the contents of many such letters. One

Figure 4.13 A macroseismic questionnaire postcard from Zafferana Etnea (Sicily), 26 December 1889 (Ufficio Centrale di Ecologia Agraria, Rome).

can find meteorological, astronomical, seismic and geomagnetic observations, advice about the design and building of observation devices, theory proofs complete with drawings, tables and diagrams, instrumental recordings and methods for the calculation of earthquake parameters. Scholars often trace connections among observations of various phenomena, such as earthquakes with meteorological displays or with astronomical phenomena, etc.

The condition of historical letter files and the different research interests of different investigators lead anyone wanting to study scientific correspondence to follow various approaches, such as (1) the study of a letter file; (2) the cross-reading of scientific correspondence between two or more scholars; or (3) the extensive study of several letter files and their subsequent crossing.

A database of the letters of the Earth scientists from 1730 to 1950

Several projects are in progress in Europe to survey and use this outstanding wealth of data. For example, the TROMOS project, 1990–2005, promoted by the *Istituto Nazionale di Geofisica e Vulcanologia* and *SGA – Storia Geofisica Ambiente* in Europe is viewable on the internet at web address http://storing.ingv.it/tromos. This project has examined an enormous quantity of letters of the earth scientists that were exchanged over the period 1730–1950 (Figures 4.14, 4.15 and 4.16). The over 14 000 letters catalogued and analysed within the scope of the TROMOS project have allowed the investigators to successfully test a comparison between letter files of meteorological, seismological and astronomical observations from

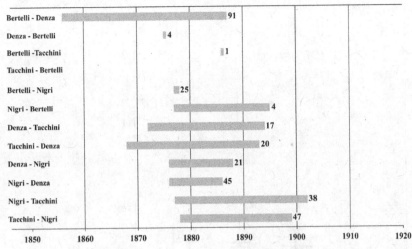

Figure 4.14 Graphical summary of the chronological distribution of the letters that Timoteo Bertelli, Francesco Denza, Vincenzo Nigri and Pietro Tacchini exchanged during the time-span of their activity. Correspondent and addressees of the letters are indicated in the axis of the ordinates (e.g. Bertelli – Denza indicates the letters sent by Bertelli to Denza, followed below by the letters sent by Denza to Bertelli). The number of letters sent is placed to the right of the strip that identifies the chronological range of each correspondence (from Ferrari, 2002).

different historical sites (http://storing.ingv.it/letters). For example, the work of scientists like Timoteo Bertelli (1826–1905) and Michele Stefano de Rossi (1834–1898), who frequently wrote to each other, is well documented in their letters. Their work led to the development of tromometers, instruments able to detect the slightest oscillations in soil. This microseismic observation technique was improved until the distribution of the standard tromometer, a pendulum whose overall length is approximately 150 cm and which is able to sense slight and slow movements of soil. It is estimated that in over 50 public and private observatories more than 300 000 tromometric measurements are preserved. Of these, 5000 have been analysed in Ferrari *et al.* (2000).

The TROMOS project has also studied the seismic instruments that were the subject of the correspondence and the thinking of the scholars who produced the letters. The intense correspondence between the people running the centres of seismological observation networks contributes to an important extent to the history of the observatories and the scientific pathways that connected different observatories. Even the most important and substantial letters, however, provide only a partial look at the topics discussed in the letters. The study of a person's correspondence reveals only half a dialogue, as though one were listening to a telephone conversation without knowing what the party is saying at the other

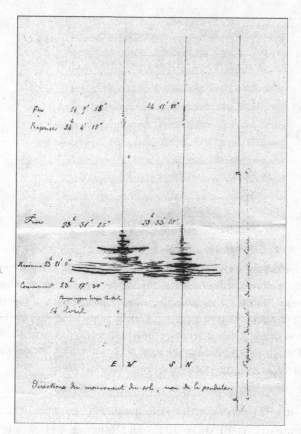

Figure 4.15 A page from a letter by Emilio Oddone, addressed to Albin Belar, the director of the observatory of Ljubljana (Slovenia), concerning the earthquake of 14 April 1895. The image reproduces the seismogram of the earthquake as recorded at Pavia (from Ferrari, 2002).

end of the line. In the case of letters, apart from the usually rare cases in which one has also obtained copies of the replies, one knows what the writer is saying but does not know the content of the addressee's letters.

A previously unknown letter from Darwin to the seismologist Perrey (1853)

The correspondence of Charles Robert Darwin (1809–1882) comprises about 14 500 letters with 2000 or so correspondents and covers the time range between 1821 and 1882, the year of his death. As is known, Darwin was in contact with the most prestigious members of the scientific community of his day, such as the geologist Charles Lyell, the zoologist Thomas Henry Huxley, the botanists Asa Gray and Joseph Dalton Hooker and the naturalist Alfred Russel Wallace. He also corresponded with numerous animal breeders and with scholars of

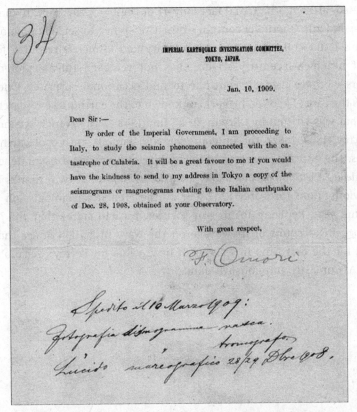

Figure 4.16 Letter by Fusakichi Omori addressed to Giulio Grablovitz, the director of the Royal Observatory in Casamicciola (Ischia-Naples), 10 January 1909. Omori asks, on behalf of the Japanese government, to receive the seismograms and magnetograms of the great earthquake of Messina in 1908 (permission kindly granted by G. Ferrari).

disciplines outside of the biological sciences, including seismologists. There is a project, started in 1985 and still in progress, devoted to publication of the complete collection of Darwin's letters (Burkhardt and Smith, 1985–). In the 23 years since this project began, so far fifteen volumes have been published, encompassing the correspondence to the year 1867. For a list of letters hitherto identified, see Burkhardt and Smith (1994).

On the website 'Darwin Correspondence Project' (http://www.darwinproject. ac.uk) an electronic version of the database of all the letters in the collection has been available since 2002, and it is updated as new finds are reported. About 5000 letters, written between 1821 and 1865, have been completely transcribed, while a summary is available for the remainder of the correspondence. By typing in the words 'earthquake' and 'tremor' into the database search engine, 122 letters are indicated, although in actual fact only 41 of these – 27 by Darwin and 14 by his correspondents – actually contain explicit reference to seismic phenomena.

Darwin's letter collection, an extraordinary and immense resource for historians of nineteenth-century science, has not yet been fully explored, and researchers who are using it continue to find occasional surprises. One such surprise is the case of a letter, hitherto unknown to the earthquake research community, that was written by Darwin on 25 July 1853 to the French seismologist Alexis Perrey (1807–1882), the indefatigable and famous author of earthquake catalogues. The letter is preserved in the archive of Perrey's correspondence, the *Biblioteca della Società Napoletana di Storia Patria*, in Naples. Darwin thanks Perrey for a previous 'note' (probably lost) sent by the French seismologist on 9 July of the same year. Replying to his interlocutor, Darwin states that the *Journal of Researches*, his account of his voyage on the Navy brig HMS *Beagle* that was published for the first time in 1839, is of little value for Perrey's seismological research (A. Comastri, unpublished data).

5

Other types of sources

5.1   Historical earthquake cartography

The historical cartography of earthquakes and tsunamis constitutes an interesting source of information about the seismic events of the past, and it represents earthquake and tsunami data in a unique and often useful way. The attempt to represent an earthquake in a conceptual scheme like a map involves both a theoretical aspect and empirical knowledge. The cartography of earthquakes is a poorly investigated discipline, which differs in important ways from general historical cartography (which is well known to historians). Fortunately, historical cartography can offer historical seismologists some background to help identify the locations of sites that have disappeared, that have changed names or that have been re-established in some way. Historical cartography provides spatial representations of a historical territory. However, the cartography of a specific earthquake or set of earthquakes can contain other fundamental information for historical seismology. Earthquake cartography plays several roles for historical seismology: (1) it is a source that documents the regional or urban geography and settlement pattern at the time of the seismic event, (2) it represents a specific document concerning the effects (especially the spatial distribution) of an earthquake or tsunami, and (3) it can be used to test theories about seismic events and their occurrences.

The oldest map about a seismic event hitherto known for the Mediterranean area is the map of Magiol dated 1564, which depicts in bright colours the effects of the earthquake which on 20 July 1564 hit the Maritime Alps (Figure 5.1). This map has been analysed from the standpoint of historical seismology by Moroni and Stucchi (1993), who have highlighted the existence of two different copies

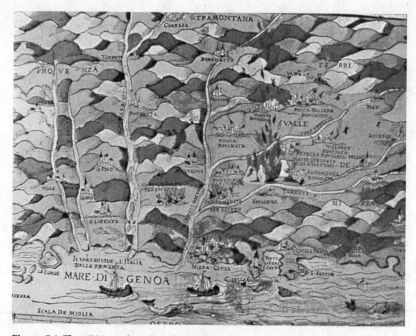

Figure 5.1 The 1564 earthquake of the Maritime Alps (southeastern France and northwestern Italy). This coloured woodcut (in the original) is the oldest existing seismic map. It shows seven damaged towns (Erlangen, Universitäts Bibliothek, Friedrich-Alexander-Universität).

of the map. This map also has the merit of being the most important source of spatial information about the earthquake that it depicts. The second most ancient seismic map hitherto known is the one that depicts the earthquake of Apulia in 1627 (Almagià, 1914). This map (Figure 5.2) too has been used by the scholars who dealt with this earthquake (Molin and Margottini, 1982; Boschi et al., 1995). As far as is known today, maps of earthquake effects must have been a little-used genre as it is not for another century before another such map is found. This map shows the effects in Palermo of the 1726 earthquake. In this case the scholar who drew it, Mongitore, was interested in depicting the distribution of damage in an urban area (Figure 5.3). As can be seen in Figure 5.3, the detailed pinpointing of the damage to the various important buildings of the city is of extraordinary detail that is of interest and impact to modern researchers. On the map, the damage is portrayed in red, with the effects notated. Several decades later, the Lisbon earthquake in 1755 aroused interest in a general study of the spatial extent of the earthquakes. One of the results of this interest was the drawing of a map that sought to depict the whole area in which that earthquake had been felt. The map extended as far as central Europe, where

Figure 5.2 Map drawn by Orazio Marinari of the damage caused by the earthquake of 1627 in Apulia (southern Italy) (in De Poardi, 1627).

the effects of a long seismic crisis that took place almost contemporaneously in the transalpine area were incorporated with the reports for the Lisbon event.

A real leap forward in the cartography of historical earthquakes was accomplished a few years later thanks to an outstanding cartographer, the monk Eliseo della Concezione, who was a physicist and mathematician of the Royal Academy of Naples. By using the data collected through a field reconnaissance made by his academic colleagues after the earthquakes in February–March 1783, Father Eliseo represented the effects of the earthquake, which he classified into three degrees of severity. On his map he indicates with one asterisk the towns that were partly damaged, with two asterisks the towns that were partly destroyed, and with three asterisks the towns that were totally destroyed (Figures 5.4 and 5.5). This map of great size (141 × 114 cm) can be considered the first one to represent the effects of an earthquake in the modern sense of the term. In 1846 Johann Jakob Nöggerath published a map that was similar, if less elaborate, to that of Father Eliseo. This was a map of the earthquake that took place in the Rhineland in the year the map was published. In this map, the area of damage was outlined with a curved line, and Nöggerath indicated a central point from which, in his opinion, the event had originated (Figure 5.6).

Figure 5.3 Damage in Palermo (Sicily) due to the earthquake on 1 September 1726. This map was engraved by Antonino Bova (in Mongitore, 1727). As an example here some buildings on the map are enlarged, and they are drawn showing the damage suffered.

From the mid nineteenth century onward the cartography of earthquakes intensified greatly, becoming increasingly careful to portray properly the empirical data but also with the aim of providing a geophysical interpretation of the event represented. Nöggerath's sober and effective map was almost forgotten. Mallet (1862) used a study of the direction of the fall of objects and the trend of cracks in seeking to calculate the point of origin of an earthquake. Unfortunately, the map of the earthquake observations that he produced was an almost undecipherable spaghetti of lines (Figure 5.7). Even so, his method was copied and its use became widespread. For example, it was adopted for the earthquake in 1884 in Andalusia (Figure 5.8) and western Liguria in 1887 (Figure 5.9). Only

Figure 5.4 Detail from the great map of Father Eliseo della Concezione, who mapped the damage in Calabria caused by the earthquakes in 1783. The author is portrayed on the right, with the specific instrument of the cartographer (map in Schiantarelli and Stile, 1784).

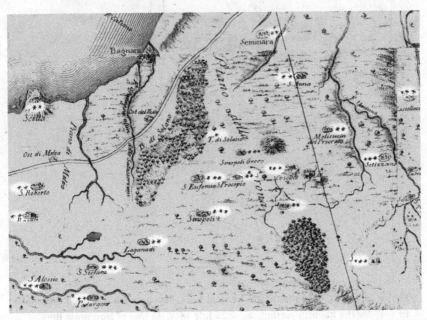

Figure 5.5 Another detail of the great map by Father Eliseo della Concezione, from which a rudimentary classification of damage due to the 1783 earthquake is shown by the number of stars next to a town, only some of which are higlighted in white (map in Schiantarelli and Stile, 1784).

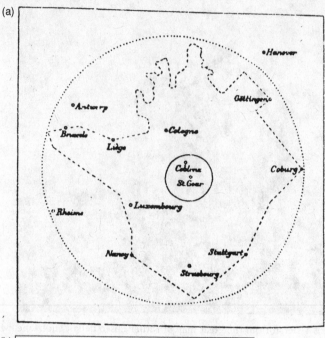

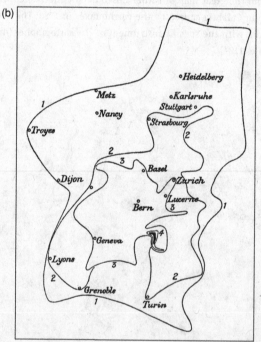

Figure 5.6 (a) Map with the first isoseismal lines, drawn by Johann Jakob Nöggerath in 1847 for the Rhenish earthquake of 29 July 1846. (b) Map of isoseismal lines of 25 July 1855 Visp valley (Switzerland) earthquake, drawn by A. Petermann from materials supplied by Georg H. O. Volger (1856) (from Davison, 1927).

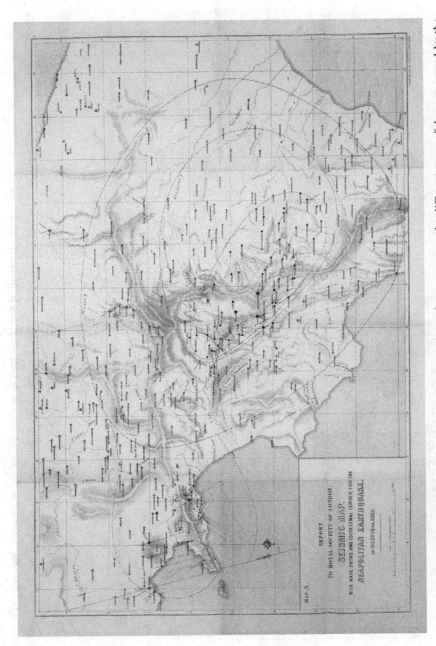

Figure 5.7 Map by Robert Mallet (1862) with the isoseismic lines that encompass the different types of damage caused by the earthquake of 16 December 1857 in the Kingdom of Naples (southern Italy).

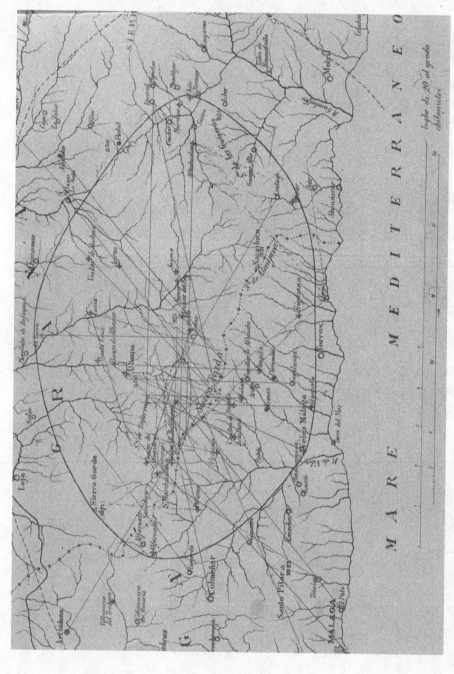

Figure 5.8 Map by Torquato Taramelli and Giuseppe Mercalli which represents the directions of the motion of the seismic waves, serving to locate the epicentre of the earthquake on 25 December 1884 in Andalusia (Spain).

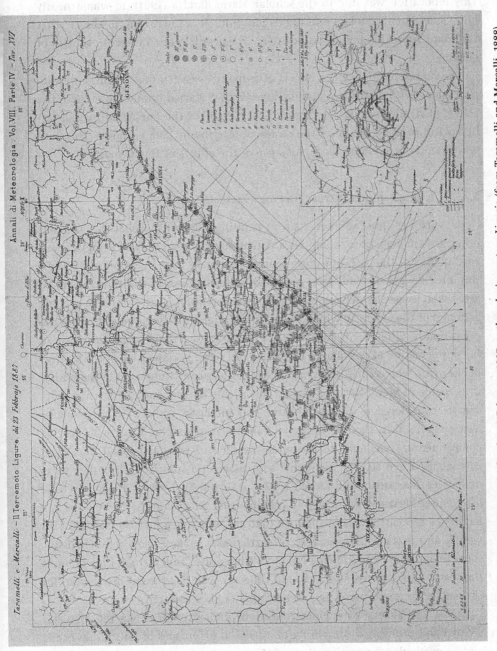

Figure 5.9 Map in the Mallet 'style' for the 23 February 1887 earthquake in western Liguria (from Taramelli and Mercalli, 1888).

at the end of the nineteenth century were the first seismic maps published that employed curved lines to indicate variations of the types of effects surveyed. This innovation was due to the scholar Mario Baratta (1901), to whom nearly all of subsequent scholars referred (Figure 5.10). It is interesting to observe that the maps from this period do not refer to epicentres but rather to 'centres of shaking'.

The representation of the effects of tsunamis went through a similar evolution. Along with instrumental data, investigators started producing maps of the effects of strong tsunamis. In Figure 5.11, a detail of the important map drawn by Baratta of the tsunami of the Straits of Messina in 1908 is shown. This map was based on a personal reconnaissance of the effects of this tsunami.

5.2   Iconographic sources (drawings, frescoes, etc.)

Iconographic sources, such as paintings, drawings, relief carvings and other artistic depictions, comprise a potentially valuable body of sources about earthquakes of the past, but these sources cannot be used in a direct way. Their use must always depend upon a critical understanding of the thought of the artist who made the depiction, of the prevailing fashions and tastes in graphic arts, and of the extent to which the artist experienced or was removed from the event that is represented. The factor of greatest critical importance for interpreting any depiction of an earthquake or tsunami scene is knowledge of the motives of the artist who created the work. These motives constitute what epistemology calls 'point of view'. Nothing carried out within the sphere of human activity can be considered a totally objective and unbiased activity. If one remains aware of this fact, then a researcher can use iconographic sources to provide complementary elements to the understanding of what happened due to an earthquake. Even so, inconographic sources are no substitute for other types of sources (Margottini and Kozák, 1992; Kozák and Ebel, 1996). Today a website (http://nisee.berkeley.edu/kozak/) is available that allows anyone to view the collection of 875 earthquake depictions that has been assembled by J. Kozák. This virtual and extraordinary gallery of earthquake images, some quite realistic, some imaginary, some touched up, but all integrated into a single collection, is the result of a sensitive and cultured collector who has turned his passion for this subject into an interesting and original field of research.

Today, the chance to explore collections and inventories via the Web has created and multiplied something of a passion for images of natural disasters, and in particular for earthquakes, of the past. A scientific use of such images is, however, definitely limited. As with photographs (see the next section), the interpretation of inconographic sources must always be conditioned by a critical

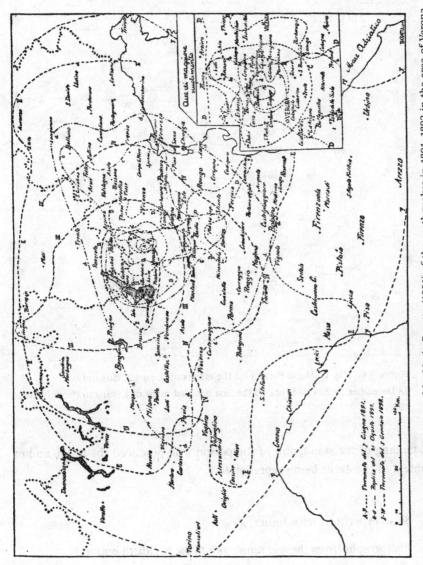

Figure 5.10 Isoseismal lines traced by Mario Baratta (1901) of the earthquakes in 1891–1892 in the area of Verona (northern Italy).

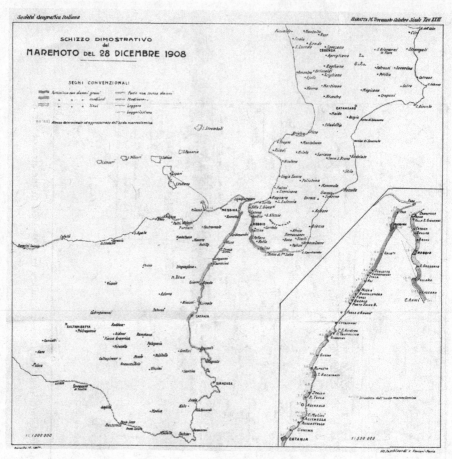

Figure 5.11 Map by Mario Baratta on the effects of the great tsunami on
28 December 1908; this is one of the first detailed maps of a tsunami (from Baratta,
1910).

understanding of the standpoint of the person who produced the image and of
the context of the event being represented.

5.3   Sources written with light

5.3.1   *Photographs (from the mid nineteenth to the twentieth century)*

Before the advent of photography in the mid nineteenth century, the
only images available of the effects of earthquakes and tsunamis were paintings,
drawings, sketches, engravings, etc. Above all in the natural sciences, visual
descriptions were disseminated by means of accurate drawings. Many things
changed thanks to the advent of photography. As soon as the *daguerreotype*

process invented by L. J. M. Daguerre in 1839 became publicized, some experimenters, mostly scientists, carried out studies to explore its technical and scientific applications. It soon became clear that the daguerreotype offered great opportunities, but it also had limitations due to the irreproducibility of the image. This was a serious obstacle to the early spread of this new process to create pictorial images. The initial difficulties were overcome with the development of the *calotype*, a fully fledged photographic process that, by taking the negative of the image, allowed for the creation of an almost unlimited number of positive images (Frizot, 1998). Within the span of a few years after this invention, photography was being used in all the fields of the sciences.

In application to earthquakes, the first ever photographic report was the one performed by the French photographer Alphonse Bernoud (see Becchetti and Ferrari, 2004). Immediately after a violent earthquake occurred on 16 December 1857 in the Kingdom of Naples (which caused about 20 000 deaths: $I_0 = XI$; $M = 7.0$), Bernoud set off for the areas hit by the disaster. He had to overcome numerous difficulties to carry out his work, difficulties both as a result of the inhospitable nature of the locations most affected by the earthquake and due to the unusual equipment he was using, which even required an armed guard to protect it from theft (Becchetti and Ferrari, 2004). Bernoud, a precocious and instinctive photo-reporter, made his way back to Naples after this reconnaissance where he had one of his studios draw up some documents that were quite shocking for their time. His photographs were immediately published with a comment by Marc Monnier in *L'Illustration*, and they moved the world. Between 28 December 1857 and the end of May 1858 Bernoud carried out four photographic campaigns, taking no fewer than 150 stereoscopic pictures.

Robert Mallet, the Irish engineer well known today for having given a name to seismology and for his contribution to different sectors of seismology (earthquake catalogues, seismic theory, seismological experimentation, etc.) also carried out between 10 and 28 February 1858 meticulous research in the area devastated by the earthquake of 16 December 1857. His method consisted of observations of the directions in which the objects and structures such as walls, etc., had fallen down. Mallet saw Bernoud's photographs in Naples, but he preferred to commission a new and more specific photographic reconnaissance that was conducted by another photographer, a man named Grellier. Mallet asked Grellier to reproduce exactly those subjects affected by the earthquake in which Mallet was most interested. In actual fact Grellier, unbeknown to Mallet, performed only a part of the commission given to him to carry out, and he sent Mallet, who was already back in London, 156 photographs (120 stereoscopic and 36 monoscopic: Figures 5.12 and 5.13), of which at least 57 were unsigned but were recognized by experts as being ones actually taken by Bernoud (see

Figure 5.12 Stereoscopic photograph by Alphonse Bernoud of the small village of Brienza (southern Italy) damaged by the earthquake on 16 December 1857 (from Mallet, 1862).

Figure 5.13 Photograph by Grellier dated 1857 of the town of Saponara (present-day Grumento Nova, southern Italy) (from Mallet, 1862).

Becchetti and Ferrari, 2004). Mallet used those photographs (today preserved at the Royal Society of London) to draft his weighty report on the 1857 earthquake. However, owing to technical constraints, only a subset of the complete set of photographs was reproduced in the form of engravings in his work (Mallet, 1862). It can be said that those first photographs assumed a special scientific relevance

Figure 5.14 A rare photograph, dated 1859, by Robert Macpherson (photograph no. 187 from the catalogue of Macpherson) of Norcia (central Italy) that was damaged by the earthquake on 22 August 1859.

in that they were commissioned by Mallet in order to document some of the effects of the earthquake on the inhabited and the natural environment.

In the first half of the nineteenth century photography had been used to document various disasters such as fires or wars, but never before 1857 had a photographer's lens pictured the effects of an earthquake with a scientific purpose. Bernoud's report along with Grellier's much less well-known one are, both for their quality and the very large number of photographs produced, a significant part of the history of international scientific photography. Another photographer working in Naples, Achille Mauri, known for having taken over Bernoud's studio and archives (he may have been aware of the 1857 earthquake report), carried out a photographic campaign on the effects of the earthquake at Casamicciola (the island of Ischia, southern Italy) on 28 July 1883 (I_0 = IX–X; M = 5.8).

Bernoud's report was the first in a long series of earthquake testimonies carried out by photographers who did not have a specific scientific aim. Among other early documenters of this kind was the Scottish photographer Robert Macpherson (1811–1872) who was based in Rome. Two views of Norcia (Umbria, central Italy) hit by the earthquake on 22 August 1859 (I_0 = IX; M = 5.8) were featured in his photographic catalogue in 1863. These photographs document in an exemplary way what Macpherson saw in that city a few days after the earthquake. His photographs document the work already under way of propping up falling walls that were damaged by the seismic shaking (Figure 5.14).

Figure 5.15 Collapses in Calabria (southern Italy) due to the earthquake of
16 November 1894 (from the Sieberg Archive of Jena).

The earthquake at Casamicciola on 28 July 1883 was photographed in an
exemplary way by Raffaello Ferretti, a photographer who worked in Naples.
Ferretti collected the photographs taken in an album laconically entitled *Casamicciola – Ruins – 1883*, documenting to a good technical level the devastation
caused by that earthquake. Of particular documentary value also is the series of
photographs of this same earthquake taken, as has already been said, by Achille
Mauri. The effects of other destructive earthquakes that took place in the world
in the 1880s and 1890s also were photographed, including for example, the
Charleston, South Carolina earthquake on 31 August 1886; the Chaman, India
earthquake on 20 December 1892; the Calabria earthquake on 16 November 1894
(Figure 5.15); and the earthquake in Assam, India on 12 June 1897.

Photographs often contain important information that was not the intended
subject as envisioned by the photographer but nevertheless is of interest to
scientists and engineers. For example, in portraying collapses and destruction
of structures, photographers have often left behind invaluable documentation
of building traditions and techniques, of the types of houses in which most

Figure 5.16 A church of Diano Castello (western Liguria, Italy) after the earthquake of 23 February 1887. Broken iron chains can be observed that had been previously inserted, perhaps to fix a detachment after a previous earthquake (from Ferrari, 1991).

people lived, and of the quality of the walls that were affected by a strong earthquake. These explicit visual records of the elements of the buildings affected by a seismic event preserve details that were seldom present in the written sources. For example, from 1887 there exists a photograph of the church of Diano Castello (Liguria, northern Italy), a village abandoned because of an earthquake that had extensively damaged it. In the photograph (Figure 5.16) one can see, apart from the great collapse, some iron chains that had been installed before the earthquake to strengthen the building. This evidence may indicate that this structure had already been shaken strongly and perhaps damaged by a previous seismic event. However, this is just conjecture since there are no written testimonies that describe an earlier earthquake that affected this church.

Photography proved to be an outstanding instrument of documentation for scientific research, and it was exploited by the earliest seismologists of the twentieth century. The archive of August Sieberg preserves, amongst others, two rather unusual photographs of the 1902 earthquake in Antigua, Guatemala (Figure 5.17). These photographs show the disastrous effects of this earthquake in a street of the capital. In another picture (Figure 5.18) some parts of a collapsed monument can be seen, and the caption points out that the damage of the 1773 earthquake evidently had not been removed and nor had the monument been

Ocos-Töcben (Guatemala) am 18.4. 1902

Figure 5.17 Collapses and ruins in Ocos (Guatemala) due to the earthquake of 1902 (from the Sieberg Archive, Jena, Germany).

restored. George R. Lawrence is probably owed the distinction of taking the first aerial photographs of the effects of an earthquake. On 5 May 1906 he flew a captive airship over the ruins of San Francisco, California caused by the earthquake on 18 April and photographed a panorama of the devastation due to the earthquake and subsequent fire (Figure 5.19).

One of the most significant seismic events that took place during the first decade of the twentieth century was the earthquake in the Straits of Messina (Italy) on 28 December 1908, which destroyed the two cities of Messina and Reggio Calabria as well as dozens of villages in the Calabria and Sicilian hinterlands ($I_0 = XI$; $M = 7.1$). The number of victims was almost 80 000, of whom probably only a few hundreds perished as a result of the great tsunami triggered by the earthquake. The seriousness of the destruction triggered a campaign that brought aid not only from throughout all of Italy but also from many other countries. A number of photographic campaigns were carried out by domestic as well as foreign photographers. One photographic campaign of special note was conducted by one of the leading photo-reporters

Zerstörung Antiquas (Guatemala) durch das Erdbeben 1773

Figure 5.18 A photograph from the beginning of the twentieth century from Antigua (Guatemala). Handwriting on the photo, perhaps by Sieberg, indicates that the damage caused by the earthquake of 1773 is visible (from the Sieberg Archive, Jena, Germany).

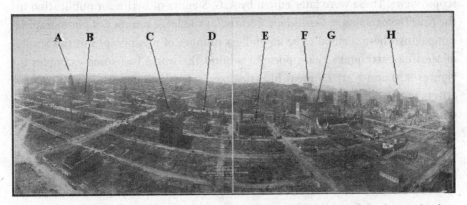

Figure 5.19 Aerial photographic taken by George R. Lawrence of the destruction in San Francisco, California due to the 1906 earthquake and fire. The photograph was taken looking north over (A) City Hall, (B) the Post Office, (C) California Casket Company, (D) Market Street, (E) the US Mint, (F) the Fairmont Hotel, (G) the ruins of the Emporium department store and (H) the Call Building (from the US Library of Congress).

Figure 5.20 Photograph of collapsed structures in Reggio Calabria (southern Italy) due to the earthquake of 28 December 1908.

of the day, Luca Comerio. Immediately following the event, he went to the area devastated by the earthquake and tsunami to document the damage. He took an enormous number of pictures, many of which were later published in many newspapers of the time. He also took a large number of very valuable stereoscopic views. These were later edited by A. G. Steglitz of Berlin for publication in the *Neue Photographische Gesellschaft*. Because of the large number of photographic campaigns, there seemed to be an endless number of photographs of the Straits of Messina earthquake that appeared around the world (see some examples in Figures 5.20 and 5.21). Of great historical and scientific interest are not just the effects of the earthquake and tsunami (well described by dozens of media, government and scientific sources), but also of the post-earthquake reconstruction phases. As can be seen in Figure 5.22 Reggio Calabria (which suffered degree X effects) was seemingly for several decades a city filled with temporary and makeshift wooden structures.

The photographic documentation of earthquakes that took place in the nineteenth century became highly popular to the point where they almost became a fashion. Postcards of various earthquakes often circulated with the phrase printed on them 'Best wishes from . . .', as though the earthquakes were a kind of sightseers' attraction. Nonetheless, they played the immediate role of arousing a new as well as widespread social awareness of seismic disasters and of the aid to the affected populations that needed to be given following an event. The

Figure 5.21 Photograph of ruins in Messina (Sicily) due to the earthquake and tsunami of 28 December 1908.

Figure 5.22 The new city of Reggio Calabria in its reconstruction phase after the earthquake on 28 December 1908.

number of postcards circulating in Europe after a strong earthquake between the end of the nineteenth century and the first half of the twentieth century is quite astonishing. For example, this kind of postcard became the object of collectors. However, from the standpoint of historical seismology, such testimonies are not always easy to use as sources of scientific information about the earthquakes and their effects. Indeed, the photographers' gaze, like that of the draughtsmen and artists before them, always selected the subject of the message. As is typical even today, what was photographed was always the most serious damage or the most dramatic views. This element, which could be said to be semiotic, should be borne in mind if one wants to make proper use of such kinds of sources.

5.3.2   Films

As regards film-making, the earliest earthquake films are certainly worth a mention. Probably the most important early films are the ones that captured the Calabria earthquake on 8 September 1905 and the effects of the San Francisco earthquake and subsequent fire on 18 April 1906. Two short documentaries were produced of the Calabria earthquake by the documentary film-makers Albertini and Santoni, while there were several films that preserved the effects of the San Francisco earthquake and fire. For the San Francisco earthquake the films primarily document the effects of the post-earthquake fire that destroyed so much of the city. From these films, scientists try to assess how much damage was caused by the earthquake shaking and how much destruction was due to the subsequent fire and fire-fighting efforts. In an attempt to stop the fire, many buildings in San Francisco were collapsed in dynamite explosions, and some of this was recorded on film. The state of a number of buildings after the earthquake but before dynamiting or burning can be seen in the still photographic record. It is this record that must be used to estimate the amount of damage in San Francisco that was caused by the earthquake itself. Comerio, the famous photographer of the Messina 1908 earthquake, was also an enterprising cinematographic documentary film-maker. He filmed at least three documentaries of both the destruction and the reconstruction of Messina. At least three other documentaries and foreign cinema newsreels also are known for the Messina earthquake. This was one of the cases where cinematography gave a new and spectacular viewpoint of the effects of a major seismic disaster.

During this same era, a seismic event in central Italy known as the Marsica or Avezzano earthquake on 13 January 1915 ($I_0 = XI; M = 7.0$) took place (Boschi et al., 1995; Guidoboni et al., 2007). It was well documented by photographs, although to a lesser extent than that of Messina. The photographic investigations may have been less extensive because Europe was at war and the conflict was already knocking on Italy's door. However, public opinion was again aroused by means of the plethora of postcards depicting the worst-hit locations. Cinematography captured the destructive effects of the 1915 Marsica earthquake on film more extensively than of the Messina earthquake. However, it must be pointed out that such a cinematographic record was more helpful in arousing compassion and aid for the earthquake victims than it was in producing a new effective and scientifically useful representation of the effects of that earthquake. The great emotional emphasis generated by this photographic genre may perhaps be due to the divide between official seismology and the mass media, which was becoming better organized and more sophisticated in its presentations even in those early decades of the twentieth century.

5.4   Unwritten sources

5.4.1   Oral sources

Some recent destructive earthquakes have occurred at unusual moments in that they immediately preceded or followed other natural or man-made destructive events (large-scale wars, epidemics and floods, in particular). In those cases where little or no information about an earthquake was formally reported, it may sometimes be possible to obtain or add to descriptive elements about the event and its effects by means of reliable direct interviews with eyewitnesses. Oral evidence can often be remarkably informative, but it must be collected and recorded using the appropriate methodology for acquiring evidence of this type, including a recorded interview and subsequent transcription. Recent seismological literature has itself paid some attention to the possibility of using oral evidence regarding the environmental effects of recent large earthquakes (Blumetti et al., 1988).

5.4.2   Ethnographic and anthropological sources

Seismotectonic phenomena have often found a place in myths and legends. Such topics and literature have largely been a subject of study for ethnographers or classical scholars, but classical seismology also has its own tradition in this field. Montessus de Ballore (1923), for instance, included earthquake-related myths and legends amongst his vast scientific interests. These give an indication of the complex relationships regarding earthquakes that have grown in time and in different cultures, either in an attempt to explain seismic events or simply in order to include them in the collective memory. Sources of this kind may contain matters of interest to historical seismology, especially regarding seismically active areas long characterised by the presence of archaic peasant cultures.

Anthropological and ethnographic elements must be the concern of specialists in order to avoid misunderstandings and mistakes. Indications of earthquakes found in sources of this kind reveal something of the complexity of the responses to earthquakes in archaic cultures in which oral culture characteristically plays a major role. The ways in which information about felt seismic effects is transmitted are not those of literal remembrances, and hence appropriate critical tools must be used to analyse this genre of information.

If the corpus of sources which in some way expresses the response of the collective imagination to seismotectonic phenomena is to be used for seismological purposes, it cannot be approached in a simplified way. A consistent ethnographic examination must be carried out, bringing to bear an understanding of the means of expression and narrative systems of archaic cultures. This examination must be carried out in such a way as to guide any interpretations and deductions that may be of use to present-day historical seismology (see also the discussion in Chapter 3 on legends as sources for historical seismology).

Tales and ancient stories from Cascadia preserve the memory of a giant seismic event (26 January 1700)

A recent as well as interesting case study using ethnographic and anthropological sources has been published by Ludwin *et al.* (2005) regarding the Cascadia subduction zone. In the 1990s, studies were performed on the primary plate-boundary fault of the Cascadia subduction zone, which separates the oceanic Juan de Fuca Plate from the continental North American Plate. This plate boundary lies parallel to the coast, from the middle of Vancouver Island to northern California, and the plate boundary comes to the surface some 80 km offshore. This subduction zone has been the subject of a number of geological studies, including that by Nelson *et al.* (1995). By combining botanical data with geomorphological elements and radiocarbon dating, Nelson *et al.* (1995) suggested that the most recent earthquake in the area must have occurred around 300 years ago. The exact time and date of this great event, i.e. 9.00 p.m. on 26 January 1700, emerged from an historical report of a tsunami in Japan that is contained in Japanese annals. There was no earthquake in Japan associated with the 1700 tsunami along the Japanese coast, and the causative earthquake was traced to a seismic event along the North American coast (Satake *et al.*, 1996; Atwater *et al.*, 2005).

American anthropologists and ethnologists have studied the languages and the traditions of the native populations of Cascadia for more than a century (specific bibliography in Ludwin *et al.*, 2005, pp. 147–8). A large collection of local stories were written or transcribed in the nineteenth century based on the oral accounts of witnesses, and these have been compiled as part of the ethnological research of the Pacific Northwest of North America. This ethnographic heritage (see for example Figure 5.23) has enabled a targeted analysis in which those texts, based mostly on piecemeal personal memories, were identified that contain mention of strong ground-shaking and of extensive sea flooding. Comparative studies of these ancient stories have highlighted different types of story.

Figure 5.23 Indian mask from mainland British Columbia (Canada), interpreted by anthropologists as the ritual face correlated to a fear of earthquakes and tsunamis (from Ludwin *et al.*, 2005, p. 143).

Some of the stories appear to contain elements of historical reality and others do not. Among the tales with apparently historical roots, some appear to contain elements that allow for the dating of the event, while others do not. The textual analysis can be linked to the spatial locations that are mentioned or implicit in the accounts. Taken together, a map of localities where the tales took place can be made. Using this information, the geographical contours of the area apparently affected by this major earthquake and tsunami was constructed. This result was then corroborated by archaeological analyses that have investigated the ancient native villages of the British Columbia, Washington and Oregon coasts. In a number of studies (Cole *et al.*, 1996; Minor and Grant, 1996; Hutchinson and McMillan, 1997; Losey, 2002) it was observed that a number of villages subsided, were flooded by tsunamis and were abandoned, and this highlights how the major tectonic earthquake and tsunami in 1700 affected the native inhabitants of this region (see also Chapter 12).

The inhabitants of Staffora Valley (Piedmont, northern Italy) re-outline the effects of two shocks, 14 and 29 June 1945, in their interviews

It is not unusual for recent earthquakes to be as hard to learn about and evaluate as an earthquake that occurred many centuries ago. No matter when an earthquake took place, the social and cultural conditions of the time determine whether the event was recorded in some form of written memoir. The case study presented here concerns two shocks which occurred on 14 and 29 June 1945 in a valley in Piedmont, Italy. The effects of the earthquakes have been well recognized on the grounds of numerous interviews with the inhabitants of the villages in the valley. In those days immediately following the Second World War, many Italian national newspapers had temporarily stopped being published, and only a few local newspapers reported some sporadic news of the earthquakes. Unfortunately, these newpaper reports do not record the full extent of the earthquake effects. Furthermore, the Italian national seismic services were suspended at the time, and since this was not a catastrophic earthquake, no bulletin was issued concerning this event. Thus, oral sources must be used in order to outline a sufficiently broad picture of the effects. Between February and March 1984 dozens of residents, parish priests, mayors and soldiers were interviewed, who at the time of the earthquake were aged between 20 and 30 years. The interviews were all diligently recorded, dated, underwritten by the interviewee and by the interviewer. Although now they are written down, the recollections that were collected are oral sources in every sense. At present, they are preserved in the library–archive of Storia Geofisica Ambiente in Bologna.

The interviewees had all been direct witnesses to the event, but they initially were reticent to give their accounts. At first the interviewers were confused by the reluctance of the witnesses, but they later realized that the interviewees did not want to specify exactly where they were on the days of the earthquake. Partisan freedom battles were still in progress in those days, and whether a person was in the village or in the mountains in those days had an important personal and social significance because it indicated whether or not the person was involved in the liberation of the Staffora Valley. Once the interviewers found ways to overcome this obstacle, the interviews became accurate and useful. One curious case is that of a parish priest, who was able to indicate with great accuracy where a crack had appeared in his church. As that crack had never been repaired, it was still visible in 1984. On the grounds of these oral sources a collection of local photographs was put together, preserved by the local inhabitants, which indirectly confirmed the importance of the effects that were caused by these events. In the *Catalogue of Strong Earthquakes in Italy* (Guidoboni et al., 2007) these earthquakes were evaluated as follows: the shock on 14 June 1945 was of $I_0 = $ VI–VII and $M = 4.9$; the one on 29 June 1945 was of $I_0 = $ VII–VIII and $M = 5.2$. The epicentres of both earthquakes were in the Staffora Valley.

6

Potential problems in historical records

6.1 Problems inherent in the historical sources

6.1.1 Potential value and availability of sources

In any time period or geographical area, the potential usefulness of sources of information is related to and derives from the different selections that have been made in the course of history, that is to say, with the actual *availability* of sources. Every country has an invaluable heritage of institutional and private sources, whose age and number vary according to that country's own particular history (see Chapter 3). Yet the institutional sources of interest to historical seismology research may not actually be preserved within the country whose seismic events are being studied. Sources may be preserved in the archives of other countries, because, as will be seen later on, the country's government may in some periods have been dominated by powers situated beyond today's national boundaries. Other possibilities are that the government may have been situated within the realm of a vast empire or it may have been influenced by military, political and colonial situations of the time.

6.1.2 Destruction and dispersion of sources

In every time period and in every geographical region, a process of involuntary selection of sources has been brought about by events that have led to the destruction or dispersion of historical records. Hence, the survival of evidence has been determined by the fortuitous effects of wars, invasions, fires, sudden disturbances in political and administrative organization, and so on. In this regard, each country has its own *source history*. Contrary to what might be expected, documents from periods close to the time of an event are not always available because short-sighted elimination of records and widespread archival chaos often make it difficult or impossible to locate and consult the documents produced by the administrative branches of the states of the time. Moreover, when a country was characterized by the simultaneous presence of different foreign populations it may well be the case that historical sources for earthquakes are still preserved in the place of origin of the foreigners who dealt with the affected territory. This situation is generally due to the presence of merchant colonies (in the medieval period), to military or political occupations, or to colonial situations in a strict sense.

If the above is true for the archival sources, i.e. those produced within the scope of the administration of a territory or a country, it is even more true for the sources that preserve the memories of individuals. For this reason annals, chronicles and diaries written in a given country can today be found in the libraries of other countries, as a result of complex events, not always known, due to the history of the people who left behind these written works as well as to the history of the individual codices and their owners. It is therefore always advisable to have direct contact with the historians who deal with these specific aspects of the historical record and who know these issues profoundly.

Why is the 1570 Ferrara (Italy) earthquake remembered in a manuscript in Zurich, Switzerland?

In the University Library of Zurich (Switzerland), there is a curious sixteenth-century manuscript noted for the first time in the field of historical seismology by Jean Vogt (personal communication). This codex contains, besides a text written in ancient German, a beautiful watercolour sketch (Figure 6.1) which portrays a rather shabby and crumbling castle and some houses pulled asunder, with their corners collapsed. At the top in the middle of the sketch there is a dragon with a long tail and a rather menacing-looking expression blowing a gust of wind. At the bottom, some small human figures seem to be following a small procession with gestures of devotion and desperation. What is this all about?

Some writing in Gothic German in a different handwriting from the one of the written text is inserted in the sketch. It identifies, without a shadow of doubt, this castle as being that of Ferrara, then one of the most important courts in Europe. The drawing is thus closely correlated to the text, which narrates in letter form, rather plainly and concisely, some aspects of the earthquake that hit Ferrara on 17 November 1570. The author might have been a Swiss soldier (perhaps an officer), one of the 150 or so mercenaries who were in the city in the pay of the Duke Alfonso II d'Este (in the city there was a building known as 'home of the German guards'). An inquisitive or perhaps even frightened witness to that strong earthquake, which damaged Ferrara extensively, the anonymous

Figure 6.1 The earthquake on 17 November 1570 in Ferrara (Italy) (from Bächtold, 1993).

mercenary thus wrote home and completed the news with a drawing, perhaps sketched out by a companion of his. It is possible that the text, which today is bound in this manuscript with other sheets in the same handwriting, was put together many years later, and the letter was sewn together with the manuscript to the memory of that event, at which our unknown soldier was present.

This is one of the many cases in which it appears to be worthwhile consulting the manuscripts of many libraries even far from the area under examination or, at any rate, being in contact with researchers from other countries. Cases of this kind show that studies based exclusively on sources retrievable in the area local to the earthquake may preclude an examination of the full extent of the basic historical data. In this case, although there are very many sources that described the earthquake of Ferrara in 1570, the codex of Zurich contains the only drawing known today of that event.

How is the historical seismicity of Algeria reconstructed?

For many ex-colonial countries the situation of the documentary stock and the sources in general is very complicated today. Destructions, dispersions and dislocations of records make historical research often very difficult. This obviously hampers research into the seismicity of the area. A case of this kind is Algeria, a country in the north of Africa with a relatively high seismicity. The last strong earthquake occurred on 21 May 2003, at 18:44 hours UT (20:44

local time) with two shocks each of magnitude 6.8 localized at 36.9° N, 3.6° E, at a superficial depth (less than 15 km). The event lasted over 20 seconds and claimed more than 2000 victims (along with over 10 000 injured). The epicentre of this earthquake was located at sea, about 7 km from the city of Boumerdes, which was heavily damaged. Many cities and villages to the east of Algiers (Zemmouri, Reghaia, Ain Taya, Dellys, Bordj Menaiel and many others) were heavily damaged, with numerous lives lost. The earthquake caused a tsunami that mainly affected the Spanish coast and the Balearic Islands, where more than 300 boats were damaged. The city of Algiers sustained little damage, but the local sailors tell of an astonishing drawing back of the sea that lasted several seconds. The focal mechanism of the earthquake calculated by various agencies (Istituto Nazionale di Geofisica e Vulcanologia, Harvard and Eidgenössische Technische Hochschule Zurich) corresponds to an almost pure reverse fault mechanism with a NE–SW strike direction. This is in agreement with the active tectonics of the region and with the location of active reverse faults along the coast.

A reliable assessment of the seismic hazard of Algiers and the other populous cities of Algeria cannot be based on historical data. Indeed, for Algeria some excellent studies are available, but the characteristics of the earthquake catalogues in these studies are quite heterogeneous. A summary of historical data, whose most ancient earthquake is that of Algiers in 1365, is in Ambraseys and Vogt (1988), who have taken up, with some authority, an earlier but interrupted tradition of historical studies. For other studies concerning the seismicity of the eighteenth and nineteenth centuries see also Vogt and Ambraseys (1991, 1992). Perrey (1848) was a pioneer in studies of the seismicity of the whole of the Maghreb, as were Chesneau (1892) and Montessus de Ballore (1892, 1906). In the late twentieth century, there were also important studies in Algeria itself. There is an excellent catalogue containing data for the twentieth century (Benouar, 1994), based on newspaper information. Furthermore, a catalogue compiled by the *Centre de Recherche en Astronomie Astrophysique et Géophysique* (CRAAG, 1994) provides a systematic and critical treatment of the literature data. More recently, Harbi *et al.* (2003) have provided a critical review of the historical data, but there are still some large chronological gaps. The historical data examined only start in the fourteenth century. The construction of a reliable historical earthquake list is problematic for both the ancient and the medieval periods, as well as for the centuries just preceding the twentieth century. Benouar (2004) presents a brief survey of previous studies regarding the largest historical earthquakes in Algeria.

Bridging this kind of information gap is very difficult and demands specific projects because for the past centuries the institutional sources are only partially found in Algerian territory. Going back in time one finds that for the French colonial period (1830–1962) there is a lack of documentation remaining in Algeria, since records were transported to France (Aix-en-Provence, Archives d'Outremer) during the crisis of the French occupation (1956–1962). The current National Archives of Algeria indeed preserve only a part of the archives produced in the period of Turkish domination from the sixteenth century until 1830 (period of the beginning of the colonial French occupation). The Ottoman archives preserved today in Algiers include state acts, financial and fiscal documents, archives of the tribunals of the *Sahiria*, and files with correspondence with the 'Sublime Porte' (the archives of the central Turkish government, preserved in Istanbul). However, their consistency with the history of the Algerian territory is modest, and these documents need to be integrated with the ones preserved at the 'Sublime Porte'. Other archives could offer supplementary materials. For example, for the seventeenth and eighteenth centuries the archives of the *Propaganda Fide* could be investigated. Other possible sources of information for this time period that could be examined are the archives of the European religious orders (such as Spanish and French Franciscans), correspondence preserved in the Foreign Office of London, or the Archives de Guerre of Paris (Château de Vincennes).

Even for the classical Arab period (that is, from the seventh century to the fifteenth century) no administrative documentation has survived. The information on the earthquakes and tsunamis of this time period can only be drawn from the chronicles and local histories of Arab writers. But the process of publishing modern transcriptions and translations of these texts is very slow; in the great libraries of the Maghreb countries there are tens of thousands of preserved manuscripts. Historical seismologists cannot replace philologists in attacking the tremendous amount of work that still needs to be done. They can, however, establish regular contacts with experts in codicology and textual analysis and develop some joint projects that satisfy the needs of both groups of researchers.

For the time periods preceding the Arab invasion, that is from late antiquity until the Byzantine domination (fourth to sixth centuries), there are no administrative sources. However, the religious literature preserves a priceless memory of a strong earthquake at Sitifis (present-day Sétif) in AD 419, in a sermon of Saint Augustine (Augustine, *Sermons*, 19.6) (reported in Lepelley, 1984). For the ancient period, when Algeria belonged to the prosperous Roman provinces of the imperial period, two inscriptions are known that mention an earthquake in AD 267 at Ad Maiores (Besseriani, a few kilometres south of Négrine). It is also possible that another earthquake affected Algeria, datable to the second century AD and which struck an area on the border with Tunisia (an inscription was found at Aunobaris, about 10 km south of Teboursouk). This complex documentary situation calls for a multidisciplinary approach and a close collaboration between historians, textual scholars, epigraphists and archaeologists on the one hand, and seismologists and palaeoseismologists on the other.

6.1.3 Lost sources only preserved in late traditions

Historical events that affected a country and its archival heritage may have led to the loss of original texts. However, these original texts may have been preserved in later traditions through copies of the original sources, even when the original sources may themselves have long since disappeared.

This situation can cause problems in evaluating historical information, as it is sometimes impossible to know what value to attribute to such late-tradition texts. There are, however, criteria of historiographic and textual analysis that should be adhered to as demanded by the individual cases. It should also be kept in mind that before the invention of printing, and thus the chance to more easily reproduce and disseminate a work, a handwritten text could only circulate in a fairly limited way and within special circles, with copying being carried out within specialist environments. Interpolations are thus often the work of later periods, and the editors of the critical editions as well as textual experts can recognize the various hands involved. Recent critical editions almost always enable the user to assess the authority of a later version of an earlier written source. In that way, even late texts can be used and can retain a certain degree of reliability.

John of Nikiu: memories of earthquakes from the Coptic to the Ethiopian language

John of Nikiu (seventh century AD) was an Egyptian bishop who wrote a universal chronicle covering the period from Adam to the immediate aftermath of the Arab conquest of Egypt (684–685). His work is thought to have been originally written in Coptic, but survives only in a rather flawed Ethiopic translation of 1601, which relies upon an intermediate Arabic text, now lost. The earlier parts of the Chronicle make use of Malalas, a Byzantine chronicler of the sixth century, with a few additions from local Egyptian sources. Notwithstanding these textual problems and the chronological distance, John of Nikiu in the Ethiopian translation is an indirect source for some earthquakes of the Mediterranean area. He mentions, together with other authors, the AD 362 and 368 earthquakes of Nicaea (Iznik, Turkey) and of Antioch (Antakya, Turkey) in 526 and 588.

His text is laconic, but contains a significant echo of the fame of these great territories, mentioned in lost sources. For example, for the territory of Antioch in October 588, John of Nikiu records:

> *Similarly in the reign of this Maurice [582–602], the city of Antioch suffered a great earthquake and was destroyed seven times. And many lands in the east were destroyed, as well as islands, and there were countless earthquake victims.*

For earthquakes so far back in time, even a passing mention is evidence that is nonetheless vitally important for the seismic history of the sites.

6.1.4 *Only one source available*

One of the basic criteria of historical seismology is, as has been seen, that of defining an earthquake of the past by using several sources that are hopefully of different types, so as to be able to evaluate the effects described from many points of view. This criterion may provide specific weight, so to speak, to the qualitative data because it allows not only for a comparison but also for a sort of calibration of the set of historical records. There are, however, cases of earthquakes attested to by just one source. What should be done in such cases? Obviously, every piece of information is vital in historical seismology, and no one can afford to ignore seismic events because they are attested to in just one source. It is necessary to try to understand why this situation exists; parallel research may highlight other sources or confirm the information gap. It should, however, be remembered that it is the *authoritativeness* of the source that allows one to accept or reject its information for a location or region, not the fact that the source itself is unique.

In textual studies it is not unusual for important texts from antiquity or the Middle Ages to come down to us today through a single text copied several times over. Naturally, such codices are invaluable, although their contents cannot be compared with other contemporary sources that may have been lost. The information concerning seismic events deriving from a single text should always be analysed on a case-by-case basis. The search for more sources should always remain a priority, although for the ancient and early medieval world a single source may represent the limit to the documentation that can be uncovered by even the most careful research.

For very well-documented eras it should not be necessary to stop at just one available source. However, above all for more remote centuries or for areas scarcely or sporadically documented, there are many cases of earthquakes attested to by only one source, without this fact undermining the data themselves. As a matter of fact, contributions of this kind must be considered irreplaceable items to be handled very carefully.

6.1.5 *Negative sources: the meaning of silence*

In the past, research into historical seismology was almost always oriented toward the identification of positive facts about the seismic history of a region, and it almost invariably neglected any analysis of the contextual territorial history. A well-argued assessment of the state of knowledge of territorial history as well as of the sources of that history makes it possible to draw more accurate conclusions as to the state of knowledge of seismic events. One should not ignore the fact that negative sources may be able to provide seismological information. A gap in knowledge resulting from a lack of information may have different meanings, depending on the types of available sources:

> *Chronicle sources*: for a given period and area, when contemporary, authoritative and reliable memorial sources, with no chronological gaps, do not record the effects of a given seismic event, they can be used as negative sources. The silence of the sources of direct information may confirm particular aspects of seismic propagation. This datum *ex silentio* can be reasonably used to indicate that a given earthquake had not been felt in that place.
>
> *Public administrative archive sources*: the significance of silence about earthquakes is different in the institutional sources. The existence of negative public sources can have different meanings: (1) the earthquake did not cause any damage; (2) damage caused by the earthquake was slight or such that public authority intervention was not felt to be required; (3) the damage may have taken place, but for whatever reason the public administration may not have intervened to repair the private building stock.

Historical research must look to determine what pertinent administrative practices actually existed and what did not. At times, the public administrations only dealt with the damage to infrastructure such as bridges, roads and military facilities. In these cases it is possible to determine in the specific archival series only whether the public buildings were damaged or not.

6.2   Problems inherent in the use of historical sources

6.2.1   Area periodization and historical source production

In order to make a proper assessment of the potential that relevant information exists for a given region, the history of the region must be taken as a starting point. Once this knowledge has been acquired by historians, it can be used to carry out a *periodization* of the area in question, in order to identify available sources capable of supplying direct or indirect information on the effects of seismic events. This long-term view makes it possible to carry out a general historical mapping of the area of concern to the historical seismology researcher. However, the researcher must be careful not to limit the locations

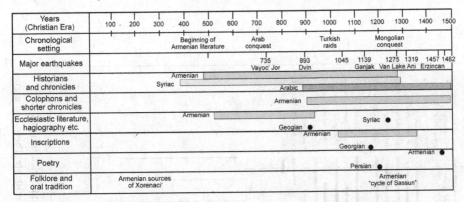

Figure 6.2 Typology of the historical sources concerning earthquakes in Armenia (Guidoboni and Traina, 1995).

where relevant records are sought only to the sites affected by the earthquakes. Generally speaking, a whole historical region or the whole country in which the region is situated should be taken as the appropriate context for evaluation, because, as a result of the circulation of information, it is possible to discover descriptive data quite a long way from the actual location of a seismic event.

The cases of Armenia and Italy

The importance of periodization for historical seismological purposes is well illustrated by the case of the Armenian region. The sources for this area were in fact produced by different cultural and linguistic traditions, reflecting the overlap of civilizations that affected the history of the Armenian region. Figure 6.2 (Guidoboni and Traina, 1995) shows the typology of the historical sources for Armenia relative to their time of production. This complex relation between time and available sources for the Armenian region led seismologists in the recent past to commit various errors of dating and locating seismic events that occurred in Armenia. This is easily understood if one considers that to have a fairly correct picture of the seismicity of Armenia one must examine and compare local and foreign sources. Armenian seismicity links two important earthquake-prone areas, the Mediterranean and the Iranian plateau, which have been traditionally considered culturally two worlds apart. To study Armenian earthquakes in a thorough and complete way one needs a broad cultural horizon (Babayan, 2006).

As for Italy, Figure 6.3 displays the particularly rich documentary availability of this ancient country. This is due to a wide and differentiated production of sources. The Italian earthquake catalogue is almost complete starting from the seventeenth century, for since that time in the whole Italian territory there

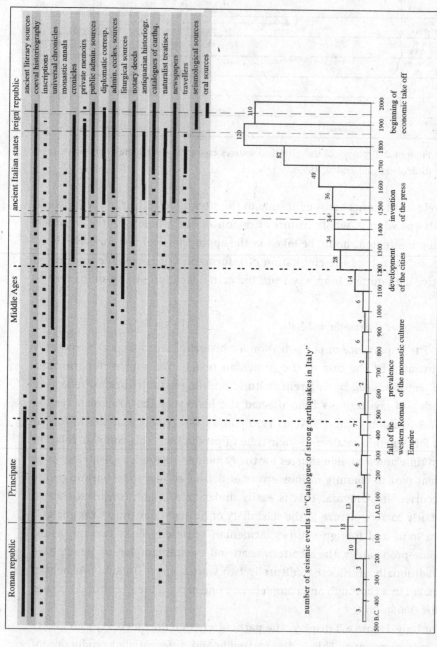

Figure 6.3 Graphical representation of the availability of historical written sources in Italy by kind and period from the fifth century BC to the twentieth century AD. The graph shows the density of different types of sources available. There was a great increase in new sources in the modern age, starting in the sixteenth century (from Boschi et al., 1995).

were produced many different and mutually independent sources, which taken together provide seismic data for a dense network of localities, even ones of very small demographic size.

The historical periodization of a region can bring to the forefront gaps in the continuity of political or administrative services that have punctuated the region's societal evolution. The ravages of war, invasions and occupations, together with significant and drastic changes of political, military or administrative regimes, the moving of centres of power from one place to another, and frontier changes, are all factors that would have directly affected the *production* of written sources.

6.2.2   Filters and selections

In every period, human *filters* have affected the selection of sources, thereby slightly or substantially reducing the potential available information for a region, or affecting the structure of the available records in particular ways. All texts of the past (including the recent past) have been (and still are) subject to a complex process of *selection*, involving both voluntary and involuntary elements. As far as earthquakes are concerned, one can identify various levels of selection responsible for filtering the transmission of historical records and thereby exerting a direct influence on their quality and quantity. Apart from the cases of the selection of sources due to the loss or destruction of the historical heritage of a country (a problem already discussed in Chapter 3) here we examine the main elements that serve to select the surviving information on the earthquakes and tsunamis of the past:

(1) the choices and motives of source authors, which affects the content and means of expression used in sources and is governed by: (a) cultural contexts; (b) subjective situations; and (c) specific institutional attitudes;

(2) the central or marginal political status of the affected area;

(3) another natural event distinct from the earthquake (hurricane, whirlwind or landslide), or an anthropic event (such as the explosion of a powder keg or something similar)

(4) housing characteristics;

(5) the concomitant occurrence of other natural or man-made destructive events that may obscure the seismic effects in the minds of the observers.

As is explained in the paragraphs that follow, these aspects are peculiar to historical seismology and contribute to the quality and the reliability of all interpretations of the data.

6.2.3   *Choices and motives of source authors*

For the authors of historical sources, the choices of what information to write down and the motives for recording that information depends not only on the characteristics of the writers' cultural contexts but also on subjective situations of the writers or on their particular points of view – personal or institutional. As for cultural contexts, it must first of all be said that this is a level of selection that goes back to the period when the evidence was written. It controls the mentality of the time and the cognitive frameworks within which the contemporaries expressed themselves. It also limits to particular cognitive contexts the perception of actual natural events. In other words, it is a selection related to the very conception of nature and naturalist and/or scientific knowledge. Within this relationship between written sources and actual events, the selection of those effects of natural events that were recorded depends on the body of knowledge and the opinions that make up the accepted and widespread cognitive framework of the area and the time period. In addition, it involves not only the interpretation of observed natural phenomena, but also the means of expression used to convey that interpretation. One of the basic principles of modern epistemology, and which also indicates the limits of knowledge, is that 'everyone sees what they know'. This consideration can help to keep attention alert so as not to misconstrue the written texts from cultural contexts that are far in the past.

The figurative use of the term 'earthquake' in Greek papyri from Egypt

Perhaps the following has happened to some seismologists. On opening up the newspaper in the morning, he or she is startled to read the first word of a headline written in block capitals: 'Earthquake at . . . the Stock Exchange' or 'in financial circles'. The figurative use of the word 'earthquake' to refer to rapid and unexpected changes is not a feature of modern language alone; it is well attested to also in the ancient world. In the past just as today, this term was used as a rhetorical device. It is not always easy to recognize its figurative use, especially if there is just one source available. Even experienced scholars have fallen into this trap and created earthquakes that had never actually occurred in the first place. This is the case of the Egyptian 'earthquakes' of 184 and 95 BC, the most ancient earthquakes that one finds in the Egyptian catalogue (Ambraseys *et al.*, 1994). However, there were no seismic events in these years in Egypt, since the word 'earthquake' was used in the sources in a figurative sense in these cases. A papyrologist (Mazza, 1998) has pointed this out. She discovered the figurative sense of the term 'earthquake' in those sources by means of a comparative analysis. A critical review of the text of the Greek papyri from which the data concerning the above two earthquakes were taken has shown that the correct meaning of the word 'earthquake' in the two contexts is really 'blackmail' and 'extortion', respectively.

The 1259 Trapani (Sicily) 'earthquake' in contemporary monastic annals

The entire tradition of historical seismological studies (Perrey, 1848; Mercalli, 1883; Baratta, 1901) has passed down news of an earthquake at Trapani (northwestern Sicily) in 1259. This record was based on an authoritative source of the time, i.e. the annals written in the monastery of Cava (*Annales Cavenses*), southern Italy. Starting in the eighteenth century, the news about this earthquake was considered to be very reliable. On the basis of the information in the Cava annals a damaging earthquake in Trapani was included in the Italian parametric earthquake catalogues in use until the 1990s. An authoritative European catalogue (Alexandre, 1990) corroborated these data with other coeval records. Apparent confirmation of this earthquake in 1259 was contained in other European monastic annals, written in the monasteries of Salzburg (Austria), Schaeftlarn (Freising, Germany) and Wittewierum (Münster, Germany). The Alexandre (1990) catalogue reports a description of the 1259 event that is broader than the other texts. This report reads:

> The earthquake cracked Mount Erice and, as they fell, the rocks damaged some houses and blocked the road, so that the monks could no longer enter the city (Trapani)

This was clearly a landslide rather than an earthquake. What was the reason for this misunderstanding? The many authors who reported this 'earthquake', unless they had copied from each other, have all read the medieval texts with a modern mentality. Indeed, for us today the word 'earthquake' has a seismological meaning. On the other hand, in Latin and in the Romance languages, the term *terrae motus* meant every conceivable 'movement of the Earth', and for different causes. It thus covered a variety of phenomena, including some of a meteorological and geomorphological nature (cyclones, landslides, etc.). Since the supposed damaged area is a territory characterized by low seismicity, a strong earthquake in 1259 would represent a very interesting element for the assessment of the seismic hazard of the Trapani area. Indeed, before being definitively deleted from the Italian catalogue, this 'earthquake' was used to draw an ad-hoc seismogenetic area on seismic hazard maps of Italy. An orange-coloured strip of locally higher earthquake hazard appears in northwestern Sicily on the seismic hazard charts of Italy produced until the end of the 1990s. This case shows that the reliability of historical records is strongly connected to their correct interpretation. It is not the historical sources themselves that are limited or unreliable, but rather mistakes arise from the interpreters who do not properly analyse the written sources.

From the preceding example, it is seen that the word 'earthquake' could also have indicated other natural events that implied a 'movement' of the Earth, such as landslides. But there are also cases in which the coeval witnesses themselves mistook landslides for earthquakes. In these cases, the wrongful interpretation of the phenomenon is codified in the source text by the original authors. For the purposes of earthquake catalogues nothing changes if the misinterpretation arises from the original author of the source rather than from a later reader; there is still a false earthquake that must be recognized and perhaps removed from existing earthquake catalogues. However, it must be noted that in some cases the error that gives rise to a false earthquake is not generated by filters where the language of the original author has been badly interpreted but rather by the cultural filters of the authors of the original source. Thus it may be that the sources themselves contain a mistaken interpretation of the phenomena, even when the phenomena were observed directly. The case that follows shows another way in which sources can be misunderstood when cultural filters are at work.

A truly diabolical 'earthquake': Issime (northern Italy), 1600

There is an 'earthquake' dated September 1600 to which seismologists had given a location at Issime (Valle d'Aosta, northern Italy, $I_0 = IX$; $M = 5.7$). Such a destructive 'earthquake' was the strongest event known in an area otherwise characterized by rather low seismicity. This posed a question to the seismologists: is this an event that has a long return period?

The basic source of the information about this event was a manuscript, preserved at the State Archives of Turin, concerning a devil being put to trial (Fabretti and Vayra, 1891), cited by Baratta (1901). The trial, as was the ecclesiastical custom in those days, was performed by an exorcist priest, called by the inhabitants of the valley because strange phenomena were taking place. What were the phenomena? From the text it can be inferred that a landslide movement was ongoing near Issime (1482 metres a.s.l.). Superficial layers of the earth were slipping. Underground noises could be heard due to the gurgling of underground water, and the great landslide produced tremors and cracks in the dwellings of the nearby villages. But in the cultural picture of the inhabitants of an isolated mountain village and an exorcist priest, everything that was underground was connected with the devil. Thus, the set of observed phenomena appeared like a diabolical earthquake, that is, due to the insidious presence of a demon, more precisely identified by the demon Astaroth. This notorious demon was tried by the priest (in the text of the trial the demon speaks in proper ecclesiastical Latin).

The judicial procedure lasted for several weeks, presumably until the landslide exhausted its capacity for movement. Astaroth was harshly sentenced and had to leave the village, and in exchange the inhabitants had to promise to build a chapel for votive offerings (Guidoboni, 1985).

This story may well arouse a certain amount of interest from the point of view of cultural anthropology and the history of the thinking of the time, but it is above all a case study for historical seismology because its proper interpretation has led to the elimination of a destructive earthquake from historical earthquake catalogues. As regards the devil, the significance of his presence can be traced back to a very ancient pre-Christian tradition of chthonic (i.e. underground) divinities. It is interesting to point out that there is also a landslide-devil connection that is attested to in medieval literature. For example, in the *Fioretti* of St Francis a great landslide is described on Mount Subiaco (central Italy) in terms of a diabolical presence.

Moving on to the analysis of subjective situations, it should be pointed out that this religious filter is influenced by the mentality and culture of the *individual* writers, in that they themselves may select specific aspects of the phenomenon they are describing for subjective and/or contingent reasons. For example, the natural philosophers and naturalists of the premodern period emphasized certain descriptive aspects of earthquakes, because these were believed to be most closely related to the theories the natural philosophers of the time were upholding. On the other hand, for the ecclesiastical writers an earthquake account was often a rhetorical or anagogic device, involving symbolic language with the effects frequently exaggerated. When these elements found their way into rhetoric and preaching, they also led to the use of some specific images of earthquakes.

From about the fifth to the tenth century AD the writers of surviving sources belonged almost exclusively to ecclesiastic institutions and monasteries which predominated over the written culture. Hence, medieval sources strongly reflect the religious conceptions of natural disasters – conceptions which tended to generalize seismic effects within their social context, omitting details thought to be superfluous in comparison with the importance of the sign which the seismic event represented. Then there was the influence of a level of intellectual indifference to events like earthquakes. The fashionable *tedium vitae* often made medieval clerics and monks pay little attention to human disasters. One therefore must arrive at the proper historical image of a region within the full scope of its real culture, going beyond the partial image which a single type of source can provide.

Even public and institutional sources (i.e. archival sources), although they do not have a subjective perspective, have equally powerful selective filters concerning the events they reference. In other words, institutional sources reflect the *modus operandi* of the social and cultural administrative systems in which they were produced. If public control structures for urban and rural property were lacking and if there was no administrative system capable of organizing reconstruction work, this may have a considerable effect on the types of source available concerning a historical earthquake. In such cases, gaps need to be filled from private documentary sources (notary deeds, wills, memoirs, letters, etc.). In many areas and for long periods, public authorities were primarily (if not solely) interested in restoring property for military or commercial reasons (fortresses, ports and bridges), or else in order to bolster the splendour and prestige of their own image. This is another point of view that may help one to assess the destructive effect of an earthquake. But the source's bias must always be kept in mind, for it is not always the case that failure to mention damage to minor private buildings is a sign that they did not suffer damage. This is an aspect of seismic effect assessment that needs to be treated with great care, because *bias* in sources of information does not affect the *degree of authority* of their content. It is up to the researcher to identify the point of view and establish the extent of its influence. For example, attributions of intensity levels based solely on institutional sources may involve the omission of other more serious damage that simply was not of concern to the institutional personnel and therefore not set down in writing.

In the past many regions in Europe and the Mediterranean area were governed by feudal powers whose authority had a private jurisdictional basis. From a source point of view, fiefs, enfeoffments and various other types of right exercised upon the resources of a region were political islands subject to an administration and a jurisdiction that were separate from those of the public domain. And for the most part, these so-called islands persisted for a long time (until the mid nineteenth century in Sicily, for example). Only the territorial history of each individual country can shed light on these administrative islands. The private archives of the great feudal families are not always available. When this results in large documentary gaps, it is necessary to use annalist, literary and naturalist sources in an attempt to find other types of independent testimony concerning earthquakes that affected these areas.

6.2.4 Historical contexts

A good knowledge of the historical territory that is to be examined is indispensable in order to place in proper context the data selected from the sources in that territory. Prior to any seismological interpretation it is indeed worthwhile correlating the selected texts with the cultural, social and economic situation of the territory being examined. In this way it is possible to detect any new sources of information and to better understand their significance.

Central or marginal status of the affected area

In many parts of the Mediterranean, many cities have enjoyed some amount of hegemony ever since ancient times, for they have been the seats of the cultural and political power to which their local countryside was subjected. In other geographical regions, the situation was different. Until very recent times, for example, urbanization was a very limited or even almost non-existent phenomenon in Armenia. In northern Italy, too, the great agricultural estates controlled by the monasteries became cultural and economic centres of a non-urban type during the early Middle Ages (seventh to tenth centuries). In the same centuries, in the southern and eastern Mediterranean large flourishing cities were developing; Cairo, Rabat, Damascus and Antioch were great, populous centres of culture and trade, each a metropolis of its day.

For historical seismology one problem that must always be faced is that of always having a rather clear territorial reference framework. It almost goes without saying that the historical sources are nearly always concentrated in the urban centres. Often, therefore, the sources that recall an earthquake cite the villages and towns affected in a very generic way, those smaller population centres being elements that did not much interest the urban-dwellers of the times. This urban-centric mindset is also true today. Indeed, people are generally more in awe of the spatially concentrated damage undergone by a city than by the sum of the sometimes much greater damage suffered by disparate villages. Geographers explain this attitude, very widespread in Western culture, in the following way: the city is such a complex and significant network of interactions that the breakdown of such a complex structure is felt emotionally in a very strong way. However, the historical seismologist is left with the problem of finding a way to learn about the full area affected by a seismic event in spite of the urban-centric attitudes of the authors of most sources.

One can proceed to attack this problem by searching for congruencies and clues. By the term *congruence* we mean the search for textual concordances contained in the sources that can be traced back to a homogeneous set of indicators about events and their effects from across the affected region. These indicators should not be in contradiction with the geographical proximity/distance relationship in which earthquakes have their strongest effects near their epicenters, with gradually lesser effects at greater distances from the causative faults. To this logical procedure *inference* can be added, albeit with due caution, in the meaning described by Musson (1998). He argues that, for example, if a location lies between two nearby damaged locations, then this third location can be *inferred* to have been damaged as well. In our opinion, however, one should make sure that the sites where damage is being inferred are to be found in homogeneous morphological and geological situations. It is important that researchers always preserve a record of their reasoning when making such damage assessments.

One way to look for congruences is through the use of appropriate symbols or annotations in historical seismology research summaries (e.g. the use of specific symbols in the cartography that represents the damage area, or the adding of notes and clarifications in recording data, in databases, catalogues and so on). It is important to keep track of all assessment processes since new historical sources might be retrieved in the future that either corroborate or undermine such assessments. The perceived importance in human terms of the area affected by an earthquake, and the hierarchy of values which town–country relationships acquired with the passing of time, inevitably influenced the written sources. This hierarchy of values was evident to contemporaries as well, and it obviously influenced in a cultural sense their reception of seismic phenomena, thus leaving its mark on the actual transmission of earthquake damage information.

Anyone studying the earthquakes that occurred during historical periods characterized by the rapid and intense growth of towns should pay special attention to both the selection and the use of sources. The development of an increasingly complex social and economic structure brings with it an increase in the number of individuals who are cultured and literate and who therefore are capable of leaving behind written testimonies. Moreover, the development of an urban economy and widespread trading activities associated with that economy enhanced the value of all the urban lands and buildings, and could therefore lead to an increased interest in and emphasis on earthquake damage when a strong earthquake occurred. In the eastern Mediterranean area and dating back as far as the fifth century AD, there were some large, highly populated cities where earthquakes could have caused thousands of deaths. The existence of a large urban area means that an intellectual class was present and capable of leaving many as well as diversified records of a destructive event like an earthquake or a tsunami. It was the quality of the intellectual life (along with the complexity and the functionality of the administrations in subsequent time periods) that played a major role in leaving sources having an information content that can be most useful historical seismology research.

There were time periods and/or geographical regions in Europe and the Mediterranean area in which ancient cities lost their privileged role or where large towns did not exist. In the Christian area large monasteries acquired a pre-eminent role in the local intellectual and administrative life, and often in the political life as well. Even where cities in the ancient tradition survived a long time (e.g. in central and southern Italy in the sixth to twelfth centuries), the importance of the great monastic centres outside the towns persisted. Elsewhere, the fortified centres (castles, citadels and fortresses) were the ones that most of all attracted concentrations of population. Thus, in regions and times with no large cities, monasteries and other centres may preserve whatever records remain of historical earthquakes and tsunamis.

Other elements that can be considered to be filters for historical sources are the systems of communication of the times. In the past, the lack of roads and transportation difficulties controlled the circulation of news of a seismic catastrophe and the organization of aid. The slowness with which news arrived at the decision-making centres worsened the impact of a destructive event on the population. The speed of relief and of the early phases of reconstruction were dramatically affected by the lack of efficient communications and transportation systems. The conditions of the survivors could have deteriorated so much as to make the number of deaths rise considerably in the days and months subsequent to an earthquake or tsunami. The lack of aid could also trigger food crises, epidemics, the loss of working capacity, prolonged states of depression and population migration.

It is worthwhile noting that situations of serious indigence might have been mentioned or omitted by the public administrative sources. Very often the bureaucratic practices that were followed after a strong earthquake developed in a way that was independent of the social situation, because the bureaucratic practices reflected the institutional situation, the administrative customs, and the degree of efficiency (or lack of it) of the bureaucracy of a given government in a given period. In most cases the institutional sources are insufficient to mirror the social reality, so it is necessary to supplement their information with the contributions from other sources. Those who are interested in knowing about the social impact of earthquakes or other extreme natural events, upon analysing the archival sources, should ask what reality the sources are mirroring. Indeed, the social reality following a destructive earthquake can be recreated through an accumulation of data. It is, therefore, always worthwhile to form a clear idea of the particular standpoints of the sources and their filters.

The self-censorship of the Bourbon high officials: on the margins of the 1857 southern Italy earthquake

On 16 December 1857 two strong shocks at 20:15 and 20:18 hours UT, and a weaker third one at 21:15, devastated a vast area of the Kingdom of Naples (then called *The Kingdom of the Two Sicilies*), which was ruled at that time by Ferdinand II of Bourbon. There were nearly 11 000 dead out of a population of around 500 000 people ($I_0 = XI$; $M = 7.0$). Over 180 towns and villages, in an area of about 20 000 km^2 (corresponding to the present-day provinces of Potenza and

Salerno in southern Italy), sustained very serious damage, and most of the houses became uninhabitable. This major disaster (Ferrari, 2004) moved the Europe of the day, which organized aid and tried to get it to the affected region. When this earthquake occurred there was strong political tension in Naples. A few months earlier, in June 1857, there had been a military expedition led by Carlo Pisacane, a revolutionary idealist who tried to stir up a mass rebellion against the Bourbons, although this attempt failed miserably. The anti-Bourbons were intellectuals, clergymen, local people and foreigners, all watched closely by the police.

After the earthquake on 16 December the institutional aid machine was set in motion. In the archive files a respectable reality appears, in which the books always balance and legal measures were passed and enforced. But part of the reality of what happened was described by a witness, a Protestant Englishman named Theophilus Roller (1861). He can be defined as a good non-governmental source. Together with one of his countrymen, he busied himself with distributing the private foreign aid. Soldiers, according to this source, arrived at Montemurro (one of the worst-hit villages) only in February 1858 and built just two or three shacks there. On the other hand, the official sources declared that 426 structures had been built. Furthermore, according to Roller, the few shacks constructed by the military were only used by the officers for their own needs. The 2000 survivors, therefore, did not benefit from any aid whatsoever.

In the two subsequent years, the effects of the Bourbon administration were dramatic. The papers of the archive slowly became silent, and the earthquake disappeared from them. However, from other non-governmental sources one learns that those who had accepted help from the English 'enemies' had to bear the brunt of the threats from the Bourbon police officials to the point that they were driven to abandon their villages. The poverty, the food shortages, and the political and economic incapacity to plan the reconstruction had thrown the population into a shocking state of penury. But none of this state of affairs is reported in the archival documents.

Housing characteristics

Another level of selection concerns housing styles, involving both the nature of buildings and the more general settlement characteristics. By relating the principal economic phases of a region to the nature of the building stock, it is possible to identify those periods when a lack of funds led to poorer-quality buildings or to inadequate or non-existent maintenance. Periods of economic growth, on the other hand, may coincide with an increase in population and may involve an increase in building activity. Only detailed research is able to shed light on the particular local characteristics of such building expansion.

Earthquakes may cause diverse effects throughout a town or city if the levels of vulnerability of the affected buildings vary depending on when they were constructed.

When an urban population rapidly increased without an expansion in its urban area, this almost inevitably led to an increase in the heights of the buildings, as it was necessary to make the most of the space within the city walls or limits. In past centuries, in the Mediterranean area, but not only there, the growth of towns has often produced conditions of great seismic risk as the number of vulnerable buildings increased. Sources dealing with earthquakes do not always make this clear, with the result that evidence of this kind must be sought by means of parallel research into the building stock, quality and distribution.

As is clearly demonstrated by seismic engineering, the form, materials, building techniques and state of preservation of buildings all contribute to determining the level of vulnerability to earthquake shaking. In an inhabited area, the buildings will display a variety of responses to earthquake shaking depending on the size and layout of each building, the materials of its construction, the strengths and weaknesses of the joints where different building elements are attached and the state of repair of the structure. In order to evaluate the effects of historical earthquakes in cities and towns with many different structures, it is usually important to make some generalizations concerning the building stock. For example, what is the prevalent building type and size? What is the most used building material (e.g., wood, stone, earth or brick)? What is the average condition of the buildings? (For further discussion on these issues, see also Chapter 11.) Historical sources rarely contain direct references to the construction variables that contribute directly to a building's earthquake response. The characteristics of an inhabited territory are nearly always only implicitly contained in the sources that specifically describe an earthquake and its effects. This is true because the authors of the source materials were substantially addressing contemporaries who were familiar with the local structures and their construction. Thus, in order for someone today to learn about the vulnerability of buildings affected by historic earthquakes, one must turn to further sources and studies.

The communication and social networks existing between the various human settlements in a territory underwent modification as the economic, political and administrative situations of a region changed. These networks could become looser or tighter, which in either case affected the production and transmission of historical sources. Earthquakes occurring in recently settled areas were very unlikely to be recorded in the written sources because the resident population was still primarily dedicated to tilling the land and because newly settled centres typically lacked an established cultured class capable of leaving a substantial collection of written records. Earthquakes occurring in periods of depopulation were also unlikely to be recorded in the written sources, either because villages had already been abandoned or because there were no suitable people around to produce the written records. It therefore follows that even strong earthquakes may not have been recorded in the written sources, even in a region or state with a long-standing written culture, because the social situation was not conducive to capturing the memory of a destructive event in a written text.

6.2.5   Wandering earthquakes: when the location is lacking

It can happen that the systematic scrutiny of sources produces information about earthquakes with exact dates but no information on the locations where the events were experienced. Such data cannot be put to immediate use, but should not on that account be abandoned. Information of this kind usually comes from premodern sources, such as *notulae*, colophons or chronicles, and they are usually reliable and valuable. Especially if the text refers to a destructive earthquake, it is worthwhile trying to attach the information to a particular area. There are various step-by-step ways of proceeding, and a competent historian can easily act as a guide. Set out below are some possible situations and solutions:

(1) if a chronicle suggests in general terms that an earthquake occurred in the region where the text was written, and if other sources are potentially available for that period, it is worthwhile searching in the archives of nearby towns;

(2) if there is absolutely no information as to the location of the reported earthquake, an attempt should be made to discover: (a) whether the text is simply transmitting the fame of a large but distant earthquake, in which case confirmatory information might be found in catalogues covering other regions or countries; (b) if the text appears to have been written in the heat of the moment by a direct witness trying to convey his great fright, it is important to know or find out where the text was written, for that place may provide an early and approximate location for the earthquake. Subsequent research could then be carried out in the area for the specific purpose of identifying other affected localities.

One should never lose track of records of this type. They can usefully be gathered into an archive of their own, a sort of container for potential research projects, to be carried out when the possibility and the occasion arise. The paths

to be followed to improve the knowledge of past earthquakes, and hence earthquake catalogues, are often neither straight nor quick.

A 'terrifying' earthquake wandering in the Mediterranean and located on the grounds of 'its' codex

An earthquake on 2 February 1438, at 12:15 hours UT, definitely happened, but where? An authoritative note in a Greek codex of the Biblioteca
Angelica in Rome (41, B.3.11, fol. 140v) preserves a memory of this event. An
anonymous scribe wrote:

> In the year 6946 [1 September 1437–31 August 1438], on 2 of the month of
> February, there was a terrifying earthquake at the seventh hour of the day.

This earthquake had the misfortune of being cited in some catalogues (Grumel,
1958; Wirth, 1966; Evangelatou-Notara, 1993) without a precise location, except
that of being in the Mediterranean area. In such cases, of course, the data cannot
be used in seismic hazard assessments even though the description suggests that
the event is of significance. What should be done then about such an event? The
fact that it is described as 'terrifying' must be considered of importance, and
a codicological search has been started to find out more about the description
itself. An attempt has been made to reconstruct the history of the codex and of
the places where it has been kept.

This notula has attracted the attention of codicologists ever since the end of
the nineteenth century. It is quoted in Allen (1890, p. 40), where the year was
given tentatively as 1028. Lampros (1910) subsequently suggested 1438. The latter date has been accepted by scholars. Neither of these first codicologists has
suggested even a hypothetical area for the earthquake mentioned in this notula,
and so its location has remained unidentified. Greek codex 41 (B.3.11) in the
Biblioteca Angelica in Rome consists of two manuscripts of works by Theodoret
(a writer of the fifth century AD). The first of these (fols. 2–55v) contains commentaries on the Old Testament, and the second (fols. 56–140) consists of the
five books of his Historia Ecclesiastica.

From an analysis of the writing, codicologists think that the first manuscript
was probably written in Basilian abbey circles in Calabria (southern Italy) in
the closing decades of the tenth century. It has been attributed by Lilla (1970,
pp. 10–13) and Follieri (1977, p. 221; 1979, pp. 324–5) to a monk named Paolo,
who was a copyist of the St Nilus school and subsequently became abbot of the
important abbey of Grottaferrata (near Rome) founded by St Nilus in 1004. At
some unknown date the codex was taken to the Abbey of Grottaferrata, where it
remained for some centuries. The second manuscript was written in the eleventh
to twelfth century. The notula was probably written when the codex was at

Grottaferrata, before it joined the Biblioteca Angelica collections. It is important to observe in this case that in the specialist literature there were precise elements that could anchor this important text for historical seismology to a location. Researchers have formulated the hypothesis that the *notula* had been written by a monk at Grottaferrata who felt the frightening earthquake of 2 February 1438. Although still hypothetical, this earthquake seems to have found its first territorial rooting. The epicentre still remains to be located, but this will be another story, because thanks to the date and the identification of the probable location of the report, more research concerning this earthquake can be carried out (Guidoboni and Comastri, 2005).

6.2.6   Concomitant damage from other natural or man-made events

Seismic effects may have been obscured by the contemporaneous action of other events, and it is not unusual in these cases for the contemporaries to have misinterpreted the effects of an earthquake. The uncertainties or inaccuracies of the original sources may also have conditioned the subsequent interpretations of the events that were chronicled. In a given inhabited environment, the effects of war, invasion, famine, epidemics and so on may have been felt to be more devastating than those of an earthquake. Other concomitant events that may have obscured the effects of seismic events or made difficult the assessment of their effects are abnormal atmospheric phenomena such as heavy rainfalls and floods, which often have a major environmental impact, or a heavy snowfall. By paying attention to these elements, historical seismological research must manage to distinguish between the various causes of reported damage and so provide a reliable picture of the proper causes and effects of the area damaged by an earthquake.

6.2.7   A damage multiplier: fire after a strong earthquake

In the past, as now, the shaking produced by a strong earthquake can cause fires, which may spread very rapidly and cause much more serious damage than that resulting from the earthquake itself. Such effects are particularly devastating where houses are chiefly made of wood. Suffice it to recall the great fire which literally devoured a large part of Lisbon after the great earthquake of 1 November 1755 or the fires at Messina after the earthquakes of 1783 and 1908. Other examples are the fire damage at San Francisco, which made the total earthquake destruction in 1906 so catastrophic, and at Tokyo in 1923. Light has been thrown on this particular damage multiplier because a great many cases are known and have been studied. It is important in such cases to carry out an in-depth analysis of descriptions of the fire, because this often leads to a better understanding of the history of the destruction and provide a clearer idea of the seismic effects that were experienced before the fire. Of course, there remains the intractable problem of distinguishing between the two types of damage in the sources, and it is understandable that the sources would describe the damage without differentiating between its two different possible causes. Only by looking at the total picture of the effects over a large area is it possible to achieve a better assessment of the impact of a post-earthquake fire as opposed to that of the initial earthquake.

Confused damage: a siege and an earthquake in the thirteenth century

On an unspecified day in September 1249, somewhere between the ninth hour and vespers, Reggio Emilia, a city in northern Italy, was struck by an earthquake which the sources define as 'great'. There is evidence for effects at Reggio Emilia in two thirteenth-century sources, both of which provide the same information. These sources are the *Liber de temporibus et aetatibus* and the *Chronica* of the authoritative Parma chronicler Salimbene de Adam. The *Liber de temporibus* is attributed to Alberto Milioli, a notary, but he was probably just the copyist. It is also known as *Memoriale Potestatum Regiensium*, a title given to it in Volume VIII of the *Rerum italicarum scriptores* by Muratori (1726), who only published the second part. The text is as follows: '1249. [. . .] *And in the month of September between the ninth hour and vespers there was a great earthquake*'. The same information is recorded in the *Chronica* of Salimbene de Adam: '*In the same year [1249] in the month of September between the ninth hour and vespers there was a great earthquake.*' According to Carlo Sigonio (1591), a Modenese historian who lived between 1520 and 1584, Modena was besieged by the Bolognese in 1249 and '*there was a sudden earthquake with a very strong shock that shattered all the roofs in the city*'.

In the earliest Bolognese chronicles, namely those of Pietro and Floriano Villola, one finds a report that the Bolognese besieged Modena for five weeks in September 1249, digging tunnels and erecting war machines to bombard the city with great quantities of stones. The Bolognese chronicles not only do not record any earthquake for that year, but they lay emphasis on the efficiency of their war machines and the destructive effect of their missiles. Sigonio does not seem to be aware of this last piece of information, and it is fairly understandable that the Bolognese chroniclers would wish to attribute the success of the siege solely to their military skill and preparation rather than to the effects of an earthquake. At the same time, however, one must give due weight to Sigonio's words, for he may have made use of lost chronicle sources – no doubt of Modenese bias – in which the earthquake was given as the only reason for the city's defeat and the damage to roofs (Guidoboni *et al.*, 2007).

Underestimated damage: bombing hides earthquake damage (Palermo 1940–1943)

An inaccurate assessment of seismic damage may occasionally result from somewhat tangled historical situations that one might describe as the result of a mingling of human events, such as war, with human reactions to natural events. In order to demonstrate how such a tangle can occur in real life, let us take a case history involving the city of Palermo.

On 15 January 1940, Palermo and some villages in northwest Sicily were struck by a medium-intensity earthquake. Many houses were severely damaged, and there were some collapses of structures. The earthquake claimed some victims, caused panic among the population and left families homeless. All in all, it was described as a typical scenario of a damaging earthquake. A few days later, the local fire brigade was ordered by the prefect of Palermo (the senior political representative of the Italian state at the local level) to carry out a detailed damage survey. This was done with scrupulous care and a fair amount of skill. Each house was inspected and described; and an assessment of the financial consequences was also made. Hundreds of cases were dealt with in the space of a few months. In Palermo, most of the older brickwork buildings were cracked to a more or less serious extent. The more recent buildings, particularly those made of reinforced concrete in the more modern parts of the city, were unscathed, except for a few slight cracks in the doors and window openings, in the dividing walls and of the ceilings. Out of the 130 000 properties existing in Palermo, 10% suffered substantial damage that compromised their stability and normal use, making necessary extensive repair work; 45% underwent slight damage, while the remaining 45% were undamaged. To this assessment of the private building stock was added a survey of damage to churches, carried out by appropriate officials.

By the time arrangements were in hand to restore the buildings, it was January 1943, and from then until June of that year, the city was repeatedly bombed as part of operations for the liberation of Italy during the Second World War. The air raids increased the damage to buildings in a city that had already suffered from the earlier earthquake. When the war was over, the new administration began compensation arrangements for war damage, which had come, of course, right on top of the earthquake damage. The earthquake was now brushed aside, the fire brigade reports were hidden away, and permission to consult them was transferred from the prefect to the ministry for internal affairs. It seems likely that the purpose of this was to facilitate access to public funds and support the local population, who had suffered one way or the other. From a seismological point of view, however, the general result is one of confusion. And it is interesting to note that even the people of Palermo almost completely forgot about the earthquake, though it is in fact a matter of great importance for their city since the earthquake was the most destructive seismic event to the city in close to its modern configuration. The 1940 earthquake is therefore a matter of great interest to those who want to understand seismic response of different sections of Palermo today (Guidoboni *et al.*, 2003a).

6.3   False and lost earthquakes

6.3.1   *False earthquakes*

This section focuses on two related cases of particular importance when assessing the seismic hazard of an area: the cases of earthquakes mistakenly assessed by the seismological tradition and of mistakes in earthquake catalogues. The assessment of a false earthquake may be based on any of a number of reasons, the most important of which are:

(1) chronological errors in dating already known earthquakes, which can lead to the duplication of the same earthquake. This is the most common mistake that generates false earthquakes in earthquake catalogues;

(2) location errors;

(3) another natural event distinct from the earthquake (hurricane, whirlwind or landslide), or an anthropic event (such as the explosion of a powder keg or something similar);

(4) misinterpretation of the historical sources that describe an earthquake, an error that may easily occur when proper interpretation requires understanding ancient cultural and linguistic contexts (see Section 6.2);

(5) incorrect interpretation of the collapses of structures that were due to other causes; the date of the collapse becomes the date of a false earthquake.

The mistaken assessment of earthquakes, either due to chronological errors or to misinterpretations of the text, may occur in particular if the analysed area was characterized by the presence of different cultures and linguistic groups, and consequently different ways of measuring date and time (see Chapter 7).

For each earthquake catalogue a history of the deleted false events needs to be documented. Here we report a few cases to illustrate the kinds of error that can be encountered in earthquake catalogues.

Dating errors: the Dvin (Armenia) earthquake in 893

The sources for the Armenian historical region belong to a number of different languages and cultures: Armenian, Syriac, Greek, Arab, Persian and Georgian. This diversity of literary, historical and cultural traditions that affected the Armenian region has caused confusion in the past insofar as it was only superficially considered by many past researchers looking into the historical seismicity of the region. On the other hand, through the application of the procedures provided by historical seismology, it has been possible to make the most of these differences in the available sources. An example of this is the earthquake that struck Dvin in AD 893, one of the strongest earthquakes in Armenia. The Armenian textual scholar Abrahamyan (1976, cols. 139–144) attempted to correct the date of the earthquake by assigning it to the year 894 on the basis of his calculation of Easter and of other chronological evidence yielded by the colophons. But this redating proved to be the result of the underevaluation of other sources not written in Armenian. Indeed, an Arab source, which may be considered to be far more accurate than the Armenian ones, reports the chronological basis for the exact calculation of the date of the earthquake, which actually occurred in 893 (Guidoboni *et al.*, 1994).

Delayed collapses: real and invented earthquakes at Constantinople

Towards midnight on 14 December 557 a violent earthquake hit the northern area of the Sea of Marmara. At Constantinople the city walls were damaged, both the ones Emperor Constantine had built in the fourth century AD and those Theodosius had built in the first half of the fifth century. Among the many churches hit was the great Christian basilica of Saint Sophia (Conant, 1939). The shocks went on for ten days. The following year, on 7 May 558, while work was in progress to repair the dome of Saint Sophia, the eastern part of the ceiling collapsed and crushed the altar and the ambo. Studies in historical seismology have understood this case correctly. But a very similar sequence of events recurred about 800 years later, and it led to a false earthquake appearing in a number of earthquake catalogues. The following is a reconstruction of the facts.

The coeval Byzantine sources attest that earthquakes struck Constantinople between October 1343 and November 1344. The latter seismic events damaged the church of Saint Sophia and were the indirect cause of the collapse of a part of the dome on 19 May 1346. Thus, the collapse did not actually coincide with any earthquake. This is the seismic sequence as reconstructed from the surviving accounts.

14 October 1343 Constantinople: earthquake whose effects are not
detailed. The source is a note written by Galaktion Madarakis
(published in Turyn, 1980, pp. 108–12).

18 October 1343 Two strong earthquakes occurred in the morning (about
07:00 hours UT, 'at the third hour') and at dusk (about 16:00
hours UT, 'first hour of the night') striking Constantinople and
the western region of the Sea of Marmara; they were felt at
Lysimachia and apparently even beyond the present-day
peninsula of Gallipoli (Thracian Chersonese) with a lesser
intensity compared with Constantinople. Probably in
concomitance with the second earthquake of 18 October 1343
there was a tsunami of great proportions. Owing to these
earthquakes at Constantinople a part of the Theodosian walls
collapsed and the maritime walls were damaged. Furthermore,
towers, buildings and churches collapsed; the eastern side of
the apse of the church of Saint Sophia was damaged. Also,
houses and walls surrounding vineyards and gardens
collapsed. So much rubble had accumulated in the streets that
people's transit was made difficult.

20 November 1343 – summer 1344 Constantinople: strong tremors. The
source is a note written by Galaktion Madarakis (published in
Turyn, 1980, pp. 108–12). The earthquakes continued for
nearly a year until the summer of 1344. Frequent tremors
again started to strike Constantinople and Thrace along the
western coast of the Sea of Marmara in the autumn of 1344.
The most destructive earthquake was the following:

6 November 1344. The earthquake occurred at around 08:00 hours UT
('fourth hour of the day'). At Constantinople the eastern side
of the apse of the church of Saint Sophia, already damaged by
the earthquakes of the previous year, slowly continued to
open, so much so that many bricks and mosaic tesseras fell
down. Also damaged was the bronze statue of the archangel St
Michael, which was located on a column in front of the
basilica of the Holy Apostles and had already been damaged
by the earthquake of 1 June 1296 (as attested to by Nicephorus
Gregoras 14.2, II, pp. 695–696).

On 19 May 1346 part of the great basilica of Saint Sophia collapsed. The
shoddy repair work to the damage suffered several years before caused the

collapse. The sources were explicit in describing the collapse and recalling its causes. Suffice it to recall Nicephorus Gregoras:

> *The eastern side of the apse of Saint Sophia, stressed and already previously under strain, slowly cracked more and more, making many bricks and mosaic tesseras fall down, until it collapsed completely.*

But this delayed collapse has found its way into seismological studies as evidence of an earthquake. In 1850 Perrey was the first to create an earthquake in 1346 based on this collapse, and after him a whole tradition of well-esteemed scholars followed suit: Downey (1955), the catalogues of Ergin *et al.* (1967), Shebalin *et al.* (1974) and Ambraseys and Finkel (1991). The epicentral intensity of this event was evaluated at degree VIII MSK. It is perhaps interesting to observe the analogy between this medieval collapse and that of the cathedral of Noto (eastern Sicily) which occurred in March 1996. Also in the case at Noto the remote cause of the collapse was the failure to provide repairs to the damage caused by a previous earthquake (13 December 1990, $I_0 = $ VII; $M = 5.4$). In the Sicilian case the problems behind the reason for the collapse were made public in a court trial, the records from which should help the future historians of earthquakes to avoid creating a false event (Guidoboni and Comastri, 2002).

Calcutta 1737: a catastrophic 'earthquake' drastically downsized in favour of a tropical storm

All pre-1990 textbooks and lists of catastrophes show the third most fatal earthquake in the world to have been an earthquake in Calcutta in 1737. It is difficult to erase such an important entry even when faced with contradictory facts, especially since the entry is repeated in many scientific lists of damaging earthquakes – for example, Dunbar *et al.* (1992). This ficticious earthquake is supposed to have occurred in September 1737 and to have killed 300 000 people. Roger Bilham submitted this event to a careful historical analysis which in view of the seismological importance that earthquake had acquired has remained exemplary (Bilham, 1994).

News of the supposed earthquake published in several European magazines were based on an anonymous report which reached Europe six months after the event reportedly took place, which later turned out to be false. Bilham looked for more reliable sources and found several, which showed that the damage and deaths, downsized to 3000, were in actual fact the result of a violent Bay of Bengal tropical cyclone and storm surge. Although earthquake shocks were mentioned in one French report, they may have been invoked as a figure of speech to describe the buffeting of the wind during the cyclone, wind that was sufficiently powerful to blow down the tower of the church.

Some key elements of the analysis by Bilham are the search for independent and authoritative sources, recourse to elements from the historical context such as demographs (in those years Calcutta had fewer than 20 000 inhabitants), the type of construction and the damage suffered, and verifying the date of the event. The last element was motivated by the fact that in the British dominions the Julian calendar was still in use. This set the Julian date of the Calcutta disaster to 30 September, corresponding to the Gregorian date of 11 October. In gazettes and correspondence there can be confusion about the date of the event if there is uncertainty about whether a date is based on the Julian or the Gregorian system.

Although Bilham did not wholly eliminate the possibility of a seismic event, he drastically downsized it and reduced its possible effects to an acceptable minimum. He proved that an earthquake of the size indicated by Dunbar et al. (1992) was quite indefensible. Thanks to historical analysis alone Bilham has show that the seismicity of the Gulf of Bengal is that of a moderate magnitude, and he brought the main event, a tropical cyclone, to the forefront. The violence of the storm caused enormous damage and the increase in rainfall raised the level of the Ganges by 40 feet (12 metres), but such events are not uncommon in the northern Bay of Bengal. The impact of this atmospheric event on local buildings (predominently made of mud and earth or very cheap bricks) was devastating. Fortunately subsequent structures in Calcutta have been built to resist damage from heavy rain, floods and extreme wind loads.

6.3.2   Lost earthquakes

Historical seismology research cannot avoid confronting the problems represented by rare strong earthquakes or prolonged seismic quiescences, nor can it avoid the problem of catalogue gaps even for the minor earthquakes. The discovery of new earthquakes (i.e. unknown to the catalogue in use and/or to the seismological tradition) or of new classes of effects (and of magnitude) for already known events constitutes a fundamental step forward in historical seismological research. Such new destructive earthquakes can be identified almost exclusively through systematic research. This category of events is the fruit of research performed by directly examining the primary historical sources. The newly detected earthquakes had been lost owing to a lack of previous research.

There are also earthquakes that were lost for other less obvious reasons, in particular during the groundwork carried out in catalogue compiling. This may have occurred due to errors, misunderstandings or failures to interpret properly historical reports by the catalogue compilers. Basically, erroneous procedures or poor skills may have led to previously known earthquakes being lost. One may be tempted to believe that such mistakes are very rare, but such cases are not infrequent and have created at times substantial losses in terms of the evaluation of local seismic hazard.

It should be remembered that correcting catalogue errors often requires more time and skill than retrieving new data from research performed *ex novo*. Here is a list of the most frequent causes of mistakes in earthquake catalogues:

(1) lack of systematic research on a territorial basis;

(2) problems of historical toponymy;

(3) earthquakes known to the nineteenth-century seismological tradition but subsequently discarded by the compilers of current catalogues without adequate checks;

(4) earthquakes known in the recent historiographic literature (such as new publications of sources or new analyses of territorial histories) but not known in the field of the historical catalogue studies, owing to a lack of contact between the two disciplinary fields;

(5) earthquakes reckoned to be of doubtful veracity by the compilers of current catalogues, who then shelved and subsequently lost them owing to a lack of adequate time or resources to investigate them further;

(6) records of isolated seismic effects not analysed more thoroughly owing to a lack of coherent and continuous historical research, with the events subsequently being forgotten;

(7) earthquakes obscured by other more important and famous seismic events as viewed by the historical and seismological tradition.

The two examples that follow are drawn from Guidoboni and Comastri (2005).

Problems between the Latin and German languages: a lost medieval Italian earthquake (eleventh century)

There is an earthquake that took place in 1046 in the Adige Valley, the sub-Alpine area of northern Italy, which was unknown to the Italian seismic catalogue tradition, but had been mentioned in Leydecker and Brüning (1988) and in Alexandre (1990). Alexandre (1990) gives the earthquake effects an intensity of IX–X degrees (MSK scale). It was, at any rate, a seismic event of considerable destructive power, and therefore of great interest for the seismicity of this area.

Before this earthquake was given a location in the Adige Valley, only one other strong earthquake was known to have occurred there, some 70 years later in 1117. The latter event, famous in all medieval Europe, was more destructive than the previous one, and affected the area south of the Adige Valley towards the Verona plain. On 9 November 1046 an earthquake struck the *Valle Tridentina*

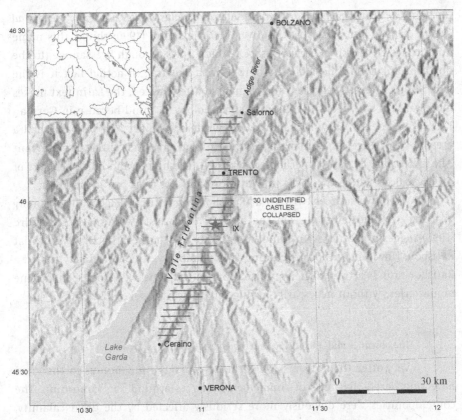

Figure 6.4 Map of the 9 November 1046 earthquake in the Valle Tridentina, northern Italy (from Guidoboni and Comastri, 2005, p. 41).

according to some twelfth-century monastic annals. In those times this term did not indicate the Trento area in a general sense (as Alexandre (1990) suggests in his map on p. 223), but rather a quite specific part of the middle Adige Valley. Indeed, the *Valle Tridentina* was the name given to the area between Salorno, about 20 km north of Trento, and the Ceraino defile, about 30 km north of Verona (see Figure 6.4). In this area more than 30 castles (fortified settlements) not specified by the annal sources collapsed in ruins (*ruerint*), killing their inhabitants in some cases. The sources are three Benedictine annals from Germany: *Annales Corbeienses* (ed. 1864, p. 39); *Annales Sancti Emmerammi* (ed. 1861, p. 571); and *Annales Ratisponenses* (ed. 1861, p. 584). The main source is the *Annales Corbeienses*, compiled between 822 and 1117 in the Benedictine monastery of St Peter at Corvey (in the diocese of Paderborn in present-day Germany). This latter record provides a decisive indication about this earthquake. It states that the earthquake brought about mountain rockfalls and that the 'Tar' River was blocked for more than ten days, during which it 'did not give a drop of water'

(*Tar insuper fluvius, montibus cadentibus interclusus, plus 10 dies nec unam guttam ad alveum dimisit*). But what was the River *Tar*? Alexandre (1990) thinks this name might refer to the River *Isarcus* (Eisack), which flows through Bressanone (to the north of Bolzano, almost on the border with Austria). In actual fact it is the River Adige, the one that the valley is named after. Indeed, the Latin text makes more sense if one keeps the German terminology in mind because in German *Tal* (= *Tar*) means *Valley*. The monk who wrote the report very probably had a greater knowledge of German than of Latin, and may well simply have meant 'the river over the valley', perhaps referring to the higher and narrower part of the river flow, where the rocks jut out more and the river bed is narrower.

The other two sources, written at Ratisbon (present-day Regensburg), simply give the earthquake a date. On the basis of these last two annals, Alexandre (1990) suggests, though with some uncertainty, that the earthquake was felt at Ratisbon. However, without excluding that possibility, it is more likely that the Ratisbon annals acquired their information via the network linking Benedictine monasteries, without necessarily having felt the earthquake *in situ*.

The obscured mid-Adriatic Sea earthquakes of 1303

Very often the transmission of the news of large and famous earthquakes have smothered the news of smaller earthquakes around the same time. The contemporaries were obviously more seriously affected by the great calamity, and so the attention of a writer usually focused more on the most profound experience. On 8 August 1303 a very strong earthquake struck the island of Crete. Besides the effects of the earthquake, a tsunami was also propagated in the southern Mediterranean as far as the coasts of Egypt and Syria.

As Crete was governed by the Venetian Republic, the destruction of more than 12 towns and castles on Crete immediately became known throughout the Latin countries, where the report was recorded in many chronicles that were written both in Latin and in the vernacular. This strong seismic event obscured a minor and completely independent earthquake that happened in the same year in Fano and Senigallia, two Italian towns on the mid-Adriatic coast. This latter earthquake caused severe damage in Fano, where a public building was completely ruined. Nevertheless, this smaller event, though mentioned in authoritative sources, ended up being forgotten as a consequence of the special attention paid by both the contemporary sources and later tradition to the Crete earthquake. It may well have been mistaken for a doublet of the great Mediterranean earthquake of 8 August 1303 and therefore eliminated from earthquake catalogues. Earthquakes that occur soon after other important seismic events are either not mentioned in the minor sources, or if they are, they run the risk of not being recognized as separate events and hence may be eliminated from catalogues.

In this case, the minor earthquake is recorded in the authoritative *Chronicon Parmense* (ed. 1902–04), and their occurrences are confirmed by the *Zibaldone da Canal*, an independent Venetian source. The *Chronicon Parmense* is an anonymous chronicle, but was almost certainly written by a notary or judge who lived in Parma between about 1270 and 1340. The other source, the *Zibaldone da Canal* (ed. 1967), is a Venetian manuscript dating to the second half of the fourteenth century. It contains miscellaneous items of information, including a short Venetian chronicle at fols. 57r–59r.

During 1303 an earthquake caused damage in the *Marca anconitana*, that is to say in an area covering a large part of the present-day Marche region in central Italy. Many towers and houses were damaged in the towns of Senigallia and Fano on the Adriatic coast. The Podestà's Palace at Fano, which had only recently been built, was badly damaged. In the same year an unspecified town in *Sclavania* was 'reduced to ruins' (*ruinata*) and was struck by a tsunami (perhaps on Croatian coast).

The sources do not provide a precise chronology for the Croatian coast event, but it can reasonably be assumed to have occurred between August and October 1303, since a reference to the earthquake in the *Chronicon Parmense* is inserted between events occurring between July and October 1303. In the text of the *Zibaldone da Canal* the earthquake seems to have occurred after the one in Crete on 8 August 1303.

6.3.3   *Effects of overestimated or underestimated earthquakes*

A particular kind of lost earthquake concerns overestimated or underestimated events. Indeed, in a number of cases multiple earthquake effects and earthquake magnitudes have been lost, an element that has altered the proper picture of the seismicity of an area. One may know the date of an event without knowing the actual size of an earthquake. If the event was large, a significant earthquake from the standpoint of the seismotectonics of the region in which it took place has been lost. The main conditions that can lead to the underestimation or overestimation of the size of an earthquake can be mainly ascribed to three factors:

(1) the influence of the conditions of the building or of the standards of living in an area affected by an earthquake. For example, when urban areas that already contained elements of decay or abandoned zones within them were hit by a strong earthquake, the damage was not perceived by the local population as being important, so descriptions

of the additional damage due to the earthquake had a very low
likelihood of being transmitted in a written text (see, for example, the
case of Palermo, in Vallerani, 1995);

(2) the influence of the large cultural centres (cities or monasteries) in the
     production of written texts; in these cases the sources, albeit direct
     and authoritative, may preserve only the modest felt effects in the
     urban area or the monastic structures due to an earthquake, while not
     reporting the more destructive effects in other areas (such as in
     outlying mountainous regions and/or in places far removed from the
     main traffic routes).

(3) an insufficient number of sources or an inadequate analysis of the
     sources and their contents.

*An underestimated earthquake because it was seen only by one city: the case of an
earthquake in 1315 in central Italy*

The 1315 earthquake had been located at L'Aquila, a city of the central Apennines, in Italian catalogues from Baratta (1901) to Camassi and Stucchi
(1997), (I_0 = VIII MCS; M = 5.2). It appeared to be one of the many medium-
impact earthquakes in the Central Apennines. Research has been carried out in
connection with work on the *Catalogue of Strong Earthquakes in Italy* (Boschi *et al.*,
1995), and palaeoseismological research has also been conducted by researchers
of the Istituto Nazionale di Geofisica. An examination of the territorial and historical background of the region was enough to arouse certain doubts about
the existence of this earthquake. L'Aquila was indeed the only cultural and
urban centre capable of recording and passing on to posterity the effects of
an earthquake. But where exactly did the seismic event occur, and how big was
it? L'Aquila was a strategically important city situated at the frontier between the
Kingdom of Naples and the Papal States. It had been founded in 1254, and the settled area continued to expand throughout the fourteenth century. In this border
region, it was one of the few inhabited places culturally capable of registering
the effects of an earthquake in a written record, because there were amongst its
inhabitants a few intellectuals (poets and historians) who could produce such a
record. It is therefore not surprising that the earthquake should become associated in chronicles solely with the city, rather than the area around it. Fresh
historical research has therefore taken as its starting point an examination of
the surrounding area.

The *contado* of L'Aquila was very large. A privilege granted in a diploma of
King Charles II of Anjou, dated 1294, shows that this area of L'Aquila included
at least 70 castles and other villages. A document dated 1317 from the Angevin
Chancery describes the earthquake as 'frightening' (*pavido*), and the way the

territory of L'Aquila was affected as 'terrible' (*terribiliter*). Soon after the earthquake, the king, Robert of Anjou, ratified a peace agreement drawn up between
the city factions 'when a frightening earthquake struck that place', and so the
king granted a pardon for crimes committed during the hostilities between the
factions. The pacification of the warring factions thus occurred at the time of
the earthquake, but the background was much broader and more dramatic than
the evidence about L'Aquila would suggest if taken on its own. For a 'frightening
earthquake' such as that referred to by the king could not simply be a modest
local shock. It is on the basis of this document – which is not contradicted by
contemporary chronicles – that the earthquake has been given a higher intensity
classification (I_{max} = IX–X; I_0 = VIII–IX). However, the epicentral area cannot
be clearly identified. Although it is difficult to give a quantitative value to our
research results, we have tried to make that task easier by establishing a notional
point within the area of L'Aquila and its 70 fortified settlements where we think
the maximum effects were felt.

The principal and most authoritative contemporary source about this event is
an order from the Naples Chancery, dated 9 March 1317, which survives in a copy
transcribed by the seventeenth-century scholar Carlo de Lellis (State Archives of
Naples). The original Angevin archive was destroyed in a fire on 30 September 1943. Hence our documentary research and interpretation of the chronicle
sources was guided by a growing suspicion that the written tradition of the city
concealed a much more destructive event. This is what the document actually
says:

> *Approval of an agreement made by certain men of L'Aquila during the
> hostilities between them; and pardon for these same men, since they were struck
> with fear at the strong shaking when a frightening earthquake soon afterwards
> struck that place in a terrible way, and they abandoned their wrongdoing and
> returned to the narrow path of their conscience. So they gave up their hatred of
> each other and established peace, each party forgiving the murders and harm
> done by the other. 9 March, in the fifteenth indiction, 1317, folio 203.*

It is a pity that we have been unable to establish whether or not the earthquake
was felt in Rome, for that would have helped locate the earthquake epicentre
since earthquakes generated in the Abruzzo Apennines are strongly felt in Rome.
But such was the situation at the time that no fourteenth-century Roman narrative sources have survived in which information about the year 1315 can be
found. In addition to this lack of narrative sources, our research revealed gaps
in administrative sources, resulting from the institutional instability in Rome
at the time. It must be remembered that all papal documentation and chronicle

material is of limited use for 1315 because the popes had by that time already been resident at Avignon for six years. Furthermore, 1315 was a particularly problematic year since the papal throne was vacant from 1314 until 1316 when John XXII was elected pope at the end of the Carpentras conclave. Research was also carried out for the Lazio region and other parts of Abruzzo, but without any positive results.

In conclusion, even if the attempt to better define this event has not had a positive result from the point of view of the historical sources, it can be argued that this is a strong earthquake of the central Apennines. A reassessment of the maximum intensity and the magnitude was performed by conjecture on the grounds of the royal document, which may not have been motivated by a small local earthquake. Thus, an exclusively logical rationale has been followed. Subsequent palaeoseismological research in the area of the Abruzzi mountains (in particular along the Ovindoli–Pezza fault-line) has shown traces of a strong earthquake between the twelfth and fourteenth centuries (D'Addezio et al., 1995; Pantosti et al., 1996)

6.3.4   Arbitrary or provisional epicentres: effects known at one site only

Even in good-quality earthquake catalogues, one will find more frequently than expected that epicentres are given a location on the basis of only a single site that is known as having suffered macroseismic effects. Such a location is necessarily arbitrary and therefore provisional. Scarcely any current catalogue indicates how many places and what intensity values were used in calculating the location of the epicentre. It is obvious that epicentres of this arbitrary kind can be hidden in earthquake catalogues, and the problematic epicentres may be known only to specialists. Statisticians who use catalogue parameters for calculating seismic hazard are not usually much concerned with the quality of the underlying historical data since they usually assume that the data are of uniform quality and accuracy. Unfortunately, they often see their contribution solely as a statistical one, and their results therefore can be misleading if the arbitrary or provisional information is not accounted for in their analyses.

Seen from the standpoint of historical seismology a fairly substantial number of epicentres in many historic earthquake catalogues are probably false, and this fact must be given due consideration because it can have an appreciable effect on seismic hazard estimates for particular areas and sites. As an example, an earthquake catalogue for Italy, while of overall good quality, contains 159 arbitrary epicentres and 48 epicentres are given a location on the basis of just two sites known to have suffered seismic effects (Gruppo di lavoro CPTI, 2004). It is admittedly the case that not all of these earthquakes that were reported at only one or two sites caused high-intensity effects since many may have been small local shocks. For these events, the paucity of reports is a consequence of their small sizes. However, for the cases where more severe effects were reported only at one or two localities, the problem of localizing the epicentre is a real one. They need to be examined and possible solutions found. The problem does not solely relate to centuries of the distant past but arises even for more recent historical times. For example, in the case of lower-intensity effects occurring in medieval times, it needs to be established whether the epicentre in an earthquake catalogue was given based on information from the only place, or from one of the few places, capable of preserving written records of seismic effects (such as a town or an important village). In these cases, the real epicentre may have been many miles away in perhaps mountainous and isolated locales for which no written records were ever made (see the case of the enigmatic earthquake of 29 November 1784 in Vogt, 1987).

How is it that earthquake catalogues have arbitrary epicentres that point to specific localities? This generally happens as a result of two basic situations that commonly arise in historical seismology research:

(1) only one text is available and it contains information about the effects at a single site;

(2) various texts are available, but they all provide information only about one and the same site.

The first case can arise due to a lack of basic research regarding the types of sources available and their geographical spread. Since the discovery of a text recording earthquake effects at a single locality may be a matter of chance, it is necessary to carry out systematic research starting at the places nearest to the known site to determine whether other information about the event can be uncovered.

The second situation mentioned in the previous paragraph may be the result of research which was indeed systematic but only covered one type of source (e.g. information solely from a chronicle or solely from public authority records). The results of the research might therefore be biased by an inadequate geographical coverage of the possible relevant sources. It may be that authoritative sources come from a single city or cultural centre (a monastery), or even that information comes from the history of a single building (a fortress, church, palace, etc.), in which case it may be sufficient to broaden both the types of sources consulted and the area in which the sources were produced. One should usually begin in this case by checking those local histories, including recent ones, which were produced within an area extending at least 20–30 or perhaps even 50 km around the site referred to in the only currently known record. It is possible that this investigation can provide some indirect indication of other effects of the earthquake to which a location is given. If so, the key to the investigation is the indication of this new locality. Once such information has been procured, new research can then be carried out in sources directly relating to the new locality. With luck, this procedure can then expand to even more localities. This procedure shows how important it is to establish a dialogue between those who are planning research in historical seismology and those who compile and edit current earthquake catalogues. The latter should not work too much in isolation if they are to produce the most accurate possible catalogues.

6.3.5   Earthquakes in areas that are silent from the standpoint of seismology

When there is no evidence of past earthquakes for some area of seismo-tectonic importance, researchers inevitably wonder whether this is due to the actual absence of seismic activity or to the absence of sources that bear witness to those earthquakes that did take place. Historical seismology, with the help of other seismological disciplines such as geology and palaeoseismology, is able to explain silent seismic areas. Although originally used in 1909 by the Japanese seismologist Omori (1909), it took a long time before the concept of *silent* earthquake areas was accepted in seismology. As is well known, Omori used this term for areas that are to be considered as seismic, since seismotectonic models indicate that earthquakes should occur in these areas in spite of the lack of reports of seismic events. Omori located a *silent seismic area* in the south central Apennines in Italy (Figure 6.5). Indeed a few years later in 1915, the destructive earthquake in the Marsica region ($I_0 =$ XI MCS; $M = 7.0$, about 30 000 deaths) corroborated his theory.

Today the analysis of silent seismic areas represents a very interesting research strategy. It seeks to locate both historical earthquakes that took place on active faults and gaps in the seismic activity of an active geological structure. Once the seismic areas are identified, those areas for which historical seismological data are not already available can be analysed in greater detail for past earthquake activity. In this case, the contribution of historical research is to delve into the historical studies concerning the area being analysed. It is important to know whether historical sources actually exist and whether they are available. Indeed, it is important to be able to account for a lack of records on past earthquakes. A lack of historical records should not be interpreted *tout court* as an indicator of a low level of seismic activity.

This research strategy, for instance, has proven to be highly relevant to Italy (Valensise and Guidoboni, 1995, 2000b). In fact, since all of the Italian active faults studied so far have generated large earthquakes with an average return period of the order of 1000 to 2000 years, it is rather unlikely that history

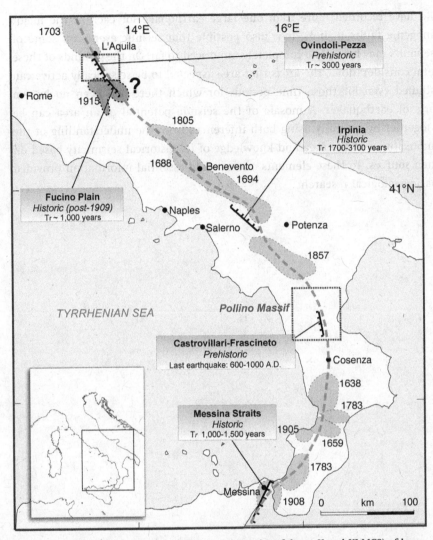

Figure 6.5 Areas of largest damage (intensity of degee X and XI MCS) of large Apennine earthquakes between 1638 and 1908 (after Omori, 1909, redrawn and modified), and average recurrence intervals for selected large Italian faults. The figure shows a remarkable alignment along the crest of Apennines and highlights two large gaps in historical seismicity (dotted areas), indicated by Omori himself after the 1908 Messina earthquake as the most seismically dangerous places of central and southern Italy. Only a few years after Omori published his work, about one-third of the Abruzzi gap was filled by the occurrence of the 1915 Fucino Plain earthquake (the intensity X area for this earthquake is indicated in a dotted line) (from Valensise and Guidoboni, 2000b).

would have recorded more than one large earthquake for each of the major seismogenic faults in Italy. It is also possible that seismic events on some of those faults may have gone completely unnoticed so far. On the grounds of these general considerations, the areas that are suspected to be seismically active can be studied even for those time periods for which there had been no known reports of earthquakes. A mosaic of the seismic potential of an area can be put together by carefully using both inferences from the understanding of the seismotectonics of the area and knowledge of its historical seismicity based on written sources. To these elements one can add essential information provided by basic geological research.

7

Determination of historical earthquakes: dates and times

7.1   The need for a common time base for earthquake catalogues

Seismologists need a common time base to study earthquake progression through time in order to estimate the rates at which earthquakes take place in a region and in order to determine the return period of strong earthquakes on a fault or other specific geological structure. It is also necessary to have a common date and time basis to relate the historical earthquake and tsunami accounts from different parts of the world in the search for a common earthquake or tsunami source for observations that are far apart and recorded by different cultures in different languages and types of documents. Today, seismologists the world over use the Gregorian calendar for dates and UT (Universal Time) for time, and these are the standards to which all historical accounts need to be related. The understanding of the calendar and time systems from different parts of the world and the capacity to convert between these and the modern date/time system are vital components for assembling modern catalogues of historical events.

Scientific consensus does not correspond to life in the everyday world; different calendars and chronological computations still coexist in the very same areas, like the Mediterranean countries, whose wide variety of dating systems both in the past and in the present represents a major challenge for seismologists. Temporal earthquake parameters, in particular the origin times of shocks, are parameters that can be easily mistaken, and virtually no earthquake catalogue is error-proof in this respect. Date and time are the main labels of any earthquake, because with enough precision they invariably are unique for each event in the catalogue of a region. Nevertheless, establishing the exact chronology of earthquakes that occurred at different historical periods and in different cultural areas can be very hard to accomplish with unquestionable reliability. As a consequence of the possibility of date and time errors, earthquake catalogues are particularly susceptible to three kinds of distortion:

(1)   earthquakes may be artificially duplicated;
(2)   separate tremors may be mistakenly conflated;
(3)   there may be a failure to recognize the nature of certain sequences of events (such as earthquakes occurring essentially simultaneously or in close succession).

In particular, the third problem (i.e. recognizing individual strong shocks within one or more seismic sequences) may be particularly difficult. Especially in the case of a seismic sequence characterized by a high number of individual shocks perceived by witnesses, the risk of discrepancies in the record of individual shocks among the various sources is high. Furthermore, when shocks were 'continuously' perceived during a seismic sequence, they were often not individually recorded. In this case direct witnesses often reported only the temporal limits within which the shocks were perceived and the approximate number of perceived shocks. Moreover, it is often the case that one or more strong aftershocks caused fresh damage in the area affected by the mainshock. This fact may not be spelled out in the written source, thus making it difficult to distinguish the effects of the individual shocks. In fact, at times direct sources only report cumulative effects since they regarded the whole sequence of earthquake as a single event. The reporting of cumulative earthquake effects may lead to an overestimation of the intensities of individual shocks in an earthquake sequence (see Chapter 8).

Conscientious historical seismological research can lead to an accurate identification of the main shocks of a seismic sequence and their related effects. This is possible only by paying great attention to the indications concerning time contained in the sources, whether those sources are ancient or recent. The best data will usually be derived from private memory, consisting mostly of chronicles and the memoirs of direct witnesses, which are more suitable for registering the sequence of individual shocks than the public memory as preserved in archives. Indeed, the latter does not pay attention to elements such as the number of single shocks but rather is concerned with the institutional process of damage examination and reconstruction. The time parameter has multiple implications in the study of an earthquake, and it is important to have an overall understanding of this aspect of historical seismology research and the problems connected with finding and interpreting it.

7.2   Dating styles and practice in ancient Mediterranean cultures

When dealing with the Mediterranean area (but it is no different for the European area as a whole) it must first of all be said that the use of different calendars which are not directly convertible into those in use today poses a constant challenge to the historical seismology researcher. Indeed, the fact is that even today different Mediterranean peoples use different calendars, and outside the scientific world societies are still a long way from arriving at a standard system of measuring time. Moreover, in the ancient world fluctuating terms of reference were the norm. In completing or reviewing an earthquake catalogue, it is thus necessary on the one hand to pay attention to chronological computations, and on the other hand to situate properly the ancient calendars in their cultural contexts.

Establishing the time at which something happened is not just a matter of arithmetic; in many cases it means making choices, attaching meanings or establishing relationships. That is why the chronological problem cannot be dealt with simply by supplying stripped down bibliographical information (which is in any case usually well known to historians, though scarcely ever to seismologists), but must also be dealt with by setting out, at least in general terms, the characteristics of the principal eras of the ancient world, thereby bringing into focus the chief problems that the various dating methods were supposed to solve. Moreover, this allows one to highlight the fact that the ancient sources are not concerned with accuracy in the sense in which the term is used today, that is to say something expressible in numerical terms. Instead, what one finds is a kind of clear vagueness or vague accuracy, if one can go so far as to use such openly paradoxical phrases. This quality is a function of the ancient cultural systems, and it creates great difficulties for modern investigators when they apply modern observational paradigms to the different cognitive worlds of the past. Even when the references provided are terminologically accurate and the modern researcher has the right tools for understanding them, there can be a fairly wide margin of error in the determination of the date or time of a seismic event that one may need to accept. Fortunately, the existence of a margin of error does not necessarily mean that the data themselves are incorrect.

7.3   Years, months and days

7.3.1   The ancient Greek and Roman worlds

In Greece, the way of counting years in terms of the cycle of the Olympiads was introduced by the Sicilian historian Timaeus in the third century

BC, but it is known that lists of winners at the Olympic games already existed by that time. The chronological system of Timaeus begins with the year 776 BC, when the Olympic games were founded. From then onwards, the games were held every four years until they were abolished at the end of the fourth century AD. From Timaeus onwards, all of the chronologies of the ancient world use the Olympiads as the basis for their calculations, and other dates are modified to fit this system. However, as the Olympic year begins in the summer, dates in Greek history often have two numerical values when expressed in terms of the Christian era. When Emperor Theodosius I issued an edict in AD 391 closing all centres of pagan worship, one of the consequences of this action was the abolition of the Olympic games three years later, although the Olympiadic dating system continued to be used in Byzantine chronology.

The days of the year were not arranged in the same way in all parts of Greece, but all the computations were based on the lunar cycle. Each month consisted of either 29 or 30 days, and there was an annual difference of seven and a half days between the lunar and the solar year. This difference was adjusted every eight years by intercalating additional months. The Athenian months were named: (1) *Hekatombion*, (2) *Metageitnion*, (3) *Boedromion*, (4) *Pyanepsion*, (5) *Maimakterion*, (6) *Poseidon*, (7) *Gamelion*, (8) *Anthesterion*, (9) *Elaphebolion*, (10) *Munychion*, (11) *Thargelion* and (12) *Skirophorion*. The intercalary month usually came after Poseidon, and was called second Poseidon. *Hekatombion*, and hence the beginning of the year, fell in the summer.

In Rome, the year always began on the first day of January. From 153 BC onwards the two consuls also took office on that date, so that the new year was identified by their names. From 222 BC until 153 BC, however, the consuls took office on 15 March. The later practice was followed until AD 541, when Flavius Basilius Junior became the last consul in Rome. In the imperial age, the year could be identified not only by means of the names of the consuls, but also by reference to the names of the emperors when accompanied by a statement of their tribunician power and imperial acclamations, as well as by the various titles that were conferred upon them.

In the Roman world, the phrase *ab Urbe condita* ('since the foundation of the city' of Rome) did not indicate an era, although the expression certainly provides us today with a valuable system for linking events chronologically. The Romans used the year of Rome for measuring the length of time that had elapsed between an occurrence and the foundation of the city, but they did not make use of this as an era because there was debate as to when the foundation of the city had actually taken place. Generally speaking, Latin historians thought the city had been founded between 759 and 748 BC. The predominant traditions were those of Cato (234–149 BC), who dated the foundation of Rome to 21 April

752 BC, and of Varro (116–27 BC), who dated it to the same day but in the year 753 BC. The latter was the dating which prevailed over all others. It had been established by Atticus (110–32 BC) in his annals, and after these were brought to public attention by Varro, this dating system acquired official status and was preferred by historians.

A list in the *Fasti Capitolini* records the number of years *ab Urbe condita*, but the figures given are one year less than those provided by Varro because the consular years have been made to fit the civic years, the former beginning on 1 January but the latter on 21 April. Thus, the consular year cuts across two civic years. That is how the years are counted in the *Fasti Capitolini*, where the foundation of the city seems to be dated to 752 BC. In the *Fasti Romani* the dating system has been preserved without gaps for 1047 consular years from Brutus to Basilius Junior. The *Fasti Consulares* of the Roman republic have come down to us today in three texts from the Augustan age: in the *Fasti Capitolini*, an inscription in the Forum which has survived in part and been added to with the aid of later calendars (the fifth century *Fasti Hydatiani*, the seventh century *Chronicon Paschale*, etc.); in Livy (59 BC – AD 17); and in Diodorus Siculus (c.90–c.20 BC) and Cassiodorus (AD c.490–c.583). All of these probably derive from a single original source.

In ancient Rome the days were counted by reference to three particular days: the Calends (*kalendae*), which was the first day of the month; the Nones (*nonae*), which was the fifth day of the month (except in the months of March, May, July and October, when it was the seventh day); and the Ides (*idus*), which was the thirteenth day of the month, except in the four months just mentioned, when it was the fifteenth day. The word *kalendae* comes from the Latin verb *calare* (to call or shout), for on the first day of the month the *pontifex* proclaimed on what day the Ides would fall. From *kalendae* comes the word *kalendarium*, which was the name given to the book that recorded the time when sums owed fell due. The *kalendarium* contained a list that indicated the days when it was permissible to administer the law (*fas*), and after it came into the public domain, each day of the year was marked 'f' (*fastus*) or 'n' (*nefastus*). Thus, the *fasti* functioned as a type of calendar. Originally, the *fasti* were a list of days in the sole custody of the *pontifex maximus*, but it became a document that was issued publicly after 304 BC.

The Nones always fell on the ninth day before the Ides (it should be kept in mind that the Romans included both the first and last days in their calculations), and was therefore either the fifth or the seventh day, depending on the month. Varro thought that *idus* was a word of Etruscan origin and that it meant the middle of the month. These three basic dates were used for calculating all the other dates in the month, the other dates being arrived at by counting how many days there were before the next base date.

7.3.2   The Hebrew calendar: the 'years of the World'

The Hebrew calendar is rather special, and it is based both on the lunar phases and on the movements of the Sun (it is called lunisolar). It repeats itself cyclically every 19 years, and each cycle is perfectly equivalent to 19 solar years. The single year may be, according to the lunar phases, common or embolismic. In the former case it is made up of 12 months for a total of 353, 354 or 355 days (defective, regular and abundant, respectively); in the latter case, it is made up of 13 lunar months for a total of 383, 384 or 385 days. The common years are 12 (the 1st, 2nd, 4th, 5th, 7th, 9th, 10th, 12th, 13th, 15th, 16th, 18th), and when interpolated with seven embolismic years (the 3rd, 6th, 8th, 11th, 14th, 17th, 19th) the result forms the 19-year cycle (known as the Metonic cycle). While being of different duration and starting in different periods (in particular, the year of the Hebrew calendar has a variable duration and thus a variable new year's day), the Hebrew calendar and the Julian and Gregorian calendar coincide every 19 years. The start of the era for the Hebrew calendar (the years are the 'years of the World' according to the Jews) is 1 *Tishri* 1 = 7 October 3761 BC. The names of the months are: (1) *Tishri*, (2) *Heshvan*, (3) *Kislev*, (4) *Tevet*, (5) *Shevat*, (6) *Adar* I (or *Adar* II), (7) *Nisan*, (8) *Iyyar*, (9) *Sivan*, (10) *Tammuz*, (11) *Av* and (12) *Elul*. The days of the week are identified by adopting the Jewish custom of giving each day an ordinal number, starting from the day after the Sabbath, a holy day for the Jews. Thus, the first day of the week corresponds to Sunday on the Western calendar.

7.3.3   The Julian calendar

In early times, the Roman year was divided into ten months, but the months *Ianuarius* and *Februarius* (January and February) were added from the fifth century BC onwards (or earlier, according to tradition). These 12 months were lunar and gave a total of 355 days, but at the time of Caesar (first century BC) the calendar was so misaligned relative to the astronomical seasons that something had to be done. Hence, in 46 BC Julius Caesar realigned the calendar with the solstices and equinoxes, and from 1 January 45 BC a new calendar with a 365-day year was created by distributing ten extra days amongst the various months of the older calendar. The Julian year and calendar was gradually introduced into all provinces by the Roman government. In the West it replaced the previous chronological systems, while in the various regions of the East, the calendar then in use was modified to fit the new system. In the Western countries of Europe the Julian calendar remained in use until the Gregorian reform of 1582. It must be remembered that in the Protestant countries of western Europe the Gregorian reform was adopted sometime after 1582 (1752 in England). The adoption of the Julian calendar in Eastern Europe is more complex. While in Poland it was substituted by the Gregorian calendar in 1586, at that time in Russia chronological computation was still based on the Byzantine dating system. The Byzantine system was used in Russia until 1699, when the Julian calendar was adopted; the latter remained in use until 1918, when the Soviet government finally adopted the Gregorian reform.

7.3.4   Regnal years: the Greek–Byzantine area

In considering the dating system based on regnal years, it must be kept in mind that until the first half of the sixth century AD, documents of the Eastern Roman Empire usually adopted the classical Roman dating system, which gave the names of the consuls in office, with or without the regnal year. In *Novella* 47 of 31 August 537, the emperor Justinian (527–565) ordered that public deeds should always be dated with the regnal year, the consular year and the indiction. From the year 541 when Flavius Basilius Junior, the last consul in the *pars Occidentalis*, was elected, it became a practice to count the years that had elapsed since he was consul (the post-consular system). When the emperor Justin II (565–578) assumed the office of consul himself in 566, it became the practice to indicate his post-consular years after the regnal years. But it must be remembered that he became consul for a second time in 568. After that date, the years may be expressed either by reference to his first or his second consulship. As time passed, however, the consular computation of the years was abandoned in favour of the exclusive use of regnal years.

The Byzantine calendar corresponds exactly to the Julian calendar except in two ways. From the sixth century onwards it makes less and less use of the system of *kalendae*, *idus* and *nonae* for computing the days of the month, preferring the use of ordinal numbers, and the year begins on 1 September, not 1 January. It must also be mentioned that up to the ninth century another type of calendar can be found alongside the Julian calendar, especially in chronicles. In most cases, this calendar was the Macedonian calendar or a variation of it (see the *Chronicon Paschale* and the *Chronographia* of Theophanes), which had come into being in Egypt and subsequently spread through Syria and Asia Minor following the conquests of Alexander the Great (334–331 BC).

7.3.5   The Christian area

Two basic elements affected the way in which Christian world eras took their form. The first was purely religious, while the second was both religious and astronomical at the same time. Those calendars which are based solely on the religious element draw a parallel between the history of the world and the six days of creation in the Bible (Genesis 1.1–32) and so make each day of the week in the Creation correspond to a thousand years in the history of the human race. The Incarnation is placed roughly in the middle of the sixth millennium, the end of which would bring the end of the human race and the beginning of eternal rest (the Epistles of Barnabas, Irenaeus, Clement of Alexandria and Hippolytus). In this system, the number 6000 assumes a powerful, sacred and eschatological significance, thereby supplying Christian chronologers with a standard measure for calculating the age of the world. The second element involved in working out world eras again derives from religious needs, but is also inextricably linked to astronomical computations. Its point of departure from the purely religious system arises from the debate about the date for the celebration of Easter, to which the entire liturgical calendar for the year is tied. Differences developed in the various Christian churches from the second to the eighth century concerning how to fix the date for the celebration of Easter, and this affected the calendars that were employed.

In AD 325, the Council of Nicaea tried to put an end to disputes about the Easter question by ordering that Easter should be celebrated on the Sunday following the first full moon after the vernal equinox. The problem with establishing a date for Easter lies in relating the story of the Passion in the Gospels, where it is dated in terms of the Jewish lunar calendar (14 *Nisan* according to the chronology of John, or 15 *Nisan* according to the Synoptic Gospels), to the liturgical requirements of Christianity, which are governed by a solar calendar.

The lunar year is approximately 11 days and 6 hours shorter than the solar year, and so, to avoid the date of Easter advancing perpetually through the days of the solar calendar, Christian chronologists from the beginning of the third century onwards began working out what are known as the *Easter tables* or *cycles*. These are a series of years (the exact number of years in an individual cycle varies; it could be 112, 84, 19, 28, 95 or 532) at the end of which the total number of days in the solar years of a cycle is the same as the total number of days in the corresponding lunar years. This is achieved by inserting into the lunar years, at regular intervals, what are known as intercalary or embolismal months (*embòlismos* = intercalation). Use is also made of other devices such as the *saltus lunae*, which eliminates the discrepancy between the course of the Sun and that of the Moon by artificially increasing the age of the Moon. In this way solar time and lunar time are realigned at the end of a cycle, and so the Easter dates established at the beginning of the cycle are taken over unchanged in later cycles. It can be maintained that an Easter cycle always underlies a world era, because the era is a multiple of the number of years in a cycle. It must be kept in mind, of course, that the calculations in this system are based on the assumption that the birth of Christ took place 5500 years after the world was supposed to have been created and that the Creation must have occurred at about the middle of the sixth millennium.

In 258, Bishop Anatolius of Laodicea worked out an Easter cycle of 19 years, which reckoned 5501 years from the Earth's creation to the coming of Christ. It was in an attempt to reform this cycle that two other important cycles for the history of Byzantine chronology came into being: the cycle of Peter of Alexandria and that first used at Constantinople in 354. Archbishop Peter adapted the cycle of Anatolius by making the first year of his own cycle correspond to the ninth of that of Anatolius, which began in 303 and coincided with the twentieth year of the reign of Diocletian. Since there were exactly 19 years between the beginning of the reign of Diocletian (284) and the end of the 19-year cycle preceding the year 303 (the year 303 is thus the first year of the next cycle), the custom was established of counting the years of a cycle from the beginning of the reign of Diocletian. If one divided the number of years which had elapsed since his reign began by 19, the remainder always gave the year in the current cycle and hence the date of Easter. That is how the *Era of Diocletian* (or *Era of the Martyrs*) came into being; it was the principal era used in the Egyptian area and in Coptic circles.

The Easter table worked out by Theophilus of Alexandria at the end of the fourth century and dedicated to the emperor Theodosius I (379–394) was based on the cycle of Archbishop Peter. Theophilus's Easter table, in turn, provided the basis for the *Alexandrian Era* which, in the version drawn up by the monk Annianus, computes 5492 years from the Creation to the birth of Christ, and begins the year on 25 March. The Alexandrian Era was also used in other parts of the Eastern Empire, especially in monastic circles (e.g. by Georgius Syncellus and Theophanes the Confessor).

During the reign of the Arian emperor Constantius II (337–361), another important modification of Anatolius of Laodicea's 19-year cycle was made at Constantinople, perhaps by a chronologist named Andreas. This new cycle was put into operation in 353, and it provided the basis for the so-called Protobyzantine Era, which begins the year at the end of March and dates the Creation to 5509 BC. This era was used, together with other dating systems, in the *Chronicon Paschale* (seventh century), but it must not be confused with the real Byzantine Era. There is disagreement among scholars about the relationship between the former and the latter. In any case, it was suggested that the era used in the *Chronicon Paschale* is simply an attempt to move the beginning of the Byzantine era to 5508.

7.3.6   *Medieval variations of the Christian era*

The Christian era was introduced by the Scythian monk Dionysius Exiguus, an abbot in Rome in the sixth century. He determined that the date

Table 7.1 *Comparison of the year computation between different dating styles and the modern calendar*

Style	Period of year	Difference from modern calendar	Adjustment
Nativity	1 Jan.–24 Dec.	years are =	years are =
	25 Dec.–31 Dec.	+ 1 year	− 1 year
Florentine	1 Jan.–24 Mar.	− 1 year	+ 1 year
	25 Mar.–31 Dec.	years are =	years are =
Pisan	1 Jan.–24 Mar.	years are =	years are =
	25 Mar.–31 Dec.	+ 1 year	− 1 year
Veneto	1 Jan.–28/29 Feb.	− 1 year	+ 1 year
	1 Mar.–31 Dec.	years are =	years are =
Easter	1 Jan.–21 Mar./24 Apr.	− 1 year	+ 1 year
	22 Mar./25 Apr.–31 Dec	years are =	years are =
Byzantine	1 Jan.–31 Aug.	years are =	years are =
	1 Sept.–31 Dec.	+ 1	− 1

of the birth of Jesus was 25 December 753 *ab Urbe condita*. The first year of the Christian era thus corresponds to the year 754 dating from the founding of Rome. This calculation is probably off by six or seven years relative to the actual date of the birth of Christ, but the precise extent of the chronological deviation is still the subject of academic debate. The use of the Christian era spread into Italy during the sixth century. By the beginning of the seventh century it had reached England, Spain and France, and in the tenth century it was known and used in the greater part of the West. Latin documents refer to the Christian era with a number of formulae: *anno incarnationis, ab incarnatione Domini, anno Domini, a nativitate Domini*, etc.

Throughout the medieval period, the date of the beginning of the year shows remarkable variations depending on geographical region and sometimes on the type of source. While all systems have the same way of indicating days and months, the date of the year for a single event can vary by a year depending on the style adopted. The following summarizes the variations in the beginning of the year in Europe (see also Table 7.1):

(1) In the present-day or modern style, the first day of the year, the Feast of the Circumcision, sometimes indicated by the formula *anno circumcisionis*, falls on 1 January; this was relatively rarely used in the Middle Ages.

(2) In the style known as that of the 'Feast of the Nativity', usually indicated with the formulae *anno a nativitate Domini* or more simply

anno Domini, the first day of the year falls on 25 December, Christmas Day, seven days earlier than in the modern calendar.

(3) In the style known as that of 'The Incarnation', usually referred to with the formulae *anno ab incarnatione Domini, Dominicae incarnationis*, the first day of the year falls on 25 March, the Feast of the Annunciation to the Virgin. Since this date is 2 months and 24 days later than the modern start of the year, the number of the year between 1 January and 24 March is one lower than that according to modern reckoning. It is also known as the Florentine style, since it was long used in Florence and other Tuscan cities, and differs from the so-called Pisan style which, although still making 25 March the first day of the year, anticipates the modern date by 9 months and 7 days.

(4) In the Veneto style or *more veneto*, sometimes indicated by the initials *m.v.*, the first day of the year falls on 1 March, postponing the modern date by 2 months. Thus, any date between 1 January and 28 or 29 February belongs to the year previous to the year according to modern reckoning. It is known as the Veneto style because it was used in Venice from the Middle Ages until as late as the fall of the Republic in 1797. It should be noted that many Venetian notaries of the thirteenth and fourteenth centuries use the formulae *ab incarnatione* or *a nativitate*, even though their year begins on 1 March.

(5) In the Easter or French style (*mos gallicanus*), the first day of the year falls at Easter on the Feast of the Resurrection, between 2 months 21 days and 3 months 24 days later than according to the modern calendar. Since the date of Easter can fall any time between 22 March and 25 April, this means that between 1 January and 21 March or 1 January and 24 April the number of the year is one lower than that calculated according to modern usage.

(6) In the Byzantine style, the first day of the year falls on 1 September, anticipating by 4 months the start according to modern usage. Dates between 1 September and 31 December thus fall one year later than the year according to modern reckoning.

7.3.7 *The Byzantine dating system*

In the Byzantine dating system (which lasted for more than 11 centuries), 5508 years were calculated to have elapsed between the Creation and the birth of Christ. This system was worked out in the closing decades of the seventh century, but it was only at the end of the tenth century that it succeeded in supplanting the other established chronological systems in the Byzantine empire. For some centuries it had to compete with other calendars which had been worked out since the third century AD in the territories of the eastern Mediterranean or, in a few cases, with calendars that went back to Hellenistic times.

When the Arabs gradually occupied the religious and cultural centres of the Fertile Crescent in the seventh century (Antioch in 636; Beirut and Jerusalem in 638; Alexandria finally conquered in 646), the chronological systems in use there had already spread to Constantinople and Byzantine Anatolia. The fact that no single dating system held sway in the Byzantine empire until the tenth century, except for the use of indictions (a cycle of 15 years) and the regnal year of the *basileus* (emperor), is reflected in chronicle usage.

The traditional Byzantine Era, in which the year began on 1 September and 5508 years had elapsed between the Creation and the birth of Christ, had certain obvious advantages over the eras mentioned above. It fitted in with the indictions and avoided the clash between the natural computation (*katà physein*) and conventional computation (*katà thésin*) implicit in the 19-year cycle of Anatolius, which neither the Alexandrian Era nor the one adopted in the *Chronicon Paschale* had succeeded in avoiding.

7.3.8   Medieval indictions

Historians date the introduction of indictions to the time of the emperor Diocletian (284–305). His fiscal reforms were based on the establishment of a close relationship between the crops a farmer was capable of producing and the quantity/quality of the cultivable land, and it was on this basis that he proceeded to regularize the requisitioning of foodstuffs to feed the army, which had been supplied in a haphazard way since the time of Trajan (AD 98–117). The word *indictio* (*epinémesis*) means fiscal distribution or imposition, and the Byzantine terms *índiktos* and *indiktió* are based upon it. At the time of Diocletian, what was probably meant by the term was the assessment of taxable property as carried out for the first time in Egypt in 287 or 297 and to be repeated at the end of each five-year period for updating purposes (Stein, 1968). In 313, during the reign of Constantine, this period of five years was extended to 15 years. The term *indiction* later lost its original fiscal sense and came to be applied to the chronological reckoning of years on a 15-year scale, so that the years were counted from one to 15 and then returned to one at the beginning of each new cycle.

The first example of a public deed with an indiction dating is a decree of Constantius II of 15 January 356, contained in the *Codex Theodosianus* (12.12.2). In 537, as we have already pointed out, Justinian made it obligatory to include the indiction year in all public deeds. At least from the fifth century onwards Byzantine indictions began on 1 September, but elsewhere in medieval Europe the beginning of the year varied. The Sienese indiction began on 8 September; the Bedan or Caesarean indiction, which was used principally in continental Europe and Italy, began on 24 September, though in the papal chancery it was only adopted at the time of Urban II (1088–1099). The most frequently used indiction from the ninth century onwards was the so-called Roman or Pontifical indiction, which began either on 25 December or, more frequently, on 1 January. This was the principal indiction used in the early Middle Ages.

7.3.9   Armenian chronological computations

The Armenian computation system is vital for understanding not only the historical earthquakes of Armenia but also those of present-day eastern Turkey, an area of abundant seismic activity, into which historical Armenia extended. In the earliest periods, the Armenians used the chronological computations of the nations to which they were politically subject or whose political influence was paramount. Even the Armenian language did not acquire literary codification until the fourth and fifth centuries AD. In Armenian circles, the earlier tradition was to write chronicles and annals in Syriac or Greek, and the authors do not seem to have adhered to a chronological system related to a calendar. As soon as Armenia was Christianized, the computations of Eusebius of Caesarea were adopted through the use of a complex system of royal and patriarchal genealogies.

It was the need to establish the dates of Christian religious festivals (especially Easter) which ensured that a fixed calendar was adopted. At first it seems likely that the Byzantine calendar was used. Later on, because Armenia became detached from the Greek world in a religious and political sense, the *great Armenian* era was introduced. Its beginning has been set at the end of the 200-year canon of Andreas, which had been worked out in Byzantium for the years 353–552 (but we know this calendar only from Armenian sources). Zoroastrian Persia had an influence on the development of the new calendar, which consisted of 12 30-day months arranged in the following order: (1) *Nawasardi*, (2) *Hori*, (3) *Sahmi*, (4) *Tre*, (5) *K'aloc'*, (6) *Arac'*, (7) *Mehekani*, (8) *Areg*, (9) *Ahekani*, (10) *Mareri*, (11) *Margac'* and (12) *Hrotic'*. To the 360 days contained in these months were added five epagomenal days (*aweleac*), in accordance with traditional Persian and Egyptian usage. The era began on 1 *Nawasardi* in year 1, which corresponded to 11 July 552 AD.

This *great Armenian* calendar was not adopted immediately. It gradually gained ground towards the end of the sixth century, and there were various attempts to modify it. Unlike the Egyptians, the Armenians did not observe the custom of adding a sixth epagomenal day every four years, with the result that their calendar had a vague year that was out of phase with the Julian calendar by one day every four years. The adoption of a 365-day cycle (i.e. one which was out of phase with the solar year by six hours every year) was always a problem, especially as regards the dates when religious feasts were to be held as prescribed by the menologists.

The *great* era was used by all Armenians until the twelfth century, when the Julian calendar was introduced by John the Deacon at the time of the patriarch (*kat'olikos*) Grigor III Pahlavuni, with the new era beginning with the year 1085. However, it was principally in northern Armenia that this *small* era caught on. Many historiographers and chroniclers made the problem more complicated by using their own chronographical computations. The situation was made even worse by the fact that the beginning of the Armenian year coincided with the Christian feast of Epiphany. Indeed, some historians have suggested that the Armenians had a double cycle, with technical computation on the one hand and historiographical computation on the other, but the fact is that any discrepancies can be attributed to individual historiographers.

In order to work out the year of the Christian era in which a year of the Armenian *great* era begins, add 551 to the Armenian year up to the year 769 inclusive. Thus 1 *Nawasardi* 769 corresponds to 1 January 1320. From the Armenian year 770 (which begins on 31 December 1320) onwards, add 550.

7.3.10   Syriac chronological computations

Syriac sources date events according to the Seleucid Era, which took its name from king Seleucus I Nicator. When he took the title of 'King of Babylonia' in 309/308 BC, Seleucus abandoned the chronology based on the regnal years of Alexander IV (the son of Alexander III the Great and Roxana) to adopt a chronology following his own regnal years, but he antedated his accession to the throne and fixed it in 311 on the first day of the year according to the Babylonian lunar calendar, i.e. *Nisan* 1 (in that year = 3 April). Later, when he was also recognized by the Greeks as *basileus*, he put back the beginning of his chronology to 312, starting to count from the beginning of the Macedonian year in the autumn on the first day of the Macedonian month Dius. Macedonian names were assigned to the Babylonian months, so that there was one calendar and two slightly different era systems (the Babylonian system for Chaldea and the Macedonian system for Syria and Asia Minor).

As in all territories conquered by the Romans, the Babylonian lunar calendar was superseded by the Julian solar calendar. Here, too, Babylonian and Macedonian names were simply assigned to the Julian months. This reform gave birth to the chronological system that was to be found in the texts of the Christian period. In practice, there are two ways of transforming a Seleucid year into a Christian year: if the date is between 1 October and 31 December, subtract 312; if it is between 1 January and 30 September, subtract 311. The month correspondences are as follows: (1) *Hyperberetaeus* = *Tishrin I* = October; (2) *Dius* = *Tishrin II* = November; (3) *Apellaeus* = *Canun I* =, December; (4) *Audynaeus* = *Canun II* = January; (5) *Peritius* = *Shebat* = February; (6) *Dystrus* = *Adar* = March; (7) *Xanthicus* = *Nisan* = April; (8) *Artemisius* = *Iyyar* = May; (9) *Daesius* = *Haziran* = June; (10) *Panemus* = *Tammuz* = July; (11) *Loüs* = *Ab* = August; (12) *Gorpiaeus* = *Elul* = September.

7.3.11   Coptic and Ethiopian chronological computation

Ethiopian culture is one of the most ancient and refined cultures of the Christian world. For the Ethiopian area there is a critical historical earthquake catalogue of high quality produced by Gouin (1979). One of the main problems Gouin had to solve was that of analysing the Ethiopian measurement system, which is particularly complex and even now well known only to a small number of scholars. The Ethiopian year is a Julian year, and its length is 364.25 solar days. In all Ethiopian chronologies, the first day of the year (1 *Meskerem*) originally corresponded to 29 August of the Julian calendar, but as Ethiopia did not adopt the Gregorian reform, 1 *Meskerem* advanced in time with respect to Gregorian dates to 11 or 12 September, as it remains at present.

Ethiopian years are grouped in cycles, which last four years each. The cycles bear the names of the Evangelists in the same order as the books in the New Testament: Matthew, Mark, Luke and John. The year of Luke is the leap year of the cycle and the extra day is added at the end of the year. In Ethiopian chronologies the leap years are those whose number divided by four leaves three as the remainder.

Ethiopian years are embolismic. They are divided into 13 months, 12 of which last 30 days, while the thirteenth lasts 5 days (6 during leap years). The names of the months are as follows: (1) *Meskerem*, (2) *Tekemt*, (3) *Hédar*, (4) *Tahsas*, (5) *Ter*, (6) *Yekatit*, (7) *Megabit*, (8) *Myazya*, (9) *Genbot*, (10) *Sené*, (11) *Hamlé*, (12) *Nahassié* and (13) *Pagoumié*.

7.3.12   Islamic chronological computation

The first year in Arab culture began on 16 July 622 in a period of religious and political transition in the Arabian Peninsula. This is what is known as the day of the *Hijra* (the Prophet's famous migration from Mecca to Medina). The first time this date was used as a point of departure was during the caliphate of 'Umar (AD 634–644), the second calliph after the death of the Prophet; 16 July 622 was called the first day of *Muharram* in year 1 (on the basis of astronomical calculations, but that day is in fact 15 July).

The Muslim year consists of 12 lunar months, each containing of 29 or 30 days. Their names are as follows (the number of days in the month is shown in brackets): (1) Muharram (30); (2) Safar (29); (3) Rabi' I (al-awwal) (30); (4) Rabi' II (al-thânî) (29); (5) Jumada I (al-ula) (30); (6) Jumada II (al-akhira) (29); (7) Rajab (30); (8) Sha'ban (29); (9) Ramadan (30); (10) Shawwal (29); (11) Dhu 'l-Qa'da (30); (12) Dhu 'l-Hijja (29; but 30 in leap years). Being based on lunar cycles, this calendar consequently runs faster than the solar calendar.

To express a date from the Muslim calendar in terms of the Gregorian calendar, it is not sufficient to carry out a scaling operation or add 622. One must take into consideration the fact that the lunar year is 11 days shorter than the solar year. This means that the number of the Muslim year must be multiplied by a term accounting for that difference, that is to say by 0.97, and then 622 must be added. Similarly, to express a Gregorian year in Muslim terms, one must subtract 622 and divide by 0.97. There are also formulas, worked out by Millosevich in 1913, for expressing the relationship between the Muslim calendar (H) and the Gregorian (G) or Julian calendar:

$$H = G - 622 + (G - 622)/32 \text{ and } G = H - (H + 622)/33.$$

These calculations, however, do not solve all the problems of relating the Muslim and Gregorian chronological systems. The fact that the months are not of equal length means that, in terms of the solar calendar, the relationship of months and days is always changing. Hence, in order to find out the exact Gregorian day and month on which a particular year begins and ends, the relevant chronological tables must be consulted (Freeman-Grenville, 1995).

In Islamic countries, or those with a predominantly Islamic population, the lunar calendar is used in different ways even today. In some of these countries, such as Saudi Arabia, the lunar calendar is used for official dating. In other countries, such as Iran and Pakistan, double dating is used. The Iranian calendar was introduced in 1925. It too is based on the Hijra, but uses a solar year. The year has twelve months and begins on 21 March (the spring equinox). Most Middle Eastern countries now use the Western calendar, with the lunar calendar being maintained for festivals and Islamic religious events.

In the ancient Arab world (i.e. before the rise of Islam) there was no precise calendar for dating events. However, the more important events were themselves used as temporal points of reference for dating later events. For example, the collapse of the Ma'rib dam in the days of the kingdom of Saba (the present-day republic of Yemen) corresponds to 120 BC; the construction of the Ka'bah (the sacred stone) in present-day Mecca corresponds to the year 1885 BC, and so on.

7.3.13   The Gregorian reform and the new calendar

During the Middle Ages differences between the Julian year and the solar year were noticed. In particular, it was observed that the former was too long (365.25), since the correct value for the tropical year is 365.242199 days. This error of 11 minutes 14 seconds per year amounted to almost 3 days in four centuries. Over many centuries, the calendar gradually became out of synchronization with the equinoxes.

Popes, monks and kings wanted to resolve this question, but no action could be taken as an accurate value of the tropical year was not available. By 1545 the vernal equinox, used in determining Easter, had moved ten days from its proper date on the calendar. When the Council of Trent met in December for the first of its sessions, it authorized Pope Paul III to take action to correct the error. But, in spite of various attempts, it was not until the election in 1572 of Pope Gregory XIII that the Julian calendar was properly reformed. Gregory XIII agreed to issue a bull that the Jesuit astronomer Christopher Clavius (1537–1612) drew up, using suggestions made by the astronomer and physician Luigi Lilio (1510–1576). The bull *Inter gravissimas*, promulgated on 24 February 1582, established that the day following the feast of St Francis (5 October) was to become 15 October. This made the civil year coincide with the solar year. In addition, Gregory established that no centennial years should be leap years unless exactly divisible by 400. In other words, three out of every four centennial years should be common years. In spite of these corrections there still exists an error of about 6 days every 10 000 years, but this means that for the next 1000 years the calendar is not in need of any noticeable correction.

The Gregorian calendar was almost immediately adopted in the Catholic countries of Europe (Spain, Portugal, France, Poland), while in most of the Protestant countries and in the Eastern European countries (Eastern Christian churches), in opposition to the papacy, it was adopted later. Among the latter, the Protestant regions of Germany and the Netherlands adopted it in 1700, Great Britain adopted the Gregorian calendar in 1752 and Sweden adopted it in 1753. In Greece it was enforced in 1923 and in Turkey in 1927. Because of these wide differences when the Gregorian calendar was adopted, it is advisable during the revision of a catalogue of historic earthquakes to check very carefully the calendar systems that were used to date the events.

Nowadays, the Gregorian calendar is almost universally used. The only historically relevant exception in modern times is that of the so-called *republican calendar*, which was established by the French revolutionary government (National Convention) on 24 November 1793. It remained in use until the end of the year 1805, when the emperor Napoleon abolished it and ordered the restoration of the Gregorian calendar. As far as earthquakes are concerned, the *republican calendar* is important for events that occurred during that period, such as the earthquake on 25 January 1799 (6 *Pluviose* An VII) in Vendée, western France, for which the records are dated in this alternative calendar. The *republican calendar* was adopted in various parts of Italy after the French revolution extended to this country.

An example of how the French revolutionary calendar affects historical seismology research is that of the earthquake at Camerino (central Italy) in 1799, where some of the sources in which the seismic event is recorded use the *republican* dating system. During the period when this earthquake occurred, the year began on the day following the autumn equinox of 1792 (22 September, anniversary of the proclamation of the French Republic).

In accordance with the principle of equality, the year in the *republican* dating system was divided into 12 months of equal length (30 days each), and every month was divided not into weeks but into in three decades of 10 days each. The names of the days were 'first day', 'second day', 'third day' and so on, until the 'tenth day'. The names of the months were given using a supposed natural principle, as they in fact were meant to reflect the natural characteristics of the weather and of the life of the countryside. Thus, there was a 'rainy month', a 'foggy month', a 'floral month', a 'vintage month', and so on. As the 12 months were composed of 30 days each, there still were 5 days (6 every four years) to be placed into the calendar. They were called the 'days of the sansculottes' (i.e. for people who did not wear 'culottes', breeches, used by the nobles), and were inserted at the end of the year.

Since the above-mentioned restoration ordered by Napoleon, the Western countries of Europe have had a unique dating system. The first time the Gregorian calendar was officially used in the Arab world of the Middle East was in Egypt, in the year 1292 of the *Hijra*, which corresponds to 1875. The problem of dating persisted after this in a few eastern European countries like Russia, as well as other countries (belonging to the ex Soviet bloc) where the Gregorian reform was adopted only in 1918. In these cases chronological mistakes typically arise due to multiple date and time conversions (Julian to Gregorian to Julian and so on) when they were applied to the dates of earthquakes in the catalogues of these areas. Such mistakes were usually caused by the uncertainty of seismologists about the earthquake data with which they were dealing. Thus, for some earthquake catalogues it may take some serious investigation to understand whether the temporal parameters were already converted to the Gregorian calendar or are still expressed in the Julian calendar. Indeed, for earthquakes that occurred before the Gregorian reform (i.e. before 1582) the years should not be converted but should remain in the form of the Julian calendar. This is necessary to maintain chronological consistency between the dates of seismic events in the earthquake catalogue and the dates of other natural or historical events that are known to have taken place around the same time.

7.3.14   *Other geographical areas and cultures*

Indian calendars

In India several eras were used in the various historical and political regions. Renou and Filliozat (1953) list some 30 eras in use in India, adopted in different areas of the country and mostly at the same time, from the sixth century BC until AD 1673. The oldest, which was generally used for centuries, is the *Kaliyuga* era, which in an approximate conversion to the Christian era, starts the counting of time from the year 3102 BC.

Within the eras, the years and the months were often counted in different ways. The largest divisions of time in the Indian cultures were fixed by the revolutions of the Moon, the Sun and even Jupiter (*Brhaspati*). The Indian calendar is based on the Moon–Sun cycle (*yuga*), which uses two systems of astronomical reference. Months are determined by lunar cycles and years by solar cycles, while Jupiter's cycles characterize cycles of 12 or 60 years. The durations of the months, the years and the longer cycles do not follow a simple numeric relation. The solar year is not a simple multiple of the lunar months, and the cycle of *Brhaspati* is not an exact cycle of the solar year. The result is that a calendar whose base is the cycles of the Moon, the Sun and *Brhaspati* needs a complex system of conventions and compensations in order to maintain a general agreement between the months, the years and the cycles determined by the orbital revolutions of the heavenly bodies that are used as references. The variations are rather complex and depend on the values indicated in the different astronomical works of reference. As is known from Western astronomy, the orbital values change if one considers the apparent movements of the stars or the average movement of the heavenly bodies.

Table 7.2 shows the name in Sanskrit of the months, which are used in corrupted forms in different localities, along with the duration of the lunisolar months.

The month (*masa*) was divided into two fortnights (*paksa*): *suklapaksa*, the clear half (*sukla*) or the period of the rising moon, and *krsnapaksa*, the dark half (*krsna*) or the period of the falling moon. Every fortnight had 15 *tithi* or lunar days, each of a duration less than that of the solar day (so that the 62nd *tithi* occurred at the same moment as the 61st solar day). Each day was indicated with its progressive number as indeed belonging to the clear or to the dark fortnight of the month. The months were grouped together two by two according to the seasons (*rtu*). Besides the lunar months and the lunar days (*tithi*), solar months and days were also in use.

The Indian solar year counts twelve solar months, and each month is defined by the 12 signs of the zodiac. The months have an unequal duration and are

Table 7.2 *The months of the Indian calendar and their duration with respect to the Western calendar*

Months	Duration in the Western calendar
(1) Caitra	(15 March–14 April)
(2) Vaisakha	(15 April–14 May)
(3) Jyaistha (or Jyestha)	(15 May–14 June)
(4) Asadha	(15 June–14 July)
(5) Sravana	(15 July–14 August)
(6) Bhadrapada (or Prausthapada)	(15 August–14 September)
(7) Asvina or Asvayuja	(15 September–14 October)
(8) Karttika	(15 October–14 November)
(9) Margasirsa (or Margasiras or Agrahayana)	(15 November–14 December)
(10) Pausa (or Taisa)	(15 December–14 January)
(11) Magha	(15 January–14 February)
(12) Phalguna	(15 February–14 March)

used in parallel with the lunar months, with which they partly overlap. The year, probably 365 or 366 days long, starts with different months according to the eras adopted, the cultures, the areas and the periods.

The complications in the Indian calendar are evident. On the one hand, the lunar month is shorter than the average solar month, so it may happen that in a solar month two different lunar months will have an end and a beginning, with precise and specific names. This also can happen for two solar months within one lunar month. Longer term complications are due to the fact that over a 25-year cycle one month must be suppressed so that the plenilunium goes back to coinciding with the months.

To compensate for the complexity of the system and the multiplicity of reference points, the dating in inscriptions and ancient manuscripts is often very detailed. They generally indicate the era, the year, the months, the fortnight, the lunar days and the days of the week. Sometimes confusing things, the year can be either the one just finished or the current one, but often this is not specified. The ambiguity can be resolved by the indication of the era. But when this is missing, the difficulties and the uncertainties in the conversion of the Indian dates to the Western calendar system cannot be overcome. Even in the Western area there has not always been the custom of indicating the Christian era (certainly for the centuries prior to the birth of Christ). Furthermore, in the past centuries within the Christian era there were also different styles of dating, depending on the areas and the periods.

The Indian eras, as has been said, number more than 30 and often must be deduced by knowing the historical, geographical and cultural context of the texts. Another element of uncertainty in the conversion of the date is because of the custom of only writing the last two digits of the date year. This omission, which is a common practice even in the West, means that at times the day, the month and the year of an event may be known, but not the century. In India the number of the year is often found to be expressed either in figures or through the name of the year in one of the *Brhaspati* cycles. In the latter case, the interpretation of the date is something of a riddle as the order number of the cycle is not normally stated.

As regards the subdivisions of the day into smaller parts, the ancient Indian system was not much different from the one used in the West. It took into account of the length of the day on the basis of daylight and was therefore variable depending on the season. Indologists note the persistence of references to water-clocks, also present in Babylonian astronomy, which subdivided the day into time segments equivalent to one-sixtieth of a day.

Today in India different some eras are still in use. The main ones are *Vikrama* and *Sakha* or *Salivahana*. The former, in use in the whole of northern India, started in 58 BC, from the first day of the clear fortnight of *Karttika* (October). The other era, used in southern India, started in AD 78 with the first day of the clear fortnight of *Caitra* (March).

A sixteenth-century Spaniard provides information about early Mexican calendars

The Jesuit José de Acosta (1540–1600) was one of the most outstanding theologians and natural philosophers of his time (see Chapter 4). He spent many years in Mexico, then known as New Spain, where he became a keen and enthusiastic observer of the native Mexicans. His first-hand observations of the Mexican calendars in use at the time provide us with evidence not only about Mexican attitudes and procedures in recording the events they considered important, but also about their attempts to adapt the preceding calendar to that of the conquistadores.

> The year was divided into eighteen months, and they [the native Mexicans] attributed twenty days to each month, thus producing a total of 360 days. The other five days required to complete the year were not attributed to any month, being counted separately and called 'empty days'. During these days people did nothing. They did not even go to the temple, simply visiting one another to pass the time; and even the priests in the temples stopped making sacrifices. Once those days were over, they started counting another year, the first month of the

new year corresponding roughly to March, when leaves start sprouting again, though they included three days from February, for the first day of their year was the twenty-sixth of February, as can be seen from their calendar.

Our calendar [i.e. the Julian calendar, used in Spain until 1582] was incorporated in theirs not without considerable calculations and adjustments, effected by the early Indios who encountered the first Spaniards. I have seen that calendar, and still have one available, for it deserves study if one wishes to appreciate the intelligence and skills of the Mexican Indios.

Each of the above eighteen months has its own name and its own painting and signs, which normally derive from the most important feast celebrated in that month, or else from the characteristics of the season concerned or from the changes brought about by the progress of the year. And all their feast days are indicated in the calendar. The weeks had thirteen days each, a day being indicated by means of a zero or small circle, and when the number of zeros reached thirteen, they started counting again: one, two, three and so on.

They also used signs to divide the years into groups of four, and they gave each year a sign. The four figures used were houses, rabbits, reeds and flint. That is what they painted, and they used them to indicate the current year by saying: this or that event occurred at so many houses or so many flints in such and such a wheel. [For this system of dating records of earthquakes in Mexican sources, see Chapter 3.]

It should be understood that their wheel, which represented a 'century', contained four 'weeks' of thirteen years each, so that each wheel represented fifty-two years. In the middle [of the wheel] they painted a sun, from which four arms or lines extended in the form of a cross as far as the circumference of the wheel, thus dividing it into four parts, each of which was identified using the same colour as that of its arm. There were four different colours: green, blue, red, and yellow; and each part had its own thirteen spaces with house, rabbit, reed or flint signs, each one representing a year, and at their side the events of that year were painted.

And so I saw that in the calendar I have described the year when the Spanish entered Mexico was indicated by a painting of a man wearing Spanish style red clothes, for that is how the first Spaniard sent by Hernán Cortés must have been dressed. (Acosta, 1590, 1608 edn, pp. 397–8.)

7.4   The measurement of the hours from the ancient world to the modern era

The Mediterranean area, with sites of ancient and diverse civilizations, has a rich and varied tradition and history of time measurement. The relevance of this history to historical seismology is that there is not always an easy or safe method for figuring out the correct temporal parameters for historical earthquakes. Each system for the measurement of the hours for any culture has implicit assumptions concerning the concept of time, and this demands great attention. Indeed, for a great many centuries time has not had any significance *per se* as a physical measure, but as the encyclopaedist Isidore of Seville (AD 560–636) wrote, its significance pertained to *human actions*. This definition involves a criterion of continuity, i.e. of actions that can be carried out within a specific time interval, whether that interval is seen in temporal or human terms. Such a concept of time is a far cry from that underlying present-day chronometric measurements, where time is conceived as a sequence of precise and exactly measurable instants which are independent of human activities. In earlier times the actions for which time was pertinent were the ones marked by monastic rules, and the succession and duration of such actions were referenced to a relatively fixed point: the nocturnal awakening of the monasteries for prayer or other activities.

For over 1500 years (since monastic computation of time became established) the hours of the day were counted from one sunset to the next. Scholars of the history of clocks have highlighted not only the mechanical discoveries that brought about a new way of measuring time but also the way of thinking that has supported this fascinating history over the centuries (Bilfinger, 1892; Landes, 1983; Dorhn-van Rossum, 1996). We shall leave the more specific aspects of the history of clocks to the authors who have dealt with the subject. Let it suffice to say that ancient clocks were regulated by means of evaluations made on the basis of normal meridians, i.e. on real solar time. Real solar time is not constant in the course of the year. It varies owing to the elliptical nature of the Earth's solar orbit and the inclination of about 23 degrees of the Earth's axis in relation to the plane on which this orbital movement occurs. Until the end of the seventeenth century, astronomers used the time of the *actual solar day*. To make the scale of time measurement based on the Earth's rotation more uniform a fictitious Sun was devised for use in time computations. Having reached an agreement on the corrections to be made between the actual solar time and time relative to this fictitious sun, astronomers as early as the beginning of the eighteenth century adopted *mean solar time*, following the example of John Flamsteed (1646–1719, the first astronomer of Greenwich, from 1675 to 1719), measuring along the notional pathway of an imaginary Sun in the sky that moves at a constant velocity along the equator. Its position corresponds to that of the real Sun only at two points of the Earth's elliptical orbit: at the apogee and the perigee, the maximum and minimum distances, respectively, between the two celestial bodies. Such a fictitious Sun crosses the meridian at regular

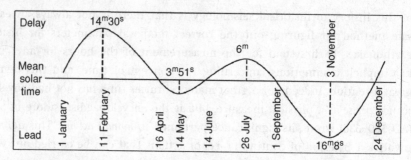

Figure 7.1 A plot of the so-called *Time Equation*, a trend during the year of the difference between true solar time and mean solar time.

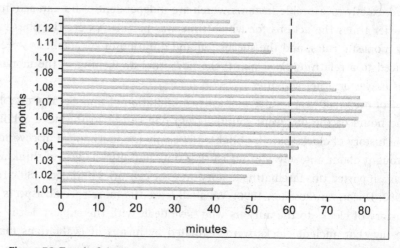

Figure 7.2 Trend of the duration of the canonical hours at the latitude of the Astronomical Observatory of Brera (Milan). From April (1.04) to September (1.09) the canonical hour has a longer duration than the equinox hour (lasting more than 60 minutes), while from October (1.10) to March (1.03) it has a shorter duration.

24-hour intervals. The difference between actual solar time and mean solar time is expressed in the so-called *Time Equation* illustrated in Figures 7.1 and 7.2.

Before the widespread acceptance of clocks based on a modern conception of time, the measurement of the hours was very imprecise compared to today. This fact has often discouraged the compilers of earthquake catalogues from investigating the specific hour at which an event took place. However, it remains very important for the study of earthquakes to identify, as best as one can, the origin time of a given seismic event in order to assign the earthquake effects at different locations to individual shocks when multiple events were known to have taken place (see the case histories below). Thus, efforts to determine the time that an earthquake in historical times took place are always strongly encouraged, provided, of course, that the interpretation from the historical sources is both direct and accurate.

Today, the scientific community agrees to report temporal parameters in terms of *Universal Time* (UT), the fixed point of reference derived from an international convention (in Washington, DC in 1884) which put the meridian of Greenwich at the zero hour. The need for a general time standard had in fact been increasingly demanded by scientists since the early nineteenth century, given the multiplicity of chronological systems existing in the world. In the same decades when the Greenwich standard was being developed, the need for standardized time was being met in Europe by the use of *Central European Mean Time* (CEMT). The process of unifying time measurements was slow and did not cover all inhabited areas. For example, in Italy in 1866 the system in operation was Rome Mean Time (*Tempo Medio di Roma*, TMR), but this was used exclusively in the cities and only for public administration purposes (offices, telegraphs and trains). Local time applied to secular routines and rural areas. Local time was marked by the pealing of bells by monks or clergymen.

The adoption of Central European Mean Time took place in Italy at the end of 1893. Before this unification, which subsequently became official and obligatory, a variety of time systems had coexisted in Italy, their differences being not only regional but cultural (reflecting the needs of the clergy, the armed forces and the ordinary people). Agreement as to the hour was somewhat rough, being based on different cultural and social needs. But for centuries church clocks and the sound of church bells ringing out the canonical hours were an established and important point of temporal reference.

For historical seismology, if earlier time systems are to be expressed in terms of UT, it is necessary to take not only the literal meaning of the sources into account but also the longitude, expressed in minutes, of the different locations, since the time references in the source must be related to the Greenwich meridian. For earthquakes that affected neighbouring states, it may happen that the earthquake catalogues of different countries list different times and epicentres of a single local earthquake owing to the lack of a uniform criterion for the calculation of time.

The scales of measurement of the hours with which man has governed his activities and which he has used to describe natural phenomena, such as earthquakes, have always referred to the apparent movement of the Sun in relation to the Earth. The time intervals between dawn and sunset and vice versa have identified day and night. The distinction between day and night has no relevance for the present-day chronometric scale, but was widely used in the past. One of the first problems for anyone who must convert time indications from historical sources into our modern unified system is the variable duration of the canonical hours, these being in practice based upon the duration of daylight, which varied according to latitude and season (as can be seen in Figure 7.2).

7.4.1   The ancient Mediterranean world

The measurement of time in the ancient world presents a challenging problem regarding the calculation of the hours of the day. The concept of the hour was for a long time unknown in the Roman world. There is no mention of hours as late as the fifth century BC in the laws of the Twelve Tables. The Roman notion of the hour (*hora*) was often imprecise, since it could indicate either the total duration of the hour or the moment at which it began. In 338 BC the term *meridies* (from which 'midday' derives) was officially introduced (Pliny the Elder, *Naturalis historia*, VII, 212: *XII tabulis ortus tantum et occasum nominantur, post aliquot annos adiectus est meridies*), and this led to the division of the day into morning and afternoon.

The hours of the day were calculated from dawn, and those of the night from dusk. In subdividing the day into hours the Romans added to the number of hours certain habitual phrases which served to specify particular moments of the day (the hour of dinner, before cockcrow, and so on). Two temporal cycles of 12 hours each were distinguished: daytime from dawn (*mane*) to sunset (*vespera*), and night-time from sunset to dawn. The two daily cycles were in their turn divided into four periods of 3 hours each. With the division of the day and the night into 12 hours each during the whole year, the daytime and the night-time hours had an unequal duration (*horae inequales* or *horae temporariae*). In the winter months the daytime hours were shorter (lasting as few as about 40 minutes) as compared with the night-time hours (lasting up to about 80 minutes). The opposite occurred in the summer months. Only at the equinoxes were the hours of the same duration, and coincided with the ones in use today. The variations in the maximum and minimum duration of the hours according to latitude, if expressed in present-day minutes, would be as follows: in Upper Egypt 67/53; at Athens: 73/47; at Rome: 76/44; in southern Germany: 80/40; and in northern England: 90/30 (see Dorhn-van Rossum, 1996, p. 19).

The day was subdivided into four intervals, each of which ended at a specific hour. In Roman towns three of these closing hours were publicly signalled by officials: the third hour (*hora tertia*), the sixth hour (*hora sexta*) and the ninth hour (*hora nona*). The four intervals spanned the time from the beginning to the end of daylight, that is to say from *mane* to *vespera* or *suprema*. Night (*nox*) also was divided into four watches (*vigiliae*), a term deriving from the shifts of guard duty in a military camp.

The day was progressively subdivided using set phrases that specified its parts in a qualitative way. This subdivision began in the third century BC: *cum die* (dawn); *diluculum* (daybreak); *mane* (the morning in general, but with augural overtones); *magis mane* (very early morning); *ad meridiem* (towards midday, i.e. indicating the end of the morning); *de meridie* (afternoon); *suprema* (sunset); *vespera* (this marked the appearance of *Vesper*, the evening star, which takes the name of Venus at night); *crepusculum* (dusk, when objects could no longer be easily distinguished); *prima fax noctis* (the moment at which the first torch was lit); *conubium* (bedtime); *media nox* (the middle of the night); *intempesta nox* (the entire middle part of the night); and *gallicinium* (cockcrow). This last term simply indicated that midnight had passed.

Certain expressions from Roman times can still be found in medieval sources, especially regarding the intervals of time close to dusk and dawn. Examples of such terms are *occasus solis* (sunset), *crepusculum* (dusk), *vesperum* (the appearance of *Vesper*, the evening star), *conticinium* (silence), *intempestum* (the cessation of all activities), *aurora* (the first light of dawn), *diluculum* (dawn) and *exortus solis* (sunrise).

It is interesting to note that in late antiquity the hour was sometimes divided into 24 parts called *scrupuli*. There is evidence for this in inscription CIL, XI, 1513: 'Peace to the deserving. Silvana who sleeps here lived for 21 years, 3 months, 4 hours and 6 *scrupoli*. She was buried on the ninth day before the Calends of July' (*Benemerenti in pace. Silvana hic dormit, vixit ann. XXI mens. II hor. IV scrupulos VI, depos. IX kal. Julias*). Unfortunately, no such precision has yet been found in the description of an earthquake.

7.4.2   The hours in the Greek–Byzantine area

Except for the use of weeks for measuring time, the Byzantine world (fifth to fifteenth centuries) in general adopted a system that was not very different from that used in the classical Roman world. In the Byzantine world the parts of the day were identified by subdividing the *nychthémeron* (period of 24 hours), which began at sunrise. In the fifth century, the day began with the *hòra próte* (the first hour), which was at sunrise; this was followed by the third hour (*hòra tríte*) towards mid-morning, the sixth hour (*hòra hékte*) at midday, the ninth hour (*hòra enáte*) towards mid-afternoon, evening (*hespéra*) around sunset and *apódeipnon* (*completorium*) after dinner. Unlike the Romans, the Byzantines apparently did not have a parallel system for subdividing the hours of the night, though they did use the term *órthros*, corresponding to the Latin *matutinum*, to indicate the period before sunrise.

7.4.3   The Christian area: canonical hours

The calculation of the hours inherited from Roman antiquity underwent some modifications by Benedictine monks in the early Middle Ages. With the diffusion of Christianity the ancient hours ended up giving their name to the phases of monastic prayer and being defined as canonical (*horae canonicae*). This system of time computation came to regulate the life of both monks and laymen with the sound of bells. Once the computation of the canonical hours had been firmly established, two new hours were added. One was the *first* hour (Prime), which coincided with sunrise (the time of which depended on the season and latitude). The introduction of this first hour caused the recital of Matins – one of the regular services in the life of the monks – to move to a later time. At the end of the day there was Compline (*compieta*, or *completorium*), which took place at sunset. This new usage caused Vespers to move to about an hour earlier, to around the eleventh hour of the day.

As the use of the canonical hours spread, the liturgical service of the night (called *nocturn*), which began at the 'middle of the night' (*ad mediam noctem*) or at 'cockcrow' (*ad galli cantum*), was unified with Matins, which, as seen above, had in its turn been brought forward by a couple of hours. The sequence of canonical hours was thus as follows: Matins, Lauds, Prime, Terce, Sext, None, Vespers and Compline. Matters are even further complicated by the fact that one must take account of the displacement of another hour. The *hora nona* gradually moved until it came to indicate midday (hence the English word 'noon'), and this led to the gradual disappearance of the *hora sexta*. This change took two or three centuries (roughly from the thirteenth century), and the speed at which it occurred depended on the geographical area. As Dorhn-van Rossum observes (1996, pp. 30–1), although prayers were normally said at the end of the three time segments of Terce, Sext and None, there was no strict adherence to a precise timetable, so prayers might begin a little earlier. In this way, the 'hours of time' (*hora quoad tempus*) and the 'hours of prayer' (*hora quoad officium*) could drift apart.

When dealing with the hours indicated in medieval monastic annals, careful attention may well be rewarded, as is seen from the following case of the great earthquake of 1117.

3 January 1117: a comparative analysis of the time of a major known earthquake brings to light two unknown seismic events

In European seismological literature the 3 January 1117 earthquake is an interesting case study, both for the sheer size of the area in which the event is recorded by twelfth century monastic sources and for the amount of damage mentioned. A new analysis of the hour system in use in the European Benedictine monasteries has led us to hypothesize three areas affected by three distinct earthquakes. Previously, all the seismological literature considered only one affected zone from a single seismic event. A single event on this date appears in the catalogues of as many as five European countries (Italy, France, Belgium,

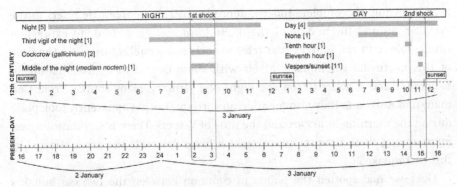

Figure 7.3 Correlation between the twelfth-century and current time systems. In grey the differentiation of the time reports from the source annals into two groups indicates the time ranges for the two main earthquakes of 3 January 1117 (from Guidoboni *et al.*, 2005). Square brackets show the number of separate texts giving information as to the time of each earthquake.

Switzerland and Portugal). For northern Italy it is listed as the largest historical earthquake. A large body of source material had been discovered by previous research (Boschi *et al.*, 1995, 2000). But only by analysing twelfth-century monastic time reckoning did it prove possible to separate the two additional individual shocks (see Figure 7.3) that had been recorded by medieval sources in a somewhat obscure fashion (Guidoboni *et al.*, 2005, 2007).

The canonical-hour system underwent changes after its first introduction (in the early Middle Ages). By detailing the nature of these changes, it was possible to relate the various time systems used by monks in their annals. The liturgical night service (called *nocturn*), which started in the 'middle of the night' (*ad mediam noctem*) or at 'cockcrow' (*ad galli cantum*), was merged with *Matins*, which was in turn brought forward by a couple of hours from dawn towards night. Thus *gallicinium* referred to late at night or the last few hours of the night itself. Apart from these variations in time reckoning, there were local differences. The sources for the 1117 earthquake are scattered throughout Europe and refer to times dependent upon the longitude and latitude where the texts were written.

The sources mentioning the night-time earthquake show significant agreement as to time references, though this is not detectable by means of linguistic analysis alone. The places where such sources originated are in present-day central and southern Germany and Austria. In Italy, only one source described this night-time earthquake as locally felt in Pisa, and this is a long annotation written by the contemporary deacon and author, Guido Pisanus. To indicate the time of the event, Guido Pisanus uses a different expression from that in the monastic texts, showing his complete independence from these sources. He refers to the event at 'around the middle of the night' (*circa mediam noctem*), an expression compatible with the sources that refer to cockcrow, enabling us to set the time of this nocturnal earthquake, albeit with some degree of approximation, to between 2 and 3 hours UT on the night of 3 January. Many sources from Germany and Austria to Italy also mention an earthquake that apparently took place during the morning hours around the time of Vespers. There is significant agreement among these time references, which helps delineate the areas affected by this earthquake.

Once we had spotted the points in common between the two earthquakes (mentioned in different linguistic terms in the sources selected; see Table 7.3), we could separate the effects across this broad region of Europe on a distance-proximity basis. A first earthquake struck southern Germany (I_0 = VII–VIII; M = 6.4) probably just after midnight. Then, some 12–13 hours later, a more violent earthquake shook northern Italy (the Verona area). This second event is the one that is familiar from the literature. It is curious to note that the size of this second event, previously rated magnitude 6.5 (Boschi *et al.*, 2000), does not decrease in estimated magnitude as one might expect when dividing up the felt effects in this area, but it actually rises to magnitude 7. The reason for this increase in the estimated magnitude is that the area where the second earthquake was most felt has proved more extensive than had been previously thought. The extent of the area where an earthquake is detected plays an appreciable part in calculating earthquake magnitude.

A third earthquake is independently attested to in sources from northwestern Tuscany (I_{max} = VII–VIII), but for this last event the epicentre and magnitude cannot be evaluated with the reports that have been investigated so far.

7.4.4   The hours in Islamic countries

In medieval Arab culture the day was divided into 24 hours, each of which had its own name: 08 *al-shuruk* (rising); 09 *al-bukur* (in the morning); 10 *al-ghadwah* (the first part of the morning); 11 *al-duhah* (the light of day); 12 *al-hajira* (midday); 13 *al-zahirah* (afternoon); 14 *al-rawah* (departure); 15 *al-'asr* (after the afternoon); 16 *al-kasr* (restrictedness); 17 *al-asil* (late afternoon, early evening); 18 *al-'isha'* (evening); 19 *al-ghurub* (sunset); 20 *al-shafak* (the remaining sunlight); 21 *al-ghasak* (dusk); 22 *al-'atmah* (darkness); 23 *al-sadaf* (the dark); 24 *al-fahm* (coal); 01 *al-zhulmah* (total darkness); 02 *al-zulfah* (the first part of the night); 03 *al-bahr* (light breathing); 04 *al-sahar* (first light); 05 *al-fajr* (daybreak); 06 *al-subh* (dawn); 07 *al-sabah* (morning). In the Middle Ages, the Arabs used sand-operated clepsydras (they existed as early as Egyptian Empire times and were used by the Romans), the quantity of sand being sufficient to measure 1 hour of the 24 into which

Table 7.3 *A comparison of sources as to the times given for the two earthquakes of 3 January 1117*

Where the sources were written[a]	Time according to the original text, with translation
First earthquake: during the night of 2–3 January 1117 – centre in southern Germany from agreement of locations of sources	
Disibodenberg* G	noctem [at night]
Freising* G	in nocte ante diem [at night before day]
Augsburg* G	semel in nocte [once at night]
Zwiefalten* G	in nocte [at night]
Melk* A	unus post tertium galli cantum [an (earthquake) after the third cockcrow]
Salzburg* A	semel in nocte [once at night]
St-Blasien* G	in galli cantu semel [once at cockcrow]
Petershausen (Constance)* G	circa tertiam noctis vigiliam [around the third vigil of the night]
Pisa I	circa mediam noctem [around the middle of the night]
Second earthquake: during the afternoon of 3 January 1117 – centre in northern Italy (Verona area) from agreement of locations of sources	
Reinhardsbrunn G	ante solis occasum [before sunset]
Aura G	ora vespertina [at the hour of Vespers]
Prague Cz	hora iam vespertinali [already at the hour of Vespers]
Bamberg G	hora vespertina [at the hour of Vespers]
Heilsbron G	advesperascente die [as the day drew towards Vespers]
Disibodenberg* G	diem [during the day]
Rheims F	ad vesperas [at Vespers]
Rheims F	ad vesperum [at Vespers]
Freising* G	ante vesperam [before Vespers]
Augsburg* G	semel in die [once during the day]
Zwiefalten* G	post vesperas [after Vespers]
Zwiefalten* G	et die [and by day]
Melk* A	hora quasi 10 [almost at the tenth hour]
Salzburg* A	semel in die [once during the day]
St-Blasien* G	ad nonam [at the ninth hour]
Petershausen (Constance)* G	undecima fere hora [almost at the eleventh hour]
Milan I	in hora vespera [at the hour of Vespers]
Cremona I	hora vesperarum [at the hour of Vespers]

[a] Places in order from north to south; asterisks *denote that an annal from that place contains a time reference for both earthquakes. Abbreviations (corresponding to the names of the present-day countries): A, Austria; B, Belgium; Cz, Czech Republic; F, France; G, Germany; I, Italy.

the day was divided. During the caliphate of Hārūn al-Rashīd (AD 786–809) a water-clock was devised which made it possible to measure time in the dark. In this device the amount of water which collected during the space of 1 hour pushed a small brass ball, causing it to fall and make a noise which indicated that an hour had passed. Since a day for the Arabs was the time between one sunset and the next, evening and night preceded daytime.

7.4.5   Hours in the modern era

The custom of counting the hours of the day from one sunset to the next, marked by the liturgical use of the *Ave Maria* in many European countries, lasted until the early decades of the nineteenth century. This style was called the *Italian manner*. In other areas such as France, the custom of starting from midnight and subdividing the day into 12 ante-meridian hours and the same number after noon was adopted from the second half of the eighteenth century. This was called the *French manner*, and was later introduced in all the areas taken over by Napoleon by the early years of the nineteenth century.

The use of these two different hour styles in historical sources is not always immediately obvious. Indeed, overlaps, local habits and parallel uses in adjacent areas were commonplace. For example, in the old Italian states before the unification of the Kingdom of Italy (i.e. before 1860) different styles were used. In the Papal States (and the Vatican from 1870) the *Italian manner* system was common, while in the neighbouring Grand Duchy of Tuscany the *French manner* was used. Hence, some late-eighteenth-century earthquakes that affected areas belonging to these two states are split into two events in old catalogues because the different hour styles of the two states made the same earthquake appear as if it were two different events. Hence, it is always advisable to check earthquake times carefully and to proceed with caution in order not to generate new catalogue errors. Only in the second half of the nineteenth century was the unification of the hour applied throughout Italy due to the spread of new means of transportation and communication, such as the railways and the telegraph.

7.5   Earthquake duration

Origin time is not the only chronological parameter for earthquakes recorded in historical sources. Estimates of the duration of the ground-shaking are sometimes given. Since accurate estimates of the duration of ground-shaking can be important guides to the size of an earthquake and perhaps how buildings responded to the earthquake shaking, the elucidation of these estimates from the historical sources is of great interest for scientific analysis of the events.

Indications of the duration of an earthquake from non-instrumental sources, in other words those derived from human estimation, often cannot be directly utilized in quantitative scientific terms. In many cases, they only provide some qualitative evaluation of the phenomenon observed. As noted by Ferrari and Marmo (1985), one of the most evident limits of unmeasured shaking duration estimates arises from the fact that all human witnesses are not sensitive to the oscillations in the ground in the same way and therefore do not constitute a homogeneous indicator of duration. Furthermore, the conditions in which such observations actually took place (for example, was the observer indoors or outdoors?) are hardly ever specified in the historical sources. However, even qualitative indications of the duration of earthquake ground-shaking enable one to get a rough idea of its actual duration. The most widespread system in the West for estimating earthquake duration was to use the length of certain prayers as a sort of unit of measurement. For example, one finds written that an earthquake lasted for a *Gloria*, an *Ave Maria* or a *Miserere* (Psalm 50).

Ferrari and Marmo (1985) undertook a study of the correlation between commonly reported prayers and current measurements of time. The duration estimates of specific prayers range between about 4 and 65 seconds. This calculation was performed on versions of these prayers that followed the popular recital custom (which differs from the monastic one). The objectivity and accuracy of this form of duration estimate comes from the fact that such prayers have been stably codified and practised following a millennium-long tradition. Even so, the precise estimate of earthquake shaking duration using this measure, as the authors themselves state, seems too uncertain for quantitative scientific purposes.

Other measures of earthquake duration that can sometimes be gleaned from written texts, the number of heartbeats, for example, are even more subjective than the duration of the recital of a specific prayer. The use of heartbeats may indicate a desire to provide time measurement accuracy for situations where time measurement instruments (such as manually controlled chronometers) were lacking. Moreover, it is worth bearing in mind that some of the durations reported in the sources clearly cannot be understood in the modern sense of earthquake duration, for in premodern sources an abnormal duration, such as half an hour or a third of an hour, is often indicated. It seems reasonable to say that in these cases the author is probably referring to a series of shocks spaced very closely in time and perceived without interruptions or with negligible intervals between individual shocks.

In Arab sources, too, there are indications of time duration, but these can be rather enigmatic in terms of their translatability into modern time standards.

Arab sources sometimes use the term *daraja* (literally 'degree') to measure the passing of time. The use of *degrees* as a time unit in the Middle East goes back to Babylonian times (eighth century BC), when they served to measure astronomical signs. At first these were organised as 15 signs arranged in an arc of a circle 18 degrees wide, but later they were reduced to 12 and placed in a 360-degree circle (419 BC). The Muslim mathematician and astronomer al-Khwarizmi (232 of the *Hijra* or 847 AD), used *al-daraja* for his astronomical records and also to measure the geographical distance between one place and another. Since 24 hours are equivalent to 360 degrees, 1 degree is equivalent to 4 minutes. This would seem to provide an accurate definition of the term *daraja* (Lane, 1867, *ad vocem*). At times, however, the term seems to have been applied to very long estimated durations for earthquake shocks or, at any rate, longer than the estimates that can be drawn from the analogous indications present in Western (Christian) sources. Confirmation of these uncertainties in the translation of such temporal indications into current chronometric systems can be found from two authors. According to Quatremère (1845, vol. II/2, pp. 216–17), *daraja* means *a short time* or *a minute*, while for Nejjar (1974, p. 85) it is equivalent to 5 *minutes*. Another way in which the Muslim sources indicate temporal duration, akin to one found in Christian sources, is that of referring to the time needed to recite certain verses of the *suras* in the Koran.

PART III PRACTICAL GUIDELINES FOR THE ANALYSIS OF HISTORICAL EARTHQUAKE DATA

8

Planning the goals of analysis of historical earthquake data

8.1 Reviewing existing earthquake and tsunami catalogues

Before the commencement of any new research in historical seismology, which obviously requires human as well as financial resources, it is essential to undertake a systematic review of the available historical earthquake catalogues. Since previous-generation catalogues are usually catalogues of catalogues, one must begin by analysing the known bibliography of the earthquake catalogue under examination. It will generally be found that each bibliographical entry refers to other texts that in turn refer to yet other texts, while the primary sources are often not cited at all. These review procedures are often laborious because they can involve the identification and critical rereading of a great number of works, but even at this early stage in historical seismology research it is possible to identify errors of time and place for individual earthquakes that have been passed on from one text to another. Even more serious misunderstandings may be discovered, for it is well known that other types of natural or man-made events, such as landslides, powder magazine explosions (before the advent of the lightning conductor), storms and tempests, etc., may sometimes be listed as earthquakes and tsunamis.

Performing a systematic review of an entire earthquake catalogue is usually very difficult if the catalogue covers many centuries of history, and such a review invariably calls for specialized research groups if it is to be completely and accurately accomplished. Such a review might take decades, whereas the urgent need of a specialized project, such as a seismic hazard estimate for the construction of an important facility, may demand the delivery of results in a much shorter time period.

The two main steps in the assessment of historical seismicity are:

(1) Making a list of earthquakes that have caused effects in the region or area that is being examined. If one wishes to examine the seismicity of a site (in order, for example, to make a hazard assessment for an urban area or the site of a high-risk power or industrial plant), the size of the area to be examined will vary depending on the geological and seismotectonic setting, and its shape may need to be asymmetric, particularly when there is a highly seismogenic zone at some distance from the site. From this work a preliminary earthquake catalogue can be assembled, with the catalogue consisting of compilations from previous catalogues.

(2) Each earthquake in this preliminary catalogue should then be the subject of a careful historical review. For reasons of time and available resources, it is not always possible to carry out thorough historical analyses reviewing the affected areas of all the known epicentres in the vicinity of a particular site if a seismic hazard assessment is being carried out. In such circumstances it may be necessary to concentrate solely on those earthquakes that caused observed or significant effects at the specific site under consideration (see the account of this methodology in Albarello and Mucciarelli, 2002).

This first review of the preliminary catalogue should provide new elements for a fresh, basic analysis of *each* earthquake, identifying the texts cited in the bibliography of each catalogue in which the event is reported. One can then compile:

- an evaluated catalogue bibliography, in which the basic texts used by the initial compilers are identified;
- a new reading of the historical records that had been utilized in the earlier catalogue compilations, which should help identify confirmed or contradictory points as well as perhaps provide additional data on the known earthquakes;
- a statistical summary of the quality of the re-evaluated bibliography;
- any new parameters that can be justified based on a historical and critical analyses.

Based on the data compiled from the review of the preliminary catalogue, a decision can then be made as to whether the review carried out so far attains the degree of detail desired in the seismic hazard analysis or whether new sources not utilized by the original catalogue compilers need to be found. In order to make this decision for a site, the following questions need answers:

(1) Is enough known about past seismic effects within some small radius of the site (perhaps 30 km or less) where active faults or strong nearby earthquakes may be an important consideration?

(2) Is enough known about historical earthquakes affecting the site area to allow for the drafting of an accurate regional map of the seismic effects of each earthquake? If, for example, epicentres have been given a location on the basis of evidence of active faults from geological analyses of the region, it would be worthwhile performing historical research *ex novo* in order to have other pointers confirming seismic activity independent of the geological data.

(3) Are the maximum known effects in the area in question due to events whose epicentral areas are clearly identifiable on a macroseismic basis, even for instrumental earthquakes?

(4) Which very large seismic events and which epicentral areas for the large events identified on the basis of historical data are nearest to the site?

After having evaluated and reassessed the entire catalogue as described above, it can be decided whether the basic information and the related parameters are sufficient to provide an adequate picture of the seismic activity of the area being examined or if instead the catalogue needs to be checked and analysed more thoroughly. If it is judged that insufficient data are available from the analysis of the preliminary catalogue, steps must be taken to collect more information. This will normally entail fresh historical research. In order to facilitate a new analysis of the events in the preliminary catalogue, it is useful to georeference every location for which there is information about each past earthquake of importance and to map the available information about each of these earthquakes. Earthquake catalogues generally only contain summary parameters such as the date, time location and estimated size of the earthquakes. In order to verify the accuracy of these summary parameters it is important to have access, for each earthquake, to the lists of the sites to which a macroseismic intensity value has been attributed (acquired perhaps from a research database, specific publications, or compiled maps). This review of a catalogue is, in the first place, a review of those descriptive data that have contributed to parameter calculation. Regardless of which macroseismic scale was used to report the earthquake intensities in a catalogue, it may well be necessary to re-evaluate the intensity assignments on a site-by-site basis starting once more from the qualitative data in the original sources. If the original reports are not accessible in the catalogues from which the event information was compiled, it will be necessary to retrieve them through systematic checking of the catalogue bibliography.

From the preliminary analysis of the intensity reports for an earthquake, it is useful to construct a map of known effects in order to evaluate:

- whether and where any important information is missing and in which locations or areas the seismic intensity information should be enhanced in order to better outline the area affected by the earthquake;
- if there are manifest contradictions in the data, such as highly destructive effects reported only a few kilometres away from much milder earthquake effects;
- what the relationship is between localities where effects of the earthquake were described and those localities existing at the time of the earthquake where no information about the earthquake is known.

8.2   The search for fresh historical data

Any search for fresh historical data will have two objectives:

(1) to extend the knowledge, in both depth and detail, about any seismic event considered to be particularly important for elucidating the seismotectonics in the area being examined;

(2) to check whether there is any particular information or knowledge gap that historical research can either fill in or at least help explain.

In both cases the research needs the contributions of expert specialists. Historical seismological research, wherever it is carried out, has its own investigation tools, hermeneutic rules and specializations (depending partly on the period of history in which seismicity is being examined). These may affect the research aims, activities and even questions the sources are expected to answer. For all these reasons, it is essential to create a dialogue, if one does not already exist, between seismologists and historians in order to maximize the potential effectiveness of the research. While research requires a financial investment, it above all needs intellectual and human resources. It is also for these reasons that the historians involved should at least have an interest in environmental history and have had some mature research experience in territorial history.

The planning of new research and the need to establish a multidisciplinary dialogue in this field are not just theoretical requirements. Rather, they are required by the particular focus of the research, because large areas and long periods of history may need to be studied to answer properly the fundamental research questions under investigation. They are also made necessary by the variety of data to be extrapolated into the time period and spatial location that is the subject of investigation. These extrapolated data often do not fall within the specific sphere of interest of historiography, although they may use the results of historiographic studies on the territory in question, as discussed in the foregoing paragraphs.

As has been seen, the sources containing information on past earthquakes may be collected in archives or may be scattered in places yet to be identified, perhaps even outside the countries whose seismic history is being established. In these cases, collaboration with professional historians is required in order to guarantee overall project cohesion as well as an appropriate research methodology. In order to avoid fragmentation of the research effort, to guide the development of the research, and to achieve practical results, it is worthwhile setting up centralized recording and control systems of the historical data insofar as the available resources and research aims allow.

Workgroups or individual researchers draw up their research strategies according to the problems they must deal with, to the spatial areas they are investigating and to the time periods being examined. Setting out on a historical quest is often a moment of great enthusiasm, but it can also be a time of uncertainty. One hardly ever knows in advance what to expect from archive work or inventory checking from library manuscripts. Even so, consistency and systemization of approach must always form the basis for acquiring new data.

8.2.1   Historical research on-line or with CD-ROM support

For almost a decade now one has been able to use computers to tap new resources, over and above the forms traditionally available, for historical research. One can browse for information on the World Wide Web and can use computer databases that contain many individual sources or more than one complete corpus of sources. Consultation of this kind is very useful in historical seismology and should not be underrated (even if historians of generations past are turning in their graves), although at the same time the advantages of these new research tools should not overestimated. New editions of ancient and medieval sources stored on CD-ROM can aid the search using keywords, dates or place-names, and hence are of great value. To use these properly, one certainly needs an exhaustive and properly graded glossary of keywords. However, greater speed of consultation should not substitute for careful direct textual analysis using the proper critical tools. It is almost never acceptable to confine research to cutting and pasting sets of phrases bearing on seismic effects lifted from historical records without a critical analysis of each source from which the words are taken. As will be seen in later chapters, knowledge of a linguistic context or a real situation beyond the text is essential if gross errors of interpretation or judgement are to be avoided.

Analytical care is also needed when working from online manuscript inventories that libraries have been committing to the World Wide Web, though one is still far from having anything that compares favourably with direct access of original sources. Nonetheless, this form of consultation is already having a considerable impact on historical seismology research. It is not only document inventories that have benefited from the coming of the digital age. When it comes to medieval, modern and contemporary archives, many archive centres now allow some collections of documents to be accessed from the Web. Original documents can thus be visually examined and reproduced at any online computer. Reading and analysing these original documents are then up to the individual, with all the ensuing needs for palaeographic, diplomatic and archival expertise. With the availability of such Web-based images, one is free to browse and study at leisure new discoveries of important records on the Web without having to visit each and every archive centre. These kinds of capabilities are in general provided only by major archives. For research in historical seismology where large spatial distances are to be covered, one should check for online consultation facilities when planning research and to use them wherever possible, since they can optimize the efficient use of time and resources. As mentioned in Chapter 3, there are many projects in progress to create and make available digital archives of historic data. In just a few years' time one may well have direct access to thousands and thousands of documents that today can be accessed only by a personal visit to the archives where they are housed.

8.3   Different research strategies for large and small earthquakes

Those earthquakes that are particularly large in magnitude and spatially extensive in their shaking effects and those others that cause slight or no damage and were felt over a small area are at the opposite ends of a spectrum that includes a vast range of events intermediate between these two extremes. In order to analyse the full spectrum of events between these two extremes of very large and very small earthquakes, particular attention must be paid to certain aspects of both research organization and the types of sources to be sought. Since the aim of much historical seismology work is to ensure that an event that is reconstructed from the historical information is as close as possible to the real one, there are certain key questions to which the research into historical seismic events invariably must seek answers.

8.3.1   Very large earthquakes

In many parts of the world, it can be very difficult to get a full measure of the damage and felt effects of large earthquakes since this can call for:

(1) the examination and control of a great bulk of documents and texts of many kinds and languages; (2) a knowledge of the various social and cultural impacts of the earthquake effects at diverse localities; and (3) a need to compare the research results with those of other research groups so as to attempt to arrive at consensus parameters to describe the observations. In the case of very large earthquakes, the following preliminary questions need to be answered:

(1) Are the affected areas in one or more than one country?

(2) Is this a single event or have a number of earthquakes close in time and place led to the establishment of a historical image in which the multiple events have become mistakenly considered as one earthquake?

The first preliminary question means that special attention must be paid to political and administrative systems, to languages, and to the whereabouts of the various research centres (archives and libraries) in which records of importance might be found. The chief challenge when there is a positive answer to this question may well lie in organizing a research collaboration, with the objective of getting workgroups in the different countries to operate in a coordinated and parallel way on the same object of study in order to bring together the various research results into a single team outcome. This means that standardized criteria for evaluating earthquake effects (including the adoption of a single macroseismic intensity scale) must be agreed on before the research is undertaken.

The second preliminary question involves a thorny problem for historical seismology: how can separate earthquakes be identified when they occur within a relatively short span of time (like hours or days)? Nowadays, the instrumental detection of shocks makes it possible to establish the precise times of occurrence of individual shocks as well as their relative sizes, but the examination of seismograms is not possible in the case of a number of historical earthquakes that occurred in a short space of time. In analysing particularly large earthquakes, it is important to have a clear understanding of the time system used by the historical sources (see Chapter 7). But that is not all that is needed. One may have clues from the historical sources that two or more earthquakes took place in quick succession but still may not be able to establish this fact with unequivocal certainty, let alone separate out the times and effects of the different shocks.

Unfortunately, many of the really large earthquakes listed in historical catalogues, especially those in the most distant past centuries, may in fact be the result of cumulative effects of multiple earthquakes that have not been disentangled. In such cases, the magnitudes assigned to such events will likely be overestimates (see the points raised from such analysis by Gasperini et al., 1999).

The epicentre assigned when multiple shocks are mistaken as a single seismic event may also be in the wrong place. If the areas of the effects of two different earthquakes overlap and the historical seismology researcher is not aware of that fact, there will be a natural tendency to place the epicentre in the middle between the two areas most affected the individual events. This could well result in a false earthquake location, such as an area where perhaps there is no active fault. Such a mistake could divert geoscience researchers if the individual shocks that caused the earthquake effects did take place on one or more active faults well away from the mistaken location for the single-earthquake scenario.

Multiple events in historic records are often very difficult to identify. By their very nature the relevant historical data will have considerable limitations. There is nothing to prevent a researcher from forming hypotheses about multiple, nearly simultaneous earthquakes and separating in some way the affected areas thought to be connected to each of the multiple events, but in such a case one must make it quite clear that this is only conjecture. It may be useful to provide dual interpretations of the data, one based on a multiple epicentre hypothesis and the other predicated on the occurrence of only a single earthquake. This would highlight the uncertainties faced when trying to document very large historical earthquakes. When faced with a really large earthquake, therefore, a researcher at least needs to acknowledge that the historical sources may refer to multiple seismic events. Different solutions to decide between a single large earthquake and contemporaneous multiple earthquakes may need to be tried for different historical cases. The imagination, skill and precision of the various researchers involved in a project must be brought into play to address these often difficult issues. When the affected area of what appears to have been a very large earthquake makes it look like a single event took place, a healthy dose of scepticism is always advisable. In such cases, the researchers should also look for support for other interpretations and hypothesis.

An event that was 'too' large: the case of the 1456 earthquake in the Kingdom of Naples

A geological knowledge of the active faults and the by now known correlation between the length of the faults and their seismogenetic power (synthesized in the magnitude parameter) can thus contribute to shedding new light on the historical data. This feedback can lead to new interpretations of the available data to try to get increasingly close to the phenomenonology of the great seismic events occurring in the past. The lack of textual elements, which allow us to identify a sequence of distinct earthquakes, can obviously discourage these attempts or make them over-fanciful. On the one hand, the attempts to 'separate' frameworks of effects caused by very great events are welcome, but on the other hand it would also be necessary to avoid making them unreliable. Explicitly formulating some hypotheses is always legitimate, as long as the hypotheses are supported by congruent, non-contradictory elements. An hypothesis or simply a doubt can be the honest result of research that is very difficult and thorough. Whoever has to evaluate the likelihood of the occurrence of a great earthquake in an area ought to take the hypotheses or uncertainties into account.

The earthquake that hit the Kingdom of Naples on 5 December 1456 is considered to be one of the greatest events in the history of Italian seismicity for the sheer size of the area affected (at least intensity VII over an area of about 27 000 km^2) and the destructiveness of the event (116 sites classified from degree IX to XI: see Boschi et al., 1995, 2000; Guidoboni et al., 2007). This great event has been at the heart of historical and geological research for many years. Thorough and rigorous historical research has brought to light a broad, authoritative and valuable corpus of sources (Figliuolo, 1988–89).

The effects caused by this great earthquake are represented as involving three regions: Molise, Campania and Apulia. The strength parameters for this event reported by past earthquake catalogues were $I_0 = XI$; $M = 7.3$. However, many geologists are convinced that this could not have been a single seismic event, given the length of area that was strongly affected (about 200 km). The hypothesis that this was really several seismic events close together in time was formulated in the 1980s (Magri and Molin, 1984, 1985; Meletti et al., 1988–89). But how can one make sense of the historical facts and the geology of the area and unwrap the available evidence into a complex seismic scenario involving several different earthquakes? The effects mapped out seem to be inextricably intertwined, and the sources, analysed over and over again, do not provide any obvious clues whereby one may separate the seismic scene into different shocks. Fortunately, a single extra-textual element has finally unlocked this situation, allowing for the formulation of a new hypothesis about the large area affected in December 1456.

The answer to this conundrum was actually right under the noses of the researchers who had been investigating the seismic reports. In retrospect, the missing element was quite obvious: where were the undamaged sites? The maps drawn to show the affected area were limited to show the locations of the damage effects as cited in the sources (Figure 8.1). Later it was considered that there may have been other areas at the time of the earthquake where the damage was slight or absent altogether. What was the spatial relationship between these undamaged locations and the strongly affected areas?

Fortunately, this region is very well documented, thanks in part to the fact that it was ruled by a centralized and spendthrift monarchy (that of Aragon) that was well enough organized to raise taxes and impose levies on the population.

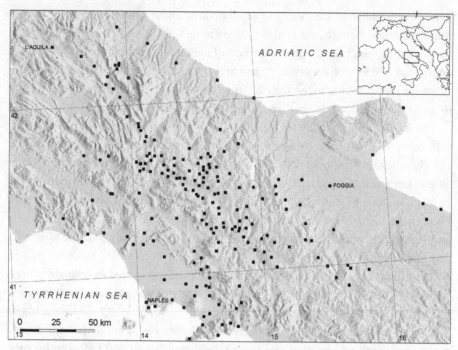

Figure 8.1 Map of the area affected by the earthquake in December 1456 (central-southern Italy). Symbol ■ indicates the damaged sites as cited by the contemporary sources.

A few years before the earthquake, in 1453, a census of the kingdom was carried out. The ruling officials thought it necessary to check that no inhabited areas had escaped taxation. Through painstaking research, all of the towns and villages mentioned in the census have now been located. These have been mapped with blacks dot on Figure 8.2, while on that same figure the villages that suffered damage have been mapped with squares. From Figure 8.2 one is able to see that the area of significant effects was not homogeneous but split into multiple regions that are interspersed with bands of undamaged villages.

As can be seen in Figure 8.2, the northwestern damage area is surrounded by black dots, which clearly delimit the damage zone. This is where the epicentre of one event is hypothetically placed (epicentre 1 in Figure 8.3). The two central damage areas in Figure 8.2, with corresponding epicentres 2 and 3 in Figure 8.3, have a joined central part, perhaps due to overlapping effects, so the separation of these two areas is somewhat arbitrary. To the north and the south of this elongated zone there are other dots that seem to indicate the limit of the greatest effects in this central area. Finally, the southeastern damage area in Figure 8.2, with a postulated epicentre indicated at 4 in Figure 8.3, is one in which the

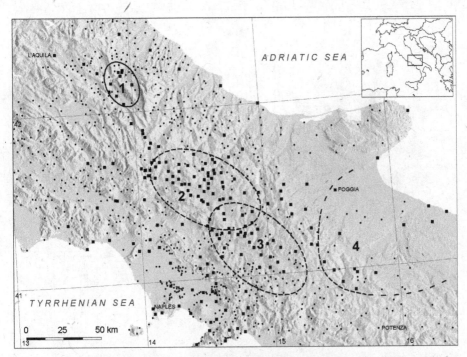

Figure 8.2 Map of towns in central Italy, existing at the time of the earthquake of 1456 and catalogued in the census of 1453. Squares represent the sites of Figure 8.1, while the black dots show other sites existing at the time of the 1456 earthquake that were not cited by the sources of the time as experiencing an earthquake. The solid and dashed lines are an interpretation of the four separate shocks, based on the distribution of the damage and felt effects (Guidoboni and Comastri, 2005).

dispersion of effects prevents one from circumscribing a well-defined epicentral area. What does all this mean? Historical seismologists have tended to be rather puzzled by this unusual spatial pattern and have wondered if it could have been due to the unusual propagation of the seismic waves from a single earthquake. Researchers in seismotectonics have alternatively interpreted the data of Figure 8.2 as evidence of great deep structures running counter to the Apennines, for which a model has already been conjectured (Fracassi and Valensise, 2004; DISS Working Group, 2005; Fracassi and Valensise, 2007) on these and other grounds. Guidoboni and Comastri (2005) were the ones who provided a new interpretation of these earthquake reports and attributed them to four different epicentres (Figure 8.3). Their reassessment of the parameters of these events' magnitudes are calculated as 5.8, 7.0 and 6.3, respectively – values which better fit the area's Apennine geology than the earlier interpretations that these reports represent a single large earthquake (Guidoboni and Comastri, 2005). Support for this multiple-event interpretation, based only on the distribution of the

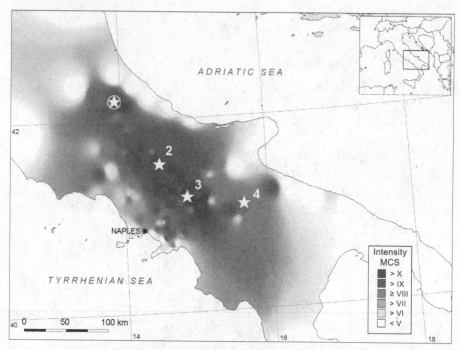

Figure 8.3 Map of the total spatial distribution of the earthquake effects in 1456, with the darker shading representing a greater intensity of ground shaking. The map gives the impression that there had been only one earthquake. The four separate epicentres hypothesized for 1456 are located by stars.

damage effects, can be obtained by mapping all the effects of this 1456 earthquake (i.e., Figure 8.3) and comparing the size of the damage area with that of other earthquakes in the same area, for example, that of 23 November 1980 (3000 dead, $I_0 = X$; $M = 6.7$) (Figure 8.4). As can be seen in Figure 8.4, the size of the area affected by the 1456 event, if it were considered a single earthquake, would be completely out of scale with the major earthquake of 1980.

The earthquake on 1 November 1755: to whom does it belong?

The earthquake that struck Portugal, Spain and Morocco on 1 November 1755 was an epoch-making event in European culture not only for the havoc it caused (including the destruction of Lisbon, completed by the ensuing fire) but also for the extent of its effects, involving several different countries and cultures. The literature on this great and long earthquake sequence (it settled down nearly three years later) is now vast. The event shook not just the Earth but also man's awareness as well; it troubled philosophers, naturalists, religious and political leaders as probably never before. Some texts by Voltaire

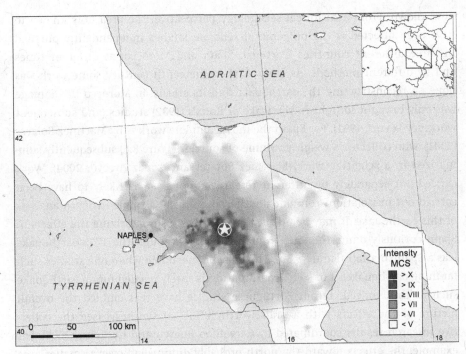

Figure 8.4 Map of the effects of the earthquake in November 1980 (I_{max} XI, magnitude 6.9) which caused the deaths of about 3000 people. On this map, the darker the colour, the greater the intensity of the earthquake ground-shaking.

managed to fix this disaster in the collective memory of the Western world. Two chapters of *Candide* and his poem on the Lisbon earthquake, which opened the famous discussion on Illuminist optimism (Tagliapietra, 2004; Poirier, 2005), memorialize this event. Just when the Enlightenment of European culture was positing a bountiful Nature, this disaster came to wipe out the surging optimism of the time. All certainty tottered, it was written, though perhaps there was some exaggeration in this report or perhaps the cultural context was *sui generis*.

But how big was this earthquake from a scientific viewpoint? The image that the contemporaries had conjured up was not a likely one. A few years after the event one academic believed that this earthquake extended across Europe, Africa and America (Ribeiro Sanchez, cited in Martínez Solares, 2001). By the twentieth century there were studies that were more thoughtful (Pereira de Sousa, 1919–28), and the reconstruction of the earthquake effects by Kendrick (1956) is fairly reliable. Thanks to the body of source material (very large in kind and quantity) and thanks also to the work of many eminent scholars (historians, geographers and geophysicists), many important aspects of this catastrophic event are now known. From a historical seismology perspective, the area now known to have been affected is the outcome of separate studies independently pursued in three different countries (Portugal, Spain and Morocco) at different times and by different methods. As early as the nineteenth century some work was done on reconstructing the earthquake and its effects. In Morocco the damage was only brought to light in the wake of Roux's (1932) studies (and subsequent work; see Levret, 1991). For Spain the most complete work is by Martínez Solares (2001), who collected a weighty volume of sources (Figure 8.5) subsequently summarized in a scientific work (Martínez Solares and Lopez Arroyo, 2004). Work carried out separately by different research groups seems likely to have been carried out in isolation from one another, preventing a single in-depth analysis of this earthquake. It may be no accident that work summarizing the effects in Spain curiously omits the data from Portugal, as though the 1755 earthquake was a Spanish affair (Figure 8.5b). This may be inevitable when one studies such far-flung earthquakes, but some coordination is surely needed for an earthquake such as this. Indeed, recent Portuguese studies have reassembled the overall picture of the effects with a praiseworthy effort at synthesis (see the collective research results coordinated by Carvalhao Buescu and Cordeiro, 2005). For example, the effects toward the north probably diminished over a smaller area than even contemporaries realized (Martínez Solares (2001) suggests that the felt earthquake shaking did not cross the Pyrenees). One would likely be correct in thinking that suggestibility probably caused earthquake effects to be claimed in far-distant areas like central Europe where the earthquake could not have been experienced. Authors in these areas no doubt mistakenly ascribed different local tremors to the great 1755 event (see the 1755–1762 seismic crisis in northwest Europe, in the very interesting research work by Alexandre and Vogt, 1994). The evidence of seiches in lakes of northern Europe may have contributed to the idea that the earthquake had been felt all over Europe, although no perceptible shaking was observed.

Recent studies have ably and insightfully brought to light the many cultural, urban-design and theoretical sides to the 1755 Lisbon event. Contributions analysing the secondary effects of a major earthquake are a great help in clarifying the earthquake as a phenomenon in its own right (see, for instance, the brilliant and very useful monograph by Poirier, 2005).

8.3.2   Widely felt earthquakes of low intensity

Special research interest exists for earthquakes that were felt over a broad area but with low-level effects, even below the threshold of damage. For example, they are typical of many parts of central Europe, from Germany to Poland or to Belgium, and have been duly recorded in the written sources. Events

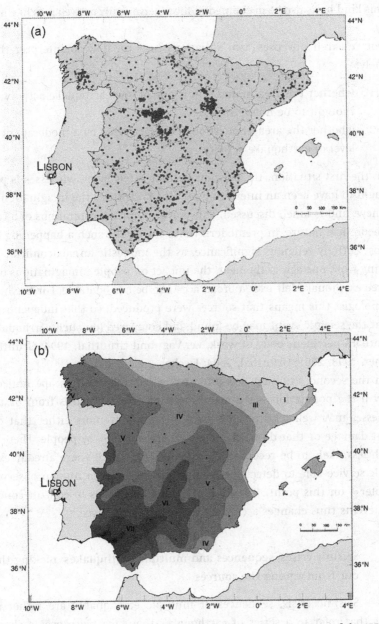

Figure 8.5 (a) Map showing sites with reports of the effects in Spain of the 1755 Lisbon earthquake. (b) An isoseismal map of Spain in the 1755 Lisbon earthquake excluding the area of Portugal.

of this kind have also been claimed in more seismically active areas like northern Italy.

For research purposes, two situations bear on how to interpret the data, namely:

(1) whether the area in question is one of low earthquake activity or is thought to be non-seismic;
(2) whether the area in question is characterized by a moderate or high level of earthquake activity.

In the first situation, the earthquake that is felt by eyewitnesses is probably thought to have been an interesting novelty. It captures the imagination, makes the news and is much discussed. This may prompt interpretations of a religious or theoretical nature. In premodern times the awe at such a happening bore an almost entirely religious significance, as the monastic annals confirm. Though varying from one age to the next, the impact on people's imaginations of a low-degree earthquake felt over a broad area can be considerable. For the historical seismologist this means that sources were produced, so that independent data about the seismic event may be found and compared (on such earthquakes, see Grünthal's pertinent body of work, i.e. Vogt and Grünthal, 1994; Grünthal and Fischer, 1998, 2001; Grünthal, 2004; Fischer et al., 2001).

In the second situation, low-intensity effects may well escape notice. Especially if the population has a memory of major earthquakes from past times, witnesses may well tend to gloss over the minor tremors either that do only slight damage or that do no damage and are only felt by people. These can be much less likely to be recorded. The picture changes if there already existed a public service able to detect the tremor by a network of human observers (see Chapter 4 on this point). The kind of source one needs to consult concerning such events thus changes according to the time and place.

8.4   Seismic crises, sequences and multiple earthquakes: picking them out from among the sources

Seismic crisis, sequences or multiple earthquakes are rather similar terms that refer to a series of earthquakes that take place over a short time period, usually associated with a single or a small number of related seismotectonic features. One or more main shocks may be distinguished and are preceded and/or followed by more minor shocks. In total, the earthquakes of such a crises can leave the casual observer with the impression of having experienced, to use a figurative image, a seismic turbulence. Not by chance, the rather literary and suggestive term 'earthquake storms' has been applied at times to indicate this type of series of seismic events (see Nur, 2002). From a terminological point of view, it can be said that the term 'crisis' implies an interpretative aspect since the various earthquakes are not seen independently from one another but rather are considered interrelated in some way. Thus, to define as a crisis a series of earthquakes that are closely spaced in time and in geographical space implies the recognition of some type of seismotectonic tie among the individual seismic events. This term attributes a significant meaning to the entire series of shocks, something which descriptions of the individual earthquakes alone may well not provide.

Many studies have shown that a large earthquake causes a redistribution of the rock stress around the fault on which the earthquake took place. The resulting stress changes may be large enough to trigger further earthquakes in nearby areas. Permanent changes in the rock stress after an earthquake are restricted to a region that is approximately comparable in size with the fault itself that experienced slip in the earthquake. Seismologists have also documented cases where a strong earthquake in one locality triggered smaller earthquakes at distant locations, even many hundreds of kilometres away, from the fault on which the large event took place. The evidence shows that it is the passage of the seismic waves, especially the S waves and surface waves, that appear to be the triggers that initiate the occurrences of the remote earthquakes. That in some circumstances earthquakes tend to cluster in time is an accepted fact; to know why it happens and to demonstrate those circumstances under which it takes place is a problem that geophysicists are still addressing. To the historical seismologist, the possibility that earthquakes can cluster calls for special attention to the separation of individual shocks, the dating of each event and the location of each earthquake epicentre. Of course, this must be done through careful analysis of each type of historical source that is used in the research, and each source must be independent and authoritative in its information on the seismic effects. This may not be so easy when a sequence of seismic events is being studied. When the areas affected by multiple earthquakes are close and the times between events are small, even the immediate sources (let alone later texts) tend to lump together or overlap the individual seismic events and their effects, or they may mention only the salient features of the activity. The sources may refer generically to earthquakes without any differentiating among individual shocks. Unlike what one might expect, the historical records here tend to show a drop in precision when there are multiple earthquakes that comprise a seismic crisis, as though the occurrences of multiple tremors close together in space and time caused some kind of information overload. Only painstaking in-depth research can break through the confusion of the information and pave the way for an accurate analysis of what really happened.

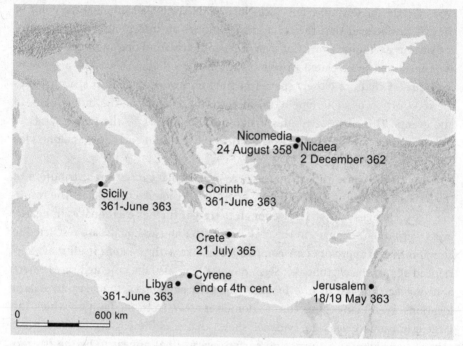

Figure 8.6 Map of the Mediterranean with epicentres from AD 358 to 365. Eight great seismic events are identified (from Guidoboni *et al.*, 1994).

The map in Figure 8.6 is perhaps of some interest in this regard. The fact that so many strong earthquakes occurred within a period of eight years (from AD 358 to 365) in the eastern part of the Mediterranean basin is worth pondering. In spite of the uniqueness of this plethora of strong events, which has no counterpart in earlier or later centuries, it is certainly worth assessing, as indeed the author of the map does, whether such an unusual burst of strong earthquake activity is coincidental or not. From a scientific point of view, there is a need to provide a physical model that accounts for the apparent migration of earthquakes occurring at such great distances and within diverse geological contexts. The problems with this burst of earthquake activity, from the point of view of historical research, are not easy ones to answer. Historical seismology researchers need to carefully address at least two points:

(1) that in this time period there really was a number of earthquakes that took place at a rate well above the long-term average recorded in the same area (and especially the average rate in the better documented recent times);

(2) that there were no cultural and historical reasons that perhaps led to overemphasizing the effects of individual earthquakes, such that even

mild earthquakes were remembered as having had very serious damage effects.

8.4.1   The seismic crisis of 1741–1744 in the central Mediterranean

On 23 June 1741 a long and complex seismic crisis started in the central Mediterranean, taking place over a vast area comprising the Greek islands of the Ionian Sea, western Greece and southern Italy (Apulia and Calabria). The crisis lasted until the spring of 1744. By means of a very vast corpus of historical sources, which has preserved highly detailed information about the seismic events and their effects, it has been possible to distinguish five important earthquakes. Amongst these earthquakes was the particularly violent one on 20 February 1743, which may in turn have been triggered by two other earthquakes. The full scope of this earthquake storm is not easily discovered in an old-style historical seismology approach, which tends to see even the most complex seismic activity as just one earthquake. Without a full understanding of the earthquake activity that occurred at this time, useful elements are lost both for the understanding of which faults are active in this region and of the full seismic hazard faced by the local populations of the region.

The interpretative history of this crisis was incorporated into that of the violent earthquake of 1743, whose epicentre was in the middle of the damage areas of two different as well as distinct earthquakes. Furthermore, a scientific bias previously had influenced the interpretation of the descriptive data, since geologists were convinced that the Apulia region, the Italian area closest to the Greek islands, was not seismically active. Before the latest analyses performed in 2005, the understanding of the earthquake activity of 1743 was the one represented in Figure 8.7, where substantial damage effects from an epicentre more than 200 km away had been interpreted as effects of the propagation of the seismic energy from the suspected source area. The new study by Guidoboni and Mariotti (2005) has enlarged the time-frame of the historical observations, highlighting a much more complex seismic scenario that is considered more realistic than the one in the older thinking. The new image of this seismic activity is shown in Figure 8.8, from which one can also glean some implications for historical seismological research:

(1)   a thorough analysis of a seismic sequence can produce some surprises that are worth knowing about;

(2)   extending the time-frame of the analysis beyond the days immediately following a seismic event is always useful;

(3)   earthquakes do not pay attention to political boundaries;

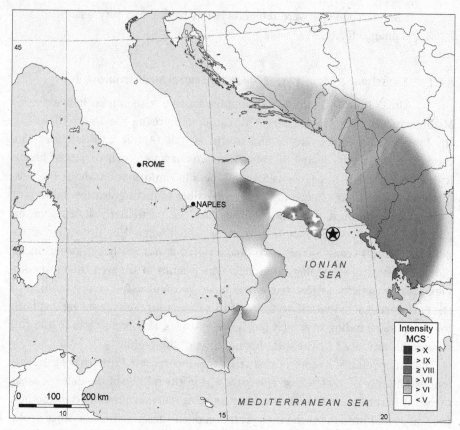

Figure 8.7 Map of the damage and felt effects of the earthquake activity of 1743 with the interpretation of only one epicentre (star). The darker the shading, the greater the macroseismic intensity.

(4) the reality of the events is always more complicated than what one manages to capture from the written sources, and so one must try to approach the written sources without bias.

On *23 June 1741* a seismic sequence started on the island of Cephalonia that lasted several months with very frequent shocks (up to 40 a day were felt). There were reported several deaths and serious damage to all the main inhabited centres of the island. At Argostoli, Asso, Lixuri and in the capital Cephalonia all the houses were damaged, with many collapsing completely and the remainder severely shaken. Serious damage also was reported to the military fortifications and the religious buildings, in particular the cathedral of San Nicolò and the church of Lixuri, which mostly had to be rebuilt.

About eight months after the start of the Cephalonia sequence, on *25 February 1742*, a strong earthquake hit the island of Zante. Many homes collapsed in the

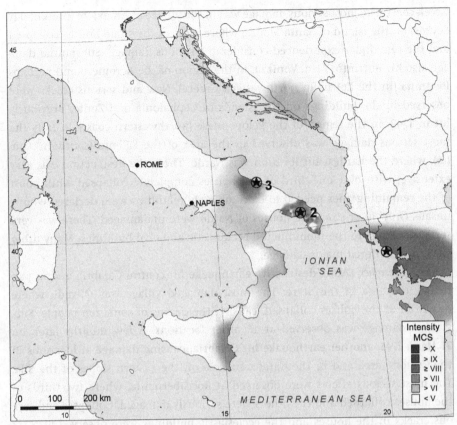

Figure 8.8 The same map as in Figure 8.7, but with an interpretation of three separate earthquakes (the stars represent the postulated epicentres) based on an analysis of the attenuation of the earthquake ground-shaking and the distribution of the macroseismic intensities.

city of Zante and in the various other inhabited areas of the island, causing the deaths of over 100 people. Besides the houses, in Zante there was serious damage to the praetorian (government) palace, the army lodgings and depots, and various churches. The new cathedral of San Marco, whose construction was still in progress, almost completely collapsed.

On *20 February 1743* an even stronger earthquake occurred. It was felt across a huge area, extending from the Peloponnese to the island of Malta and all the way up to northern Italy. This earthquake was perceived by the majority of the first-hand witnesses in areas far removed from the damage region as a sequence of three violent shocks that succeeded one another in a span of time between 7 and 20 minutes, according to the sources. In the Greek area, the earthquake hit the islands of Santa Maura and Corfu and locations adjacent to the western

Greek coast. The most serious damage was in the village of Amaxichi (present-day Levkás) on the island of Santa Maura, where most of the stone houses collapsed and the remainder experienced considerable serious damage. Substantial damage also hit Butrintò, Arta, Vonizza, in the region of the Xeromero and perhaps Ioannina (in the Epirus in northwestern Greece). New and serious cracks were observed in the buildings on the islands of Cephalonia and Zante, previously hit in 1742, in the region of the Peloponnese (northwestern coast). In Italy the most serious damage was observed in the area of the Salento (southern Apulia), where the hardest hit location was Nardò. There the destruction was very extensive, with over one-third of the houses completely collapsed while most of the remaining ones needed to be completely rebuilt or were declared wholly unsafe. Only 30 or so of the houses of Nardò were undamaged. There was very serious damage to the monumental ecclesiastic and civil buildings. Many other structures suffered major damage.

On 7 December 1743, a destructive earthquake hit central Calabria and in particular the area of the Serre. The most damaged village was Olivadi, where nearly all of the houses collapsed, causing the deaths of some ten people. Substantial damage was observed at 16 other locations. A few months later, on 21 March 1744, another earthquake hit Calabria, causing damage at locations in the Crotone area and in the Catanzaro area on the eastern slopes of the Sila. The most serious effects were observed at Roccabernarda, where two-thirds of the houses collapsed and the others were seriously damaged. Collapses and serious cracks in the houses and the ecclesiastic buildings were observed at a six other locations, amongst which was the city of Catanzaro where several churches were seriously damaged or half-destroyed. Less serious damage was observed in a further 12 villages.

A map of all the earthquakes described here is shown in Figure 8.9.

8.5   Foreshocks and aftershocks: why targeted research is useful

The forerunners to many major earthquakes, small preceding earthquakes called foreshocks, can provide useful clues about a large earthquake. They can indicate that a population was alerted about earthquake activity and perhaps even left their houses, thus explaining, for example, a low death toll despite extensive destruction in the mainshock. Likewise, the aftershocks following a large earthquake can hold important information, for as tremor follows tremor with some perhaps approaching the intensity of the mainshock, the damage may worsen during the days or weeks of the greatest aftershock activity. Such points should be borne in mind for at least two reasons:

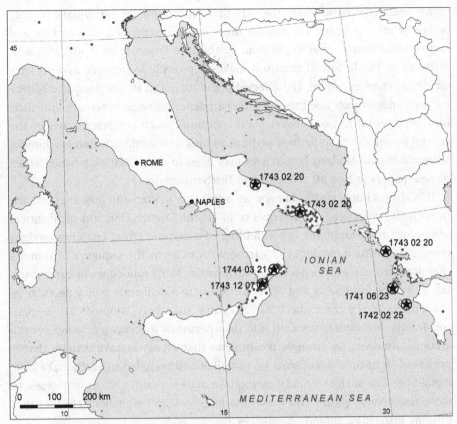

Figure 8.9 Map of the seismic crisis in 1743–1744 (south Italy, Greece) with all the epicentres located with stars and dates. There are seven earthquakes with serious damage.

(1) the initial scope of the damage due to the main earthquake may increase in a short space of time by as much as two degrees of intensity due to strong aftershocks. Such a record from past events may be of interest to present-day rescue-workers who are faced with search and rescue efforts after a modern strong earthquake in an area prone to strong aftershocks;

(2) if the aftershocks added to the damage from the mainshock, sources written as little as a month later usually will refer to cumulative effects without differentiating what effects were caused by the main earthquake. By contrast, sources written just after the main tremor (letters, newspaper reports, etc.) may err on the low side of reporting the total damage due to the entire earthquake sequence.

For reasons such as these, records that report smaller earthquake activity before or after a strong earthquake must not be overlooked. The archive and institutional sources may capture an earthquake sequence as it unfolds, especially when the historical situation makes it possible to produce and preserve written exchanges among the affected population and its rescuers. The historical seismologist may need to sift correspondence between governors and their local representatives, or between local communities and representatives of the central power, for time periods perhaps as long as months after an earthquake sequence began. Modern researchers may need to check diaries, private letters or newspapers to find all references to further tremors.

It is always important to know *when* a text was written and how that text fits in with the chronology of a sequence of seismic events. One aim of historical seismology is to highlight the unfolding of a seismic crisis, enabling modern researchers using a catalogue of earthquakes to know the sequence of tremors that preceded and followed a strong earthquake. Many catalogues in current use fail to list the foreshocks and aftershocks due to a deliberate policy decision by the authors of the catalogue. This policy may have been adopted due to space constraints, incomplete research into the aftershock sequences, or other considerations. However, we strongly recommend that all earthquake activity that is preserved in historic documents be compiled and included in earthquake catalogues, because such secondary earthquake material can be of major interest to some researchers. Figure 8.10 shows the foreshock and aftershock chronologies of some historic earthquake sequences.

8.6   Epicentres at sea or on land?

When an earthquake affects coastal areas, a tough problem always arises. As the data on the damage and felt effects almost invariably come only from the onshore areas, one is inevitably faced with the possibility of having an incomplete seismic scenario. Where could the epicentre be? The temptation to place an epicentre at an arbitrary location should be avoided. Indeed, any objective system for the estimation of the location parameters of a historical earthquake is always better than a subjective assignment of a source location using reasoning that is not transparent and may even be scientifically flawed. However, any system of calculation that only uses the data from specific localities where the earthquake caused damage or was felt, and therefore were on the mainland or on inhabited islands, tends to be biased toward putting the calculated earthquake epicentre onshore or very near the onshore area. In every seismological research group, there is no shortage of discussions and divergences of opinions about how to most accurately estimate the location of the active

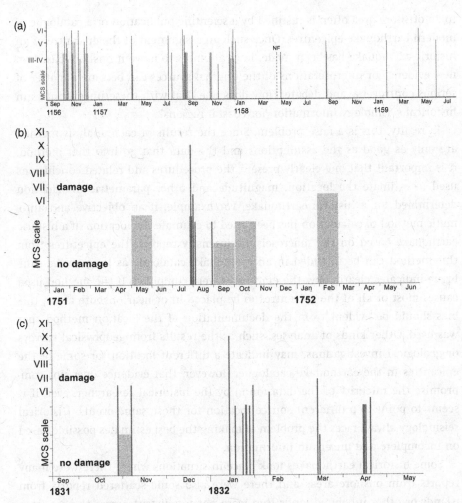

Figure 8.10 (a) Foreshocks and aftershocks of the seismic crisis in 1156–1159, felt at Damascus and described by an eyewitness (from Guidoboni and Comastri, 2005). (b) Simplified trend of the earthquake sequence around the time of the earthquake of 27 July 1751 ($I_0 = X$; $M = 6.3$) that hit Umbria (central Italy). This event sequence is reconstructed on the basis of the direct sources (from Guidoboni *et al.*, 2007). (c) Foreshocks and aftershocks of the long seismic crisis in 1832 in northern Italy from descriptions of the direct sources (from Guidoboni *et al.*, 2007).

earthquake fault when presented with such macroseismic data. On the other hand, if the geological data provide some indications that an offshore geological structure may have been one on which earthquake faulting might have occurred, even if the evidence is highly debatable and not well supported, it is often the case that the epicentres of nearby historic earthquakes are assigned to that offshore structure. In these cases, the assignment of a historic epicentre to an offshore area often is justified by a scientific publication or a catalogue of modern earthquake epicentres. Once such an assignment of the epicentre of an historic earthquake has been made, later attempts to move it onshore based on new evidence or interpretations of the original sources can become a matter of serious controversy and debate. How does one deal with these uncertainties in historical seismology information for coastal regions?

In reality, this is a false problem. Since the results of each analysis method are only as good as the assumptions and the data that go into that method, it is important that one clearly present the procedures and related conclusions used to estimate the location, magnitude and other parametric information determined for a historic earthquake. For example, if an objective and auto-matic method of calculation has been used to estimate the location of a historic earthquake based on the macroseismic intensity reports, the epicentres from this method can be included in an earthquake catalogue as long as the cata-logue indicates clearly how the epicentres were computed. If the method used causes most or all of the epicentres to be placed in or near onshore areas, this bias should be evident from the documentation of the location method that was used. Other kinds of analyses, such as the results from geophysical surveys or geological investigations, may indicate a different location for some of the epicentres in the earthquake catalogue. However, that evidence does not com-promise the integrity of the data found by the historical researcher, even if it seems to point to a different source location for those same events. Historical seismology always faces the problem of making the best estimates possible based on incomplete and uncertain information.

Some historical earthquakes took place in situations where there were many reports from onshore areas and there were also some scattered reports from islands or other inhabited lands that were not too distant across the sea from the main continental area where the event was experienced. From the stand-point of historical research, for these cases one can try to collect data capable of enclosing the area of the strongest seismic effects. As populations expanded with time into offshore areas, it can happen that earthquakes that occurred in that area in earlier time periods appear to have arisen from different locations from more recent historical earthquake or from the seismically active areas known today from instrumental earthquake monitoring. For example, Figures 8.11 and 8.12 show maps of the macroseismic intensities derived from two different time periods. In both cases, the seismic events were experienced in northwestern Italy and southeastern France, one in 1887 and one in 1963. For the 1887 event, the locations where the earthquake is reported are restricted almost exclusively to the continental areas of Italy and France, and the epicentre is located just onshore. For the 1963 event, the locations from which earthquake reports were received are distributed across a much wider range of coastal areas as well as

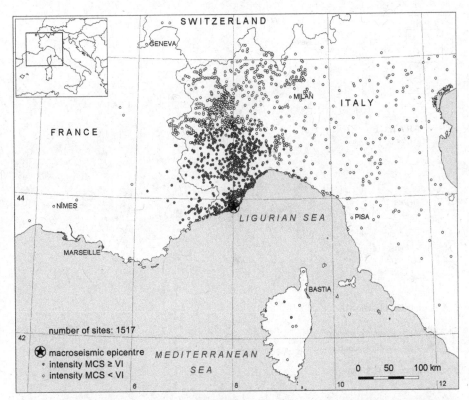

Figure 8.11 Effects of the earthquake on 23 February 1887, which hit southern France and western Italy ($I_0 = IX$; $M = 6.3$). A calculation of the epicentre locates the event on the coast (from Guidoboni *et al.*, 2007).

from offshore islands. In this case, the estimation of the epicentre is made using a geographically more extensive set of data, and the epicentre is located at sea. This example should not be taken to mean that all onshore epicentres in coastal areas are biased due to a lack of data from the offshore region. Natural phenomena like earthquakes follow variable and complex patterns, and understanding them fully is an objective that can never be achieved with the incomplete data with which seismological research must work. Perhaps the most important lesson to be learned is that one should not draw hasty conclusions based on the data at hand but rather welcome all data that can help answer the research questions being addressed.

8.7   The completeness of an earthquake catalogue: some general considerations from the historical point of view

One special problem in historical seismology is that of having to deal with *missing* information. While it is assumed that there is a certain stability

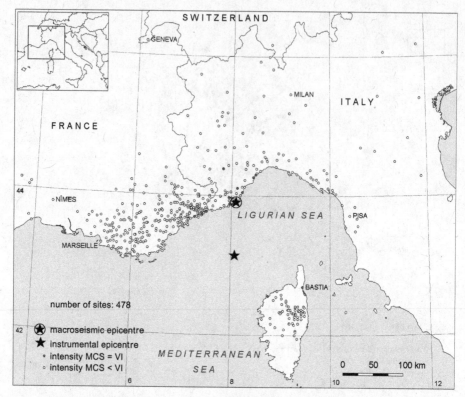

Figure 8.12 Effects of the earthquake on 19 July 1963, which hit southern France and western Italy ($I_0 = 6$, $M_e = 5.5$). A calculation of the epicentre locates the event at sea (from Guidoboni *et al.*, 2007).

in the long-term spatial and temporal patterns of seismic activity, it is known that average return times of strong earthquakes in an area can vary considerably even on a single fault. Furthermore, many scientific analyses require a complete knowledge of all earthquakes above some magnitude going back as far in time as possible. Thus, the level of *completeness* of an earthquake catalogue, by which we mean our confidence that all earthquakes above some specified magnitude that took place in an area are known and contained in the catalogue, is an important issue for the scientist as well as the historian.

A destructive natural phenomenon across a territory, such as a strong earthquake, will probably lead in the course of time to the production of written evidence of the event, provided that certain conditions are fulfilled:

(1) the affected area was occupied by man;
(2) even within different cognitive frameworks, the earthquake was recognized as such and not mistaken for some other phenomenon;

(3) someone had a reason to write about it;

(4) the relevant written texts have survived.

As far as written texts are concerned, one can say that there is a general tendency for information gaps to grow bigger the further one goes back in time, as there is an increased likelihood that texts will have been destroyed or lost. Obviously, for various reasons, there are also geographical regions that are poorly documented even for recent periods.

Analyses carried out in historical seismology should make it possible to achieve positive new results that improve various aspects of the understanding of past earthquakes. First of all, the parameters for known seismic events may be revised and corrected. At this level of analysis the assessment of new earthquake information does not immediately need to match geological information about the area being studied. New results can also be achieved in the classification of the known historical records. This should allow for an improved evaluation of the quality of the earthquake catalogue as a source of information for a site or area. Another way that knowledge of past earthquakes can be improved is by evaluating the availability of historical records for particular periods of time. Analyses of this type seek to identify those time periods for which historical information is not available. Finally, an important new result might well lie in how one assesses the principal forms of selection which controlled source production and preservation, both by chronological period and in relation to the earthquake data (i.e. in particular, regarding earthquake chronologies and even sizes) in the available earthquake catalogues.

The seismic hazard estimate for a particular site is based on a multiplicity of factors that emerge from different types of research. The term *seismic hazard* is used to mean the probability that the following proposition is true: 'Within a period of time t, the site will be subject to a seismic shock of strength I (expressed in terms of intensity, acceleration, or some other measure of the strength of the earthquake effects at the site) which is greater than or equal to some reference parameter I_0.' The I_0 and t parameters are independent variables and are based on the protection strategies devised in a non-scientific (i.e. political or social) context. In most cases, the amount of information available is only sufficient to make very rough seismic hazard estimates for a particular site, and the reliability of such estimates can only be determined *a posteriori* or by means of retrospective forecasts based on the site's seismic history. In this sense, the deterministic estimate becomes probabilistic in terms of the reliability of the prediction.

A second approach for making a seismic hazard estimate is based on the possibility of using the site's past seismic history to identify certain useful regularities in the pattern of earthquakes that are known to have taken place. The seismic hazard estimate is generated using a phenomenological model capable of reproducing the past seismicity of the site in a way that extrapolates the past seismic activity into the future (in terms of probability). In this sense, the model's physical plausibility proves to be less important than its phenomenological validity.

Finally, there is a third approach, which involves making use in a multidisciplinary way of what is known about great historical earthquakes, together with the catalogued activity of active faults and recorded instrumental data. However, this more sophisticated and more complex model only leads to positive results if thorough independence is established between the various types of seismological and geological research involved, especially regarding the classification of active faults and the historical seismology analyses. The sites of the seismic effects in a particular area must be studied independently of the knowledge of active faults. Knowledge that an active fault is present should not necessarily sway the assigned location of a historical epicentre before the whole field of effects from past earthquakes has been reconstructed in an objective and critical manner. Furthermore, a knowledge of the seismic history of a site is crucially important in establishing phenomenological models and in validating seismic hazard estimates, by whatever means those hazard values have been reached. The degree of reliability of the available seismic history (i.e. its level of completeness) determines the degree of reliability of the seismic hazard estimates.

The required completeness level for an earthquake catalogue varies according to the method adopted for achieving a seismic hazard estimate. Generally speaking, the most stringent requirements are those related to a combined statistical and phenomenological approach in which the seismic history has the dual role of making a phenomenological model possible and validating its utility for forecasting purposes.

9

Processing historical records

9.1   The validation of historical data

In order to make proper as well as transparent use of the results of historical research, the data must be submitted to a *validation process*. Such a validation process involves setting up a basic record that makes it possible to lay down the main elements needed to *classify* a text.

The basic critical points required to classify a text should address the following issues:

(1) *Author*   Who was he, when did he live, where was he writing from, and why was he writing (in other words, who, when, where, why)?

(2) *Title of the text*   This is not a problem for printed texts; however, if dealing with archival documents, one needs to be very careful. It is useful to list each selected document separately in a bibliography according to the document type (e.g. a plea, a benefice, a letter requesting tax reductions, a technical survey, etc.). This kind of classification can prove to be very useful during any stage of preliminary data analysis (during which a decision may be made as to whether or not to carry on with the research) and to give advice about the documents to those who wish to use them to perform further research. By clearly highlighting the prevalent type of information in the documents, researchers can better understand the points of view that were used in the reconstruction of the earthquake and its effects.

(3) *Date*   This is not a problem for printed texts, although one needs to be careful to include the date when a text was written, especially if it is an edition. For example, if a chronicle was written sometime between

1670 and 1690 but published in a critical edition in 1997, the relevant bibliographical reference must show both dates.

For the date of a memorial-style manuscript, one must be sure to include reference to the date expressed in the text. When this is missing, one should generally date the manuscript using the information contained in the inventories available in the library that preserves the manuscript or information in the codicological literature, when available.

Archival documents (i.e. administrative acts, the documents of a chancellery, etc.) have always included the date of the document because they were part of the official procedures. Thus, no problems should arise in dating such documents.

It may also be worthwhile remembering that the date in a file should be copied and referenced in the format in which it is found in the original text, especially for time periods prior to the Gregorian calendar reform. It is necessary to convert all the dates to the current Gregorian calendar only in such cases where sources from different countries with different calendars (such as the Julian or Gregorian calendars) are used for the *same earthquake*. In order to preserve transparency in the processing of a text, converted dates should be distinguished in a special way, such as by using square brackets or other notational indicators.

The time of an event, such as the hour when it took place, should be reproduced from a document by maintaining the original wording of the text and giving a clear explanation or indication of any interpretations or conversions of the original time report.

(4) *Shelf-mark and place of conservation*   It may be quite obvious to point out how essential it is to ensure accuracy in indicating the exact shelf-mark of the manuscript and the folios or pages containing the selected texts. Accuracy is necessary for quality historiographic work and must in no way be overlooked.

It is important for historical seismology research to have a complete picture of the research in progress or of that already performed for a specific location or territory, since it is very useful to be able to map out the archives, libraries and other sites where research has already been carried out. To create such an overall picture, it is enough to insert the name of the library and its location into the bibliographical file in a research document (whether it is a manuscript or a printed text). Indeed, many sources may have already been published and,

when possible, it is always best to use those in the critical editions
that are available.

(5) *Geographical coverage*   It is advisable to keep some kind of *address book*
of the various research sites (archives and libraries) of interest in the
project. Making a map of these sites can help one keep tabs on the
geographical coverage of the historical data.

(6) *Contents*   It is very good practice to select those texts that are relevant,
to transcribe them in the original language in which they were
created, and to follow the transcription with a translation of the text.
However, these operations, apparently very simple, can contain some
pitfalls. The biggest potential pitfall concerns the accuracy of the
transcription and the text selection. As regards the creation of an
exact transcription, today digital photography or microfilm can be a
significant help since they make it possible to consult the original
freely, above all if it is handwritten, and check with the original in
case of any doubts about its contents. As regards to translating the
text, it is advisable to select more than just those parts of the text that
strictly concern the earthquake being examined so as not to lose the
overall sense of the text that is being used. Above all, for ancient and
medieval sources, preserving the contextual parts of the narratives can
help one obtain information such as a better understanding of the
date of the event, a topographic reference, or the author's way of
narrating events.

(7) *Systematic card-indexing*   Each selected text can contain pieces of
information for one or more earthquakes or tsunamis. All of the
earthquake and tsunami notations in a source should be card-indexed
so as to preserve in its original form the information concerning
seismic events not being examined at that moment. In areas where
the effects of frequent earthquakes are often felt, systematic
card-indexing allows one to maximize the results from a textual
analysis. Indeed, one can also gather the effects of far-off or unknown
earthquakes in this way. But apart from that, from the standpoint of
textual criticism one can get an overall picture of the researcher's
particular interest in the earthquake source and his own particular
way of dealing with the source.

A well-indexed historical record makes it possible to gain a general overview
of the research, as well as to preserve detailed information about the seismic
events described. As pointed out above, it is crucial to preserve historical records

in their complete and original form, even after a detailed index of the contents of the historical records has been created.

A historical source or an erudite historiographic work may contain information concerning various historical earthquakes and tsunamis, but not all the information will necessarily have the same critical value. For example, a chronicle that preserves the memory of events that had occurred over a very long range of time can be considered a direct source only for the events that had happened just before or at the time that the work was written. For earlier events, such as those that predate the lifetime of the author, this kind of work must be considered as an indirect (secondary) source. It is important to understand which other sources such a chronicle relied upon for its information and whether these earlier sources are still available or lost. A text covering a long time period cannot therefore be considered to be a source in itself; it is only a primary source for those events that its author or his contemporaries had directly witnessed.

The critical distinction between primary and secondary sources must be based on historiographic requirements, and all historical seismological researchers must pay attention to these requirements precisely because their work inevitably makes use of the historiographic method. In research practice it may happen that important information concerning a seismic event is contained in secondary sources. Of course, this information cannot be ignored or discarded. The reference to the seismic event in the source might have been derived from prior sources that are now lost, or it may have been a spurious addition that the author thought relevant or interesting. It is not always easy to distinguish between these extreme and opposite cases, especially if there is not the means or the time to make specific in-depth analyses. Of course, it is worthwhile to accumulate and retain the information of the secondary sources, but indications of their lesser value should be preserved as well. Thus, a correct classification of the texts is necessary to highlight such aspects and to highlight the need for other future research into the primary sources.

The construction of a database can make it easy to manage the basic qualitative historical data and to organize them. There are many examples of such databases now available on the Internet (see websites such as www.ingv.it, www.sisfrance.fr, etc.).

9.2   Classifying a list of references

Classifying a list of references about an earthquake (that is, the list of the basic texts that have been used to define the earthquake's spatial extent and effects) means giving each text a value according to precise historiographic criteria. For *each individual* seismic event a data set needs to established, but each individual element of the data set will have its own informational weight and importance. Classifying a bibliography thus means clarifying the available information and placing it on a scale of importance.

Although text types may vary a great deal in different countries, certain key criteria can be identified, though they may be subject to minor modifications in accordance with particular historiographic situations. Whichever procedure is adopted, however, it is useful to bear in mind some basic historiographic categories. Value codes have a very precise meaning for historical researchers, who are the major players in the research into earthquakes of the historical past. The assignment of values codes allows one to easily check the databases available and to obtain statistics on each type of text that contributes to outlining the image of the earthquakes to be analysed.

9.2.1   An example of the classification of a list of references with a value code

The following classification scheme is an example, one among the many that can be made to classify a bibliography. A good classification system should make it easy for a researcher to comprehend the nature of the data that are available. In fact, each country has different problems to solve with the data set for seismic events derived from the available sources. The aim of any classification scheme is that it be simple and easy to apply. In the classification system suggested here we refer back to the definitions of various terms that we have given in the previous chapters of this book.

S1   *Source*, that is direct source. A text written by a reliable contemporary witness, living geographically close to the event described. This evidence must be sound and authoritative. The source may be a public, ecclesiastical or private archive document; a literary work, memoir or inscription; a private journal (if letters or first-hand reports are involved); reports from a naturalist (first-hand records); or scientific data (macroseismic postcards, first-hand records made in the field, etc.).

S2   *Source*, that is indirect or secondary source. A text written by an author who is not himself a witness, but who draws upon authoritative primary sources. These sources may be chronicle sources that are particularly authoritative, although they are not contemporary.

NS   *Negative source*. Sometimes it is important to highlight the silence of the direct sources that were supposed to provide information.

HS   *Historiographic studies*. This category includes specific contributions by historians on historical earthquakes or on contextual aspects of economic, demographic and social history, etc., which can be used to improve the analysis of the effects of an earthquake. These studies may

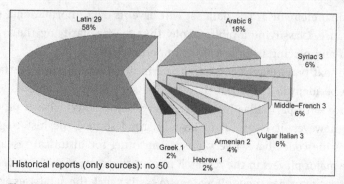

Figure 9.1 Type and number of historical sources containing information on the earthquake of 29 June 1170 in Syria. Noteworthy is the large number of Latin texts (29, 58%) in comparison with sources of other languages, i.e. Arabic, Syriac, Greek, etc. (from Guidoboni *et al.*, 2004).

cite sources, contain assessments of research to be performed, provide important outlines of research to be carried out, point out important and specialist indications on the sources and their type, and possibly may contain a history of the territory where the earthquakes occurred. Special attention must also be addressed to the local historiography, which is often not based on direct sources.

GI   Generic information

C   *Catalogues.* To simplify, with this codex two types of texts can be classified. One is the earthquake lists compiled from an erudite, literary or naturalist standpoint, and while the other is catalogues in a more modern sense. The latter are lists of earthquakes compiled for scientific purposes dating from the nineteenth century onwards. They may be based on primary sources or on cited works, or they may not provide any reference to the sources used. They may be descriptive or parametric (generally speaking starting from the mid twentieth century). Obviously, if a country has a rich and extensive tradition with an erudite premodern list of earthquakes, such a list is relevant to this particular kind of text.

EC   *Early catalogues.*

B   *Bulletins.* Publications by specialist bodies, based on their scientific observations or recordings that include tables of measurements and either whole or partial series of data recordings. Such publications can appear in a variety of contexts (such as acts of academies, articles, etc.). Observations of seismic effects recorded in observatory registers are categorized as S1 or S2, as appropriate.

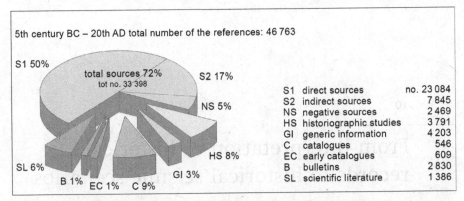

5th century BC – 20th AD total number of the references: 46 763

		no.
S1	direct sources	23 084
S2	indirect sources	7 845
NS	negative sources	2 469
HS	historiographic studies	3 791
GI	generic information	4 203
C	catalogues	546
EC	early catalogues	609
B	bulletins	2 830
SL	scientific literature	1 386

Figure 9.2 Example of the comprehensive bibliographic composition of a current catalogue of historical earthquakes, the *Catalogue of Strong Earthquakes in Italy* (CFTI4Med, Guidoboni *et al.*, 2007). The various types of information have contributed to the descriptive image or to the critique of the sources of the various seismic events analysed, differently depending on the related historical period. Over 70% of the overall bibliography is made up of historical sources.

SL   *Scientific Literature*. Includes contemporary studies carried out within a particular theory-based cognitive and interpretative framework, from the nineteenth century onwards, and recent studies on historical earthquakes.

Once a bibliography has been prepared by attributing to each title a *value code*, one can obtain a schematic picture of the *information quality* that underpins the reconstruction of an historical earthquake. If there are few direct sources and the information mostly comes from secondary sources, historiographic texts or old earthquake catalogues, one must then decide whether more dedicated historical research into the earthquake is warranted and practical. Figure 9.1 shows an example of the classification of the sources for a single earthquake, while Figure 9.2 shows an example of the classification of the sources for an earthquake catalogue.

From interpretation of historical records to historical seismic scenarios

10.1 Constructing seismic scenarios: a painstaking montage of different elements

A wise rule for limiting error when interpreting historical records of earthquakes and tsunamis is to relate the source texts closely to their historical, cultural and social contexts. How successful this operation will be depends on expert knowledge of the texts concerned and of their contexts. Historical seismology, as has been discussed, is based on qualitative data that can be easily misread, can often appear to be deceptively clear, but are never exhaustive enough in their content. The interpretation of the historical data selected to investigate historical earthquakes and tsunamis should therefore use specific interpretation tools as much as possible, paying particular attention to the following:

(1) the calendars and the related chronological systems in use in the various areas and periods where earthquakes were experienced, of which we have provided a summary in Chapter 7;

(2) an analysis of terminology and linguistic expressions;

(3) toponyms, microtoponyms and historical geotoponyms, in order to learn about their variations over time;

(4) the administrative and political history of the area of the earthquake reports, in order to define accurately the control of the areas and localities being studied in terms of their local and central political power; this is critical when an earthquake affects areas in several states or regions;

(5) historical population data to evaluate the impact of a given earthquake on the population, especially regarding the numbers of dead, injured and homeless.

Knowledge of or the ability to evaluate each of these points need not be confined to one person. The aim of any research effort must be to foster collaboration between the various kinds of experts who engage in historical research. The goal of any such collaboration is to improve the assessment of an accurate seismic scenario for each historic earthquake that is being analysed.

10.2   Terminology and modes of expression

When historical analysis addresses a long period of time, it may happen that the relevant sources for a single region are written in different languages. This calls for contributions from multiple specialists in order to ensure that the original data are properly preserved, transcribed and translated. Furthermore, even for sources with a common language, authors of different social and cultural classes will adopt different modes of expression concerning events that they write about. Sometimes subtly different terminologies while at other times strongly specific expressions may have been used in the same time period by writers in the bureaucratic, religious, naturalist or technical fields. Only by paying proper attention to these linguistic factors of the source material is it possible to avoid misunderstandings of the information in the sources.

As already mentioned, one element that must be taken into account when interpreting sources is understanding the purpose of the author in writing the account. When due attention is paid to this element, the data become easier to understand and interpret. At times, the users of historical sources have the feeling that the texts found are unsatisfactory, riddled with gaps or lacking informational significance, but this usually results from an inadequate appreciation of the relationship between the questions the researcher is asking, the texts being investigated and the nature of the historical document under examination. A historical source cannot be considered exhaustive or poor in itself but only in relation to the research questions one is asking. Indeed, the texts that one finds and selects today to learn about the earthquakes and tsunamis of the past were written and produced for reasons other than those for which they are now being consulted. Every source may contain only pieces or fragments of an answer to the questions of the researcher, which means that if one is going to establish the effects of an earthquake in a particular historical region, one needs multiple pieces of evidence that take into account a variety of points of view of the event being researched.

In Chapter 4 we pointed out the multiple meanings that the term earthquake has had throughout history and the different meanings that may be associated with this term in historical sources. In historical records, the term earthquake has denoted not only a seismic event in the modern sense but also the concept of a strong sudden shock due, for example, to a burst of lightning, to a gunpowder explosion, or to a landslide. In this chapter we examine the words frequently used in historic sources to qualitatively describe an earthquake shock or succession of shocks. The terms that we examine here belong to the common language, and are not scientific terms. However, our purpose is to indicate what scientific information can be gleaned from the terminology used in the written accounts.

A cursory analysis of recurrent terms used to describe an earthquake can help us to better understand the text available. A triangle of interconnected elements is formed by the historical text, its analysis and the researcher's needs. On one side of the triangle is the author's viewpoint, while on another is the factual reality of the event described in the historic report. The third side of the triangle is a set of *inferential strategies* (as they are defined by communications experts) which should allow the researcher to distinguish between the two levels of comprehension, that of the *literal text*, on the one hand, and the *facts*, on the other hand. This interaction between written report and reality also occurs when one reads a newspaper. In this case, although there is no direct relationship between the reader and the facts that actually took place, these facts are likely to be part of a complex reality that is only partly known to the reader. Moreover, in order to be correctly informed by a written text from the past concerning environmental and habitational situations that have changed (especially situations that no longer exist), one must accept at least the one basic requirement that there is no substantial difference between the author's desired and implied contents (because they are known to the contemporary recipients) and the literal meaning of the text.

10.2.1 Lexica

The words used in the past to describe earthquakes and tsunamis have their own specific originality and can be defined as a *lexicon*. Together they comprise something less than a dictionary, in which the contents are systematically arranged, and something more than a plurality of minor individual definitions. Obviously, the text type plays an important role (see Chapter 3), because it influences the style of writing. For instance, an administrative text and a book of memories necessarily use different styles, so that typology, quality and quantity of the information change as well.

If twelfth-century Latin sources are considered, for example, one notes that only a minority of the texts actually use terms that convey subjective factors (such as fear or fright). More frequent is the use of terms concerning the size of an event, such as 'large' (in Latin *magnus*) and its superlatives 'largest' (*maximus*) and 'enormous' (*ingens*). For example, a comparative analysis of the use of the term *maximus* in about ten medieval earthquakes of the Latin area has shown that this term is used to describe a very destructive earthquake, whose damage is attested to by other sources.

From the start of the modern era (that is from the sixteenth century) the description of earthquakes is in general more tied to the perception of effects, which are then also described in a more precise and complex way than previously. Generally, there is an increase in the use of terms that stress the *emotional perception* (such as 'frightening' or 'terrible') of the seismic event and its effects. This development cannot be traced back in a simple evolutionary line, but it probably arose both with the increasing awareness of a writer's own perceptions and with new paradigms and forms in scientific thinking (see Chapter 4).

Another point of note is that often the same shock could have several evaluations in the sources. Indeed, one can find different sources that define the same seismic event occurring in the same place as being *small*, *strong*, *very strong*, etc. These inconsistencies in judgement, which are part of an individual's idiolect, are hardly surprising as they are really subjective data in the sources. Only a set of independent sources, in other words those not derived from one another, allow one to assess the factual elements of the information in the sources. Sometimes the information in the sources may simply be poorly presented and even contradictory due to subjective perception of those who wrote the reports.

When examined in detail, one can distil the terminology used for earthquake and tsunami descriptions in pre-contemporary sources, irrespective of the individual authors and the historical period, into a well-defined and limited vocabulary that was borrowed from the common language of the time (i.e. descriptions of the noise of the earthquake, the type of shock, etc.). The selection of terminology may have answered the needs of objectivity, such as the writer's desire to refer to common experiences. As far back as the eighteenth century one can observe a tendency to give some terms (*weak*, *strong*, *violent*, etc.) predetermined meanings that are not entirely subjective but rather convey some generally understood connotations. These terms become more frequent in scientific and technical sources (assessments, official reports, scientific texts, etc.) as these kinds of reports became more abundantly produced. The need to give some categorization to the qualitative terms pertaining to the perception of a shock led to the formation of the first macroseismic intensity scales for scientific use in the nineteenth century.

Regarding the concept of single words conveying the emotional aspects of a seismic event, it can be said that the words themselves have not changed in any appreciable way with time. When writing about earthquakes and tsunamis, authors of all time periods use words that express fear, awe, shock and surprise. But as has been said, these are not the expressions that inform historical seismology research when it seeks to evaluate earthquake effects.

It is also not infrequent to be confronted with texts that are intrinsically rather generic. What should one do when a text speaks of *ruins*, but does not say of what; *damage* without saying of what type or to which buildings; *collapses* but without saying whether these are total or partial and how they are distributed among all the structures of the city or settlement? Depending on the time period and the availability of sources, it is worthwhile making sure that such generic statements do not form the sole characterization of the research results. Rather, starting from these generic terms, it is important to engage in a further research effort to uncover more clear-cut descriptions to shed light on the meaning of these generic terms. The shift from generic information (albeit from an authoritative source) to specific information can only be achieved by having recourse to other sources.

There are cases in which some terms have had a culturally shared meaning that is different from what the term connotes in the modern language. For example, in many past earthquake accounts the word 'collapse', if not otherwise specified, often indicates only a partial collapse. In older records it is hardly ever a mere exaggeration (though that can sometimes be found in sources), but instead it has the implied meaning of the partial destruction of a structure. Indeed, in texts based principally on personal memory there is a tendency to stress not so much losses to a physical structure itself but rather the *loss of functionality* of a building or defensive structure. Especially in ancient and medieval sources (but not unusually also in later ones), if an earthquake damaged buildings having a strong functional and symbolic value (such as places of worship, defensive city walls, military fortresses, etc.) the trend was to remember their collapse as a cessation of functionality. Neglecting to appreciate this aspect of historical descriptions of earthquake damage can swing the macroseismic intensity estimate of local effects by one or even two degrees.

When one reads an account that states that 'the (city) walls collapsed', one should not forget that in the Mediterranean area until a few centuries ago the partial collapses of city walls often led neighbouring enemies to besiege the affected city. Attacking a defenceless town was so frequent in the past that a Benedictine chronicle of the ninth century (*Chronicon Vulturnense*) recalls, as an example for posterity's sake, the case of a Saracen commander who did not make an attack. This was Abu Massar, who was preparing to attack Isernia, a small town in southern Italy. Having discovered that Isernia had just been struck by a strong earthquake (that of June 847, $I_0 = $ VIII; $M = 5.6$), he avoided ransacking the town and uttered the historic phrase, 'The Lord of all things is enraged with it (Isernia), and who am I to add salt to the wounds?'. A more common example took place in 1638 after the earthquake of 27 March ($I_0 = $ XI; $M = 7.0$) when the partial collapse of the walls of Nicastro (present-day Lamezia Terme in Calabria) encouraged the subsequent assault by Turkish galleons, who took advantage of the destruction due to the earthquake. In another case, towards the end of the eighteenth century it was written that in Rimini (a papal city on the Adriatic coast), after the earthquake on 25 December 1786 (I_0 = VIII; M = 5.6) that had damaged and opened up some parts of the city walls, pirate incursions were greatly feared. In those days raids by pirates, who sailed upstream from the mouths of the rivers, were not infrequent events, and the earthquake suddenly had greatly compromised one of the most important city defences against enemy attacks. Thus, one can attach more precise meanings to words that describe earthquake effect by bearing in mind the historical contexts of the sources.

As in any technical field, specialized and often obscure terms were used in the historical building sector. Master masons, architect–builders and specialized technicians of centuries past used such terms when they drafted reports and assessments of damage and set out the reconstruction to be carried out, as requested by princes, rulers of various ranks and proprietors (see Chapter 3). In these types of sources, fully fledged professional terminology (almost a jargon) was used, and it may often be properly understood only by historians of architecture or by architects familiar with brick buildings. In these cases, collaboration with those who have expertise in these subjects may be vital in order not to miss or misunderstand important information about the specific earthquake effects on structures.

10.3   Place-names, administrative boundaries, frontiers and their changes

Historical areas that have been settled for a long period of time and have longstanding written cultures will have undergone substantial territorial changes in the course of time, with consequent developments in their administrative and political systems, language, and cultural and power centres. All these factors must be taken into consideration in analysing historical seismicity, whether one's point of departure is a specific area or a corpus of sources. For example, in certain Mediterranean regions there are places that have changed their name three or four times. These variations must obviously be recognized and associated with their present-day names. If places are georeferenced (i.e. given geographical coordinates in degrees, minutes and seconds), it becomes quite simple to identify historical sites in a uniform way.

Since the frontiers of present-day countries are often substantially different from those of ancient times, one must take into account in the interpretation of historic sources the ways in which frontiers have moved, as well as the administrative and political centres which controlled a particular region over a period of time. Frontiers are a particularly relevant variable, partly because earthquakes often encompass areas that at the time of the seismic event were, and in most cases still are, divided into different political, cultural and linguistic territories. This is the case of transfrontier earthquakes, which represent a difficult case study for historical seismology, in particular regarding the retrieval of direct historical sources. This topic is so important that it was the exact subject of one particular research project (Albini et al., 1995). Earthquake sources, and in particular administrative documents, usually provide only partial information about the extent of seismic effects in the case of transfrontier earthquakes. In fact, if the damage zone of an earthquake in historical time was split between more than one political region, information provided by documentary sources will be split according to the administrative division of the affected area, and archive documents will have to be retrieved in different countries. Furthermore, earthquakes sources will be produced within different cognitive models in different political and cultural regions. A similar tendency can be found even in non-institutional accounts. Due to linguistic and cultural barriers, chroniclers of one country often fail to provide accurate accounts for a foreign damage zone, but rather they often concentrate on the effects of the earthquake in the cultural area of the author of the account.

How changing boundaries and territories influence research and the use of sources: the cases of Italy and Armenia

An example of the relevance of historical frontiers and the way they have changed is the territory of present-day Italy, which is made up of a number of different earlier states. In the sixteenth century it consisted of 18 large and small states (Figure 10.1), each with its own administrative and fiscal procedures and its own governmental culture. Each state also had different types of documentation in recording the damage caused by destructive earthquakes and different financial procedures and possibilities regarding post-earthquake reconstruction. Such a situation, seen in terms of historical research, means that different types of archive research must be undertaken for different geographic areas. Some of the old states are now regions (e.g. Tuscany), while others covered what are now three or more regions (e.g. the Papal States, which included the Lazio, Umbria, Marche, Abruzzo and Emilia–Romagna regions). One therefore needs to know how these various states ran their administrations if one is to track down sources describing the effects of strong earthquakes within the states or at their boundaries. Although Italy is a country with long-term habitational stability, its administrative and cultural traditions are extremely varied with both place and time.

A second example concerns historical Armenia. The present-day Republic of Armenia is the result of a sort of subtraction of territories where Armenian

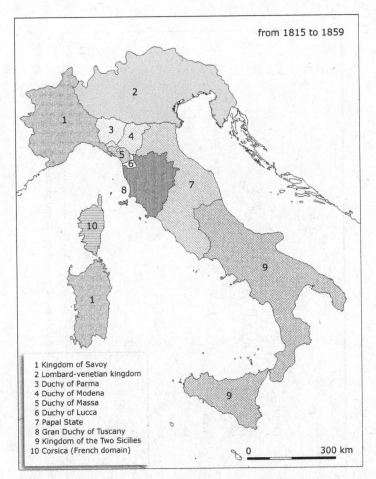

from 1815 to 1859

1 Kingdom of Savoy
2 Lombard-venetian kingdom
3 Duchy of Parma
4 Duchy of Modena
5 Duchy of Massa
6 Duchy of Lucca
7 Papal State
8 Gran Duchy of Tuscany
9 Kingdom of the Two Sicilies
10 Corsica (French domain)

0   300 km

Figure 10.1 Map of Italy before its unification in 1860. The territory was comprised of various reigns, principalities and duchies (from Boschi *et al.*, 1995).

culture had been present (Figure 10.2). The fact is that in the past there were no proper state boundaries but simply a large territory occupying a vast plateau at an elevation of between 1000 and 2000 m to the south of the Great Caucasus mountain chain. This plateau region acted as an important hub for trade between Asia and Europe. Its particular geographical position encouraged commercial and cultural encounters and exchanges over the centuries but also led to conflict with neighbouring powers and to military occupation. Persians, Byzantines, Arabs, Turks and Ottomans entered the Armenian territory in succession and held sway there. But in spite of their different cultures, religions, languages and value systems, the Armenians have always maintained strong ethnic, linguistic and cultural continuity in the teeth of complex and often tragic historical experiences. That is why, up to the early twentieth century, Armenian historical

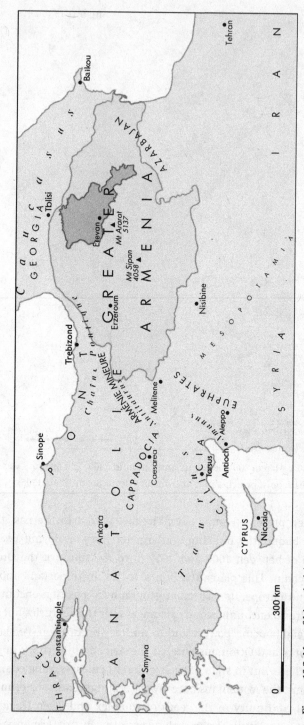

Figure 10.2 Map of historical (greater Armenia in light grey) and present-day Armenia (dark grey) (from Mutafian and Van Lauve, 2001).

sources may contain evidence for historical earthquakes in what is now eastern Turkey (where Ani, one of their ancient capitals, was situated), as well as in the eastern Mediterranean, thanks to their cultural and commercial presence there. The Armenian experience illustrates how different cultural and political influences can operate over time in a territory.

10.4   Territorial factors in seismic scenarios

In order to assess seismic effects in inhabited areas, it is useful to examine various aspects of the social, economic and demographic contexts of the areas that experienced the earthquake. The following considerations are of particular importance towards improving assessments of earthquake effects:

(1) population density of the localities concerned;
(2) chief characteristics of historical buildings (materials used and widely adopted building techniques);
(3) general characteristics of the local economy (living standards, economic expansion, crises, negative situations, etc.);
(4) dominant characteristics of the local culture (predominant religious groups, attachment to the locality).

The above list highlights important *historical variables* that may have a determining influence on both the overall picture of the effects and the historical image of the earthquake as conveyed by the sources. If research is to be organized in an efficient way, it is useful to have available such general, basic instruments as will permit comprehensive assessments of these historical variables. A general work-plan, which includes a database, can help one to rationalize the research process and to organize the material progressively selected for analysis. In order to achieve this, it is worthwhile to begin by establishing a simple, rational record of the documents being researched. Simplicity and rationality can be decided by each research group according to its own experience. When possible, texts should be electronically stored, and reproductions (photocopies, microfilms or digital scans) of original documents made. It is not advisable to record and save summaries of individual sources (whether descriptive catalogues or primary sources), because summarizing is always a subjective process and there is the inevitable risk of omitting details that may later turn out to be important. Only complete texts should be taken into account, and they must be studied in their original language. Detailed accounts of the prior work carried out by individual historical researchers are extremely useful, because they make it possible to note negative as well as positive data that have emerged during their research. The value nowadays of a piece of seismological research into historical

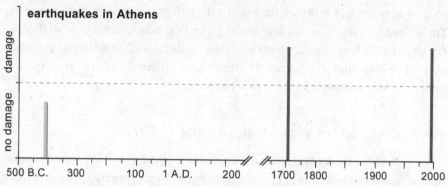

Figure 10.3 Earthquakes in Athens from 500 BC to AD 2000, according to the current state of historical knowledge.

earthquakes depends partly on a description of the processes by which its results are achieved.

Athens and Florence: two capitals of culture but two opposite cases of seismic history

Periods when towns were abandoned or suffered sudden population decreases may be indications of an economic crisis serious enough to obscure the effects of even a strong earthquake. Despite being the depository of one of the most ancient and developed written cultures, Athens suffers from a great gap in historical information on the earthquakes that have struck it over time. On 7 September 1999 an earthquake of magnitude 5.9 hit Athens, causing 140 deaths and serious damage. The event was localized in the western suburbs of the city where no active faulting had been mapped previously. The earthquake was centred in the Fili Valley, along the foothills of Parnitha near the small town of Fili and west of the Athens suburb Ano Liossia. This event surprised Greek seismologists, who called it, in the earliest phase, 'unexpected' (Pavlides *et al.*, 1999). The Fili Valley had experienced no historical earthquakes. A similar magnitude earthquake struck in 1705, but the epicentre for this old event is not well constrained. It may have been in the Oropos–Plataees on an E–W trending fault 30 km north of Athens. In any case the 1999 earthquake is the largest known quake in the Athens metropolitan area (Figure 10.3).

The 1999 earthquake caused geophysicists and seismologists immediately to get to work looking for the geological structure to explain the location of this unexpected earthquake. A few years later, analyses of instrumental data and field observations highlighted a fault about 5 km long on which the earthquake in 1999 took place (Ganas *et al.*, 2001; Pavlides *et al.*, 2002). It was discovered that the initial unexpectedness of the 1999 earthquake was due to the near-complete lack of a historical memory on this area's seismicity. Such an information gap is primarily due to the history itself of the city and its changing cultural and political role from the splendours of the classical age in Greek history to the present time.

The decline of Athens as a cultural centre took place over 15 centuries, starting from the middle of the second century AD when Greece became a Roman province. In the second half of the third century there are some testimonies to a disrupted state of the city, but the causes of the dispute are unknown. Even without specific indications about the disruptions, historians have attributed this situation to barbarian invasions almost as though it were a commonplace fact. With this lack of certainty about the cause of the distress of the city, one cannot exclude the possibility that destructive earthquakes had occurred. At the end of the fourth century the Roman Empire was subdivided and Greece was integrated into the Eastern Empire, thus entering the Byzantine area of influence. For that period and for subsequent centuries historians tend to attribute the ruins of the great Athenian classical structures to invasions and devastations (first due to the Goths, then to the Slavs and the Bulgarians). It is possible that modern research focused on medieval urban archaeology might provide new information about how these structures fared during this historical period. From the thirteenth century the Venetians, Genoese and Catalans fought over control of Greece. In 1456 (that is, immediately after the fall of Constantinople in 1453) Greece was occupied by the Turks and became part of the Ottoman Empire until the start of the nineteenth century. Athens was in that period reduced in influence to something like a village. Only in 1822 did Athens reassume its autonomy, after which it was proclaimed the capital of the new Greek nation in 1834.

Florence represents a case opposite to that of Athens. The continuity of power and stability of its administrative and political centre have enabled Florence to have a virtually continuous written history from the twelfth century until today. The high economic level of this city has allowed the development of an extensive and wealthy culture, capable of producing numerous sources and different typologies. For these reasons, today considerable detail is known concerning the earthquake history that Florence has experienced for eight consecutive centuries (Figures 10.4 and 10.5). The historical testimonies allow the evaluation of earthquakes that have affected Florence. For this time period, there was no earthquake that exceeded the VII degree of intensity in Florence (Crespellani et al., 1992; Guidoboni and Ferrari, 1995; Molin and Paciello, 1999). Although the seismicity of Florence and surrounding areas is of a medium-low level, the value of the architectural, artistic and economic assets exposed is very high, and so the seismic risk in this renowned city also is very high.

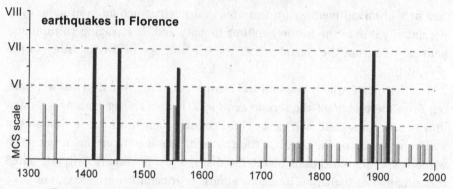

Figure 10.4 Earthquakes in Florence from AD 1300 to 2000, attested to in the direct historical sources.

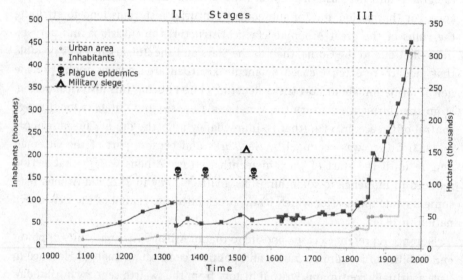

Figure 10.5 Relationship between the number of inhabitants and the extent of the urban area of Florence from the twelfth century to 2000. Until the nineteenth century Florence kept growing within its fourteenth-century circle of walls by filling spaces that had never been built on before. The stages indicated on the chart are: I, circle of walls of the twelfth century; II, circle of walls of the fourteenth century; III, additions outside the walls. From stage II to stage III the urban area was the same, but the graph takes into account the area within the walls that was actually built up. The declines in the demographic trend were due to plague epidemics in 1348, 1417, 1529 and 1630; in 1529 it was due also to a military siege (from Guidoboni and Ferrari, 1995).

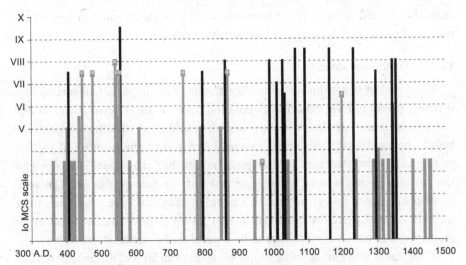

Figure 10.6 Earthquakes at Constantinople/Istanbul from AD 300 to 1500. The continuity of the locally written direct sources allows for an extensive and presumably quite complete earthquake history for this city (elaborated after Guidoboni *et al.*, 1994 and Guidoboni and Comastri, 2005).

10.4.1 Continuity of the historical sources, continuity of data: the case of Constantinople/Istanbul

The Byzantine Empire lasted for over a millennium (from the fourth century AD to 1453). Its inhabitants had the perception that the empire was so immortal and solid that it would never be overcome, even in spite of numerous and powerful attacks by the enemy populations over time. As the capital of a great empire, Constantinople was always blessed with an intellectual class capable of leaving a written legacy concerning major occurrences in the city. For this reason it is possible today to compile a fairly complete picture of the earthquakes (both strong and mild) that hit the city before the Turkish conquest (1453) when it was renamed Istanbul (Figure 10.6). Information on the past earthquakes that affected the city come only from literary and memorialistic records and not from archival records, because following the Turkish conquest all the Byzantine archives were lost (excluding those from some monasteries). In spite of this gap in the written sources, there is continuity in the historical records that pertain to the city. However, there are some aspects of the historical records that must be appreciated. On the one hand, the more recent historical information must be interpreted from Turkish sources, which continue into modern times. On the other hand, until at least the seventeenth century the urban area to which the Byzantine and the Turkish sources relate had a much smaller spatial extent than the current metropolis of Istanbul, and the style of construction of the buildings of the city has evolved significantly with time.

10.4.2 Population scales

In order to get the best possible estimate of the impact of an earthquake, it is very useful to have historical population data for the affected localities and to know both the population density and the number of buildings in existence before the earthquake. Historical sources often record the number of victims or the number of damaged or collapsed houses, but they usually do not give the percentage of the total number of residents or houses affected. The total number of persons and structures in a population centre were considered to be well known and therefore belonged to that group of facts that were taken for granted at the time but can prove to be a problem for modern researcher.

If the total number of inhabitants of a community before an earthquake is known, other information can be deduced from this number. For example, resident population estimates can lead, via appropriate indices, to the number of houses in the community. These indices are derived by historians of demography on the basis of the average size of resident families or of the principal occupations of the residents. Families may have been more or less extended, depending on the place, time period and local economy. Conversely, if one knows the total number of houses, one can arrive at the approximate size of the resident population. Fortunately, even approximate figures can help one to arrive at an understanding of the probable impact of a seismic event. In order to track down population data one needs to consult specialized bibliographies and sources other than those that recorded the information about the earthquake. Only some kinds of technical and/or scientific sources (such as damage reports or certain reports to governments) will contain data that relate the number of inhabitants to the quantity and nature of the damage by building or damage type.

Since making numerical estimates for populations and housing stocks at particular periods and in particular places can be difficult, it is necessary to rely as far as possible on information supplied by the most reputable specialized studies to determine this information. Changes in population density in long-settled cities can give one an idea of the degree of comprehensiveness of a historical earthquake catalogue. If a site has been continuously inhabited over a long time period, it will usually have given rise to written sources, in which case one can assume, for example, that any very destructive events like earthquakes or tsunamis during periods of continuous habitation would not totally escape report. An example of reported earthquake data as related to population data is shown in Figure 10.7 for the 1887 earthquake in Liguria, Italy.

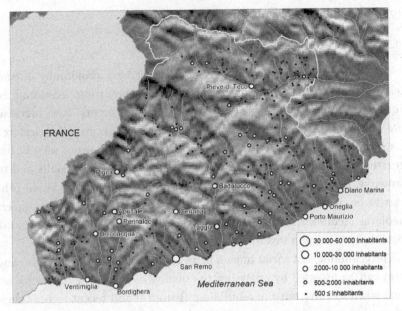

7a

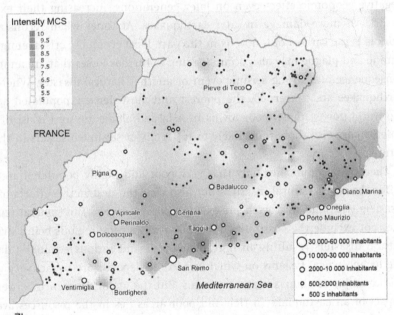

7b

Figure 10.7 (a) Distribution of the population in Liguria (northwestern Italy) according to the 1881 census. (b) The effects of the 1887 earthquake relative to the distribution of the population in Liguria according to the 1881 census. The earthquake predominantly hit the most populated inhabited centres, situated along the coast, where there was already a largely tourist-based economy. In the years following this earthquake, the human settlement network and the distribution of the population changed significantly: many inhabitants moved from the mountains towards the coast, abandoning the villages that were unlikely to have the resources for repairing the damage suffered.

10.5   Human impact

The major seismic disasters of the past have profoundly affected the living standards of the victims. In areas with a precarious economic equilibrium, seismic disasters, like other extreme natural events, have increased the rate of impoverishment in the short and medium terms. Historical research has shown that this kind of poverty results from the loss not only of building structures and means of production, but also of knowledge, skills and trading networks. Furthermore, such disasters have not infrequently caused inhabited sites to be abandoned. This was a common phenomenon in the ancient and medieval Mediterranean world, and it persists to this day, often resulting in social isolation and a loss of cultural identity.

The economic and social impact of an earthquake over the medium and long term can be assessed by analysing the quality of reconstruction work and the time taken to carry out the rebuilding. Historical and recent records show that when reconstruction is slow and funds are largely or totally lacking, there is a negative economic effect even on later generations, increasing their exposure to risks of more damage in later earthquakes. At times when reconstruction work is being carried out, now as in the past, local economic crises, emigration, famine and plague may also occur, leading to further losses that are an indirect consequence of a strong seismic event or another natural disaster. Where large earthquakes are concerned, the poorest and most densely populated areas in the world (for example, poor towns in Asia and along the west coast of South America) are also those most likely to suffer major seismic disasters in the near future.

There is no doubt not only that very poor and highly populated areas will suffer the worst natural disasters but also that the destructive effects of these disasters will tend to worsen pre-existing poverty levels. This general tendency does not exclude the possibility of economically strong regions being struck by natural disasters resulting in high death tolls and serious economic damage. The technological systems on which urban life depends are in fact very vulnerable to major earthquakes and tsunamis. But the major difference between the impact of an earthquake in rich and poor areas lies in the resources available for rebuilding operations and the time required to effect those operations.

The severe impact of a modern earthquake on a poor population: the 2005 Kashmir earthquake

On 8 October 2005 at 3:50 hours UT (8:50 Pakistan standard time) an earthquake of M 7.6 was centred in mountainous Kashmir, an area of disputed ownership between Pakistan, India and China. A number of cities and villages were severely damaged, with at least 87 000 dead in areas controlled by Pakistan, at least 1800 dead in areas controlled by India, and several deaths were reported in Afghanistan. It is estimated that over 2 million people were made homeless by the earthquake. The earthquake took place in a relatively heavily populated area of the western Himalayas, and landslides induced by the earthquake shaking along mountainsides took out or blocked many of the roads in the area of the region affected by the event. The losses to the road system effectively isolated the area most severely damaged by the earthquake and stranded many of the homeless. This significantly reduced the amount of aid that could be moved into the damage area following the event, and it also made it very difficult to move the injured to towns with hospitals and the homeless population to localities with temporary housing.

The poverty of the area affected by the earthquake added to the disaster in several ways. The typical building was constructed of cement blocks held together by mud mixed with just a little mortar. The walls were not able to resist the strong, sideways ground motions generated by the earthquake, and as a consequence most of the buildings in the area of strong ground-shaking crumbled into jumbled piles of concrete, tile and stone. Many people who were unfortunate enough to have been in these buildings when the earthquake took place were crushed as the buildings collapsed. After the earthquake there was very little heavy equipment available in the affected areas to move the debris that trapped survivors in collapsed buildings, and so most initial search and rescue efforts were undertaken using only hand tools. Following the earthquake, food was in short supply, and the loss of roads made feeding the population a long-term problem. Many residents refused to leave their properties after the earthquake, even if their homes were totally destroyed. In this region, there is no formal documentation or governmental administration of land ownership, but rather the persons who live on the land are considered its owners. If a family leaves its property and a new family moves onto that property, the new family becomes the new owners of the land. Thus, for most residents, a decision to leave their land to move into villages with temporary housing for displaced persons meant that they would permanently lose their rights to their property. For this reason, many people, even those in the more inaccessible and harsher higher elevations, spent the autumn and winter months camped out on their own land with little shelter and uncertain access to food and other necessities. Airlifts had trouble delivering sufficient aid to the population that was distributed throughout the affected area, and snowfall in the higher elevations further hindered relief efforts. Cases of pneumonia and upper respiratory infections were widespread during the cold weather, and some deaths were reported due to the cold winter weather and to disease.

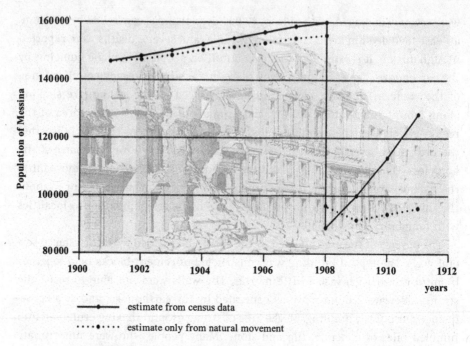

estimate from census data

estimate only from natural movement

Figure 10.8 Population of Messina (Sicily) from 1900 to 1912. The dramatic fall in the population was due to the earthquake on 28 December 1908. The fast recovery was due to immigration for the reconstruction work (from Restifo, 1995).

10.5.1   Dead, injured and homeless

A large earthquake has a direct, immediate impact on a population. It causes a rent in the social fabric of a community, the impact varying with the violence of the event. The impacts of the earthquake are quicker or slower to heal depending on the economic condition of the area where the seismic event strikes. Loss of human life is the direct immediate result of a strong earthquake. As in a military defeat, from the number of victims one can gauge the degree of impact of the earthquake or tsunami (e.g. Figure 10.8).

With most earthquakes the number of victims may depend on the time of day when the earthquake strikes. If the event takes place at night when most are at home, it could be worse than during the day when many are at work. In the common agricultural setting of the past, being at work mainly meant being in the fields, a situation of very low risk when the earth starts to shake. If an earthquake happens on a day of religious worship or on some religious holiday, it poses a special risk if it takes place when people are gathered in church to pray. When a strong earthquake catches worshippers in a stone or brick church that is vulnerable to earthquake damage, the collapse of a single building could well cause hundreds of deaths. When an earthquake causes damage that is spread over a wide spatial area, the toll of deaths and injuries tends not to be controlled by a few individual catastrophic collapses. With buildings razed to the ground throughout many towns in a region due to an earthquake, a large death toll simply confirms that many manmade structures succumbed to the earthquake shaking and crashed down onto their inhabitants.

Among medium-sized earthquakes there may be a great range in casualty figures. The death toll affords no direct yardstick of the local macroseismic intensity or the magnitude of the earthquake. The damage reported in moderate earthquakes does not correlate with the loss of life. For this reason, there is no scale of intensity that refers to the number of casualties. Of course, where a single source for a past earthquake gives a single indication that the victims were an extremely high proportion of the inhabitants of a locality, one can reasonably assume that the casualty report can be equated to a higher degree of macroseismic intensity. It may sound like splitting hairs, but only when specific details are known about the time of an earthquake, the percentage of the total population that died and the types of buildings in which they were killed, do the numbers of those who died actually acquire specific significance for historical seismology.

In evaluating historical earthquake reports one needs to ascertain, if possible, whether the victims were evenly spread over the territory affected or were concentrated in a few localities. The latter ought to make a researcher pay special attention to the report. Why were there so many deaths in such concentrated or selected localities? Were there special seismic effects or concomitant events (like a landslide or the bursting of a dam or dyke) that led to the number of deaths? Was the seismic effect somehow locally amplified? Also, there are known cases of earthquakes that caused demographic turbulence due to large numbers of persons permanently leaving the affected area following an earthquake, with dire consequences for the survivors and for the task of rebuilding. Some recent earthquakes provide such scenarios and give insight into how these disasters may have taken place in the past.

A glance at Table 10.1 may give the mistaken impression that earthquakes are becoming more violently destructive and deadly than in the past, with the possible exception of earthquakes in China. This is an artefact that is due to at least three important factors:

(1) Demographic scales have changed with time. In the last 1000 years the overall population has grown enormously so that the absolute death toll in more recent earthquakes is inevitably higher than it was in events long ago.

Table 10.1 *Earthquakes causing at least 10 000 victims from the year AD 1000 to today*

Year Month Day	Deaths (000s)	Latitude	Longitude	Country	Area	Source[a]
1008 04 27	16	34.6	47.4	Iran	Dinevar	MR01
1038 01 09	23	35.0	110.5	China	Shensi	MR01
1042 08 21	50	34.0	36.0	Syria	Palmyra, Baalbek	MR01
1042 11 04	40	38.1	46.3	Iran	Tabriz	MR01
1057 03 24	25	38.5	118.0	China	Chilhi, BoHai	MR01
1068 03 18	20	28.5	36.7	Israel	Eilat	MR01
1068 05 29	15	32.0	34.8	Israel	Ramla	MR01
1157 08 15	20	35.1	36.3	Syria	Hama	MR01
1170 06 29	20	35.3	36.3	Syria/Lebanon		GC
1202 05 20	30	33.7	35.9	Lebanon	Bekaa	MR01
1254 10 11	15	40.0	39.0	Turkey	Erzincan	MR01
1268 04 17	60	37.0	35.0	Turkey	Cilicia	MR01
1290 09 27	100	38.0	120.0	China	Bohai, Shantun	MR01
1293 05 27	22	35.9	133.5	Japan	Kamakura	MR01
1303 09 17	200	36.0	111.0	China	Linfen	MR01
1336 10 21	25	34.7	59.7	Iran	Khwaf	MR01
1348 01 25	10	46.5	13.0	Austria/Italy	Carinthia/Friuli	B
1405 11 23	30	36.5	59.0	Iran	Nishapur	MR01
1456 12 05	12	41.5	14.5	Italy	Campania	B
1458	32	39.5	39.0	Turkey	Erzincan	B
1498 09 20	41	34.1	138.2	Japan	Tokai	MR01
1499 07 17	20	25.3	100.3	China	Yunnan, Weishan	MR01
1509 09 14	13	41.0	28.8	Turkey	Tsurlu, Istanbul	Utsu
1531 01 26	30	39.0	−8.0	Portugal	Lisbon	MR01
1556 01 23	830	34.5	109.7	China	Shensi	MR01
1598	60	40.6	35.8	Turkey	Amasya, Black Sea	MR01
1622 10 25	150	36.5	106.3	China	Guyuan, Ningxia	MR01
1638 03 27	10	39.1	16.3	Italy	Calabria	B
1641 02 05	13	37.9	46.1	Iran	Tabriz	Utsu
1648 06 21	30	41.0	28.6	Turkey	W Istanbul	MR01
1653 02 23	15	38.3	27.1	Turkey	Izmir	Utsu
1654 07 21	31	34.5	106.0	China	Gansu, Tianshui	MR01
1667 11 18	12	37.2	57.5	Iran	Shirvan	Utsu
1668 01 04	80	40.5	48.5	Azerbaijan	Shemaka	Utsu
1668 07 25	50	34.5	119.0	China	SW Chingtao	MR01
1679 09 02	45	40.0	117.0	China	NE Beijing	MR01
1679 06 14	30	40.0	44.8	Armenia	Garni	GH
1688 06 05	10	41.3	14.6	Italy	Campania	B
1688 07 05	30	38.3	27.1	Turkey	Izmir	B

Table 10.1 (*cont.*)

Year Month Day	Deaths (000s)	Latitude	Longitude	Country	Area	Source[a]
1693 01 11	54	37.5	15.2	Italy	Catania	MR01, B
1695 05 18	53	36.5	111.5	China	Linfen, Shanxi	MR01, Utsu
1702 02 25	12	37.7	29.0	Turkey	Denizli	MR01
1703 01 14	10	42.5	13.0	Italy	Laga Mts	B
1716 02 03	20	36.2	2.8	Algeria	Medea	MR01
1718 06 19	75	35.0	103.0	China	C Kansu	MR01, Utsu
1721 04 26	40	37.9	46.7	Iran	SE Tabriz	MR01
1731 11 30	100	40.0	116.2	China	NW Beijing	MR01
1739 01 03	50	38.5	106.5	China	Ningxia, Yinchuan	MR01
1746 10 28	18	−12.0	−77.0	Peru	Lima, Callao	MR01
1755 11 01	30	39.0	−8.5	Portugal	Lisbon	MR01
1759 10 30	20	33.1	35.6	Syria	Baalbek	MR01
1771 04 24	12	26.0	128.0	Japan	Nansei-shoto	MR01
1773 06 03	20	14.6	−91.2	Guatemala	Santiago	Utsu
1780 01 08	50	38.2	46.0	Iran	Tabriz	MR01
1783 02 05	31	38.5	16.0	Italy	Calabria	B
1784 07 18	12	39.5	40.2	Turkey	Elmali	MR01
1789 05 28	51	38.8	39.5	Turkey	Elazig	MR01
1792 05 21	15	32.8	130.3	Japan	Nagasaki	Utsu
1797 02 04	40	1.0	−79.0	Ecuador	Quito	MR01
1794 12 04	16	10.5	−64.5	Venezuela	Cumana	Utsu
1812 03 26	20	10.0	−67.0	Venezuela	Caracas	MR01
1815 10 23	37	34.8	111.2	China	Pinglu, Shanxi	MR01
1815 11 27	10	8.0	115.2	Indonesia	Bali	Utsu
1822 08 13	20	36.7	36.9	Syria	Aleppo	Utsu
1824 06 25	20	29.8	52.4	Iran	Shiraz	MR01
1850 09 12	300	27.8	102.3	China	Sichuan	MR01, Utsu
1854 12 23	31	34.1	137.8	Japan	Nankaido	MR01
1857 12 16	11	40.4	15.9	Italy	Basilicata	B
1861 03 21	18	33.0	−69.0	Argentina	Mendoza	MR01
1868 08 13	25	−18.5	−71.0	Chile	Arica	Utsu
1868 08 16	40	0.5	−78.0	Ecuador	Ibarra	MR01
1875 05 18	16	8.5	71.0	Venezuela	Merida	MR01
1879 07 01	30	33.2	104.7	China	Gansu, Wudu	MR01
1883 10 15	15	38.3	26.3	Turkey	Izmir, Cesme	MR01
1883 10 15	15	38.4	26.1	Turkey	E Erzincan	MR01
1893 11 17	15	37.0	58.4	Iran	Quchan	MR01
1896 06 15	27	39.5	144.0	Japan	Sanriku	MR01, Utsu

(*cont.*)

Table 10.1 (*cont.*)

Year Month Day	Deaths (000s)	Latitude	Longitude	Country	Area	Source[a]
1905 04 04	19	33.0	76.0	India	Kangra	MR01
1906 08 17	20	−33.0	−72.0	Chile	Valparaíso	MR01
1907 10 21	12	38.0	69.0	Tadjikistan	Dushanbe	MR01
1908 12 28	80	38.0	15.5	Italy	Messina	B
1915 01 13	33	42.0	13.7	Italy	Avezzano	B
1917 01 21	15	8.5	115.0	Indonesia	Bali	MR01
1920 12 16	235	36.6	105.3	China	Haiyuan	MR01
1927 05 23	40	37.4	102.3	China	Gansu, Gulang	MR01
1923 09 01	143	35.4	139.1	Japan	Tokyo	MR01
1931 08 11	10	47.1	89.8	China	Xinjiang, Fuyun	Utsu
1932 12 25	77	39.8	96.7	China	Kansu	MR01
1934 01 15	11	26.5	86.5	India	Bihar	Utsu
1935 05 30	60	28.9	66.2	Pakistan	Quetta	MR01
1939 01 25	28	−36.2	−72.2	Chile	Chillan	MR01
1939 12 26	33	39.8	39.5	Turkey	Erzincan	MR01
1948 10 05	20	37.5	58.0	Turkmenistan	Ashkhabad	MR01
1949 07 10	12	39.2	70.8	Tadjikistan	Khait	Utsu
1960 02 29	13	30.5	−9.6	Morocco	Agadir	MR01
1962 09 01	12	35.6	49.8	Iran	Buyin Zara	MR01
1968 08 31	12	34.0	59.0	Iran	Dasht-e-Bayaz	MR01
1970 01 04	16	24.1	102.5	China	Yunnan, Tonghai	MR01
1970 05 31	67	−9.2	−78.8	Peru	Yungay	MR01
1972 12 23	11	12.0	−86.0	Nicaragua	Managua	MR01
1976 02 04	23	15.3	−89.1	Guatemala	Guatemala City	MR01
1976 07 28	290	39.6	117.9	China	Tangshan	MR01
1978 09 16	20	33.3	57.4	Iran	Tabas	MR01
1988 12 07	25	40.9	44.1	Armenia	Spitak	MR01
1990 06 20	40	37.0	49.2	Iran	Rudbar	MR01
1993 09 29	10	18.1	76.5	India	Latur	Utsu
1999 08 17	17	40.8	30.0	Turkey	Izmit	MR01
2001 01 26	100	23.4	70.2	India	Bhuj	MR01
2003 12 26	40	29.0	58.3	Iran	Bam	JAJ
2004 12 26	283	3.3	95.9	Sumatra	Bandar Aceh	NEIC
2005 10 07	86	34.5	73.6	Pakistan	Kashmir	NEIC
2008 05 12	69	30.9	103.3	China	Sichuan	NEIC

[a] MR01, MunichRe; Utsu, IASPEI, 2002; B, Boschi *et al.*, 2000; GC, Guidoboni and Comastri, 2005; GH, Guidoboni *et al.*, 2003b; JAJ, Jackson, 2006; NEIC, USGS National Earthquake Information Center.

Source: The table may contain gaps of knowledge, the data in all likelihood erring on the low side. These data were compiled and kindly made available by Professor James A. Jackson (Cambridge University).

(2) Movement from distributed population settlement patterns to more concentrated urban centres is a major factor in demographic evolution in many parts of the world. Cities are becoming metropolises, markedly increasing the density of inhabitation and hence increasing the seismic risk in those urban areas.

(3) Economic differences between countries or within individual countries are large and are increasing to a great extent. Falling standards of living in poorer countries mean a lowering of past levels of security against natural hazards such as earthquakes.

Together, these items mean that the impact of major earthquakes on human population on a global basis will tend to get worse in the future, especially in those areas that are the poorest and most overpopulated. The absolute death tolls in future earthquakes can only be expected to rise compared to what has been experienced in the past.

10.5.2 Rebuilding

In areas frequently damaged by earthquakes, rebuilding operations may pose a special challenge for the historian. The need for rebuilding is a recurring socio-economic burden affecting not only the standard of living among the populations involved but also the quality of the buildings that the population used. Of course, the quality of building refurbishment following an earthquake will bear significantly on the severity of the effects of subsequent earthquakes. Careful analysis of the evidence of rebuilding throws important light on possible past earthquake scenarios.

Before the days of national communities presupposing a bond of solidarity, with rare exceptions the planned intervention of a central power after a destructive earthquake was a low-key affair. Efforts were usually confined to restoring certain categories of buildings such as military, religious or public structures. There was little or no public money available for private citizens to rebuild or otherwise cope with the effects of a seismic event. Assessment of damage often led to situations where the central government passed down responsibility for repair and reconstruction to local administrations, and therefore assessment was influenced by the competing desires of differing interests. The central administration sought to give as little aid as possible, while the local community's goal was to get as much aid as possible from outside the community. Whatever the ultimate financial decree by the central administration, the estimate of damage formed a kind of bargaining process between wealthy private owners and public administration as well as between local and central administrations. In administrative history the reality of this bargaining process explains the various guarantees demanded by all governments on damage estimates and the gaps between how the local community evaluated its damage and what the central government's own or hired assessors concluded. Divergences of opinions often gave rise to interminable paperwork, which can be rediscovered today through archive research. The disputes that this paperwork can reveal might have been resolved by the intervention of local men of influence or of trusted officials from the central government, but such efforts at compromise might have come to nothing when war, revolt or other setbacks stood in the way of attempts at negotiation. In general, those who were looking for government aid following an earthquake were seen as seeking something that did not rightfully belong to them. No legal text (then or now, as it happens) has upheld the right to be indemnified for the need to rebuild one's own property following an unexpected natural disaster like an earthquake.

The burden of reconstruction in the past normally was borne by the individual inhabitants. This led to the result that every owner, wealthy or not, rebuilt according to the means that he possessed. This still happens in many parts of the world today. Except among the elite ranks of the prosperous, rebuilding was usually carried out on a shoestring budget with whatever resources the average individual could scrape together.

In Europe, with very few exceptions until the end of the nineteenth century, public intervention consisted in granting partial tax relief but not in making any form of budgetary allowance. However, in times of economic prosperity or during times of special administrative and social reform, some major earthquakes were followed by properly planned reconstruction. Examples of planned reconstruction include the rise of the baroque towns of eastern Sicily after the large earthquake of January 1693, the rebuilding of Lisbon after the earthquake of 1755 or the reconstruction of San Francisco after the 1906 earthquake and fire. Government reconstruction schemes formerly did not entail the financing of small property owners but simply served political ends. They were meant to ensure law and order, to promote town planning and to impose some order of priority in the access to labour, the containment of labour and material costs (invariably an ineffective enterprise), and the purchase of new sites if the reconstruction plan involved relocating the worst-hit settlements.

Reconstruction following a proper plan is a rare event indeed. This is especially true if one thinks of the countless number of rebuilding projects that mainstream history has forgotten but seismological history has a duty to record. In most cases these projects were carried out with the scanty resources of survivors and hardly any public assistance. From the standpoint of historical seismology, the public administrations that did become involved in reconstruction produced a great wealth of files and documents, affording many details of the effects of seismic events, which can now be used to gain knowledge of those earthquakes.

Different administrative praxis lead to differing patterns of reconstruction in one and the same earthquake (1661): two cases in Italy compared

On 22 March 1661 a violent earthquake ($I_0 = IX$; $M = 6.1$) struck a broad swath of the Tuscany and Romagna Apennines in northern Italy, causing damage in over 40 towns and villages. Part of the area affected belonged to the Papal States, while the rest was part of the Grand Duchy of Tuscany. The homeless in the former area received no aid except for minor tax relief. No surveys of the damage were performed, with only some generic estimates made of the monetary losses (hence of little use as sources for a historical seismologist). The lack of aid led to voluntary emigration by the artisan class, which caused a period of economic depression. The area under the Grand Duchy of Tuscany was administered quite differently. Careful surveys of the earthquake effects were conducted (first-rate material for historical seismology), and then the central administration made financial loans at 4% interest (deemed a low rate of interest at the time), repayable over a lengthy period. The granting of easy mortgage terms, pursued almost exclusively by the Grand Duchy, was the most enlightened policy in this region and a novelty on the European scene. The policy enabled local communities to rebuild their houses with a healthy circulation of liquid cash. The long-term scheme of repayment to the public administration bound the inhabitants to their land (some debts were not paid off until the mid eighteenth century), which prevented neglect and dereliction of buildings – one widespread social bane following many earthquakes. What is more, the exchequer of the Grand Duchy itself accrued some benefit from this rebuilding scheme. To the historian of seismology the Tuscan documentation yields the information that out of 4560 houses, the building patrimony of 15 villages, 27% (1234 houses) were seriously damaged, and in some settlements (i.e. Rocca San Casciano and Galeata) as many as 76% of the dwellings were damaged.

The weight of a badly executed prior reconstruction aggravated the damage of the 1786 earthquake in Rimini

On Christmas Day, 25 December 1786, the Adriatic coast of the region of Romagna was hit by a medium intensity earthquake ($I_0 = VIII$; $M = 5.6$), which caused a large amount of serious damage to the city of Rimini (today a very popular seaside resort). The effects on the buildings were the subject of two expert examinations, performed independently by two skilled architects: Camillo Morigia and Giuseppe Valadier. At the time the latter was starting his career at the court of Rome. The analyses of these two experts (weighty handwritten volumes thousands of pages long) represent quite an exceptional case from the archival standpoint. They were produced as a result of an administrative rivalry between the central management of power (the Papal Court of Rome) and local government management, which was entrusted to the powerful Cardinal Colonna. A detailed analysis of these two examinations not only allows one to evaluate the local response to the earthquake but also to discern an image of the damage on the basis of two expert judgements. Although the picture of the damage was quite serious, a general reconstruction project was not drafted by the government for Rimini, and the initiative for reconstruction was left entirely to private parties. Perhaps also for this reason no important urban developments following the earthquake were undertaken, but rather the owners of the damaged houses saw only to repairing their own homes, which to the experts seemed to be more serious than at first expected.

Among the causes that contributed to worsening the damage, Camillo Morigia identified a general lack of overall care and attention in the quality of the minor civil buildings, but above all he pointed to the 'ill-repaired damage' of a previous earthquake, one on 14 April 1672 (I_0 = VIII; M = 5.6). Based on his examination, he observed that 'little attention and skill' had been put into repairing the cracks from the earlier event and that damaged walls had not been tied securely to the rest of the structure. Thus, the load-bearing walls had not been fully restored, and the cracks in those walls, even very large ones, had only been filled with lime or gypsum and concealed beneath new plaster. Furthermore, the architect Morigia had found many walls supported by temporary metal chains or by wooden props, which over a century before had been placed temporarily to avoid the collapse of leaning walls. These had been left in place as the consolidation work had never been carried out. For the repairs to the damage due to the 1672 earthquake, cheap materials had been used. This material had been recovered from demolitions of other structures or were river cobbles joined together with very poor mortar and held together with small wooden keys. In the large civil buildings (public and private buildings) Morigia observed that the quality of the load-bearing walls was excellent; however, the dividing walls were too weak and had been built with bricks laid upon the narrowest possible base. Even the ceilings, made from organic material and gypsum, were very fragile, as were the roofs. The wooden rafters supporting the ceilings were short, barely spanning the load-bearing walls. The rafters could easily fall out of their small sockets in the walls in even a small horizontal shaking, and this could account for the inward collapses of the roofs of many of the houses.

Giuseppe Valadier, after visiting the houses in Rimini damaged by the 1786 earthquake, also stressed the inadequate care that was taken in repairing the damage from the earthquake in 1672. He foresaw, with great disappointment, that even after this new earthquake in 1786 not everyone would carry out the work necessary to ensure the stability of their houses. He wrote:

> In the examinations I wrote down house by house I noted only the most visible and serious damage and also calculated their costs [. . .]. But it will happen that some will claim to have spent more, and others less. This will happen because some, with the pretext of the earthquake, will want to have a new house made for them and will spend more, while others will repair them poorly, and will spend less. This had already happened with the earthquake on April 14, 1672. Those constructions that were then badly repaired suffered greatly as a result of this latest earthquake. (Valadier, 1786.)

Reconstruction from thousands of documents: the great archive of the Giunta di Cassa Sacra *(State Archives of Catanzaro) throws light on the earthquakes of 1783*

In February–March 1783 a long and devastating seismic sequence struck Calabria (southern Italy), with five violent earthquakes and hundreds of lesser tremors. The cumulative toll was dire. Destruction was accompanied by widespread upheaval of the ground and havoc to the water system. In area it extended across the whole of southern Calabria. Reports indicate that 182 villages suffered well-nigh total destruction. In the worst-hit parts of Calabria, there were more than 35 000 dead out of a population of a little more than 400 000 (8% of the whole resident population). The enormous extent of the damage involved centres of importance to political, economic and military life in the Kingdom of Naples, such as Reggio Calabria and Catanzaro. The sheer range of the area destroyed meant that the institutions and organs of the central state had to step in for relief and reconstruction efforts.

The earthquake was an opportunity to try to redistribute resources, land ownership in particular. In June 1784 the Bourbon government set in motion a complex procedure to expropriate uncultivated lands owned by the Catholic Church or by local barons. By administering and selling off such land assets, the government intended to raise funds with which to rebuild whole settlements from scratch. At the same time, through the land redistribution the government aimed to launch a new phase of development in areas that had languished for centuries on the fringe of the kingdom's social and economic life. To carry out this ambitious process of property reconstruction, the Bourbons set up a special office called the *Giunta di Cassa Sacra* (State Archives of Catanzaro), which is nowadays represented by a major collection of documents. In Naples, where the King was residing, the office of *Corrispondenza con la Cassa Sacra* was set up, to manage the direct control of the procedures. The plan to repair the damage to the social fabric by the earthquake failed. The factors responsible for the backward-ness and poverty of the social and economic life in Calabria had been diagnosed by intellectuals and reformers of the time. These factors were perceived to be the lingering feudal structure (83% of Calabria was under baronial power), the lack of economic initiative by the major land-owners, the decline of the arable, woodland and pasture resources across the territory, the onerous demands of the central tax system, and, by no means least, the weakness of the State in the face of the barons (Placanica, 1970, 1979, 1985a). Every attempt at land expro-priation by the Bourbon government was met with appeal after appeal, ending in a bureaucratic spiral that frustrated efforts at reform or drastically curtailed their effect. Thousands of archive entries testify to the difficulty of rebuilding, yet they also preserve a clear and detailed picture of the damage incurred by the various local communities – deeds and documents of exceptional descriptive value for historical seismology.

In the years that followed the earthquake, much of the agricultural popu-lation drifted towards the bigger inhabited centres, an exodus that worsened the already grave shortage of manpower farming in those outlying parts of the region. Not even in the big towns did the social and economic life pick up to the level prior to 1783. Delays in the rebuilding programme, and the setbacks that went with the delays, added to the difficulty of getting the population to move back to the outlying areas, depressing the economy still further. In some major centres the long sequence of main shock and aftershocks and the havoc that ensued brought a deep crisis of productivity. Palmi suffered a major reces-sion in its wool and silk industries, and after the earthquake there was a fever epidemic that exacerbated the economic downturn. At Pizzo few died under the earthquake rubble but the epidemic that followed caused many deaths among the population camped in shacks along the beaches. Some centres like Seminara, Oppido Mamertina and Briatico were decimated by another wave of fever in the summer of 1783 which, of itself, caused nearly 19 000 deaths (from Boschi et al., 2000; Guidoboni et al., 2007).

10.5.3 Abandonments

In the southern Mediterranean area (places like Sicily, Calabria and some parts of Algeria) there are settlements that have been abandoned after strong earthquakes, some cases in times long past and some more recently. Abandon-ment has been a frequent reaction in the Mediterranean basin after a seismic catastrophe ever since ancient times (e.g. Figure 10.9). In ancient times aban-donment was motivated by a culture that judged a place that had suffered an earthquake, or a place where phenomena known as prodigies had taken place, to be considered uninhabitable. The people would leave the location devastated by the earthquake and would rebuild their villages elsewhere, perhaps with a

Figure 10.9 Ruins of villages abandoned due to earthquakes: the remains of Occhiolà (eastern Sicily), destroyed in January 1693 (from Dufour and Raymond, 1994).

new name. Sometimes rather than rebuilding the homeless inhabitants might have scattered to other locations. The result was that settlements that were damaged by earthquakes were immediately abandoned, and those localities remained abandoned. The abandonment did not only come about as a direct result of the damage suffered, but also arose due to circumstances such as the area's poor economic prospects following the earthquake or the concomitant occurrence of other unfavourable events like wars or changes in key trade routes. Abandoned sites would be plundered of all materials that could be used for the new buildings. An earthquake could be one of the factors contributing to the abandonment of a settlement, although the most important reason may have been the search for new communication routes to improve trade or to find new market opportunities. Moreover, throughout the nineteenth century, the sheer bulk of rubble resulting from a strong earthquake may have discouraged the removal of the debris and therefore reconstruction in the same place. It was typical that the debris was burnt to eliminate corpses trapped in the ruins so as to avoid the risk of epidemics.

Until the end of the nineteenth century, the area under consideration where a new settlement would be rebuilt following devastation by an earthquake depended on the social and political rank of the persons who were overseeing the rebuilding. If the inhabitants were the ones to make this choice – in cases in which the central administration took on no direct role in the matter – the new site would generally be situated no more than a few hundred metres, at most perhaps a kilometre, from the old site. In this way the materials from

Figure 10.10 Ruins of Gibellina (Sicily) due to the earthquake on 15 January 1968 in the Belice Valley. The remains of the abandoned village were covered in white cement, made by the sculptor Alberto Burri in 1985–89. These are the first seismic ruins that have become a monument in the Mediterranean area.

the old town could be easily recovered and reused (this was the case of Basilicata, southern Italy, after the earthquake disaster in the Agri Valley in 1857; see Mallet, 1862). Instead, if the parties responsible for choosing a new site had more ambitious economic or developmental plans in mind, the distance from the old site to the new settlement site could be as much as 10 km, as was the case in Noto (eastern Sicily) after the 1693 earthquakes (Tobriner, 1982). The relocations of towns have led to permanent changes in place names and have substantially changed the cultural landscape, the road network and the local trade routes. The abandoned sites due to earthquake devastation represent a sub-group within that of the abandoned sites for a number of other reasons, such as changes in trade routes, epidemics and above all wars (see the case of the Calabria, southern Italy, in Teti, 2004). From the standpoint of historical seismology, the relocation of villages must be taken into account. Indeed, when seismic scenarios are reconstructed it is important to locate the earthquake effects at the sites that actually experienced them and not at the new locations of the reconstructed sites, especially if the new site is as far as 10 km away from the location where the earthquake damage took place (for example, Figures 10.10 and 10.11).

10.6   The effects of earthquakes on construction practices

The effects of earthquakes on buildings depend on a variety of factors that are often difficult to differentiate from the written reports of earthquake damage itself. The type and amount of damage to a structure is determined

Figure 10.11 Bird's-eye view of Gibilmanna (Sicily) after the earthquake on 15 January 1968.

not only by the energy of the earthquake shaking but also by the particular geological condition of the site and by the structural and non-structural characteristics of the building. The susceptibility of a building to sustaining damage in earthquake shaking depends on the morphology and foundation type of the structure, how it was originally built and subsequently modified, and on its state of preservation and maintenance. In other words, a series of correlated and interdependent factors are involved in controlling how structures are damaged when the ground shakes strongly, and this can lead to differences in seismic effects across small areas (spanning thousands or even hundreds of metres). This is true even for buildings that are ostensibly made of the same materials and erected using the same techniques. If a seismic intensity scale such as the MSK or EMS scale is being applied to categorize the earthquake effects, a classification of building construction types, taking into consideration the vulnerability of each building to earthquake shaking, is required in order to assign the proper macroseismic intensity to the damage at a site. It is therefore necessary to establish a general periodization of the materials and building techniques that existed at the time of an earthquake in the area under examination.

10.6.1 Building materials and techniques: identifying the predominant characteristics by area and period

According to earthquake engineers, the extent of damage caused to a structure by earthquake shaking may be controlled by one or more of the following factors: poor quality of the materials employed, badly executed

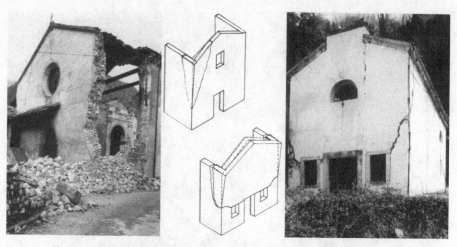

Figure 10.12 Earthquake effects on particular buildings. Shown here are some small churches of Friuli (northeastern Italy) damaged by the earthquake in 1976. Left: the church of San Giovanni at Venzone; right: the church of Osoppo. For both churches, a detailed engineering analysis has shown the various mechanisms by which these churches were damaged (from Doglioni *et al.*, 1994).

building elements and details, asymmetrical buildings, the grouping together of the buildings, and similar construction details and defects. These factors can be found in many building types in poorer parts of the world, and they are surprisingly frequent even in well-developed countries. Of these factors, the kind and quality of building materials and the state of preservation of a building are particularly important elements that can predispose a seismic disaster when strong earthquake shaking takes place. Buildings constructed of stone, brick or clay can be particularly vulnerable to earthquake shaking, and the layouts and sizes of the structures made of these materials can be quite different depending on whether they were built in urban centres or in the countryside, and in mountainous localities or in coastal locations (Figure 10.12). Furthermore, the type and quality of building materials can change radically from one locality to the next, especially if there is a great change in topography (say from a river valley to a high mountain location).

Very often, the user of seismic catalogues is not given a full view of the distribution of the types and compositions of the buildings that were affected by a historical earthquake. In an attempt to highlight some aspects of this subject, we summarize here the most commonly used materials in historical buildings. While these materials are still used in contemporary buildings, their use was much more widespread in the past than it is today. The quality and state of conservation of buildings is generally linked to economic levels. This relationship is not the determining factor in how any single building may have fared in a past earthquake, but may be taken as a general indicator of how a building was probably constructed. More attention to building materials and techniques and a greater awareness of their contribution to the amount and severity of seismic damage can help provide a more complete picture of how strongly a settled area was shaken by a historical seismic event. Seismologists and earthquake engineers might try to make direct use of this information to construct a detailed map of the variations in the strength of the ground-shaking across an area that comprised a city or of a number of distributed villages, or they might want to add flexibility to macroseismic intensity scales to adapt them to the different types and conditions of buildings from different epochs and historical contexts.

As is known, *brick* is a very ancient building material for common structures, and houses made of brick are widespread in many parts of the world. Because of the pervasive use of clay brick even into modern times and its significant vulnerability to earthquake shaking, brick structures represent perhaps the most studied and best-known type of damaged buildings that have been studied by earthquake engineers after strong seismic shocks. Scattered throughout the flat areas or concentrated in the populous villages of the plains, the brick house is a characteristic historical feature of the flat-land and mid-hillside landscape in many parts of the world, from the small village to the city. It tends to be found in less hilly areas, where the earth can be worked more easily and where at least some flat land can be found as a suitable base upon which the structure can be built.

Buildings of *stone* are often considered noble or special structures, and in ancient buildings stone can be synonymous with a high-quality building material. Stone buildings are often found from the tops of mountains to mid-hillside sites, places where it is easiest to use the local stone for building materials. Because stone is heavy and difficult to work, its use generally results in the creation of simple structures, structural shapes and grouping systems. For openings like doors and windows in stone buildings, the relatively low ability of stone to resist flexure means that the weight of the wall directly above an opening must be prevented from resting directly on the stone lintel, especially if the lintel has a considerable span. Stratagems such as relieving arches or simpler devices such as triangular elements are therefore used to deviate the heavy wall loads away from a door or window opening. The collection and shaping of the stone used in stone buildings can demand that specialized techniques be applied to the building materials. Where there were no local or regional major quarries that could provide building stone, there were usually small quarrying excavations near the towns where the stone structures were being constructed, and the products of the quarry were almost exclusively used for local requirements.

In many remote areas, it was the peasants themselves who collected or quarried the stones for the structures, and they would lay them without any particular assistance. In most structures in these areas, the stone was either squared using simple tools or was left completely coarse. The stones were selected for placement in the walls according to their shapes. The most even and square were used in crucial areas such as corners, jambs, lintels and hearths, while less regular pieces were used for the walls. Different ways of working the stones led to different types of surface and construction systems. The most widely used method to construct stone buildings was the cavity wall, with the even-structured solid wall was less frequently employed. The outer walls of stone structures may not be plastered, while the inner walls were nearly always plastered with lime and sand. Thus, the surfaces of a masonry structure varied according to the shapes of the stones used and the finish applied to the building elements. It is worth mentioning that in the same town different stone buildings can be found with different types of external surfaces:

- unworked stones and pebbles of different sizes held together by a large quantity of mortar (particularly in isolated houses on rises)
- very roughly hewn stones of different sizes
- unevenly squared and different-sized stones
- stones squared into the same shape and size (usually for the most prestigious buildings).

In some isolated regions where the population had only very limited financial resources, *earth* materials such as *clay* may be used as a low-cost building material to construct housing. This situation was found in regions where materials such as stone and wood were scarce and where bricks were not economical because they required processing and firing in kilns. A mixture of earth, shredded straw, water and finely crushed stone was fairly widely used by peasants to build their low-cost houses in open countryside. The mixture of earth was shaped in moulds or simply hewn into clods and laid by the peasant according to procedures handed down through generations. The mixture used to build the walls directly over the foundations was prepared in a specially dug pit or within the perimeter of the building. There are many drawbacks to housing made of clay or earth. It has a damp interior, requires annual maintenance, and has a relatively brief life span (usually no more than 50 or 60 years). Wherever practical, homes of earth construction were eventually abandoned in favour buildings made of more long-lasting and wholesome materials. However, there is no doubt that at times in the past buildings made of earth were common in very many areas, both throughout and outside the Mediterranean region.

Adobe houses are made of sun-dried earth, and they can be found in different varieties throughout almost the entire world. Adobe brick houses were found where the use of timber was very limited, either because of building traditions or because locally there were no large woods to supply timber as a building material. Adobe construction has played a major role in nearly all of the great seismic disasters of past history, and even today major losses of life in earthquakes often involved the failures of adobe buildings. In the past adobe was used much more than it is today, and therefore it is a type of construction material that must not be forgotten when researching the reasons for a major loss of life in a historic earthquake. Adobe is one of the weakest building materials when shaken by an earthquakes, and so it sustains damage at a lower level of earthquake shaking, and therefore macroseismic intensity, than almost any other building material. In many regions it was the practice to add straw or similar kinds of natural materials to the adobe bricks to strengthen them. The straw has the added advantage that it offers protection, though only partially, from damage due to water leaking through the adobe bricks, as the straw fibres encourage the evaporation of excessive water, thereby limiting the erosional effects of the water on the clay. An inconvenience due to the use of vegetable materials to strengthen adobe that cannot be overlooked is the risk of rot and parasites.

In adobe structures, horizontal elements such as ceilings and the framing that supports the roof were generally constructed from natural materials such as wood, reeds and straw. These materials may be used to support adobe elements or they may be used alone for these important structural members. Ceilings were generally formed from a structure made of large trusses (usually roughly hewn trunks of oak or chestnut) which were inserted into the adobe masonry.

The case of Italy: the ancient roots of different building cultures existing in the north and in the south

Those studying historical earthquakes in Italy cannot help coming across major construction differences between the northern and southern parts of the country. While the south was a region where construction materials consisted of stone (squared or not, depending on the areas and the economic conditions), of tufa and of brick, the north for centuries was characterized by wooden buildings or mixed materials, stone and wood or cobble and wood, and only in the fifteenth to sixteenth centuries did the use of bricks become widespread.

The roots of the use of these different construction materials can be traced back to the sixth century AD and the invasion of the Longobards. This was a time when there were two areas with different cultures that were settled in the country, the *Longobardia* and the *Romania*. There are some territories that were first subject to the Longobards and then later to the Franks, while there were other territories that were not influenced by these first groups but rather fell under a continuous Byzantine dominion (the *Romanoi*). In Longobardy there prevailed a population pattern of scattered rural settlements organized in large farms. The structures on these farms were mostly built of wood, sometimes mixed with other cheap building materials. The rationale for this construction tradition reflected the need to maximize the use of the simplest building materials available in the territory. Naturally, thanks to the presence of large wooded areas, great use was made of wood, while in the inner Apennine areas the use of rough-hewn local stone lasted for a long time. The widespread use of wood, which is believed to be more accentuated in the Longobard period as compared with the previous one of the Roman Empire, offered the advantage of requiring labour that was skilled only in carpentry, as were most Longobardian peasants. It was quite feasible to have wood as the exclusive material of a building, which could consist of vertically or horizontally intersecting planks that lay directly upon the ground, on a brick wall or on a foundation of boards. Sometimes wood served as the framework of walls that were infilled with dry clay, with clay and straw, or with cobblestones. From the written sources it can be observed that wood also was utilized in the structures for roofing (*scandolae*). The practical nature of these buildings and their stable use over time can also be attributed to the fact that such houses could be dismantled and their components reused in other sites. This type of movement of an entire structure often became necessary when agricultural contracts expired (Galetti, 1985). In the cities of the Longobard area simple houses made of wood and other cheap materials coexisted alongside public and civil buildings and places of worship for which stone was the preferred building material. These larger structures often reused components of ancient stone structures, effectively using the earlier stone remains as though they were on-site quarries (Cagiano de Azevedo, 1974; Ward-Perkins, 1984).

In central and southern Italy with its Byzantine influence, even if those areas also were intermixed here and there with some distinctive Frankish enclaves, different building characteristics are observed compared to those in northern Italy. The countryside was characterized by centralized settlements of a more urban nature. The walls of the houses were built with clay, with cobblestones and clay, or with fragments of brick and mortar. In the cities there was a great availability of bricks for reuse when older structures were abandoned or razed. These materials comprised the lower parts of the buildings. The upper parts of most structures generally were made of *tufa*, a lighter material that is widely available in this region. In those areas where Byzantine territories (which later became

Figure 10.13 Drawing of an arch with slippage of the central stone in the church of Santa Maria del Soccorso at Leghorn (Tuscany) caused by the earthquake on 14 August 1846. This sketch was drawn by Pilla (1846).

autonomous) and Longobard territories existed contemporaneously, there was an interesting coexistence of building materials and different building customs. This was due in part to the different economic conditions and in part to the different ways in which labour was organized in the two distinct territories. From the basic characteristics that were unique to the Longobards and to the Byzantines, there later developed some particular building techniques that arose from the earlier construction traditions, and these techniques continued to characterize northern and southern Italy almost until the nineteenth century.

Observations about seismic effects on a brick building: Leghorn, 1846

The earthquake of 14 August 1846 caused damage that became a sort of observational workshop for scholars, who were interested in and had been identifying the traces of seismic shaking on monuments. Leopoldo Pilla (1805–1848), a Neapolitan geologist at the University of Pisa, made an important field observation concerning the effects of the 1846 earthquake after questioning many eyewitnesses and recording their accounts in great detail. He noticed that the horizontal movement (in fact, the movement of the ground in any earthquake is complex and has a number of different components) imposed by the impact of a strong shock on an arched structure could, in a matter of seconds, split the arches open and close them up again, leaving a lowering of the central part of the arch as tangible evidence of this momentary occurrence (Figure 10.13). Pilla made sketches to show how the central part of the arches had been displaced in this way. He described this effect at the church of Santa Maria del Soccorso at Leghorn, for which he took down the evidence of the slippage of the keystone of an arch according to the report provided by a squarer who was inside the building when the strongest shock was felt,

I was just sitting quietly as I usually do at that rest time when I suddenly heard a noise like a gust of wind. I looked up at the roof and at that very moment it shook and I thought I saw a piece of stone fall down. I stood up in fright and ran towards the nave, and at that moment all the vaulting above the opposite aisle rose up, the arches split open and the sun suddenly shone in below the aisle. (Pilla, 1846, p. 39.)

The lowering of the arch keystones in the church of Santa Maria del Soccorso can no longer be detected. The church interior has been plastered and the keystones stuccoed. Observations similar to those of Pilla had already been made in structures damaged by the 1755 earthquake at Lisbon in Portugal. A witness in 1755 declared to the Royal Society of London that he had seen houses split open from top to bottom and close up again without leaving any trace of the split.

10.6.2 Sources that mention collapses or rebuilding without explicit reference to an earthquake as the cause: how to use them?

Historical sources occasionally mention the collapse of or damage to important buildings, such as places of worship, military or defence works (including town walls), or socially prominent edifices like patrician houses, towers, belfries or minarets, and yet, for cultural and/or religious reasons which sometimes escape us today, the cause of the damage may not be explicitly mentioned in the source. This raises a problem for historical seismology, which may be losing potential information on the specific damaging effects at a locality due to an earthquake. Why does this occur and how should historical seismology research deal with such sources?

In many sources, while the cause of a collapse is completely ignored, there is a wealth of detail on dates and places associated with the collapse. Likewise, restoration work may be authoritatively documented, yet no mention is made of the reason for the damage that is being repaired. Such a grey area about the causes of damage that is being repaired is puzzling. Was there an earthquake that affected the structures? Was it war damage? Did the buildings collapse through old age or for reasons of instability? In the absence of further evidence, no acceptable theory can be formed for these remote and unverifiable contexts. It is advisable, therefore, to play it safe and not to use such data as evidence of historical earthquakes unless careful research manages to throw light on the cause of the destruction. One should not forget that buildings may collapse through simple faults of construction or poor upkeep.

Inscriptions recalling collapses or reconstructions with no explicit
reference to earthquakes

Seven inscriptions found at Thubursicum Numidarum (present-day Khémissa, or Henchir Khamissa, about 30 km southwest of Souk-Ahras, Algeria), the first one datable to between AD 355 and 361, and the others to the reign of Julian (November 361 – 26 June 363) recall the work carried out for the reconstruction of the city's *forum novum* (*Inscriptions Latines de l'Algérie*, 1.1275, 1229, 1247, 1274, 1276, 1285, 1286). Three of the inscriptions (1.1229, 1247, 1274) mention ruins or collapsed statues, thereby referring to a sudden and destructive event, although they do not contain any explicit mention of the cause of such collapses. Could it be an earthquake? Very likely, but when did it happen? According to Lepelley (1984, p. 484, n. 73) a shock must have hit this city before AD 355, taking as the term *ante quem* the dating of the inscription 1.1275, in which the future Emperor Julian is remembered, appointed *Caesar* in Gallia in that very same year.

When the inscriptions neither specify the cause nor the year when the destruction actually occurred endless debates begin as to the use and interpretation of such data. The most well-known debate in this field, and the one that has perhaps most of all affected the results, is the one relating to the effects of the tsunami and the earthquake on AD 21 July 365, which paved the way to a *querelle* that has been going on ever since the 1970s and which is not over yet. Ten inscriptions have been correlated to this great event, which originated in Crete, by Di Vita (1964, 1982), Beschaouch (1975) and Rebuffat (1980); 16 inscriptions have been collected by Rebuffat (1980) and attest to the intense building activity carried out by Publilius Caeionius Caecina Albinus during his governorship in Numidia (a region roughly corresponding to the northeastern part of present-day Algeria), that is between the years 364 and 367. According to the hypothesis of Rebuffat (1980, pp. 323–324, on the grounds of vol. 1 of *The Prosopography of the Later Roman Empire*, Jones *et al.*, 1971), nearly all of the 16 inscriptions supposedly testify to the great restoration work carried out to repair the damage caused by the AD 365 earthquake, which, again according to Rebuffat, is believed to have struck this area of Africa as well, in present-day northeastern Algeria. This interpretation has supported a 'hyper-evaluation' of the energy of the AD 365 earthquake; indeed, the distance between Cirta (present-day Constantine) – the city which three of the 16 inscriptions of the dossier of Publilius Caeionius Caecina Albinus refer to – and southern Crete, is over 1500 km.

Lepelley (1984), after an examination of the sources relating to mid fourth century northern Africa in which there is no mention of the AD 365 earthquake, objected that the intense building activity observable in the years 364–367 can be put down to the favourable building laws passed by the Emperor Julian in 362 rather than to the effects of a destructive earthquake (*Codex Theodosianus*, 10.3.1; *Codex Iustinianus*, 11.70.1; Ammianus Marcellinus, 25.4.15). Those disposition indeed provided for the return of the *vectigalia* (indirect taxations) to the various cities, which were supposed to use them for the maintenance and restoration of public buildings. Lepelley points out that the rich epigraphic dossier he had collected 'does not provide information as to any earthquakes that could have occurred in Africa in that period. It can in no way be excluded that some of the major destruction mentioned (such as at *Thubursicum Numidarum*, before 355) did indeed have a seismic origin. But nothing attests to the ruins caused by an earthquake in 365 [. . .].' However, this well-known epigrapher does not exclude the possibility that 'the ruins of the 4th century detected could well, at least in part, be due to this seismic activity'.

Lepelley's overall indecisiveness has been superseded in regard to the site of the ancient Roman city of Cuicul (present-day Djemila, in Algeria) by Blanchard-Lemée (1984), who convincingly confuted Rebuffat's arguments, demonstrating that such inscriptions are independent from the seismic event of AD 365.

Cases of this kind are not infrequent in historical seismology. A case study has concerned 21 inscriptions that attest to the building interventions of Fabius Maximus and Autonius Iustinianus, governors of the province of Samnium around AD 346–357. Three (or four) of these clearly mention the earthquake, attested to by other sources as well, which hit the area of Benevento, Isernia and Caserta (southern Italy) in AD 346 (Camodeca, 1971; Buonocore, 1992; Guidoboni *et al.*, 1994). The other 16 inscriptions, mostly fragmentary, recall Fabius Maximus as a 'restorer' or 'founder' of city walls or public buildings, and do not explicitly mention the earthquake as being the cause behind his interventions. Those sources contain references to precise locations and can therefore potentially broaden the area of the known effects. The temptation is thus comprehensible, to the point that these inscriptions have been correlated to the earthquake of AD 346, albeit with some cautionary note (Galadini and Galli, 2004): the area of the effects thus identified seemed to be very similar to that of a strong earthquake in this area, which occurred in 1805 ($I_0 = X$; $M = 6.6$), and has thus given rise to further analyses as to the fault and the return times. Interpreting is certainly a legitimate process; nevertheless, if the reader is well informed, he or she can contest such interpretations and draw up different hypotheses.

10.6.3   Collapses and their cultural contexts

The great basilica of St John Lateran in Rome collapsed completely (*ad terram de toto*) in the year 896. The event is recorded authoritatively in the *Liber Pontificalis*, and in the sixteenth century the collapse was attributed to an earthquake (Sigonio, 1580). Belated and over-imaginative interpretation of the source texts by learned historians in this case obscured the precarious stability of an illustrious building, although the original source was not intended to deliberately mislead future historians of seismology. The Lateran church was restored at the time of Pope Sergius III (904–911). Centuries later, when Rome was struck by a very violent earthquake in the central Apennines in September 1349, this same basilica suffered partial collapse, this time limited to the façade. From this report one can conclude that some of the structural problems of the earlier building had been overcome since major components of the restored basilica survived the earthquake.

Sometimes official documents show an explicit awareness of the unstable state of large buildings. Curiously, one does find instances of such rationality in medieval pontifical documents (see below for Clement VI's letter of 7 July 1351). However, in some cases the collapse of a building is cloaked in fantasy and miracle. Sometimes it may be inferred that the narrator is concealing or altering the true explanation of the physical cause of the destruction and rather is playing up a religious or ideological significance of the incident. A clear example of such narrative practice is found in the *Chronicon Salernitanum*, written in the tenth century by an anonymous Benedictine monk from Salerno (southern Italy). Records show that in 870–872 the Saracen tyrant Abdila died in the neighbourhood of Salerno, crushed by a beam falling from a church ceiling as he was attempting to ravish a maiden. The narrator here interposes his own opinion that God intervened lest the Saracens should later deny that Abdila's death was due to divine cause and claim it was pure accident. The monk wrote that such a false interpretation might be given credit by the most frequent fact 'that churches are seen to collapse through old age'. To prove it was God's expressed will to punish Abdila, and not the general dilapidated state of church buildings at the time that caused the incident, the author of the source wrote that the church walls fell apart separately and that not far from the altar a great crack opened. The yawning crevasse and the collapse by stages might lead one to envisage an earthquake scene, perhaps due to a closely spaced series of tremors. But there is lacking what a judge – in this case the seismological historian – should demand of a witness, namely an explicit statement. Hence in cases where the cause of a collapse is either not mentioned or is unduly slanted by an author's ideology, failing other source material it is best to suspend judgement rather than to start filling an earthquake catalogue with possible seismic events that may later prove unfounded (Westerbergh, 1956).

Problems of stability and precariousness as joint causes of a collapse: the awareness of a medieval pope

In September 1349 a strong earthquake (conjectured to be a cluster of at least four violent seismic events: Guidoboni and Comastri, 2005) struck central

Italy, damaging Rome among other cities and towns. Churches, houses and papal strongholds collapsed or suffered damage. Work was set in motion almost at once to restore the large religious buildings. Less than two years later, in July 1351, Clement VI's chancery in Avignon, the provisional see of the Papacy at the time, urged the faithful in a circular letter to donate to reconstructing the basilica of San Paolo, a church that had been seriously damaged by the earthquake. The bell tower had fallen and a collapsing column in the nave had brought down part of the roof. Many sources of the period describe the event in general or exaggerated terms. Even the account by the great poet Francesco Petrarca, who had personally visited Rome shortly after the earthquake, is summary and unclear about what exactly happened: 'the one (the church) dedicated to the Apostle Paul fell to the ground' (F. Petrarca, *Familiarum rerum*, XI, 7). Petrarca's statement hints at a total collapse. It is interesting to see from a papal letter that in technical circles at the time it was known, and perhaps widely known, that the destruction of this old building, precipitated by the earthquake, had arisen from a combination of causes and that repair work would need to remedy pre-existing faults in the state of repair of the church. This is quite clearly stated in one report:

> the venerable basilica [. . .] shaken in many parts by the earthquake some days ago and demolished or damaged at certain points calls to be repaired all the more urgently in that [. . .] the ancient basilica structure was known anyway and for other reasons to need the benefit of new repair work. (Vatican Secret Archives, Registra Vaticana, Clement VI, vol. 145.)

10.7   Effects in towns: constructing an urban seismic scenario of the past

From what has been said above, it may be clear that the level of detail necessary to localize the effects of an earthquake in an urban area is usually only possible for periods that are not too far back in time. Even with the best conditions of preservation of the documentary assets, it is difficult to go back beyond the seventeenth century in the European area for detailed accounts of earthquake damage, although in some cases (such as in Italy) the sixteenth century can also be reached. The most useful documentation is the institutional material. As will be seen at greater length in Chapter 11 the actions of central administrations following a destructive earthquake in granting tax concessions or financing repairs required precise reports of the damage suffered. In many cases such reports were not entrusted to experts but rather to government clerks. Archive research can bring to light some pleasant surprises. Lists of buildings with descriptions of the damage suffered, as well as reports by architects and engineers or tax officials on damage to specific structures, can be found by hunting down the administrative files created in the wake of seismic destruction. Not infrequently the estimates of the damage suffered were only made from an economic standpoint, and in a cursory way, without a detailed description of the effects themselves. In these cases, the documentation cannot be used for detailed seismological analysis. The more homogeneous the documentation used (that is, belonging to a single archival type) the more likely it is that the data can be usefully compared to garner an image of exactly what damage was experienced due to the seismic event.

The earthquakes in a treasure house of art: the case of Florence

The urban and social life of Florence, a treasure trove of art, has been extraordinarily well documented for over 700 years. Even so, the overall long-term picture of the effects of seismic events on Florence's artistic and monumental heritage is still an unknown quantity. The different types of sources available (administrative documents, reliable chronicles, technical and scientific reports, etc.) enable one to paint a fairly complete picture of the effects of earthquakes on the city. Historical seismology seeks to evaluate this knowledge and make available the information concerning earthquakes contained in the historical sources. Among the many earthquakes felt in Florence from the eleventh century onwards, three caused the greatest damage to the city (degree VII MCS): these are the earthquakes that occurred in 1414 (I_0 = VII–VIII; M = 5.6), 1453 (I_0 = VII–VIII; M = 5.3) and 1895 (I_0 = VIII; M = 5.4) (Figure 10.14). There were also nine earthquakes that caused damage in Florence (degree VI or VI–VII MCS) in 1345 (two shocks), 1399, 1542, 1554, 1729, 1770, 1873 and 1919 (Guidoboni and Ferrari, 1995; Guidoboni et al., 2007). The epicentre of the earthquake at Colfiorito (Umbria) in 1997 (I_0 = VIII; M = 5.7) was about 160 km from Florence, but it caused damage to the Ponte alle Grazie in the city (uplifting several slabs of the porphyry paving on the bridge).

The high value of Florence's historic buildings (some of which are now museums) points to a very high seismic risk that is exposed to a seismic hazard level that is relatively low but not inconsequential. Moreover, the scenarios of effects of past earthquakes display some common characteristic elements: (1) an elevated ground-shaking response in the vicinity of the River Arno, which flows through the city; and (2) a high degree of vulnerability in many buildings which can make them more susceptible to seismic shaking, largely due to a lack of basic maintenance work. Shaking effects inside buildings also has led to the destruction of some works of art in past earthquakes. After the earthquake of 1895, house-to-house inspections were carried out by city council technicians, and descriptions of the effects of the earthquake shaking were recorded.

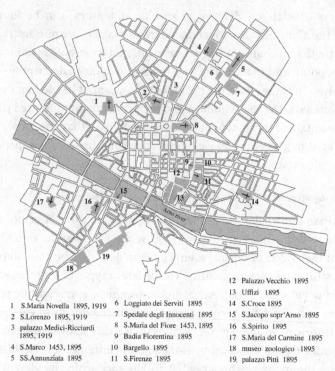

1 S.Maria Novella 1895, 1919
2 S.Lorenzo 1895, 1919
3 palazzo Medici-Ricciardi
   1895, 1919
4 S.Marco 1453, 1895
5 SS.Annunziata 1895

6 Loggiato dei Serviti 1895
7 Spedale degli Innocenti 1895
8 S.Maria del Fiore 1453, 1895
9 Badia Fiorentina 1895
10 Bargello 1895
11 S.Firenze 1895

12 Palazzo Vecchio 1895
13 Uffizi 1895
14 S.Croce 1895
15 S.Jacopo sopr'Arno 1895
16 S.Spirito 1895
17 S.Maria del Carmine 1895
18 museo zoologico 1895
19 palazzo Pitti 1895

Figure 10.14 Map of the historical centre of Florence: buildings that were damaged by the three earthquakes of 1453, 1895 and 1919. The damages caused by the 1895 earthquake are located on the map on the basis of the contemporary technical documents.

These reports were processed and maps of the damage were related to the local geology (Vannucci *et al.*, 1999a; Boccaletti *et al.*, 2001). From this information, there is now a better understanding concerning which parts of the city are more at risk in earthquake shaking. Furthermore, Florence is not the only city in Italy for which studies of the site-specific response to earthquake shaking of different parts of the city (even different buildings) have been carried out. Other such cities include Melfi (Figure 10.15), Bologna (Figure 10.16) and Palermo (Figures 10.17 and 10.18).

10.8   Effects on the natural environment

If the scenario of a past earthquake is to be comprehensive and complete, a description of the effects of the event on the natural environment is required along with that for the built environment. As a matter of course, earthquake catalogues ought to contain relevant information on this category of effects when such effects are mentioned in the historical sources. Although they are usually overlooked, afforded only sporadic attention or mentioned only in relation to a

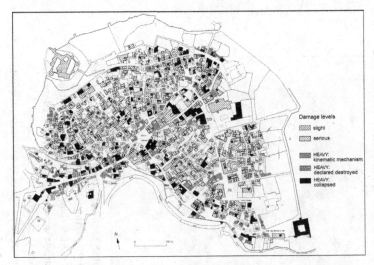

Figure 10.15 Map of the historic centre of Melfi. The buildings damaged by the earthquake on 23 July 1930 are located and classified on the basis of the contemporary technical documents (Gizzi and Masini, 2004).

few specific events, changes in the natural environment due to an earthquake are not a secondary part of a seismic scenario. Rather, they contain vital information of great interest to seismologists who want to know about past earthquake activity. The correct recording of historical data should therefore also include complete descriptions of the effects of earthquake on the natural environment (see the interesting considerations and data in Vogt, 1977a, 1977b, 1984a, 1984b, 1993). It is thus worthwhile to prepare a classification scheme of such effects, where the classification includes the location and characteristics of the effect. A preliminary classification can indeed be rather detailed, in accordance with present-day cognitive categories. However, the greatest problem in dealing with studies of changes in the natural environment due to earthquakes remains that of recognising such phenomena in the historical descriptions.

10.8.1   Classification of effects on the environment: an example

In the database of the *Catalogue of Strong Earthquakes in Italy* (Boschi *et al.*, 2000; Guidoboni *et al.*, 2007) there are over 2000 descriptions of the effects of earthquakes on the natural environment. Such descriptions are contained in full on the on-line version of that catalogue, and in fact this is the first catalogue to contain these data presented in a systematic fashion. Altogether, such data represent an informative package of statistical substance, but considering the chronological span of investigation of almost 2500 years, it obviously contains many uncertainties and explanations of the reports that are included.

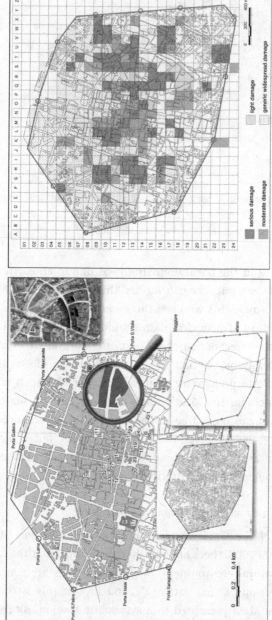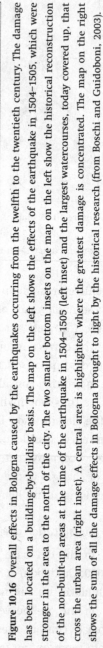

Figure 10.16 Overall effects in Bologna caused by the earthquakes occurring from the twelfth to the twentieth century. The damage has been located on a building-by-building basis. The map on the left shows the effects of the earthquake in 1504–1505, which were stronger in the area to the north of the city. The two smaller bottom insets on the map on the left show the historical reconstruction of the non-built-up areas at the time of the earthquake in 1504–1505 (left inset) and the largest watercourses, today covered up, that cross the urban area (right inset). A central area is highlighted where the greatest damage is concentrated. The map on the right shows the sum of all the damage effects in Bologna brought to light by the historical research (from Boschi and Guidoboni, 2003).

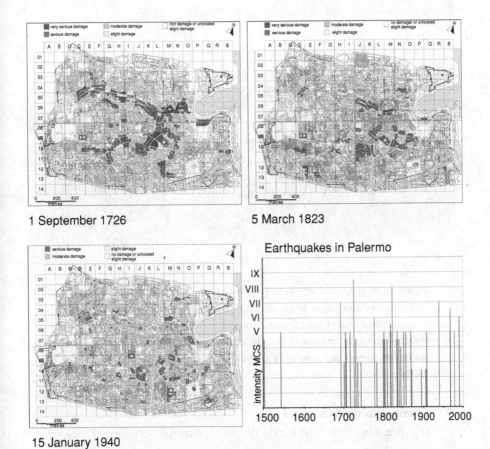

1 September 1726

5 March 1823

15 January 1940

Earthquakes in Palermo

Figure 10.17 Scenarios of the effects caused in the historical centre of Palermo by three of the strongest earthquakes of the last few centuries, documented by direct sources. The damage has been located on a building-by-building scale (from Guidoboni *et al.*, 2003a). The lower right graph shows the trend of earthquakes known in Palermo from 1500 to 2000.

Indeed, the data gathered in this database come from different cultures and have been generated within different cognitive frameworks whose language is clearly far removed from that of the present day. When the catalogue is used, these factors will certainly raise some queries as to the general reliability of the data gathered, problems in their initial interpretation and their usefulness in increasing the knowledge of the earthquakes to which they refer. In practice, six families of effects have been identified, within which there are further subdivisions intended to help situate the described phenomenon in the historical records. The classification scheme is as follows (from Valensise and Guidoboni, 2000a):

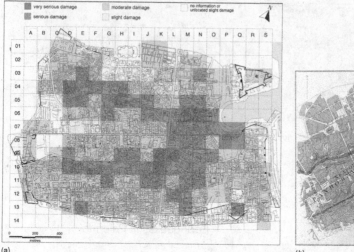

(a)   (b)

Figure 10.18 (a) Overall effects in Palermo of the three earthquakes of 1726, 1823 and 1940 (see also Figure 10.17). An area of serious effects is indicated by a circumscribed area of the current historical centre. (b) Historical map of the city of Palermo, the ancient harbour and the courses of the two waterways Papireto and Maltempo, buried and covered in the seventeenth century (from Guidoboni *et al.*, 2003a).

(A) **Changes in the landscape**
   (1) Breaking up of the earth, fractures, fissures, splitting of rock or soil
   (2) Subsidence, slumps and landslips
   (3) Ground uplifts
   (4) Cavity collapses or effects on the structures of caves
   (5) Landslides, earth flows, mud flows, ground failing
   (6) Landslides

(B) **Watercourses**
   (1) Increase in the flow of watercourses
   (2) Decrease in the flow of watercourses
   (3) Overflow: land turns into marshland
   (4) Alluvium, filling up with unconsolidated earth
   (5) Changes in river beds, deviations, interruptions
   (6) Burst river banks, flooding
   (7) Water clouding or colour changes

(C) **Lake basins**
   (1) Appearance or disappearance of lake basins
   (2) Clouding of water or colour changes in lake basins

    (3) Variations in the water level of lake basins

    (4) Flooding of lake basins

(D) **Subterranean streams**

    (1) Changes in spring flows

    (2) Appearance or disappearance of springs

    (3) Clouding of water in springs

    (4) Changes in water levels of wells

    (5) Water or mud outflow from the earth or liquefaction

    (6) Changes in water temperatures

    (7) Changes in water chemistry

(E) **Marine or coastal environment**

    (1) General changes to the coastline

    (2) Erosion or the moving back of the coastline

    (3) Seaquake or tsunami

(F) **Others**

    (1) Death of fish and other organisms

    (2) Exhaling or emission of gases, eruptions of mud / mud cones

    (3) Sulphurous exhalations

    (4) Electrical or magnetic phenomena

    (5) Visible light phenomena

10.8.2 *Some observations from a historical point of view*

In the ancient and late Greek and Latin world, the effects of earthquakes on the natural environment left a deep impression on the minds of observers, who were imbued with feelings of the sacred and magical with which natural phenomena were perceived. People at the time regarded the changes as mysterious because of their idea that nature was unchangeable. Indeed, effects on the natural environment are often recorded in the sources with such emphasis that the direct cause of the effects fades into the background, so that today's reader may sometimes wonder whether the phenomenon described was actually caused by an earthquake or by some other phenomenon. At times, such descriptions are unequivocal, and for that reason those descriptions are all the more important to modern researchers.

The new lakes at Apamea after the earthquake of 88 BC

Nicolaus Damascenus was a Greek-speaking writer of the first century BC. He was much struck by changes to the natural environment brought about by an earthquake in 88 BC near the city of Apamea in Phrygia (now Dinar, in Turkey). The earthquake is also mentioned by Strabo (12.8.18). Nicolaus writes that:

lakes which did not exist before appeared in the area, as well as rivers, and while some springs gushed forth as a result of the shaking, others disappeared [. . .] Brackish, pale blue water also flowed out.

He then continues with a description that is hard for us to explain today, since Apamea was about 150 km from the sea:

in spite of the great distance from the coast, the surrounding area became full of shells and fish. (Nicolaus Damascenus, *FGrHist* 90 F 74.)

Evidence concerning the early Middle Ages was mainly preserved in monastic sources or was produced by intellectuals in church circles. The evidence in these sources is important because it is all but unique and often provides information about areas outside the urban centres. One might think that this permitted a greater number of direct observations of the effects of earthquakes on the natural environment to be made, but the culture that permeated the production of such observations at the time did not encourage the explicit and direct observation of these phenomena.

A large number of descriptions of environmental effects in the medieval texts have proved to be inexact or mistakenly correlated to earthquakes. Language often envelops such phenomena within an aura of exaggeration, for example claiming that mountains split in two, that the belching of smoke from the Earth was seen or that howling noises were heard. The interpretation of this type of strongly figurative language within our present cognitive models is not a simple matter. Generally speaking, an accurate modern understanding of the meaning of often fantastic descriptions only becomes possible when the effects that were probably being described are repeated in the same areas and in later periods when better observations of the phenomena are recorded. Thus, the *splitting* of a mountain may be accounted for as a reference to the formation of fractures in undulating land that is sliding downhill, *smoke* to the dust stirred up by strong shaking on dry hillsides (a phenomenon commonly reported today in strong earthquake shaking), and *howling* can be recognized as the perception of seismic waves at low acoustic frequencies, possibly amplified by the local soil conditions. In fact, roaring or thunderous noises are commonly reported from the ground near the epicentres of strong earthquakes, while higher-pitched sounds (such as cracks and explosive bangs) are perceived near the epicentres of smaller-magnitude earthquakes (magnitude 2 to 4). Such sounds are not surprising since the seismic waves in the rock can convert to acoustic waves in the air that can be heard if they contain audible frequencies. It must be concluded from this discussion that what seem to be sometimes fanciful accounts of natural effects that accompany an earthquake may well be exaggerated or misinterpreted phenomena that really did accompany the earthquake. In some cases, the medieval chroniclers paid great attention to effects of earthquakes on the natural environment, and sometimes they described the effects with surprising accuracy.

Landslides and lakes caused by earthquakes: a 'Chinese box' in the medieval texts

On 30 April 1279, in the afternoon (14 hours UT), a strong earthquake struck an extensive area of central Italy (the Umbria and Marche regions). Since there is evidence for the event in many independent contemporary sources, it has been possible to carry out a good reconstruction of the area of notable effects, which comprised 14 localities (I_0 = IX; M = 6.5). The principal sources that describe this event are authoritative chronicles and annals, some from Italy and others from central Europe, which exist because news of the earthquake reached quite distant monasteries (see Guidoboni and Comastri, 2005). Apart from direct references to places, the sources also refer to effects on the natural environment. Such effects probably made a considerable impact on the imagination of contemporaries, but their location cannot be immediately pinpointed as the sources do not supply explicit information about the locations of the effects. The writers' way of communicating is rather like a set of Chinese boxes, in that place-name data are concealed within a series of territorial allusions. Their way of proceeding is to supply an identifiable toponym, and then make some hidden references, i.e. without a location, and then provide some identifiable elements from which the location might be guessed.

Obviously, this way of recounting events was not the result of a desire to play semantic games with the readers of years to come, but rather it arises from the very nature of the medieval system of communication. Word-of-mouth transmission of information involved various stages where the information was passed from one person to the next, which progressively allowed distortions to become incorporated into the account before the piece of information became distilled in a written text. During the transmission stages some items of information, such as precise geographical references, were lost, while others were added, such as qualitative judgements about the events described. The latter always tended to be strongly emphasized (a custom that still is followed today). One can observe this evolution in the passages set out below, which have been translated from medieval Latin. The texts, in their original version, are preceded by some lines in which the authors describe the destruction in their city and reveal, through place names, that an area of about 3000 km^2, situated in the present-day regions of Marche and Umbria, was affected. The geographical references are highlighted in *italics*; the cryptic terms are *italic underlined*.

Chronica S. Petri Erfordensis moderna

Nocera was situated on *a mountain*, and *another mountain* was in front of
it; and *a castle* was situated between either mountain; there were
exactly five hundred guests here, and it was called *Serravalle*. These
mountains came against each other and covered that castle that lay in
the middle, with all the people who were inside it. And now it is so
flat, as though there had never been any building there. And close to
those *two cities* there are *other castles*, in which many men died as a
result of the same earthquake.

Salimbene de Adam, Cronica

Equally *three mountains*, between which lay two lakes and a castle, [. . .]
were alternately joined together and then detached; and
a lake and a river, whose bursting of the banks had produced lakes,
were completely dried up. And in a short time *all the castles* that were
to be found in those mountainous parts withstood great damage. Even
a castle was completely destroyed.

Annales Polonorum

Even *two mountains* joined into one; between these there flowed *a river*
that was therefore enclosed between the mountains, which not having
any bed, formed *a lake* around it, up to seventy miles and more,
overcoming everything that lay in its path. Even *some castles* situated
on the mountains, as the earth opened for the earthquakes that
continued for 15 days, the same castles collapsed with the mountains;
some huge mountains were reduced to fields, entirely flattened out to the
ground, and it (the earthquake) caused a terrible massacre among the
people.

It is only possible to identify the terms indicated if one knows the historical
nature of this area in the thirteenth century. For example, there were many
castles (small fortified villages) between Camerino and Nocera (see the map in
Figure 10.19). The higher part of the present-day town of Serravalle di Chienti
is still called the Castle of *Serravalle*. It lies at the opening of the Chienti River
valley, at a short distance from the pass that separated the hills from the plains
of Colfiorito (epicentre of the earthquake in September 1997, also known as
the Assisi earthquake, $I_0 = VIII; M = 5.7$). In the higher part of Serravalle the
remains of a castle are visible even today, and that is probably the one mentioned
in the medieval texts. Some portions of the walls of that castle date back to the
late thirteenth century. The Chienti is the *river* that has been identified as the
one mentioned only as '*river*' in the sources. This river flows through the village

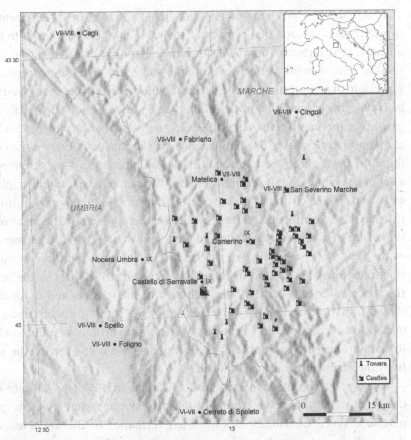

Figure 10.19 Map of the towers and the castles that existed at the time of the earthquake in 1279, not cited in the contemporary sources, in relation to the earthquake. The degrees of intensity (MCS scale) are related to the sites cited in the medieval texts as being damaged.

of Serravalle. When the Chienti River was in full flow in the medieval period, it formed marshy areas that were used and exploited by the local rural economy. This area of surface water cannot be detected today because it was drained in the fifteenth century. The collapse of the castles in the mountains can be interpreted as due to a landslide. The flattened mountains ('*reduced to fields*') are to be interpreted as textual hyperbole. But the landslide, which probably worsened a hydrological situation that had not yet been brought under control, really did take place. A group of geologists has recently identified it by means of field reconnaissance and cartographic observations. Thus, the anonymous monks who wrote these apparently fanciful and cryptic descriptions were not making up stories at all. One really can take advantage of their enthralling riddles to discern a scenario of the effects of a strong earthquake.

It was in the early part of the modern era (sixteenth to eighteenth centuries) that the observation of natural environmental effects started to become an important factor in improving the understanding of earthquakes. Descriptions of environmental effects became the subject of discussions regarding the real origin of seismic phenomena. Natural philosophers, scholars and naturalists began to take special note of landslides, large surface strata slips, emissions of gas flames (perhaps methane), changes in the behaviour of underground water and atmospheric effects like lights and a red sky. Such observations can also be found in town chronicles, personal diaries and gazettes. What attracted most attention was the overall seismic scenario. The symbolic and moral significance of an earthquake obviously did not fail to appear in contemporary culture, but it tended to be confined to certain particular literary genres. The great observational leap towards more accurate and more immediately perceptive descriptions of the effects of earthquakes on the natural environment within current scientific frameworks took place in the eighteenth century, clearly as the result of the major advancements that were achieved in the sciences and in naturalistic thought. In the previous century descriptions of the effects of strong earthquakes had already shown a growing awareness of earthquake-induced changes in the natural environment, and this awareness tends to grow in the sources from the eighteenth century onwards.

It was after the major earthquakes of Lisbon in 1755 (see Chapter 6) and Calabria in 1783 that the repertory of seismic environmental effects acquired a new importance. Landslides, rock splittings, formation or altering of lakes, landslips and chasms were described and often drawn with great accuracy (Figure 10.20). It can be said that until the end of the eighteenth century, the measurements supplied by various authors as to the size, duration and location of such effects are nearly always approximate and very crude. Even if the units of measurement of the different historical systems in use in the various areas are correctly converted into current ones, there may still remain exaggerations or incongruencies in the descriptions that were recorded. It is possible that more accurate measures were mentioned in the original sources or that reliable conversion systems to compare these measurements and modify them have not been found. Furthermore, there may be difficulties in determining the correct locations of the effects described. In such cases it is useful to keep separate in the catalogue database the generic descriptions of the effects from the ones that can be located accurately (i.e. with their geographic coordinates). In other words, it is better not to locate an environmental effect rather than to locate it haphazardly based on inadequate information.

With the development of positivist sciences, increasing attention was paid to seismic effects on the natural environment. During the nineteenth century a

Figure 10.20 One of the 70 plates drawn by Pompeo Schiantarelli and Ignazio Stile (1784) to represent the effects of the earthquakes in Calabria in 1783 in the villages and in the natural environment. This is Plate no. 26 drawn by Pompeo Schiantarelli and shows the lakes formed by the earthquake close to Polistena (Reggio Calabria).

truly naturalistic and geological approach to the study of earthquakes was set out in great detail. The types of damage to buildings also began to be correlated systematically with the types of ground on which the structures were built. This correlation led to increased attention on the part of informed observers (such as trained seismologists, engineers and geologists) to the relationships of the observations of damage with the local terrain and the natural environment in general.

Many of the nineteenth-century observations of earthquake effects contain a remarkable wealth of detail and add important information about the earthquake itself. There are numerous cases of this new attention, such as the scientific report on the earthquake of 16 December 1857 in the Kingdom of Naples, made by Robert Mallet (1862) for the Royal Society, or for the case of the earthquake in France and Italy on 23 February 1887, for which a panel of scientific experts (Issel, 1888; Taramelli and Mercalli, 1888) has left some highly detailed descriptions of the effects of the earthquake on the terrain. Such reports were addressed to the most important scientific academies of the day (such as the Academy of Sciences in Paris or the Accademia dei Lincei in Rome), where such reports were read, discussed and published.

The attention given at this time to the observed effects of an earthquake on the environment were observed shows a newfound awareness in which confidence and caution reveal an attentive and competent scientific framework in which investigators worked. For example, the nineteenth-century researchers recorded observations of elements that today, in a different cognitive framework, allow modern seismologists to suggest a source mechanism for the Calabria earthquake on 16 November 1894 ($I_0 = IX$; $M = 6.1$). An Italian scientist, Riccò (1907), described as follows the instrumental measurement of a variation in sea level measured at Punta Faro (the southernmost tip of Calabria, southern Italy):

> The next morning, at 8:30 hours, the needle indicated a level slightly lower than that of the previous day at the same time; and afterwards, on starting the instrument's clock and paper, it traced a similar line to that of the day before, but somewhat lower, corresponding to a sea-level 5 or 6 centimetres lower; a longer study is needed (which cannot be done here) to see whether this small variation really depends on the relative lowering of the sea-level, or instead on a steady rising of the mareograph or of the ground on which it is installed caused by the earthquake, or if it is a question of the mechanism being moved as a result of the quake, or if it is simply a question of the change in the influence of the sun and the moon and of the marine and air currents on the tide, which is more likely. (Riccò, 1907.)

Within the end of the nineteenth century and the start of the twentieth century, observations of the effects of large earthquakes on the environment became an integral part of scientific observation in the current sense of the term. Observations of the changes in the coastline caused by the earthquake at San Francisco in 1906 (Lawson, 1908) and at Messina in 1908 (Baratta, 1910), and the detailed description of the ground fractures caused by the earthquakes in the Pakistan/Afghanistan region in 1892 (McMahon, 1897) and in the Marsica region (central Italy) in 1915 were ahead of their time. These are pioneering descriptions in the field of observational seismology. The importance of historical descriptions of environmental effects basically lies in their ability to illustrate the distribution of those areas that are unstable in earthquake ground-shaking and to quantify the levels of shaking that trigger these effects. When dealing with historical accounts of seismically induced changes in the natural landscape, a correct reading of the reported descriptions and identification of the triggering processes is generally difficult and potentially ambiguous, and so great care must be taken with these reports.

10.8.3 *Effects on the natural environment and macroseismic intensity scales*

Over the last 15 years attention has grown significantly on how to classify the effects of earthquakes on the natural environment. The Mercalli Scale assigns degree VII to the initial collapse of river banks, degree VIII to the formation of cracks and to liquefaction phenomena in unconsolidated ground, and degree X to the falling of 'whole boulders', going on to the 'large-scale devastation' described in the definition of degree XII. Mercalli intensities IX to XII form a range of descriptions that is more reminiscent of the collective imagination of catastrophes than a rigorous scientific classification. This also applies to similar observations in the MSK scale, the MM scale and the scale used by the Japan Meteorological Agency. This matter has been the subject of different analyses and points of view, including that of Serva (1994) and of Grünthal (1993), the latter being expressed within the framework of the definition of the European Macroseismic Scale (EMS: 1992). These analyses share the basic conviction that earthquake effects on the natural environment vary a great deal from place to place for the same level of ground shaking. Nevertheless, whilst Serva (1994) concluded by suggesting that such observations could still be used to evaluate the severity of the effects of past earthquakes as long as they are measured beforehand against the effects found in buildings, scepticism and concern seem to prevail among the compilers of the EMS, who fear that the inclusion of these effects on the natural environment in the framework of earthquake evaluation may generate more doubts than it actually dispels.

Yet there are different opinions and the discussions continue. These is no doubt that the systematic collection of the seismic effects in the natural environment is precious and allows us to glean elements not included in the degrees of intensity. In this regard, a specialized workgroup has been set up, the International Union for Quaternary Research (INQUA), which has developed this strand of inquiry for over eight years (coordinated by L. Serva). From this research the Environmental Seismic Intensity Scale 2007 (ESI) (Guerrieri and Vittori, 2007, pp. 7–13) has taken shape. This systematizes and classifies a series of elements that can be correlated with the seismic activity. (For some applications of a previous version of this scale see Papathanassiou and Pavlides, 2007; Fokaefs and Papadopoulos, 2007; Tatevossian, 2007; Serva *et al.*, 2007.)

The effects are distinguished into *primary effects*, that is ruptures, uplifts, escarpments produced by the active faults; and the *secondary effects*, that is landslides, movements of rocky blocks, ground collapses, liquefaction, tidal waves, etc.

Descriptions of the breaking of the ground surface, of fractures with the downdrop of one side as compared with the other and of subsidence and the formation of marshland, particularly if spread out over some or many kilometres, all represent clues about permanent surface deformations induced by an earthquake. Such evidence may also have great practical importance, such as in the planning of high-risk industrial plants or large-scale public works that could be seriously damaged by any permanent ground displacement. Nevertheless, all over the world observations regarding this category of earthquake effects have enthralled generations of historians, naturalists and geologists, above all due to their relative rarity and the unparalleled quality of the direct testimonies of endogenic activity.

10.9   Identifying faulting and liquefaction features in historical accounts

It is almost never easy to identify *faulting* and *liquefaction* phenomena in historical sources. The former involves a fracture or system of fractures along which the rock is moved in opposite directions on either side of the fault. The movements can be vertical, horizontal or some combination. The slip of the rock on either side of the fault can vary from a few centimetres in a small earthquake to a few tens of metres in a very major earthquake. Also, the larger the earthquake, the greater the fault length. Surface faulting that has only a few centimetres of slip usually extends only a few kilometres along the fault. On the other hand, a very major earthquake with a very large fault slip will typically occur on a fault that spans a few hundred kilometres or more.

Liquefaction is a surface seismic effect (to a depth of 100 metres at the most) where the ground-shaking of soils consisting of small water-soaked granules (sand or finer) undergo an increase in water pressure, leading to a loss of strength in the soil itself. The loss of strength in the soil causes it to behave like a liquid rather than like a solid, and this behaviour is called liquefaction. Liquefaction occurs when the ground-shaking exceeds a threshold level and typically takes place in a time period that is no more than probably 10–20 seconds. Liquefaction happens in surface or subsurface soil layers where there are fine or very fine sand deposits that are relatively unconsolidated and are highly saturated with water. Furthermore, sand may lose its cohesion even if it is not fully saturated with fluids. Liquefaction can manifest itself in several ways. It may be associated with sudden landslides, where soils atop liquefied layers slide downhill on otherwise stable slopes, as happened in the Turnegan Heights in Anchorage, Alaska in the 1964 earthquake. Lateral spreading, where a soil layer spreads out along a very shallow slope (as small as 1 degree), is a liquefaction effect that often occurs along shorelines. Sometimes subsurface liquefied layers can erupt explosively, throwing sand and mud like a small volcano onto the surface of the earth in what is called a *sand blow*. Sand blows can be as much as 100 metres across. When a soil undergoes liquefaction, it can lose its ability to support loads such as buildings or other structures, which then start to sink and tilt into the soil. These seismic effects are much feared even today, because even proper anti-earthquake building techniques can be rendered ineffective if the foundation of a building is compromised by liquefaction.

Available historical data on surface earthquake faulting and on soil liquefaction are extremely rare in comparison with what is known about other environmental seismic effects, such as landslides. The reason for this lack of data is fairly obvious, since an appreciation of both faulting and liquefaction as well as their relation to seismic shaking are comparatively recent additions (within the past 100 or so years) to our knowledge of earthquakes. Consequently, when these phenomena might have been observed in the past, they tended to escape notice because they fell outside theories about the origins of earthquakes and were perhaps considered as marginal ancillary phenomena. Since the epistemological principle still holds good, that 'everyone sees what he knows', it is only by chance and perhaps by indirect inference that one might be lucky enough to find evidence of these two types of effects in historical texts. It must also be kept in mind that faulting features can be confused both with ordinary fissures that naturally form in some types of terrain or with fissures and cracks associated with non-seismic landslips.

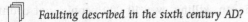 *Faulting described in the sixth century AD?*

Marcellinus Comes was a Latin writer from Dardania, a region corresponding to southern Serbia and present-day northern Macedonia (a map is in Croke, 2001, p. 50). Very litttle is known about Marcellinus Comes, apart from the fact that he lived in the fifth and sixth centuries AD. In his *Chronicon* he described the effects of an earthquake (or several earthquakes) that occurred in the year AD 518. It seems likely that he witnessed these effects himself or, at any rate, used some accounts of first-hand witnesses (on this earthquake see Ambraseys, 1971; Shebalin *et al.*, 1974; Comninakis and Papazachos, 1982; Guidoboni *et al.*, 1994; Papazachos and Papazachos, 1997).

After having mentioned destruction and damage at 24 villages, the ruins of Scupus (now Skopje) and extensive rockfalls from the mountains, Marcellinus describes apparent faulting, of which he also provides some measurements. He also provides the well-identified location of some thermal effects apparently associated with the earthquake:

> *Most of the mountains of this entire province were split by this earthquake, stones were torn from their outcrops, and trees uprooted. Through a deep chasm thirty miles [c.43.5 km] long and twelve feet [c.3.6 m] wide the quake provided a*

> burial ground for a large number of the inhabitants fleeing from the ruins of
> the fortresses and defensive positions, also fleeing from the attacks of enemies.
>   In one fort of the Gavisa region, called Sarnontum, the earth's channels were
> dislocated and after boiling up like a burning furnace disgorged a fiery shower
> that had built up for a long time on the inside. (Marcellinus Comes
> translation by Croke, 1995.)

It is not simple to decide what use can be made of these data today, or how exactly to interpret them. However, at a first level of approximation, this description confirms that this was a large seismic event. On 26 July 1963, the same area was hit by a strong earthquake, which caused more than 1000 deaths and 3000 injured (M = 6.1), according to Papazachos and Papazachos (1997).

A thirteenth-century monk travelling along the North Anatolian fault

In the year 1253 a Franciscan missionary, William of Rubruck, set off on a long journey that took him all the way to the court of the Great Khan of the Mongols. Having left Constantinople in May 1253, at the end of December of the same year he arrived at the camp of the Great Khan near Karakorum, the ancient capital of the Mongol empire. During his return journey, after having crossed Central Asia and the Caucasus, in February 1255 William of Rubruck found himself travelling through eastern Anatolia. Just four months earlier, on 11 October 1254, that area had been hit hard by a very violent earthquake which had completely wiped out the city of Erzincan, killing over 10 000 people, not counting those called 'poor' whose number was probably not even known before the earthquake. William of Rubruck left a detailed account of his journey through Mongolia in Latin, a sort of travelogue entitled *Itinerarium*, in which the idiosyncratic traveller described those effects of the earthquake that he had seen at first-hand:

> [. . .] we arrived at a certain castle they call Camath [Kemah]. Here the
> Euphrates winds southwards, towards Alapia [Aleppo]. But, once we had crossed
> the river, we headed towards a very tall mountain, through thick snow, to the
> west. But in that year there was such a great earthquake that in one city, that
> they call Arsengen [Erzincan], ten thousand people perished, known by their
> names, except the poor, of whom there was no news. Continuing on horseback
> for three days, we saw a crack in the earth, that had been opened by the
> earthquake, and heaps of earth that had slid down from the mountains and
> had filled in the valleys.

As far as is known, William of Rubruck's account is one of the earliest descriptions of what was probably surficial faulting. At that time he could not know that what he had followed for three days on horseback was the surface trace of a segment of the North Anatolian fault, which stretches for several hundred kilometres across northern Turkey. The entire North Anatolian fault is more than 1000 km long. This fault had undoubtedly generated the great earthquake of 1254, one of the many that in the course of history have periodically devastated this part of Turkey (the most recent occurred in 1939; its magnitude was close to 8.0 and was among the strongest ever recorded in the Mediterranean area).

William's chronicle is an example of how even a text dating back nearly eight centuries, and written by an author with a completely different purpose in mind, can provide unique as well as useful information for historical seismology studies. This source was first brought to light by Ambraseys and Melville (1995). According to them, the effects on the ground observed by William of Rubruck extended for about 50 km along a segment of the North Anatolian fault west of Erzincan, but in their opinion the surface faulting in fact probably extended all the way to Erzincan and the area of the greatest destruction, and so for an overall length of about 150 km. This length of surface earthquake rupturing would make the earthquake of October 1254 one of the most violent ever to have occurred along this fault, similar to that of 1939.

An uplift in 1638 in Calabria (southern Italy)?

Despite the use of fairly general terms like 'splitting' and 'fissures' in their descriptions, some authors may perhaps be referring to much more complex phenomena that potentially provide important information about a seismic source. An example can be seen in the descriptions of the effects on the natural environment of the earthquakes in Calabria in March 1638. A report of a '60-mile long' fracture, with an approximately '3-palm high' difference between the two sides the fracture, makes one think of the surface effects of a great fault (according to Galli and Bosi, 2004). The description is certainly inaccurate in its measurements, but confirms that attention to earthquake effects may be present in a variety of cultural frameworks. In this case the context is that of the occurrence of calamities, foreshadowed by sinister signs (red skies) and comets. The following passage is recorded by A. di Somma (1641):

> From the edge of Policastro up to the furthest part of the mountain called Sila, towards the north, the ground sank to a depth of three palms on one side, over a distance of sixty miles, and stretched out in a straight furrow, and what appears even more astonishing, it spread into the lowest valleys as much as into the highest mountains.

Of the two phenomena, faulting and liquefaction, perhaps the latter made the deeper impression on persons in the past who observed the phenomenon.

Figure 10.21 The imagined effects of liquefaction in the little house that was built obliquely by Pirro Ligorio c.1552, in the great historical garden of Bomarzo near Viterbo, central Italy.

Even in the ancient world, there are reports of buildings tilting, apparently due to earthquake liquefaction, and these effects are preserved, so to speak, within the cultural category of prodigies. An example is an earthquake of the first century BC. Obsequens (45), on the basis of Livy's lost books, records that '*an earthquake caused houses to collapse in ruins in Picenum, while in some cases the shaking of the foundations left them standing but leaning at an angle*'. It may seem strange that in the sixteenth century a baroque representation of a house leaning at a dangerous angle was constructed in order to make anyone who entered feel the sensation of having his topographical perception upset. The building in question is the *casino* designed by Pirro Ligorio (a famous architect and intellectual of the papal court) in the great garden of Bomarzo (near Viterbo, central Italy). The similarity between this artistic achievement and subsidence due to liquefaction effects (which he perhaps observed) is rather impressive (see Figure 10.21; Figure 10.22 shows a very similar situation from San Francisco in 1906).

It is not unusual to find a description of liquefaction effects on buildings, in the ground or across a landscape, but these descriptions are often in very general terms that lead to more doubt than certainty. Sometimes one comes across rather curious descriptions of liquefaction effects on the ground. For

Figure 10.22 A house in San Francisco (California), which truly remained slanted as a result of the effect of liquefaction in the earthquake of 18 April 1906.

example, after the large 1783 earthquakes in Calabria (southern Italy), many effects were observed that may be interpreted as due to liquefaction. Today one can observe them from detailed drawings made and published by two members of the Royal Academy of Science in Naples. Pompeo Schiantarelli and Ignazio Stile (1784) were the persons who produced this report; examples from among the dozens of tables that they produced are shown in Figures 10.23 and 10.24.

An earthquake in the Pisa area in 1846 ($I_0 = IX$; $M = 5.9$) caused some sand funnels that were observed and drawn by contemporaries (Figure 10.25). In a number of fields, small cones were noticed with water issuing from them. Leopoldo Pilla (1846) saw them and wrote:

> the shaking (caused by the earthquake) had produced deep cracks which brought to the surface trickles of underground water which had previously been trapped inside layers of earth.

It seems that liquefaction caused some small sand blows, which brought to the surface fine sediments and mud that formed small cones. Water issuing from sand blows is a common report in modern earthquakes.

Figure 10.23 Ground deformation effects caused by the earthquakes of 1783 in Calabria documented by Schiantarelli and Stile (Plate no. 31) (Schiantarelli and Stile, 1784).

Figure 10.24 Effects of liquefaction in Calabria due to the earthquakes of 1783 (Plate no. 47) (Schiantarelli and Stile, 1784).

Fig. III.

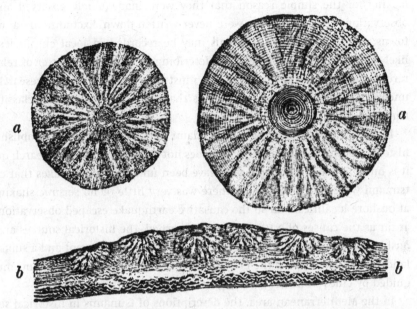

Figure 10.25 Drawing, from above and in section, of the sand funnels observed after the earthquake in Tuscany on 14 August 1846. The author is Leopoldo Pilla (1846), a geologist at Pisa University, who directly observed these features.

10.10   Tsunamis: loss of data and descriptive uniformity

The historical sources that describe a tsunami hardly ever have a comprehensive perception of the dimensions of the event with which they are concerned. Rather, the historical sources concentrate on the most obvious of the local manifestations of a tsunami, such as receding water (which premodern sources considered to be a prodigy) and the appearance of fish on the uncovered sandy seabed (often interpreted as a positive prodigy), followed by the dangerous return of the flooding sea. When tsunamis were not particularly violent or extensive, the descriptions are very incomplete, and the effects described may nowadays be confused with those of severe sea storms. In addition to vagueness in many historical tsunami descriptions, there may also be a considerable scarcity of data that even intensive research cannot always resolve. It must be remembered that coastal areas were much less inhabited in the past than they are today. In recent decades tourism has completely transformed our understanding of what a coastal area represents, and this has brought visitors and inhabitants to places which, in the Mediterranean area, were considered uninhabitable as little as a century ago. Therefore, if local tsunamis did occur, it can be assumed that for many centuries sources capable of recording them are lacking for the simple reason that they were inadequately observed and any observations that were made were never written down. Excluding data on port towns, where damage by tsunamis may be recorded in local chronicles, it is likely that the existence of sources describing tsunamis is a matter of relatively rare chance. On the other hand, one must always be aware that research may uncover information about sea storms that have been mistakenly classified as tsunamis.

In many cases, the earthquake–tsunami link cannot easily be established for historical events, and that difficulty does not always depend on research quality. It is quite possible that there may have been undersea earthquakes that caused tsunami waves but from which there was very little or no seismic shaking felt at onshore localities, and so the causative earthquake escaped observation. And as far as the causes of a tsunami are concerned, the historical sources are obviously unable to distinguish between an undersea seismic event and a submarine landslide. In such cases, the classification of the cause of the tsunami must be guided by clues from the local seismotectonics and marine geology.

In the Mediterranean area, the descriptions of tsunamis in historical sources are typologically similar no matter when the sources were written. One commonly finds an indication of how far the water penetrated inland, and this is often the only quantitative evidence given. In the case of stronger tsunamis, a certain amount of descriptive exaggeration is naturally to be expected, especially concerning the height of the waves and the damage caused by the inundation. Such events caused great panic, and typically they remained in the collective memory for centuries.

The lexis used in historical sources used to describe tsunami effects (e.g. Figure 10.26) is very similar to that used to describe flooding, tidal waves or the effects of tropical storms like hurricanes and cyclones. Terms that are used today to describe the power and range of the anomalous upward or downward movement of the sea come from a modern understanding of how sea level evolves with time. Changes in sea level and tidal variations began to be observed regularly in ports no earlier than the second half of the eighteenth century. Prior to that time, it was sailors who supplied descriptions and evidence of anomalous sea waves and of other dangers associated with the sea that they encountered.

A great Aegean tsunami in the fifth century BC, as told by the ancient Greek historians

Thucydides (fifth century BC) records that in the summer of 426 BC, an earthquake struck the Isthmus of Corinth area and the city of Orobiae in

Figure 10.26 Contemporary depiction of the tsunami in the Straits of Messina (southern Italy) that occurred in March 1783.

northwest Euboea. The earthquake itself was accompanied by an extensive seismic sea wave, which affected Orobiae, the island of Atalante off the coast of Locris and the Aegean island of Peparethus:

> During the following summer, the Peloponnesians and their allies, led by Agis son of Archidamus, king of Sparta, advanced as far as the Isthmus, with the intention of invading Attica. But a great many earthquakes occurred, causing them to turn back again, and no invasion took place. It was at about the time of the earthquakes that the sea at Orobiae in Euboea receded from what was then the shore line, only to rise up again in a great wave and engulf part of the city. Part of the water subsided and part continued to flood the land; so that what was once land is now sea. And those who did not manage to flee to high ground in time were killed. A similar flood occurred at the island of Atalante, off the coast of Opuntian Locris. It carried away part of the Athenian fort there, and wrecked one of two ships that had been drawn up on the shore. The sea also receded at Peparethus [off the northeast coast of Euboea], but there was no flood; and an earthquake knocked down part of the fort, the prytaneum and a few houses. In my opinion, the cause of the phenomenon was this: where the earthquake was most violent, the sea receded and was then pushed back with even greater violence, thus bringing about a flood. Such a thing would not have happened without an earthquake. (Thucydides, 3.89.1–5.)

Demetrius of Callatis (end of the third century BC), source of the Strabo account, writes that the earthquake struck on both sides of the Gulf of Malia:

> *A large part of Echinus and Phalara and Heraclea Trachinea was reduced to ruins, but the settlement of Phalara was destroyed right to its foundations. Something similar also occurred at Lamia and Larissa. Scarphe, too, was razed to the ground, and no fewer than seventeen hundred people were buried in the ruins. There were more than half that number of victims at Thronium. A triple sea-wave rose up, one part of which was carried towards Tarphe and Thronium, a second to Thermopylae and a third into the plain as far as Daphnus in Phocis. River sources dried up for several days, and the Sperchius changed its course, making roads navigable. The Boagrius was carried down a different ravine, and many areas of Alope, Cynus and Opus were seriously damaged, while Oeum, the castle above Opus, was completely destroyed. Part of the walls collapsed at Elatea, and during the celebration of the Thesmophoria at Alponus, twenty-five girls ran up into one of the towers at the harbour to get a better view, and when it collapsed they too were thrown into the sea. It is also said that the central part of Atalante near Euboea was split open to the extent that ships could pass through, and some of the plains were flooded as far as twenty stades [c.4 km] inland, and a trireme was lifted out of the docks and deposited over the wall.*

It may be of interest to note that on 20 and 27 April 1894 (8 and 15 April in the Julian calendar, which was still used in Greece at that time), the same area of Locris was struck by two very severe earthquakes, whose effects were very similar to those of the 426 BC earthquake (Skuphos, 1894): in the region of the Gulf of Atalandi a peninsula became an island, just as happened in the same area in 426 BC. The historical sources are drawn from Guidoboni *et al.* (1994).

10.10.1 *The most famous and discussed tsunami in the Mediterranean: 21 July 365 AD*

The earthquake and tsunami of 21 July 365 AD which occurred in the central Mediterranean represents a particular historic case that for more than 40 years has stimulated and promoted discussions between historians, epigraphists and archaeologists, and more recently, also geophysicists. Many elements together make this event a special case from the analytical methodology point of view. When different disciplines communicate about the same event, it is never easy to find common ground from the different points of view. If you also add to this the fact that from the viewpoints of the various disciplines, there are often more hypotheses than certainties, and that the limit between the former and the latter is sometimes not clear or is rarely obvious, one might expect that the results will be discussed for a long time yet.

Why should such a situation be problematic for the tsunami of 365? Indeed, it is useful to analyse this case also to show that the choice between different hypotheses does not give us a certain result, but another hypothesis, which might be more or less supported by the data, according to the knowledge in play and the congruence of the data. This case illustrates that it is not easy to go forward in this type of research. Often the protagonists in the debate over the 365 event have been so impassioned by their ideas and their methods of analysis that a scientific discussion almost becomes a personal attack. To understand the situation, we need to distinguish between the contributions of the data that have arrived in this case from four different disciplines: *history*, *epigraphy*, *archaeology* and *geophysics*. If one of the disciplines lays claim to having some sort of primary interpretation over the others, this can make it impossible to render the data congruent and to understand with the necessary openness an event in the past that cannot be restructured without mediation across the competence of various disciplines.

The history of the AD 365 earthquake and tsunami

At dawn on 21 July 365, an important natural event occurred: an earthquake that resulted in a strong tsunami which involved the whole Mediterranean area. The precise date and the memory of the event has been transmitted through time by the written historical sources to come down to us. The sources that are independent and non-controversial amongst themselves were written between the end of the fourth and the sixth centuries. These include:

(1) Athanasius, Bishop of Alexandria (295–373 AD), in one of his *Festal Letters*, which have been preserved in a summary form in an eighth-century Syriac text (*Index to the Festal Letters*, no. 37).

(2) Ammianus Marcellinus (c.330–c.400), a major historian of late antiquity. Born in Antioch, he lived in Rome from 378, where he wrote his works in Latin.

(3) Jerome (c.347–419), one of the most important Fathers of the Latin Church. Born in Dalmatia, he studied in Rome and in 384 went to Palestine. He wrote of the earthquake and tsunami of 365 in three works:

(3a) in his continuation of the *Chronicon* of Eusebius of Caesarea (which he translated from Greek into Latin), written about 380;

(3b) in his *Commentary on Isaiah* (*Commentariorum in Esaiam libri*), recalling an episode from his childhood. He remembered an inhabitant of Areopolis (present-day Rabba, Jordan) referred to the fall of the walls of this city because of an earthquake.

According to Jerome, this earthquake should have occurred in concurrence with the tsunami of 365. The effects of the earthquake for Areopolis could however refer to another earthquake different from that of 365, which occurred in 363 in the area of Palestine;

(3c)　in his *Life of St Hilarion* (*Vita sancti Hilarionis*), he spoke of the effects of a tsunami in the locality of Epidaurus (present-day Cavtat, Croatia; 15 km south of Dubrovnik).

(4)　*Consularia Constantinopolitana* (fifth century), a Latin chronicle that lists the names of Roman consuls from 509 BC to AD 468, with numerous historical notes inserted; this information comes from the end of the fourth century, and it refers only to the tsunami.

(5)　*Fasti Vindobonenses priores* and *Fasti Vindobonenses posteriores* (late sixth–seventh century), two Latin chronicles which (as with source 4) only noted the tsunami; for both of these sources the information probably comes from a common source of the fourth century. In the *Fasti Vindobonenses posteriores* the tsunami was incorrectly dated as 21 July 363.

(6)　John Cassian (c.360–after 430), born in Scythia and died in Rome. First a monk and then a bishop in Egypt. It is likely that his description of the Nile Delta region in AD 399 was based on personal experience (*Conlationes*, 11.3).

(7)　Socrates Scholasticus (c.380–c.440), a Byzantine historian, born in Constantinople. His *Ecclesiastical History* (*Historia ecclesiastica*) in seven books continues the works of Eusebius from 305 to the year 439.

(8)　Sozomen (first half of the fifth century), a Byzantine historian and protagonist of the great debates between the then predominant Christian culture and an embattled pagan culture. In Constantinople, around 443–450, he wrote an *Ecclesiastical History*, in nine books, which covered the years between 323 and 425. Sozomen incorrectly dated the earthquake and tsunami to the times of Emperor Julian (361–363).

(9)　A Coptic text of the sixth century written by Constantine, Bishop of Assiut, in Egypt (*Encomia in Athanasium duo*).

Some of these main sources provide us with three different views of the tsunami effects at Alexandria and in the area of the Nile Delta:

(i)　descriptions of the phenomenon (sources 1 and 2; see also Kelly, 2004);

(ii)　commemoration of the event as a disaster, giving its place in time by means of a religious rite known as the day of 'Fear' or 'Terror' (sources 8 and 9);

(iii) a description of permanent effects on the natural environment in the
        eastern part of the Nile Delta, written about 30 years after the
        tsunami (source 6).

Although the details are sketchy, one thing that is clear is that the sea rolled
back, exposing the seabed, at which point people could see 'various kinds of
animals stuck in the mud, and wide valleys and mountains'. Many ships were
stranded and people walked about freely on the seabed 'to pick up the fish'.
Then the waters rose up again and rushed violently onto the land, causing
many buildings to collapse. At Alexandria, boats were washed out of the water
onto the rooftops.

Furthermore, one can form an idea of what took place in the Nile Delta
from an eyewitness account. In telling of his travels to Panephysis (near present-
day El Manzala) John Cassian (source 6) wrote:

> Its lands have been covered by the sea, like most of the surrounding area, which
> was previously so fertile [. . .] The sea was disturbed by a sudden earthquake
> and broke its bounds, destroying almost every village and covering what used
> to be fertile lands with salt marshes [. . .]. In these places the many towns
> perched on the higher hills were abandoned by their inhabitants, and the hills
> were turned into islands by the flood, thus providing the solitude desired by
> holy men seeking seclusion.

In breaking through the sandbanks, the sea had invaded the area of Panephysis,
transforming it into a lagoon where there must have been a depression.

The two controversial sources that remain at the centre of discussions
between scholars are Libanius and Zosimus:

(1) Libanius (314–c.393), a Greek orator and rhetorician from Antioch who
      was a close friend of Emperor Julian. His numerous works passed to us
      as 64 speeches, 51 declamations and 1607 letters, and they are of great
      importance in our reconstruction of the cultural life of the eastern
      Roman Emperor in the fourth century AD. One section (292–293) of
      the Epitaph for the death of Emperor Julian (Oration 18) contains
      numerous references to places and areas affected by seismic events
      and other natural disasters that hit the Mediterranean basin;
      according to this text, many cities in Palestine, all of those in Libya,
      most in Sicily, all in the Greek islands except one (Athens), Nicaea, and
      finally the 'loveliest of our cities' (referring maybe to Nicomedia or to
      Antioch). The dating of this oration, and consequently of the seismic
      phenomena recorded therein, is at the centre of a lively debate
      between the specialists that has been going on for tens of years.

According to Jacques and Bousquet (1984, p. 433), and in particular according to the analysis of Henry (1985, pp. 60–61), Libanius wrote the *Epitaph* for Julian in the first months of 365 AD, that is, *before* the earthquake and the famous tsunami of 21 July 365; the tsunami is not explicitly recorded by Libanius. The final historical event mentioned by Libanius was the barbarian invasion of Goths, Sarmatians and Celts that crossed the Rhine, an episode datable to the start of January 365, and which thus represents the latest *terminus post quem* for the dating of the *Epitaph*. According to this analysis, the *Epitaph* would therefore precede the events of 21 July 365, by about six months. This interpretation was also supported by Fatouros *et al.* (2002, pp. 132–133). Nuffelen (2006), however, recently proposed a dating of the *Epitaph* to after October 368 (so after the earthquake that hit Nicaea on 11 October of that year), supporting a hypothesis formulate previously by Sievers (1868, p. 253) and by Bidez (1930, p. 336), and then discussed. An intermediate dating is supported by Felgentreu (2004), who indicated that *Oration* 18 was written around the middle of AD 366. According to the interpretation of Henry (1985, pp. 38 and 59), the earthquakes mentioned in the *Epitaph* of Libanius were considered as portents (*omina*) of the death of Julian, and therefore should be dated before 26 June 363, the day the Emperor Julian died, following his injury reported during the military expedition against the Persians.

(2) Zosimus (second half of the fifth to the beginning of the sixth century), a Byzantine historian, wrote his *New History* (*Historia Nova*) in six books, which went as far as AD 410. In one section of this work (4.18.1–2), somewhat similar to the above-mentioned section of Libanius, Zosimus mentioned strong earthquakes that hit Crete, the Peloponnese and the rest of Greece; Athens and Attica were, however, spared. Zosimus also does not refer to the effects of a tsunami. The earthquakes were dated to after the death of Emperor Valentinian I, i.e. after 17 November 375. The chronology of Zosimus is any case rather problematic: in another section of his work (5.6.3) which again briefly deals with earthquakes in Greece, we find that he wrote that they occurred 'during the rule of Valentinian'. Inevitably, scholars are also divided regarding the interpretation of the controversial chronology of Zosimus. Jacques and Bousquet (1984, p. 436) take the date after the death of Valentinian I given by Zosimus as being correct. Henry (1985, p. 48), however, indicates that Zosimus, a pagan historian who was very hostile towards Christianity, altered the dates of the earthquakes. Henry therefore believed that the earthquake in Greece

mentioned by Zosimus (and Libanius) must be dated to before June 363, and that the reference to Crete can probably be linked to the earthquake of 21 July 365.

Ex silentio historical sources

Lepelley (1984, 1994) makes an important point about northwest Africa that deserves our consideration. He points out that we have an exceptionally rich corpus of literary and epigraphic information for the fourth century, which should allow us to identify direct or indirect references to any damage that was caused by the earthquake of AD 365. There are two particular literary sources which Lepelley uses to show, *ex silentio*, that northwest Africa did not suffer any damage from this earthquake: a work written in 366–367 by Optatus, Bishop of Milevis (present-day Mila in Algeria), against the Donatist schismatics, and the *Confessions* of Saint Augustine, who was living at Thagaste at the time (present-day Souk-Ahras in Algeria). This latter provides a very accurate diary of events.

Epigraphy on the AD 365 earthquake

This field of study, wrongly considered as secondary to history and archaeology, joined the debate on the effects of the earthquake of 365 following a series of inscriptions from north Africa, where the construction and restorations mentioned that are datable between 364 and 378 have been positioned in relation to the earthquake of 365 (Beschaouch, 1975; Rebuffat, 1980; Di Vita, 1982). Also in this case, the criticism of Lepelley (1984) was substantially for exclusion on the basis of the effects of 365. Lepelley examined 76 inscriptions relating to the whole of north Africa that mentioned construction or restoration work of public municipal buildings that were dated between 364 and 383, i.e. between the start of the reign of Valentinian I (February 364) and the death of Gratian (August 383). His critical examination led him firmly to exclude the possibility that these inscriptions can be used as evidence of damage resulting from the AD 365 earthquake, as he believed that the construction work referred to in them can be explained in other ways. In particular, there were the propitious laws coming from Emperor Julian in 362 that saw the restitution of regular taxes (*vectigalia*) to the various cities that had to be used for the maintenance and restoration of public buildings. For Numidia (now northern Algeria), historians of later antiquity noted the intense building activities carried out by the Governor Publilius Caeionius Caecina Albinus in the period 364–367. Analogous critical considerations were developed by Blanchard-Lemée (1984) regarding the presumed seismic damage suffered by the locality of Cuicul (Djemila, 35 km northeast of Sétif, Algeria). A Greek Christian inscription found in the necropolis of Cyrene (today close to Shahhat, in northern Libya) might also be associated with the 365 earthquake (see Bacchielli, 1995), despite problematic dating (see Chapter 3).

Archaeology of the AD 365 earthquake

It is well known that many archaeologists have explained collapses that they have discovered as having been caused by this earthquake, and this game of 'hunt the earthquake' has become particularly popular in recent years. It is therefore worth tracing the archaeological history of the earthquake, now that some of the arguments have been settled.

If we look at the bibliography of research into the archaeological aspects of this earthquake of 21 July 365, we notice certain salient characteristics. In particular, there was a radical turning point in research in 1964, when Antonino Di Vita expressed the cautious opinion that the damage to Mausoleum B at Sabratha had been caused not so much by Austoriani raids (as had been thought until then) as by the 365 earthquake (Di Vita, 1990). It is to Di Vita's credit that he used this case to make possible the initiation of wide-ranging discussions among archaeologists and historians, for he was the first to indicate the extraordinary importance of archaeological data as sources for historical seismology. When considering in general the archaeological aspects of the proposal, the following points must be kept in mind:

(1) The chronological data provided by this archaeological research do not permit a dating *ad annum*, even though the archaeological timescales are based on the year; they can determine the *post* or *ante quem* limits, and therefore the chronological span.

(2) Even though the chronological data may be sound in themselves, they may be conditioned by the interpretation of the archaeologist or the person who published the excavation (as happened in the case of Kenrick (1986), who altered the interpretation of those who carried out the excavation).

(3) A substantial change in archaeological excavation methodology took place in the 1960s.

After Di Vita had made his initial suggestion, he himself tracked down evidence of the earthquake in other cities, and he was followed by many other scholars, who on the one hand provided chronological evidence of a quite satisfactory kind, and on the other claimed to find earthquake damage where in fact there was no evidence of destruction by natural catastrophe at all. Excavations at Gortyna (in the southern–central part of Crete) directed by Di Vita revealed collapses of important structures that were datable to about 365 (Guidoboni et al., 1994, p. 272). Later, Stiros and Papageorgiou (2001) analysed archaeological excavations at Kissamos, which is in the western area of Crete that experienced an uplift of around 9 m (see below). Collapses identified in the archaeological excavations of Tripolitania, northern Libya, were also attributed to effects of the earthquake of 365 (Goodchild, 1966–67; Bacchielli, 1995).

Geophysics of the AD 365 earthquake and tsunami

The uplift of the western end of Crete (which reached a maximum of around 10 m), which had already been reported on by Captain Thomas Spratt in 1851 (Spratt, 1865), opened a new round of interest. This started from the research of Pirazzoli and Thommeret (1977), and was then developed further in Pirazzoli et al. (1982, 1992, 1996). Did this awakening explain the historic scenario?

From recent modelling of the tsunami of 365, two interpretations have been published that are not so very different: Lorito et al. (2008) and Shaw et al. (2008). Both of these studies assumed that the uplift of the western end of Crete was co-seismic; the earthquake and the uplift happened therefore 'instantaneously'. In the first study, which concentrated on a simulation of the risk scenario of the tsunami for southern Italy, the earthquake of 365 was considered the 'maximum credible earthquake' for the Hellenic Arc source zone, one of the three potential source zones considered in this study. Lorito et al. (2008) hypothesized that the fault that generated the earthquake and the tsunami of 365 had a length of 130 km and a down-dip width of 86 km; the estimated magnitude (M_w) was 8.4. With these elements at hand, the authors worked out a predictive scenario for a possible tsunami.

The tsunami simulated by Lorito et al. (2008) predicted waves of a height greater than 1 m (with the largest peaks of 5 m at some non-specified places) along the coasts of northern Africa. Waves up to about 2 m would reach the coastal areas around Syracuse (eastern Sicily) and the Catanzaro coast (Calabria). For Egypt, an area for which we have available descriptions of the effects from written sources, Lorito et al. (2008) did not provide quantitative estimates, but indicated only that 'probably as a result of edge waves, significant energy was trapped and carried along the coast of Egypt'.

Shaw et al. (2008) described a model of the tsunami of 365 again starting from the uplift of western Crete of up to 10 m, which the authors believed occurred concomitant to the 365 earthquake on the basis of newly calibrated carbon-14 dates from accelerator mass spectrometry. Previously, Price et al. (2002) had proposed a different dating of this uplift of western Crete, as around AD 480–500, for which reason it could not be considered to be directly linked to the earthquake of 365.

In contrast to Lorito et al. (2008), Shaw et al. (2008) believed that the fault that generated the earthquake and tsunami of 365 was not located in the subduction interface beneath Crete, but on a fault dipping within the overriding plate of 100 km in length and 90 ± 15 km wide, and the magnitude estimate was put at 8.4 ± 0.1.

Regarding Alexandria, the simulation of Shaw *et al.* (2008) showed that off the shore of the city waves were generated that reached ~0.6 m in the open ocean. However, because of the many non-linear effects of the nearshore bathymetry, it would be difficult to calculate the run-up in ancient Alexandria, also considering the change in the local conditions since 365. Anyway, the authors assumed according to analogy with the tsunami of Sumatra in 2004, that the onshore effects would be devastating.

These new studies provide important progress for the geophysical understanding of the event of 365; nevertheless, the propagation models of the tsunami must be improved in future in order to be able to simulate more accurately the effects of the run-up in a single locality. This will allow more accurate comparisons with the descriptions from the historical sources, which are necessarily limited and partial, and to finally obtain more complete knowledge of this tsunami from late antiquity, which is still important today for us to understand the hazard levels of the Mediterranean from the point of view of possible tsunamis.

Finally, it is perhaps relevant to remember that only one other earthquake in Crete, from our knowledge base of today, caused a tsunami that was very similar to that of 365; this occurred about a thousand years later, to be precise on 8 August 1303 (Guidoboni and Comastri, 1997, 2005). In this case again, the earthquake was located to Crete, and in particular to its central–eastern area. The effects were widely witnessed in Venetian sources (at that time, Venice had dominance over the island). The tsunami that was generated hit Alexandria in particular, as given credence by numerous and authoritative Arab sources. Another large and destructive earthquake in Crete occurred on 12 October 1856 (Ambraseys *et al.*, 1994); however, in this last case, no tsunami hit Alexandria. It is this overall picture that needs a unifying explanation, which probably remains to be formulated.

In conclusion: the debate on the historic effects of the earthquake and tsunami of 365 is polarized between two prevalent positions:

• Guidoboni *et al.* (1994, pp. 267–274) presented a critical framework of the written, epigraphic and archaeological sources, and chose for themselves a picture of the earthquake as restricted to Crete, with the effects of the tsunami being towards the Egyptian coast, to the south of Crete, as in Alexandria, and also more to the northeast, with the lagoon of El Manzala; a strong involvement was also hypothesized for the coast of Libya, although this was not specifically supported by the written sources, with few effects in the central Adriatic. For Sicily, relevant shaking effects can be excluded, and therefore damage to buildings. These indications are confirmed by the archaeological study regarding an earthquake in the Straits of Messina that occurred halfway through the fourth century AD, not long before that of 365 (Guidoboni et al., 2000).

- Stiros (2001) presented a revision of the historical, archaeological and epigraphic sources, and of the geophysical knowledge of the area of Crete and opted for the 'universal catastrophe', which as a result of various aspects took the critical debate back to ten years previously. He outlined a framework of effects resulting from the shaking and the tsunami, spread across the whole of the Mediterranean, including north Africa, Sicily and the Adriatic. This position he has taken is in favour of the most catastrophic evaluation that has arisen from the study of this event over the last 50 years. Stiros interpreted the event of 365 as exceptional and anomolous.

If it is assumed that no single discipline has the primary interpretation, the problem that remains is the unifying of the various data. After a few tens of years of critical history, the aspect that still remains to be discussed could well be the geophysical interpretation. Are the faults indicated by Lorito et al. (2008) and in Shaw et al. (2008) the only possibilities? Has this concentration of the attention exclusively on the uplift of western Crete perhaps obscured other aspects of the problem? To doubt, and to discuss such doubts, is necessary for the continual potential revisions of the scientific data, which cannot be a result of a preconceived position.

10.11   Earthquake effects on a regional scale: outlining a complex seismic scenario

A *seismic scenario* represents the assembly of different elements (analysed in this chapter) that contribute to highlighting a wide range of effects rooted in a well-characterized historical context. The creation of a seismic scenario is the destination point for research into historical seismology and should not be delegated to people from other disciplinary sectors (geologists, geophysicists, seismologists or others), and certainly in any case to people external to the research performed. Indeed, the historical seismologist is not just a historian but a historian who also has a profound knowledge of the physics of the earthquake.

He or she must know which questions need answering when outlining a seismic scenario that can be used in the scientific field.

As a historian, the researcher in historical seismology is capable of collecting and selecting the sources, interpreting them, and relating them to the economic, demographic, cultural and administrative characteristics of the area being examined and of the historical period of the sources. From this point of view, historical seismology requires specializations for periods and areas to maximize the available knowledge that can be brought to bear on the research. The historical seismologist knows how to use some basic tools that allow him or her to use macroseismic intensity scales and thus to classify the seismic effects with common and rational criteria (on the use of the intensity scales, see in particular Chapter 12). In the phase of assembly of the data, each inhabited site historically identified must be (1) located with precise geographical coordinates determined from the qualitative information provided by the historical research, and (2) classified with a degree of the intensity of the effects. The density of this network of classified sites is nearly always directly proportional to the extensiveness and the depth of the basic historical research. Other elements not directly included in the descriptions of the different levels of macroseismic intensity also form part of a seismic scenario, such as (3) those environmental effects, especially those that can be located and mapped; (4) the inhabited centres abandoned after an earthquake, as an element of the anthropic impact on a given society.

The operation of synthesis of the data from historical seismology research is nearly always depicted by thematic maps, but these maps are not the final result of the historical seismology research. Indeed, thanks to the ease with which today one can organize a computer database, all historical research (the exegesis of the sources, the textual criticism and the knowledge of the historical contexts) can be archived and made available in text format as an indispensable medium for reading and understanding an historical seismic scenario and its parameters. A scenario of a past earthquake, which in its most immediate aspect can be understood as a simple map and/or a list of sites evaluated with degrees of intensity, is in actual fact a series of informative layers from which one can infer the quality of the data themselves. This accumulation of information ultimately is intended to be used in geophysical and seismotectonic studies, together with other scientific data (including instrumental data). Many seismic scenarios relating to the same area form a sort of seismic *anamnesis*. Examples of reconstructed seismic scenarios are depicted in Figures 10.27 to 10.29.

Site effects

With the term *site effects* we refer to localized anomalies in the seismic effects in a given location, smaller areas of either particularly intense or particularly mitigated earthquake felt or damage reports within a much larger

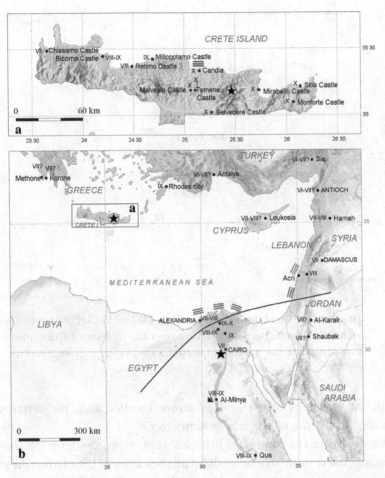

Figure 10.27 The earthquake at Crete on 8 August 1303. (a) This map locates the epicentre (shown by the star) estimated on the basis of the primary historical data. It also shows the macroseismic intensity values on the island of Crete. (b) Map of earthquake effects throughout the Mediterranean ascribed to the earthquake in 1303. The line divides the felt reports into areas of the two different earthquakes, according to El-Sayed *et al.* (2000). This figure is from Guidoboni and Comastri (2005).

surrounding region where more mutually consistent earthquake effects were experienced. Except in cases where structures are built directly on the fault that was active during an earthquake, earthquake damage is caused by the local ground-shaking due to the passage of the seismic waves generated on the earthquake fault. The seismic waves that reach the ground surface at a site can be strongly affected by the seismological properties of the rock and soils directly beneath the site. These seismological properties can alter the time duration, the strength and the predominant frequencies of the ground-shaking relative to other sites with different soil and rock properties. In areas where there are

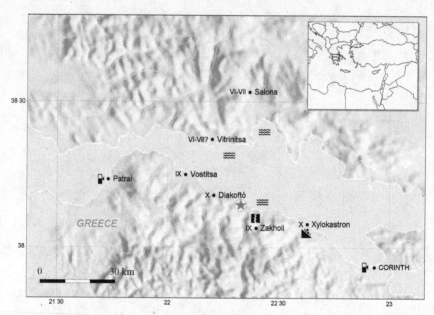

Figure 10.28 Map of the earthquake of July 1402 in the Gulf of Corinth (Greece) with the effects on the natural environment also represented (from Guidoboni and Comastri, 2005). The star shows the estimated epicentre of this earthquake.

significant variations in the geology across a settled area, the corresponding variations in the site effects can be quite strong.

When used in the context of historical seismology, the term *site effect* most frequently refers to an area of localized amplification of the earthquake ground-shaking, with the concomitant damage that stronger shaking can cause. This often occurs at localities where the ground motion has been modified due to some particular forms of topographic relief, such as on the summits of hills or mountains. In such cases the local effects are more concentrated and more severe than those in the nearby inhabited centres, at times even by one or more degrees of macroseismic intensity. Unfortunately, it is not always easy to recognize the site effects for the earthquakes of the past. Quite often the land that is nestled on mountains may undergo landslides or the slipping of soil or rocks caused by an earthquake. While these are elements that aggravate the damage due to an earthquake, they are distinct from the site effects as defined here. A similar problem differentiating whether or not site effects came into play in an earthquake arise when one deals with direct sources that describe the damaged structures as a 'heap of rubble'. Was such a report describing the collapse of a poorly constructed building, a localized site effect, or the effects of particularly strong earthquake shaking? With few sources at one's disposal or if only a quick

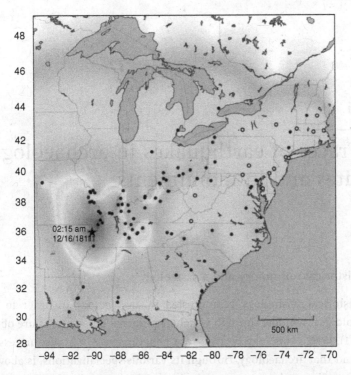

PERCEIVED SHAKING	Not felt	Weak	Leight	Moderate	Strong	Very strong	Servere	Violent	Extreme
POTENTIAL DAMAGE	None	None	None	Very light	Light	Moderate	Moderate/Heavy	Heavy	Very heavy
PEAK ACC.(%G)	< .17	.17-1.4	14-3.9	3.9-9.2	9.2-18	18-34	34-65	65-124	>124
PEAK VEL.(cm/s)	< 0.1	0.1-1.1	1.1-3.4	3.4-8.1	8.1-16	16-31	31-60	60-116	>116
INSTRUMENTAL INTENSITY	I	II-III	IV	V	VI	VII	VIII	IX	X+

Figure 10.29 One of the strong earthquakes at New Madrid in 1811 (from Hough, 2004). The MMI values are based on a reinterpretation of the original felt reports from towns that reported ground-shaking. The filled circles indicate the locations where MMI values are available, while open circles indicate locations where Hough (2004) assigned a value of unfelt.

historical analysis is carried out, it may not be possible to distinguish between the different causes of the damage. Hence, it is worthwhile, when investigating whether destructive local effects are evidenced, to ensure that the sources attesting to them are as numerous, authoritative and independent as possible, so as to give the effects assessment a reasonable degree of certainty. In order to better highlight suspected site effects it can also be useful to compare, if possible, the effects of various earthquakes in the same places. If significantly greater damage than expected is reported at some localities due to site effects, a similar pattern of reports should be observed for other earthquakes.

11

Traces of earthquakes in archaeological sites and in monuments

11.1   Historical seismology and archaeology

Historical seismology is interested in archaeology in order to widen the chronological window through which earthquake observations are obtained (Figure 11.1). While this may be true in a general sense, it is also necessary to bear in mind that archaeology, through its various specializations, is above all a method of reading and interpreting the history of geographical regions, of built structures (or their remaining ruins), and of manmade artefacts. Archaeological sources thus potentially allow one to become acquainted with an unwritten history, which can also highlight, amongst the many elements that can be brought out, direct traces of destructive events such as earthquakes and tsunamis.

Interest in the relationship between archaeology and earthquakes is not a recent development. As far back as the mid nineteenth century the convergence of these two different topics was already attracting attention in the Mediterranean area owing to the frequency and intensity of the earthquakes and to the extensive presence of ancient remains. Monuments were the first kind of archaeological objects that were studied from this standpoint. De Rossi (1874) and Lanciani (1918) did so for Rome, Willis (1928) for Palestine and Sieberg (1932b) for ancient Egypt. But a significant growth in interest only occurred starting in the early 1970s. A specific seismological orientation in archaeological studies began with the research by Karcz and Kafri (1978, 1981) and Nur and Reches (1979) for Israel. This turning point established *archaeoseismology* as a sector potentially capable of answering questions raised by seismology (Nur *et al.*, 1989; Stiros and Jones, 1996; Guidoboni *et al.*, 2000; Nur, 2002). But in spite of the development of this field, archaeological literature on earthquakes has remained relatively indifferent to the methods and objectives of this subject, although there is no

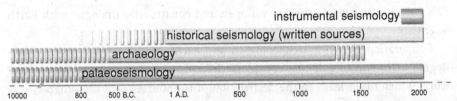

Figure 11.1 Time frames for which information concerning past earthquakes may be recovered based on research by different disciplines. Archaeology allows for a widening of the time-frame for observations of past earthquakes relative to that for historical seismology.

shortage of contributions of an excellent standard. In our opinion, an authoritative and shared multidisciplinary field of historical seismology using archaeological data has not yet truly arisen. There are still different sets of scientific interests and points of view that characterise the various geographic areas of research into this topic as well as viewpoints of the authors who have written on this topic (MacGuire *et al.*, 2000).

11.1.1 Why is it difficult to use archaeological sources in historical seismology?

Using archaeological sources to trace destructive earthquakes is not an easy matter. One of the main reasons for this arises from the methods of archaeological work itself, which have converged on homogeneous research elements of good accuracy only in the past twenty years. Unfortunately, archaeological data may be hurried in their acquisition and may be affected by elements that are not easy to control, often due to contingent reasons such as an urgency to act before a site is covered or destroyed, a paucity of resources to support the research work, etc. This is particularly true for the discarded chronological layers at an archaeological site, because they are deemed to be of little interest from a current research perspective. This means that if an archaeologist is interested in a particular historical period, for example the classical Roman period, he or she may well be unlikely to analyse with the same care and attention the remains from preceding or subsequent time periods. Only by the application of a rigorous stratigraphic method at all levels of an archaeological site can one guarantee that the data have been systematically acquired using correct procedures. One should not overlook the fact that in the Mediterranean area until the 1980s stratigraphic excavation was not as widespread as it is today because the medieval layer was considered by classical archaeologists to be of virtually no interest at all. Furthermore, the culture of classically trained archaeologists can at times make it difficult to plan a work strategy oriented towards surveying data of interest for another scientific discipline such as seismology, and this culture can also discourage an open and constructive dialogue with Earth scientists.

Cooperation between archaeologists and earth scientists takes on a certain degree of importance if one considers the nature itself of the archaeological work, which often leads to the destruction of the sources that are progressively being brought to light and examined. This is because the excavation work proceeds by removing contiguous chronological layers, moving backwards in time back to the layers where there are no traces of human presence. The stratigraphy and the surveys (drawings, notes and films or digital photographs) thus become the only remains of the layers that were progressively identified and removed. Hence, the capacity to ask oneself questions *before* the excavations very much influences the likelihood of detecting data of relevance for historical seismology.

From the more strictly archaeological standpoint, one needs to bear in mind that the indicators of seismic activity provide the simultaneous clues and ambiguities typical of every piece of archaeological evidence. An explanation of how to find a layer of a particular age from amongst the full extent of the deposits can be hypothesized and inferred, but it cannot be proved deterministically, as will be shown in more detail below. Sometimes archaeologists have resorted to the hypothesis of seismic activity to explain certain situations they have uncovered in their archaeological records. There can be cases in which there were some macro-indicators of a catastrophic event, such as collapses, perhaps even with the presence of skeletons, detected in prestigious or important structures such as villas, thermal spas, places of worship, theatres or city walls. An earthquake could be one explanation of how this archaeological situation formed.

In most of the cases attested to in the archaeological literature where a seismic event is conjectured, the earthquake is situated in a chronological phase subsequent to the abandonment of a site. Indeed, the indicators of seismic activity are obviously easier to identify where human action has not intervened to recover, reconstruct or restore the topographic and structural order of a damaged site (Figure 11.2). In this way, however, a whole set of ancient seismic evidence may be overlooked. For example, the data relating to rural settlements and to the poorer buildings, which characterize the organizational set-up of the outlying territories away from an urban centre, nearly always remain beyond the scope of an archaeological analysis. The different responses that ancient societies gave to a seismic disaster in terms of reconstruction have also escaped the attention of most archaeologists.

In the theoretical and field research formalized by Anglo-Saxon archaeology (*site formation processes*: Wood and Johnson, 1978; Butzer, 1982) the traces of a destructive earthquake are examined only as something which confuses the stratigraphy, and the only element surveyed is that of cracks in the ground.

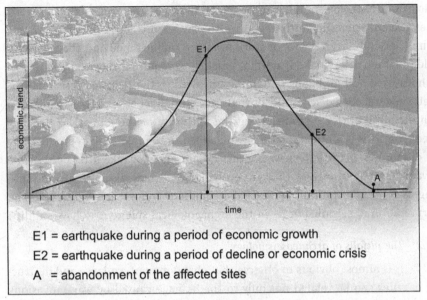

E1 = earthquake during a period of economic growth

E2 = earthquake during a period of decline or economic crisis

A = abandonment of the affected sites

Figure 11.2 The potential visibility of seismic effects at archaeological sites varies depending on when in the economic cycle of a settlement an earthquake took place. The highest likelihood that the effects of an earthquake can be observed in the archaeological remains is for situation A (abandonment of the site). An earthquake that occurred for situation E1 (earthquake during a period of economic growth) is least likely to be found at an archaeological site, while an earthquake for situation E2 (earthquake during a period of decline or economic crisis) is more likely than situation E1 but less likely than situation A to be observed.

Any cracks due to an earthquake can indeed subsequently be filled in by allochthonous material due to the activity of natural agents, usually water. The result, in strictly stratigraphic terms, consists of (1) the formation of deposits containing materials of secondary deposits; (2) the horizontal or vertical movements of material (both sedimentary and archaeological relics) in the stratification; and (3) the creation of possible false chronological and functional associations in the materials contained in the deposits. These are important effects in the archaeological field, which only a careful stratigraphic reading can detect. Such evidence – when not diagnosed and appropriately placed in the stratigraphic sequence – can divert the interpretative phase of the sequence. Unfortunately, archaeologists usually fail to arrange carefully and accurately the seismic indicators to be observed within the stratigraphy of a site. From the archaeological literature it emerges that this sort of methodological inattention (with the exception of a few cases) has seldom been bridged even by a multidisciplinary dialogue. This situation could be improved if the application of archaeological data to seismology was always carried out by a variety of specialists, such as geologists, archaeologists and historians, and in some cases even engineers, work-ing as a team. Scientific journals could also play a more active role by arranging for multidisciplinary reviews of work awaiting publication. And the reviewers could, for their part, make it normal practice to require a clear explanation of methods used and problems remaining unsolved in the reported research. It is a well-known fact that scientific journals sometimes act as a kind of standard of scientific authenticity by transferring the authority of the journal to the pub-lished data. This is all the more likely to happen in the case of data produced by new disciplines, because reviewers either do not know or take little heed of the relevant hermeneutic rules that are evolving when new disciplines come into being. A thorough multidisciplinary dialogue is therefore indispensable if the interpretation of seismic effects in the archaeological sphere is to have validity.

11.1.2 The pitfalls of archaeoseismology

It is almost obvious to observe that abandonment, collapse and destruc-tion in cities can be related not only to the damage caused by wars, invasions or political and social unrest, but also to the decay of the building stock owing to a lack of maintenance during periods of economic crisis or major cultural and/or political changes. Particularly dramatic, from the standpoint of the analy-sis of damage effects on structures, may be the traces of some types of warfaring technologies (such as catapults and, in the late medieval period, mines). Further-more, ancient restorations and reconstructions may be connected to the revival of a site after warlike destructive events or as a result of demographic expan-sion, variations in economic activity, functional changes owing to social factors, or changes in the roadways. Deformation or collapse of architectural structures may also be caused by the morphology of the ground (altimetry, regularity, etc.) coupled with geotechnical problems in the underlying soils. For example, the frequency of structural modifications detectable in an ancient monumental building (such as a theatre) may suggest a problem with lack of cohesion of the layers upon which its structural foundations are set or with instability of the fill material under and around the structure. Strong earthquake ground-shaking may accentuate static stability problems that were perhaps pre-existing in a structure. The cases of the theatres at Hierapolis in Turkey (De Bernardi Ferrero, 1993) and Stobi in Macedonia (Gebhard, 1996) are examples of this type of complex problem. The effects of earthquakes can also be confused with those of other natural phenomena, such as landslides and landslips, which can be caused by earthquake shaking but also can arise due to non-seismic causes such as excessive rainfall.

In spite of all these conceptual and practical pitfalls, it is possible that the effects of an earthquake can be positively identified and analysed at an

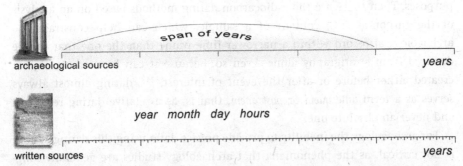

Figure 11.3 The time resolution of archaeological finds is usually no better than a span of several years, while ancient written historical can be used to resolve the day (and perhaps even time of day) when an earthquake or tsunami took place.

archaeological site. After such an identification has been reliably made, another problem usually must be tackled, namely finding an accurate date for the occurrence of the seismic event.

11.1.3   Dates: a key problem in earthquake identification

One of the core problems of archaeoseismology is that of dating the seismic or tsunami events that affected a site. Indeed, the traces of seismic activity at an archaeological site must be dated in order to make any sense for historical seismology and for the earthquake information to be usable in other research contexts. It is important to bear in mind that the written records and the archaeological sources are interpreted using two completely different timescales (Figure 11.3), due to the different nature of the records themselves. The written records may refer not only to the year, the month and the day in which the narrated event took place, but also to smaller fractions of time, such as *hours* and *minutes*. As accurate measurements of minutes have only been available for the past 200 years, generally it is the *hour* that is used as the smallest unit of time for historical dating over the long term.

An archaeological date is not based on an absolutely known calendar, but rather on a relative system of references, represented by the stratigraphic sequence in which the observed archaeological record is contained. Every piece of evidence at the archaeological site is constrained to a time that lies within a range *before* and *after* the event. This means that the archaeological dating can only be indicated as within a period lasting either *years*, *decades* or *centuries*, according to the anthropic contexts and the methods used. There are some records, such as ceramics, that can be used quite precisely as chronological indicators. Even coins can identify a restricted period of time when an event must have occurred, and they provide a valuable term *post quem* for dating purposes. Then there are the radiocarbon dating methods based on an analysis of the carbon-14 (^{14}C) content of a sample, which can help constrain the archaeological record within a narrower time frame than the one that can be deduced from stratigraphy alone. Even so, because it can be used on objects created either before or after the event of interest, ^{14}C dating almost always serves as a term *ante quem* or *post quem*, that is, as a relative dating reference and never an absolute one.

For archaeology the inability to provide precise dates using this dating system is not critical, as the phenomena that archaeology studies are generally long-lasting, such as the history of a site or a territory, the evolution of particular techniques (such as the extraction of metals) or trade routes, or changes in human settlements, etc. However, precise dates from the standard archaeological dating system become very critical when the events needing to be dated lasted just a few seconds or minutes, as earthquakes do. Indeed, in the space of years, decades or centuries, several destructive earthquakes may have occurred and affected an archaeological site.

The dating problem becomes even more complex and critical when archaeologists, seismologists or geophysicists date seismic traces found at different sites in very large archaeological territories where the sites are far from each other. In these cases, it is important that the dating of a seismic event be made in an absolute way (that is, by stating the year, the month and the day of the event). How is the conversion from relative to absolute dating undertaken? Generally, it is done by referring to contemporaneous written sources, that is, by combining the archaeological record with the dating of a clear-cut seismic event recalled by written records, which can be dated using a more accurately known timescale. The chances of grouping together the archaeological traces of earthquakes at different sites that are in actual fact independent into a single seismic event is very high. Conversely, the chance of using written sources to date a seismic event with archaeological evidence is very rare, as we shall see later on.

From an epistemological standpoint, a combinatory method correlating written and archaeological sources can lead to a sort of circular reasoning. This problem can arise when the written sources are used to provide an absolute chronology for the archaeological sources, while the archaeological information is presented as the effects caused by a precise historical earthquake, whose written sources, however, either do not specify the damage or else do so only partially or generically. The danger is that the archaeological sites with traces of collapses become interpolated into the seismic scenario of the ancient written records (which themselves are invariably incomplete in some way). When this happens there is a risk of attributing to a single earthquake attested to by the written sources the effects caused by several earthquakes that may have occurred in the space of several decades and with epicentres that may even be geographically rather far apart. This type of interpretation may lead to a series of scientific deductions that are completely mistaken.

The use of written sources for the absolute dating of the seismic effects of an archaeological site should thus be carried out with great caution. For the wholly exceptional case of Pompeii (see below) and perhaps for very few others in the Mediterranean area, one can use a direct relationship between archaeological records and historical sources. It is thus always worthwhile preserving the dating system of the archaeological sources (the time-frame) and, should it be necessary and relevant to the research, bring together the different types of sources (historical and archaeological), whilst keeping a prudent hermeneutic distance from the various explicit interpretative hypotheses.

Types of damage detectable from ancient restorations of buildings: the case of Pompeii after the earthquake in AD 62

The eruption of Vesuvius in AD 79 buried the cities of Pompeii, Herculaneum and Oplontis in southern Italy beneath 6–12 metres of ash and lava. After more than 250 years of archaeological research – indeed up to 1748 the memory itself of these cities had been lost – some priceless evidence has emerged. Today in Pompeii one can be moved by the emotion of seeing what a Roman city was like. One can not only see the houses, the courtyards, the temples, the squares and the casts of the people fleeing from the explosion of Vesuvius, but also the effects of a strong earthquake that had damaged Pompeii 17 years before the eruption. Hurriedly abandoned builders' tools and buckets of lime show a city rather like a building site, where people were busy working to repair damage and renovate houses. The earthquake on 5 February 62 had caused extensive collapses in Pompeii and in other centres of the western area of the Campania region. Seneca and Tacitus recall this earthquake, the former with great detail in the sixth book of the *Naturales quaestiones*, the latter more concisely in the *Annales* (15.22.1).

Pompeii is a case, perhaps the only one in the world, of the unquestionably correct use of historical and archaeological data. Indeed, there is no doubt that in the 17 years that had elapsed between the recorded earthquake and the eruption of Vesuvius no other destructive seismic event had struck the city. In this case the sources are both detailed and authoritative. Thus, one can effectively use a reading of the ancient building restorations (just finished or in progress in AD 79) to understand the effects of the earthquake in AD 62, together with what is contained in the written records. The archaeological observation in this case plays the role of a time capsule. Indeed, the damage and the restorations remained set in the volcano's ashes exactly as they were on that tragic day on the 24 August 79 when the eruption of Vesuvius buried the cities.

By observing the buildings brought to light by the archaeologists (Adam, 1989) one can notice that the cracks due to the earthquake were repaired with materials recovered from the ruins caused by the seismic event, something that today makes the repairs to the cracks easy to recognize. The reuse of the rubble was systematic. The collapsed walls had become heaps of rubble that were mixed together with the fragments of the roof tiles, and this rubble was subsequently recovered and reused. All of these fragments together represent a sort of tracing for recognizing the restorations. For example, when in the earthquake walls came apart and leaned at an angle that was not so great as to cause their collapse, they were not necessarily demolished but rather were often reinforced and brought back into use as part of the refurbished building. In many cases the wall surfaces were preserved in spite of deformations that were even considerable at times. New walls were built to be perfectly plumb, and so the joints between this new construction and the damaged pre-existing building elements can be easily recognized, especially where the deformation is more substantial. Deformations and weakening in the walls were repaired without resorting to demolition by supporting them with brick buttresses or earth embankments. The floor beams of upper floors were also supported in this way, and walls that did not lean too much were made more stable by giving them a new and larger surface area. Damaged or disconnected walls were given new masonry so as to strengthen the pre-existing structure and to avoid demolishing it (Figure 11.4).

One finds that in order to preserve a structure that had experienced earthquake damage, the safest means was thought to be to completely wall up the previous doors and windows. Indeed, that is what one can see in many houses and shops in Pompeii (a custom later applied for centuries in similar situations). Such work can be seen both in the restored parts of structures after the seismic events and in the modifications that were made to structures owing to changes in ownership or in the use of a building. The restorations are most evident if the walls had not yet received, or else had lost, their outer wall coating that was meant to give them a uniform and pleasing finished appearance. A white plaster was applied as the final outer covering to eliminate all traces of changes to the walls. The corners of the walls were generally highly vulnerable to having sustained damage in the earthquake, something that is commonly observed even in modern buildings shaken by strong earthquakes. Seneca wrote about this, specifically with regard to Pompeii:

> We have seen the corners of buildings split open and then move back into place. There are some buildings, in fact, which are poorly set in their foundations and

Figure 11.4 The damage to the city of Pompeii due to the earthquake in AD 62 can be highlighted by using a computer to remove the very evident restorations. These two views show a corner of Orpheus's house (see the city plan VI, 14, 20) rebuilt after the earthquake (left image) using some of the original limestone blocks with layers of tiles and bricks. The image in the right shows the damage to the structure with the restoration work removed.

were made by negligent and slovenly builders, and yet have become firmer by being repeatedly shaken in an earthquake. (6.30.4–5).

After an earthquake, the corners of a building may be reconstructed in one of a few different ways, according to the inhabitants' economic resources. Most of the restorations involved the use of recycled materials. In other cases, the restorations were performed with greater care by inserting sturdy courses of bricks arranged in an orderly fashion and tied to the adjacent walls (Figure 11.5).

The columns, vertical supporting elements omnipresent in Roman architecture, did not undergo any direct damage due to the earthquake shaking in AD 62, but their shifting or toppling was the cause of other numerous collapses of the roofs of the peristyles and arcades. It should suffice to get some idea about this point by considering that the Forum of Pompeii, as can be seen today, is preserved exactly as it was in AD 79. A private builder had replaced the previous columns made out of volcanic tuff with columns made of brickwork, in which a variety of materials can be found.

Figure 11.5 Evidence of damage in Pompeii caused by the earthquake in AD 62 deduced from restorations to a building (left: the restored building; right, the building as it would have appeared without any restoration). The corner and side wall of a domus of Samnite times was reconstructed in *opus mixtum* and in *opus incertum* (ancient Roman construction techniques). The bond between the old and new walls would have been consolidated by plastering, but that had not yet been applied when the work was buried by the eruption of Vesuvius in AD 79 (Adam, 1989).

There were probably also many deaths due to the earthquake, and many families were probably made homeless as a result of the earthquake damage. Some persons – as is known from the ancient sources – emigrated because there was no economic aid for the reconstruction of the cities. Normally, the Emperor provided such aid, but the rebuilding of Rome following the great fire two years later (AD 64) and the construction of the *Domus Aurea* absorbed the financial resources of Nero's administration. The houses of Pompeii that were left uninhabited after the earthquake were in one way or another used by the survivors. One can also hypothesize that there were changes in the trades of the building owners, who might have become shopkeepers or craftsmen after the earthquake. In any case, an archaeological reading of the buildings brought to light that the city was going through a serious crisis, with changes in fortune or status, which the inhabitants responded to in various ways by providing different solutions as they adapted to the new conditions.

11.1.4   Isolated ancient collapses: how should this archaeological evidence be used?

There are now numerous attempts in the literature to associate to single buildings, to their remains, or to an entire site some seismic effects where earthquakes are suspected or hypothesized to have taken place. Generally speaking, some collapses (as well as landslides and ancient restorations) have been highlighted as having been caused by a seismic catastrophe. In order to be able

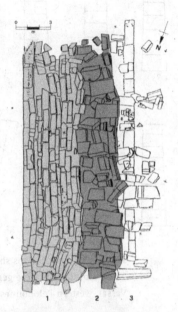

Figure 11.6 Hierapolis (Turkey): overhead photograph (left) and sketch map (right) of the collapse of the house of Frontinus caused by an earthquake in the early seventh century AD (from Guidoboni *et al.*, 1994).

to evaluate the scientific meaning of such isolated findings it is important to contextualize the evidence within the seismic history of the area being examined. If the archaeological indications of the traces of an earthquake concern an area with a high level of seismic activity, of which much is already known from historical seismology and present-day geoscience research, it is necessary to recognize that the archaeological finding might not add any particularly new scientific value to the understanding of the local earthquake activity. From the seismological standpoint the result may be a confirmation of the knowledge already acquired with other cognitive instruments, in particular from historical seismology (Figure 11.6).

If the archaeological evidence of possible earthquake effects concerns an area not believed to be seismic (i.e. is silent or has a very low level of seismicity), it is clear that the documented damage in the archaeological record is a fact that, if appropriately checked and verified, can cause a reinterpretation of the seismotectonics of the area where the damage is located (see Galli and Galadini (2003) for the collapse of the temple of Hercules at Campochiaro (Molise, southern Italy), and the case of the wall of Vaste in Apulia). But even in cases in which

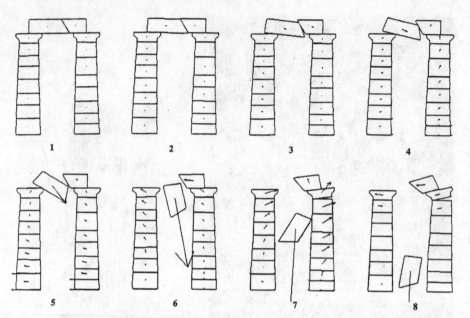

Figure 11.7 Simulations showing the seismic response of megalithic monuments. These simulations were generated to study the damage to the temple of Apollo at Vassai, western Peloponnese, Greece (from Papastamatiou and Psycharis, 1996).

the archaeological damage data come from very seismically active areas, this generalization does not mean that the observations of possible earthquake damage are unimportant. Indeed, the analysis of collapses or landslides due to earthquakes may imply a seismic source located very close to the archaeological investigation site (which in itself is a precise datum), and this may help improve the knowledge of the seismic indicators in the archaeological record and thus help direct subsequent analyses. In particular, if such data can be processed quantitatively (for example, with numerical engineering simulations of the ground motions that may have caused the observed damage), one may be able to generate models, albeit simplified, of the seismic responses of ancient structures (Papastamatiou and Psycharis, 1996) (Figure 11.7). Such structural analyses may help address the problem of how to preserve ancient monuments in archaeological areas of high seismic hazard (Figures 11.8 and 11.9).

A great collapse in a silent seismic zone: the case of Vaste
(Salento, southern Apulia, Italy)

The Salento area (i.e. the Italian peninsula that extends towards the Ionian Sea) is traditionally considered by geologists as one of the least seismically active areas in Italy. This characteristic was included in all the seismic norms of the nineteenth century, as well as by the national reclassification in 2004.

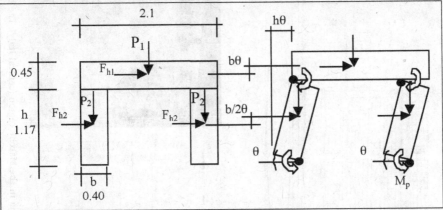

Figure 11.8 A deformed tomb atop a necropolis in Hierapolis (Turkey). The top photographs show the cracking and dislocations of the stones due to earthquake shaking, and the bottom sketches show calculations of the response of the tomb to a strong lateral seismic motion (elaboration by Professor Paolo Riva, in Guidoboni, 2002).

However, during that reclassification a great collapse was identified and studied, which reopened the discussion on the interpretation of the possibility of strong earthquakes in this area. The substantial archaeological evidence has catalysed recent geological observations, as well as the formulation of new seismotectonic models for the area, which suggest that even this Italian region might be subject to destructive earthquakes, although they might have long return periods. The Italian seismic hazard norms show a low potential hazard for the towns and villages of the Salento, which is thus one of the very few areas of Italy to be officially considered as almost aseismic (Sardinia is another).

The new data come from the archaeological site of Vaste (Lecce). This site was an important centre for the Messapic civilization, an ancient population that had settled in Apulia in the Iron Age and is attested to up until the Roman conquest. The period when they flourished was between the sixth and the third century BC. Very important elements for the ancient Vaste site were fortification

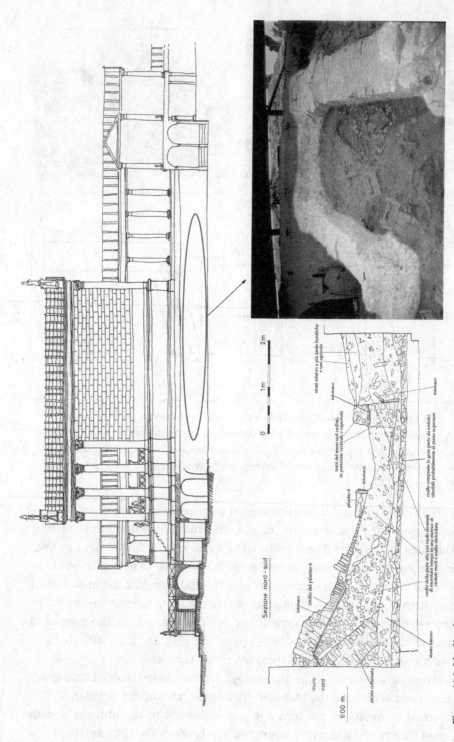

Figure 11.9 Idealized reconstruction (top) of the temple of Urbisaglia in the Marche region (central Italy). The bottom right and left panels show a cross-section and photograph of the great collapse of this structure, evidence that emerged from the excavations in 1998.

walls, built between the fourth and the third centuries BC. They were 3.35 km long and enclosed an area of about 77 hectares. These megalithic walls are up to 7 m thick at some points; they were about 6.5 m high and are made up of blocks measuring 1.50 × 0.30 × 0.60 m. The walls were surrounded by a 4-m-wide road that was kept free of buildings. Only on the eastern side of Vaste, that is in the seawards direction, was there a low wall consisting of 1-m-wide blocks designed to prevent the approach of siege-trains. Part of the collapse of the walls was observed in previous excavations (Lamboley, 1998), but damage that was previously unrecognized along other stretches of the eastern fortifications has been brought to light during the activities for the creation of the Archaeological Park of Vaste (scientific supervision by Francesco D'Andria, the director of the School of Specialization in Archaeology, University of Lecce).

The excavations performed in the period 2003–2004 have brought to light the collapse of the block walls for a length of over 20 m. As many as ten rows of blocks seem to be in a position of collapse that preserves, on the ground, their original upright arrangement. We have asked ourselves whether such a collapse may be congruent with a strong seismic event having its epicentre in the Salento (Figure 11.10). The topographical and archaeological context allows one to suggest a date for the collapse around the second century BC, when the Messapic cities were depopulated following the Roman conquest, although the structures had not yet been demolished in order to recycle the materials. The damaged wall face presents itself lying on the ground, as it could have been configured following a rigid rotation of the wall on its foundation blocks. Such a movement is compatible with the action of a significant horizontal push acting upon the wall. The cause of such a push could be linked to (1) the action of the wall's push on the face, (2) the action of the roots of plants that, over the centuries, have managed to take root in the protrusion adjacent to the wall, (3) human intervention and, lastly, (4) a seismic event.

Among the causes listed above, the only one capable of generating a collapse of the kind observed on site would seem to be a seismic event. Indeed, according to the engineers who analysed this site (Professor P. Riva of the Bergamo University and Professor P. Marini of the Milan University-Politecnico, in an unpublished report) the push that moved the stones may presumably have caused, besides the rotation, a sliding of the base of the stones. If this was the case, the most likely configuration of the displaced facing would be characterized by a stacking of the stone blocks near the base, but this is not observed. The action of roots has been discarded, as roots typically have a localized action, which would lead to a local opening of the wall face rather than a rigid rotation of a large part of it. Even human action has been reckoned to be highly unlikely, given the substantial wall mass. Finally, the possibility of the collapse

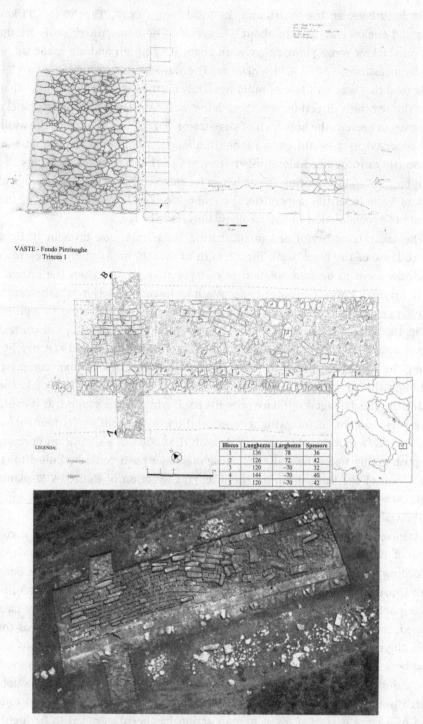

VASTE - Fondo Pizzinaghe
Trincea I

LEGENDA:

Blocco	Lunghezza	Larghezza	Spessore
1	136	78	36
2	126	72	42
3	120	~70	32
4	144	~70	40
5	120	~70	42

Figure 11.10 Three views of the collapsed great megalithic wall at Vaste (Apulia, southern Italy) from the fifth century BC. The top and middle panels show elevation and map sketches collapse, and the bottom is an overhead photograph of the site (F. D' Andria, Scuola di specializzazione in archeologia classica e medievale, Lecce University).

due the forces of gravity has been discarded, as such a collapse, not linked to the onset of a significant horizontal push, would have meant a stacking up of stone blocks near the base of the wall, without it being possible to identify the preferential direction of the fall. Hence, the seismic hypothesis remains. The analytical estimates made by the engineers is that the minimum value of the horizontal acceleration of the ground required to bring about the overturning of the wall face is equal to about 0.12–0.15 g.

Among the great variety of research techniques that have application to archaeoseismology is surveying comparatively slow seismotectonic phenomena, such as elevation changes of coastlines and slow movements on fault slopes (for example: Stiros, 1988; Papageorgiou and Stiros, 1996; Spondilis, 1996). There is also research addressed to determining the return periods of great earthquakes that can impact archaeoseismological investigations. When the area investigated preserves numerous traces of human presence for several millennia, then archaeoseismology becomes an element of palaeoseismological research. However, in these cases the search for archaeological evidence concerning great earthquakes must come to terms with the need to find many data that are compatible for the same area, that is to say, traces that can be interpreted as seismic effects and that are datable to the same time period. The imprecision of archaeological dating becomes a problem for palaeoseismological interpretations if reference must be made to excavations at different sites, since then it is likely impossible to obtain an absolute dating with sufficient precision to link to a single event the various pieces of seismic evidence. The minimum unit of time for archaeological dating is normally a few years to a decade, and more than one strong earthquake could have occurred in that time. Except in extremely unusual circumstances an archaeological investigation cannot distinguish a period of time of less than a year. It is also the case that a great earthquake (i.e. one with a magnitude above 7.5–8.0) will have widespread consequences, and hence cause significant damage at multiple sites. If such an event is suspected, one is obviously entitled to proceed by means of hypotheses and to try to interpret the available data in the light of their territorial context. If significant damage at only one site is known, the usual conundrum is that the damage could be due to a strong (but not great) earthquake near the site or to a truly great earthquake centred much farther away. Only by finding many damaged localities at considerable distances can one be sure that the earthquake was a great one. But even in cases where such an interpretation seems best, it is still always necessary to make clear that this is an *interpretation*, because archaeological sources cannot establish an absolutely contemporaneous relationship.

It is also a fact that things can be complicated from a seismologist's point of view. Take the cases of very great earthquakes like the magnitude 9.0 earthquake that affected the Indian Ocean region of northern Indonesia on 26 December 2004. The major damage effects from ground-shaking took place on the northern part of the island of Sumatra and on the islands (Andaman and Nicobar) to the north. A thousand years from now, an archaeologist who finds evidence of damage and destruction from this earthquake on northern Sumatra might underestimate the magnitude of this event if he only looks for damage to the south on the heavily populated island of Sumatra. Since he would find relatively little damage, he might conclude that the earthquake was not as big as it truly was. He may not realize or may not discover (because no archaeologist looked or because no remains exist) that the key information concerning the size of the 2004 earthquake was damage on the sparsely populated islands to the north. This example illustrates one of the limitations of archaeology's contribution, and it is one that is almost impossible to overcome. Fortunately, a thousand years from now one can hope that there will be some source of information about current times, other than that of archaeological finds, which will enable future historians of seismicity to appreciate the enormous dimensions of the 2004 Sumatra earthquake.

For historical seismology purposes, one normally must look to the set of information already gathered and brought to light by archaeologists and their investigations (even using the data of several generations of archaeologists) to identify the indications of a seismic event (indications such as collapses, abandoned sites, a decrease in the size of urban areas, the move of a necropolis, etc.). This use of archaeological evidence is more akin to making a *seismological interpretation* of archaeological data gathered for óther purposes than it is to engage in archaeological research *sensu stricto*. Owing to the variety of approaches that have been employed in the past (proven by a wealth of literary sources), the obvious problems in finding and interpreting its data, and the contradictions and the inevitable grey areas in its research results created by the lack of a single shared method, archaeoseismology should perhaps be defined more as a field of seismological questions addressed to archaeology than as a neo-discipline in a strict sense. The efficacy of a multidisciplinary approach involving historical seismology and archaeology is beyond doubt, but its effectiveness is directly proportional to the quality of the dialogue that can be established between archaeologists, historians of seismicity and geoscientists. Such a dialogue should be based upon the correct definition of the *questions* that researchers from each discipline want to answer. From both the archaeological and the geophysical standpoint, the results of combining forces will certainly be more reliable if each can avoid encroaching upon the skills and tools used by the other but rather move beyond the one disciplinary field into a truly cooperative research effort.

11.2   Traces of earthquakes in historical construction and monuments

11.2.1   *Human responses reflected in construction and reconstruction*

Reconstruction after destructive earthquakes has modified buildings, changed the profiles of towers and bell towers, and caused churches, abbeys, mansion and castles to be redesigned or transformed. But what about houses? Not infrequently entire villages and towns have been redesigned following strong earthquakes, and at times even the settlement network of an area has changed. For example, sometimes town centres have been abandoned while new towns have arisen. The choices of new sites for rebuilding in past centuries has nearly always been controlled by the locations of major roads and trade routes, and sometimes the destruction due to an earthquake created an opportunity for economic renewal and new development.

In the areas with high seismicity, building, rebuilding and repairing the damage from seismic events were recurring activities. Coping with the aftermath of a strong earthquake would make significant economic and social demands on the affected population, but it was necessary to meet these demands if the population was to get back to a normal life. All of the building activity that developed after a damaging or destructive earthquake has over time represented a historical activity of great importance because it has not only required substantial resources but also has required that local building expertise and techniques be employed. In the areas where the earthquakes frequently caused damage, those involved in construction and reconstruction have also tried to determine measures that, when applied to local construction, would mitigate effects or prevent the effects of strong earthquakes. When earthquakes did take place, some local experts had a chance to evaluate whether or not earlier measures to mitigate damage from earthquakes had been effective. Above all, the prestigious historical buildings were very often the subject of particular technical attention, both for their symbolic value and for the high cost needed to rebuild them. Major public buildings and monuments were indeed structures that, more than any other, were expected to be very longlasting. Technical information on the responses of most structures to a strong earthquake of the past are hardly ever found in the architectural treatises of the time, because these treatises were mostly concerned with empirical details implemented by the local builders on single and specific buildings. For this reason, comprehensive theories on the responses of most buildings to earthquake shaking were only formulated in relatively recent historic times (i.e. the last two centuries).

There are several different ways in which damage from one or more historic earthquakes to a specific building or to a set of buildings can be deduced:

(1) from *written documentation* that describes reconstruction and
     restoration work on a structure. Those historic written sources that are
     available for some areas and time periods represent a set of research
     sources that are the primary concern of historians. From these sources
     one can generally detect that there was seismic damage to a structure,
     with that damage no longer visible today due to the reconstruction
     work (see below the case of the cathedral of Florence). There are also
     cases, although much rarer, in which the damage described in historic
     sources is still visible today (see below the case of the cathedral of
     Syracuse in Sicily). Finally, sometimes a comparison between what the
     written sources attest to and what direct observation of a structure
     reveals can help the researcher calibrate the meaning and accuracy of
     a written text, which can provide a better understanding of the nature
     and limits of the contemporary sources (see below the cases of a great
     medieval fortress in Syria and of the cathedral of Cremona, northern
     Italy);

(2) from *direct observations of the repairs on a building that lead one back to the
     written sources*. From an architectural and engineering analysis of the
     restoration work that was carried out on a building, it may well be
     possible to approximately date the work and there develop a research
     plan aimed at uncovering written sources concerning the repair work.
     This type of research requires a collaboration between historians and
     architects. Relating the restoration work to the date of the seismic
     event that caused the damage might be difficult, since restorations can
     be made some time after an earthquake. Sometimes restoration work
     is simply hidden inside the building by subsequent works and
     forgotten (see below the cases of the sanctuary of San Luca in Bologna);

(3) from *direct observations on a historical building to detect earthquake effects
     not contained in written sources*. This is a type of investigation that
     requires the input of architects and engineers. These types of
     observations can make more complete the analysis of the seismicity of
     an area and can allow one to identify indications of unknown
     earthquakes. This kind of work clearly has a relationship with
     archaeoseismology (see below the case of Italy).

(4) from *observations on the typical architectural features* of buildings in a
     seismically active area. It is possible that some unusual construction
     practices in some areas, especially for some specific types of buildings,
     may be early attempts at earthquake-resistant engineering (see below
     the case of the Armenian churches).

11.2.2   Traces of earthquakes in historical buildings and monuments

In geographical areas with no or few documented written sources, ascertaining seismic effects on local historical buildings or monuments can play an importance role for helping determine the seismic hazard of those areas. Documentation of seismic effects on structures or monuments in these areas helps define seismic hazard in several ways:

(1)   as an indirect testimony to significant earthquake activity in the area;
(2)   as an indication of construction practices in the history of a structure or monument and how this may affect its vulnerability to earthquake shaking;
(3)   as data points for estimating the seismic behaviour of structures in the area.

There are several different kinds of traces of effects that one can expect to find today in a structure or monument that was damaged by an earthquake. Those effects that may still be directly visible are lesions, cracks, offsetting of upright structural members, and slippages of keystones in the arches. One might infer from restoration work that was carried out on a building or monument that one or more of these types of effects were repaired at some point in the past. For example, one might discover that walls were built in different periods with different masonry materials or that reinforcements made of wood or iron, supporting arches, buttressed walls and carp walls, sloped arches to prevent collapses, etc., were later added to a structure that had been deformed in a way consistent with damage from earthquake shaking. Furthermore, ancient builders may have inserted into a building during a reconstruction phase some devices or improvements with the aim of protecting the building from future damage due to earthquakes. In these cases one might notice that some corners of load-bearing stone walls were reinforced with clamps or ties or that load-bearing walls were made thicker than the traditional load-bearing walls of the area. These measures were undertaken to counteract weaknesses that led to the earthquake damage in the structure.

Traces of deformations and damage of historic structures due to earthquakes of the past are indicators of the seismic behaviour of those structures. How historic structures behaved in past earthquake shaking in many cases is not easily determined, because the buildings of the past are very different from present-day buildings for which mathematical models of their behaviour in earthquake shaking are generally known. If one compares historical buildings that were made of stone, brick, wood or some combination of these materials with modern structures that are made of reinforced concrete or steel, one sees that the buildings differ in terms not only of the materials with which they are made but also of the conceptual processes of design and engineering that went into the creation of the building. D'Agostino (2002, 2003) argues that historical buildings can be classified as *constructions* based on the methods used to create them, while current buildings can be classified as *structures* because of the practices used to design and erect them. These terms highlight two different worlds of thought on how to create a building. *Construction* has been defined by D'Agostino (2005) as a unified practice in which a building is erected using empirical handicraft expertise and following some precise *rules of the art* that guides all design, engineering and construction phases. The characteristics of *construction* arise from and interact with the natural environment and the cultural setting where the building is located. In historical *constructions* the materials used for the building and the ways in which those materials are assembled and bonded together are rooted in the availability of material and in the knowledge of local builders and artisans. This knowledge, or the *rules of the art*, has been gained through generations of testing and experience, and has been shared over time from one generation to the next (see, for instance, Cairoli Giuliani *et al.*, 2007; D'Agostino and Bellomo, 2007). A *structure*, on the other hand, is designed and engineered to fulfil exclusively static functions that are governed by the laws of structural mechanics. In a *structure*, the parts of the building are connected together by means of the assembly of industrial components. The structural calculations used for the building design and engineering follow from a rational and theoretical conception of nature. These calculations use simplified mathematical models, which are analytically formulated or are computer derived, to guide the process used to engineer a building. Today this is a highly specialized practice that is regulated by rigid normative prescriptions.

Historical *constructions* belong to the world of empiricism and experimentation by trial and error, an approach that contrasts with that of modern structural engineering. The latter, however, today tends to interpret historical *constructions* through its own schemes. Modern engineering can use mathematical tools and powerful computers to calculate quantitatively the static behaviour of historical *constructions*. From this kind of analysis of historical *constructions* the fundamental mechanical bases upon which the *rules of the art* are based can be deciphered. In theory it should be possible to verify the scientific validity of the ancient *rules of the art*. However, engineers are still far from reaching this goal because such analyses require resources and levels of effort that are rather substantial. So far, only a few specific projects have undergone such an analysis (e.g. gothic cathedrals). For this reason, the descriptions in historic sources and the direct observations of damage in historical buildings still contribute important information concerning the behaviour of a building during an earthquake. Such knowledge is vital to guide modern work to preserve historic buildings for the future. Thus, even today the traces of seismic effects on monumental buildings or on sets of traditional constructions remain vital data for those who are dealing with the preservation of monuments or traditional building in a seismically active area.

With a view to historical seismology, the direct observations of possible earthquake damage on historical buildings or monuments can help complement details found in written sources. Where written sources are absent, these observations from the archaeological data may be the only indications of a local earthquake effect. It is important to point out, however, that the effects of an earthquake observed on the single building are not sufficient to assign a macroseismic intensity value to the site. This is because the degrees of a scale of intensity are not intended to refer to the earthquake effects on just one building but rather to the effects on a statistically significant set of buildings in an inhabited location. In the cases for which it is worthwhile assigning some intensity values based solely on the damage to one or a very small number of buildings (perhaps owing to the lack of written sources), it is advisable to make macroseismic intensity estimates only as being indicative of the *presumed* effects to the site. Any such presumed intensity assignment should be indicated with a specific symbol (or something similar) in the database assembled as part of a historical seismology research project. The symbol or other indication should make clear that this intensity value was determined in a very different way from those based on written sources (this topic has also been addressed in Chapter 4).

The following examples illustrate different situations that arise when damaged buildings are evaluated for historical seismology purposes.

From written sources to a building

Seismic effects that are no longer visible: the case of the cathedral of Florence

Written sources constitute an irreplaceable memory of the seismic effects on a monument because in most cases the effects of an earthquake are no longer recognizable on the body of the building. Repairs carried out subsequent to seismic damage may have completely wiped out any visible traces of the damage that was suffered by the structure. This is the case of the cathedral of Florence, Santa Maria del Fiore, a very famous building built upon a previous Romanesque building of the eleventh century. The reconstruction of the original building started in 1296 and continued until 1436, the year in which the great cupola designed and created by Brunelleschi (1377–1446) was closed. This cupola is 84 m high and 42 m in diameter. This church suffered slight damage in the earthquakes of 1453, 1685 and 1895. This information allows researchers today to evaluate whether the great cracks that are visible today in the church are due to the effects of these past earthquakes or are due to the stabilization and dilation of the building materials under the weight of the building's dome and walls.

In 1453 the cathedral of Florence was still commonly known by its original name, Santa Reparata (the name of the Romanesque basilica). In the recently finished building, the earthquake of 28 September 1453 (I_0 = VII–VIII; M = 5.8) caused some stones to fall off the corners of the lateral nave towards the parsonage and some lesions opened up in the cupola. It is not known exactly when the two cracks that can be seen today appeared, but they had certainly existed from very early on. The strong earthquake of 1453 probably contributed to the settling process of the building, but the sources consulted do not report any trace of this.

When on 20 July 1685 another earthquake hit Florence, with intensity effects of degree VI, alarm concerning the condition of the dome grew among the local population, who worried that large visible cracks portended the imminent collapse and ruin of the cathedral. The seriousness of these fears is apparent in the decision of the Grand Duke Cosimo III de' Medici to have a special analysis of the dome and the building carried out. The analysis was performed on 18 June 1696 by Carlo Fontana, together with some masons and suppliers to the Fabbrica del Duomo, amongst whom was Giovan Battista Nelli. The original document containing their report is preserved at the Archivio della Fabbrica del Duomo (*Suppliche, Rescritti ed ordini del Governo*, 1687–1703, filza 18) and was published by Del Moro (1895). From this report it can be inferred that the major cracks in the dome were, as has already been said, pre-existing. The surveyors did not observe any alarming situation for the stability of the building in general and the dome in particular, although they did observe that the seventeenth tab, placed in order to monitor the natural movements of the dome, had opened on its two sides by several tenths of a millimetre in width. This width is well established because the surveyors wrote that 'a sheet of paper' could have been slipped into the opening.

The earthquake of 18 May 1895 (I_0 = VIII; M = 5.4) widely damaged the cathedral, although the damage was superficial rather than destructive. A thorough post-earthquake reconnaissance was made by an engineer of the time (Bassani, 1895–1902). Bassani's report provides a rather accurate picture of the effects of the earthquake (Figure 11.11), giving an idea of the way in which this building responded to the seismic input. The chain of the first archway on the western side of the building broke, separating the central nave from the northern one. The chain, situated above the window frame was 16.60 m long, had a section of 70 × 50 mm, and had an overall weight of about 500 kg. It broke in the

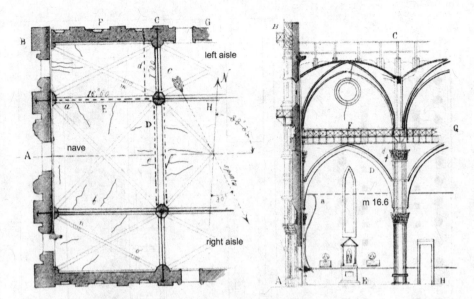

Figure 11.11 Earthquake effects at the cathedral of Florence in Italy. On the left is a sketch of the survey after the earthquake in 1895 showing the locations of surveyed cracks in the ceiling of the structure. On the right is a sketch of cracks to a window and of the breakage of a chain in the first arch on the western side of the church (from Bassani 1895–1902).

vicinity of the point where it was fitted into the first pillar of the archway, and as it curved over, it fell near the pillar against the internal wall of the façade. The old cracks mentioned above grew slightly larger. Marble peepholes in the northern wall broke, as well as a peephole associated with an ancient crack that goes along the circular holes of the dome. There was some falling of plaster and stucco, and some roof tiles detached and fell. The cross atop the spire of the lantern of the dome bent over in a northward direction. On the outside, some pieces of the green marble tarsia fell from the façade, and the cross placed at the top of the façade bent over. Also, a crack appeared in the rosettes of one of the windows.

Seismic effects that can still be recognized: the case of the cathedral of Syracuse

The cathedral of Syracuse (Sicily), named after Santa Maria delle Colonne, dates back to the seventh century AD, when the ancient Greek temple of Athena was transformed and adapted by Christianity. The building underwent several construction phases as a result of this transformation (Figure 11.12a). The building was also shaken by at least two strong earthquakes (in 1542 and 1693),

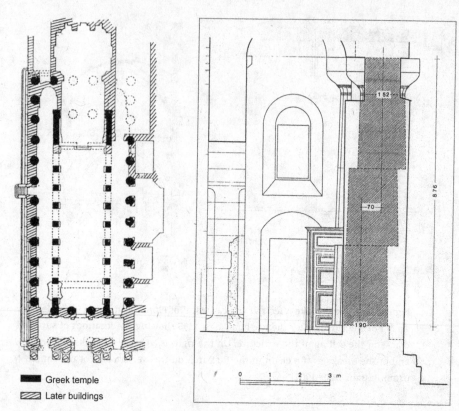

Figure 11.12 (a) Plan of the cathedral of Syracuse that was built incorporating the ancient Greek temple of Athena. (b) Deformation of one of the columns from the ancient temple that was incorporated into the northern wall of the cathedral. Owing to the earthquake on 10 December 1542 the column went off its axis by about 70 cm (survey of J.-P. Adam).

of which the exact damage caused to this very old construction is known. The earthquake on 10 December 1542 ($I_0 = X$; $M = 6.8$) shook and devastated this church. On the inside the serious damage it suffered was described in scholarly historiography and is still visible today. The colonnade fitted into the left lateral wall was greatly disjointed. The drums that contained a column slid and were so greatly askew that it was necessary to build a bulky reinforcement wall to offset this damage. For this reason the columns were then built into this thickened wall, and so today only some parts of the capitals of the columns still emerge into view. J.-P. Adam performed a survey of the original damage, and he calculated that the offsetting of the columns is about 70 cm (Figure 11.12b). Even the great tambours that make up the Doric columns of the nave underwent a strong push. The axial deviation is more evident as one gets close to the entrance of the small Byzantine apse, the only one surviving of the three primitive Byzantine apses (Agnello, 1930). The medieval façade, built in the Norman period (eleventh to twelfth century), was seriously damaged by the earthquake in January 1693, one of the most violent in eastern Sicily ($I_0 = XI$; $M = 7.4$).

Calibrating the written sources with direct observations of a building

The case of a great medieval fortress in Syria (twelfth century)

The Crak des Chevaliers (now Qalat al-Hisn, in western Syria) was in the twelfth and thirteenth centuries one of the most important fortified castles of its time. It was an imposing structure of dark basalt built on a spur of Gebel Alawi, about 750 m above sea level. It dominated the road to the Crusader port of Valénie/Banyas, and thus controlled the coastal road which ran north from Tripoli to the Principality of Antioch (Molin, 2001). This fortification is referred to in the Muslim sources as Hisn al-Akrad. The Muslim sources that recall the effects of the great earthquake on 29 June 1170 say that this fortress suffered the complete collapse of its walls. The texts describe this great castle as 'collapsed in the waves of the earthquake' (Abu Shama), and that of its walls there 'remained no trace' (Sibt Ibn al-Jawzi). It is understandable that the Muslim sources may have exaggerated the devastation because the castle was an enemy fortification. This is part of the normal bias in sources when they contain subjective judgements. But how much had those witnesses overstated the actual damage? In this case, direct observation of the castle has allowed modern researchers to evaluate the reports preserved in the sources. According to Molin (2001) the Crak des Chevaliers underwent profound transformations following the destruction caused by various earthquakes, including that of 1170 and others which occurred in the early decades of the thirteenth century. This famous and imposing fortress rises high even today, and it is so well preserved as to be considered the greatest medieval fortress in the world. Direct observation of its walls shows that the construction underwent some radical modifications and reconstructions carried out from the last decades of the twelfth century to around beginning of the thirteenth century. However, the fortress clearly did not undergo total destruction, as the written sources indicate.

A 'collapsed' cathedral: the case of Cremona (northern Italy) – 1117

The earthquake on 3 January 1117 is considered by the seismologists to be the strongest seismic event of northern Italy since the beginning of historical time. The interest in this great event has thus been rather lively, and there has been much intense research on this earthquake. Cremona was one of the most important cities struck by this event. Unfortunately, information on the local effects in Cremona only concerns the city's cathedral, the church of Santa

Maria Assunta, one of the greatest religious buildings in northern Italy. The construction of this great Romanesque building had started on a pre-existsing palaeo-Christian church in 1107, and it is very likely not to have been completed (or else had been finished for no more than about one year) when the earthquake struck. Two sources recall this destructive event: the *Annales Cremonenses*, a text compiled from the third quarter of the twelfth century onwards, and its *Supplementum*, a text based on old local annals, now lost. Both the *Annales* and the *Supplementum* date the earthquake to 1116, because they adopt the Incarnation dating style. While the *Annales* only recall the earthquake, in the typically concise and enigmatic style of this kind of text, the *Supplementum* also recalls the effects suffered by the cathedral:

> "In the year of Our Lord 1116 [for 1117], in the 10th indiction, on the 3rd day before the Nones of January [January 3]. There was a great earthquake, which caused the cathedral of Cremona to collapse, and the body of the confessor Imerio lay for a long time in the ruins" (Anno Domini MCXVI, indictione X, III Nonarum Ian. Terremotus magnus fuit, propter quem maior Cremonensis ecclesia corruit, et corpus confessoris Ymerii diu latuit sub ruina).

Subsequent chronicles and the local histories of Cremona recall the collapse of the cathedral as a complete collapse, only attributed to the violence of the earthquake. In actual fact, an analysis of the building has highlighted the fact that the construction of the 1107–1117 phase was solid in its side walls but presented a serious construction fault in the central nave. This nave rested upon a slender trabeation and upon columns that were too thin to hold the weight of the mass above. In this case, the building tells a story that is somewhat different from that in the written sources, and study of the physical building helps calibrate the precise meaning of the textual information. In 1129, fresh restoration and rebuilding work was started on the cathedral, and this work ended in 1190. In the following centuries new Gothic, Renaissance and Baroque integrations were added to the twelfth century building (Porter, 1915–17).

From the building to the written sources

Stories of forgotten restorations: what is an iron ring doing around the cupola of a sanctuary?

On a hill about 300 m above sea level, above the plain where the city of Bologna (northern Italy) rises, there is the famous sanctuary of the Madonna of San Luca, built at the end of the seventeenth century upon a twelfth-century church that had been remodelled. In the first few decades of the eighteenth century the Bolognese nobility had decided to build a larger and more imposing church to replace their earlier sanctuary, and they commissioned the work to

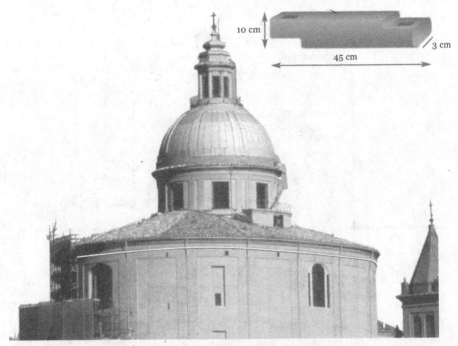

Figure 11.13 Sanctuary of the Madonna of San Luca (Bologna, Italy). Example of a metallic element used in the great belt encircling the elliptical wall of the sanctuary, applied to avoid the collapse of the walls and further damage wreaked by earthquakes.

a renowned architect, Carlo Francesco Dotti. He built an elliptical church, surmounted by a great cupola. Just a few years after the inauguration of the great sanctuary, in 1779 Bologna was hit by a seismic sequence lasting nearly a year. In the church two symmetrical vertical lesions appeared in the elliptical walls, and more lesions opened with the earthquakes on 6 February 1780, 4 October 1834 and 24 January 1881. These earthquakes, albeit not very strong (effects in Bologna not above degree VII of macroseismic intensity), were strong enough to spread alarm among the directors responsible for the safety of this great and symbolic religious building.

Recent restoration and maintenance work to the sanctuary walls have revealed that the church is girdled by a great iron ring, made up of metallic components each one measuring 45 cm, 10 cm high and 3 cm thick (Figure 11.13). What is this iron ring doing there? A plaque bears these words: *Cinta di ferro – 1882* ('Ringed in iron'). Specific research was started in the archival sources of the day concerning the church and the public works performed in that time period, and this has brought to light the technical examinations and analyses of the engineers of the time concerning this sanctuary. It was discovered that the

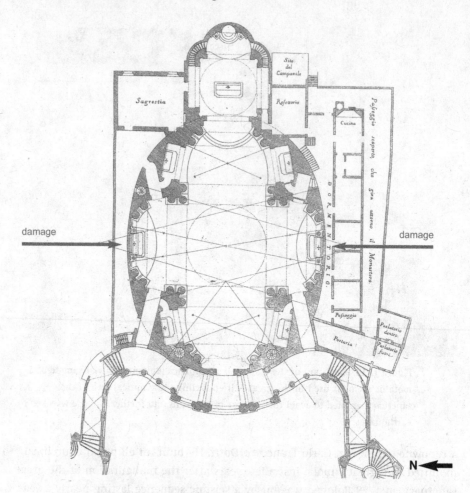

Figure 11.14 Plan of the sanctuary of the Madonna of San Luca, Bologna (Italy), and indications (arrows) showing where the cracks in the building are located.

cracks caused by the 1834 earthquake were closed in a makeshift manner with stones and stucco, but that they reopened in the subsequent 1881 earthquake (Figure 11.14). To close these opening cracks (from 2 mm to 5 mm) once and for all, it was decided to rebuild part of the walls (for about 45 cm from each side of the lesions) and consolidate them from the outside by inserting sturdy iron keys. Furthermore, it was decided to encircle the great dome from the outside in order to stabilize it and make it stronger in the event of future earthquakes. An observation of the building stimulated a search for written sources. In this case the data have created a complementary effort involving historians and architects, and this has led to a detailed understanding of the damage suffered by this church and the remedies implemented by the builders of the time (Ciuccarelli, 2003).

From observations on the typical architectural features of a seismic area

The Armenian churches: an extraordinary aesthetic style as an antiseismic consolidation

In areas that are strongly seismically active, at times architectural solutions to problems like earthquakes have become consolidated into the common construction practice of the region, and this practice has significantly modified the pre-existing construction models in the region. From this point of view, the Armenian area offers some very interesting observations. Armenian medieval ecclesiastic architecture derives from the Byzantine tradition, from which it has inherited the basilica form (that is, the naves) and the overall central structure and dome, although later substantially modified. According to historians of Armenian architecture (Cuneo, 1968; Rocchi, 1978) it can be argued that Armenian churches from the fifth century AD onwards have satisfied the need to last for all time. This characteristic may be ascertained through an examination of the oldest primitive churches or from observations of collapses. According to Rocchi (1978) the elements developed to increase the longevity of these important structures are believed to be: a reduced size of the building, very important buttresses or pilaster strips, polygonal absidal basins, diagonal brickwork and the triangular niches. The central plans require at least four absidal basins, which is a building element with a semi-cupped roof. Through these modifications the ancient Armenian builders changed the shape of their earlier churches. For example, the cathedral of Talin, which originally had a basilica plan, was transformed into a circular plan, and this was also the case for the great church of St Etchmiadzin. Many other such examples are listed in studies on Armenian architecture (Zarian, 1992; Kasangian, 1981; Cuneo, 1988). In addition to these elements high tambours with a cupola, which emerge from the centre of the architectural structure, were commonly added. All of these elements seem to confirm the intention to strengthen the building by reducing its surface plan and increasing its height (Figure 11.15). The formal beauty of these distinctive architectural features has long been investigated and annotated by the art historians. Here we consider only how those elements are driven by the need to counteract the effects of strong earthquakes.

11.2.3   *Around the concept of* anomaly

Since the 1980s the European Centre of Ravello (Naples) has been the promoter of unique studies and research for Europe and for the Mediterranean area, centred around the concept of *local seismic culture* and *anomaly* (for a review

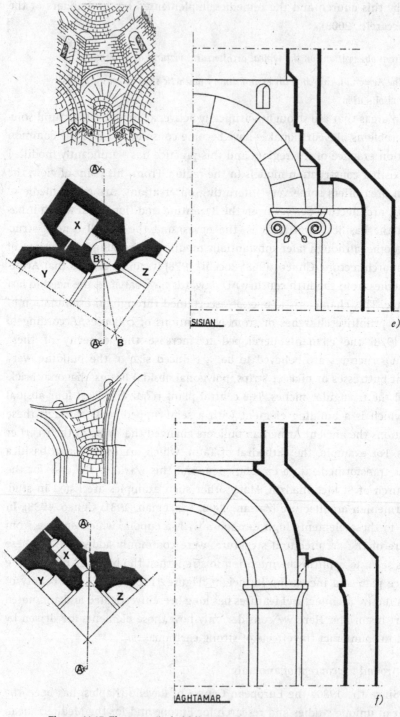

Figure 11.15 Elements (in bold lines or shading) of Armenian architecture that are interpreted as keys to help strengthen the structure against earthquake shaking (from Rocchi, 1978).

of the cases analysed and the issues dealt with, see Ferrigni *et al.*, 2005). The main hypothesis of this work – at times later modified and also rather confused in the course of the years and in terms of the various contributions – is that historical construction in seismically active areas is the result of a specific response to earthquakes. The local building culture makes use of techniques, materials and formal solutions to specific problems in an effort to cope with the threat of earthquakes. Such a culture can be observed in the *ex novo* construction phase or in the phase of repairs after a strong earthquake. Above all, in repairs following an earthquake, elements in the building construction can be observed that are *anomalous* relative to the prevailing construction style to which the buildings belong. Such anomalies, according to the experts who have brought them to light, are easily recognized and can be used as an indicator of the experience of past strong seismicity. In other words, wherever construction anomalies are detected, an earthquake is thought to have taken place there. For example, an arcade walled up to consolidate a building bears witness to a previous collapse. Another example is a ground floor doorway that has been closed off and replaced by a staircase, built adjacent to an external wall, that leads to a new door opening on the floor above. This configuration could arise when the load-bearing wall was seriously offset by a strong earthquake, necessitating the ground floor doorway be sealed to help support the upper floors of the structure. Still another example is that of a bell tower that has been lowered and widened as opposed to the proportions normally used in the prevailing style for that type of structure. Compared to the stylistic norm, such a tower might appear deformed, i.e. anomalous. This last example has been highlighted in a study of Spanish Baroque towers in Mexico, where the shorter and wider structures are clearly anomalous relative to their European counterparts and are thought to bear witness to an anti-seismic intention (Fernández, 1990).

The concept of *anomaly* is quite stimulating and original, although it is important that it not be exaggerated. Indeed, quite often economic considerations and technical shortcuts have led to stylistic variations in buildings that appear anomalous, but which in actual fact were not directly stimulated by the occurrences of past earthquakes. For these reasons, caution in applying the concepts of local seismic building culture and anomaly is a necessity, especially where written sources are lacking.

Doubled archways in a medieval church (northern Italy)

The basilica of Santo Stefano in Verona is one of the oldest churches of the city. This building was built in the fifth century AD on top of an ancient bishops' cemetery; it was nearly destroyed by Theodoric (sixth century) and rebuilt in the eighth century. The inside still preserves the sixth-century perimeter walls

Figure 11.16 The church of Santo Stefano in Verona (Italy). Note the difference in the brickwork of the walls between the outside of the nave, dating back to the early Middle Ages, and the *tiburio* (little tower), constructed after the earthquake in 1117.

(a unique case for Verona), and they are of a barbaric simplicity. The building has three naves, divided by rudimentary pillars. Inside the transept there are the archways where the weight of the *tiburio* (an octagonal body with two orders of windows) is offloaded. This archway is *doubled*, that is, it is formed of two adjacent arches (Figures 11.16 and 11.17). One arch is painted with Romanesque geometric motifs and is clearly the older of the two. The other is painted with ceiling roses in a twelfth-century style. The two arches are separated by a slit. What does this configuration of arches mean? The arches support a newer architectural body that evidently weighs more than the previous one that it replaced. Indeed, the new tiburio, which differs from the rest of the walls, was built of bricks, while the church is made of stone, cobbles and tufa. Even the façade of the tiburio is contemporary. But why build a new tiburio? It is very likely that we are faced with a restoration of the damage caused in this church by the earthquake on 3 January 1117 (I_0 = IX; M = 6.8), an important seismic event in northern Italy. This major earthquake, the largest known in this part of Italy as has already been noted, is widely attested to by the written sources. However, no source refers to damage to specific churches in Verona. Shedding light upon the exact effects in individual buildings would be an important contribution towards a better understanding this great seismic event.

Figure 11.17 Two inside arches in the area of the transept of the church of Santo Stefano in Verona (Italy). The ancient arch, painted with geometric shapes, was flanked after the earthquake of 1117 with a sturdy arch (painted with rosettes). There is a slight gap between the two arches.

According to Marchini (1978), unlike many other Verona churches that were rebuilt from scratch, this basilica does not seem to have undergone very serious damage. Both the modest height of the basilica and the presence in it of buttresses and vaults in the small lateral naves probably improved the connections of the building and limited the damage in earthquakes. The previous tiburio is quite likely to have collapsed onto the roof, and perhaps the old façade was also damaged. What we can hypothesize today, on the grounds of the anomaly of the double arch, is therefore a partial collapse, mended within a rather short time after the earthquake.

The deformed porch of an ancient mountain Tuscan hermitage

In the Tuscan Apennines in the vicinity of the village of Minucciano, there is an ancient hermitage, dedicated to the Madonna del Soccorso (The Madonna of Aid). The current form of this hermitage dates back to the fourteenth century. The building has a very simple shape, almost a parallelepiped. In the façade there is an arcade where one of the archways is clearly asymmetric and some other architectural elements display an anomalous dislocation. The pillars of the arcade and the windows indeed appear to be crooked in relation to the axis of the façade. These are anomalies that do not seem intended by the builder and that today appear the result of a changing relationship between the building and its foundation. What does this mean? Perhaps there has been a sliding in progress, due for example to a landslide. However, the geologists' surveys have clearly ruled out such a possibility. In actual fact, the building had the misadventure of being built directly upon an important fault in this area, a fault that has been identified in the ground between the villages of Castiglione and Morfino (Vannucci, 2000). The hermitage is situated perpendicular to the direction in which the fault moves. Thus, the two parts of the building on either side of the fault have moved in different directions over time, each one moving together with the portion of the grounds on which it was built. The difference in level between the two parts of the building, as is shown by Figure 11.18, is about 30 cm. The dislocation that has deformed this building must be geologically very recent. A stone under the arcade recalls the earthquake of 1837, although the timing of the deformation relative to this earthquake is unclear.

11.2.4   Anti-seismic houses in the Western world: food for thought

The history of the anti-seismic construction in the Western world is a fascinating multidisciplinary topic, embracing historical seismology, the history of architecture, and engineering. It is a subject that is rich in ideas but that still must be systematically developed and investigated. Even so, there is no lack of competent contributions, although they are relatively small in number. It is only at the end of nineteenth century and in the twentieth century, as is well known, that scientific methods based on mathematics and physics actually began to be employed by the architectural and engineering disciplines to attack the problem of defining earthquake resistant construction practices.

Little is known about the projects or attempts prior to the nineteenth century to design anti-seismic buildings to withstand earthquake shaking. Although it is reasonable to suppose that in every part of the world human societies have always sought to avert the damage caused by earthquakes, it took a long time before specific projects intended to produce an earthquake-resistant house began to be undertaken in the Western world. According to the scholars who have looked at this early work, the first projects that explicitly attempted to create dwellings that could withstand earthquake shaking are thought to be dated to around the mid eighteenth century, immediately after the earthquakes of Lisbon in 1755 and Calabria in 1783 (Tobriner, 1983). In the Portuguese area, the so-called *gaiola* house was designed at the urging by the Secretary of State, future marquis of Pombal, and in the Italian area the *casa baraccata*, perhaps a further development of the Portuguese *gaiola*, was created. These projects are complete with drawings and theoretical explanations, and both are based on the

(a)

(b)

Figure 11.18 (a) Façade of the hermitage of the Madonna del Soccorso, near Minucciano (Tuscany). (b) Scheme of the differential movements that the façade has undergone. The difference in level on either side of the arch is 30 cm (from Vannucci, 2000).

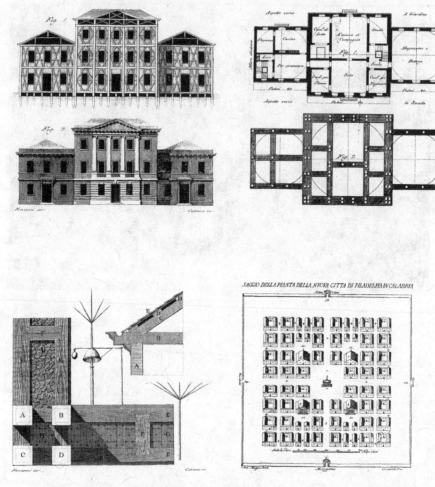

Figure 11.19 The famous earthquake-resistant house of Giovanni Vivenzio, 1784. The elevations on show the wooden elements inserted into the perimeter walls (top) and the finished building (bottom). The plans on the right show the scheme of the wooden frames (top) and the finished building (bottom) for the ground floor.

idea that wood is the material most suited to withstanding an earthquake. In these houses, poles or wooden rafters were used to form a strong interconnection between all the vertical and horizontal components of the house (Figure 11.19). It is interesting to note that in Calabria some of these so-called *case baraccate* survived the earthquake in 1908 in the Straits of Messina (I_0 = XI; M = 7.1) and were used to shelter homeless earthquake victims.

The *gaiola* and *casa baraccata* houses were not the first projects in western Europe to consider how buildings can withstand earthquakes. In 1781, the problem was addressed in a specific work by Francesco Milizia (1725–1793), the most influential theoretician of Italian classicism in architecture. He published a treatise on civil architecture that later became very popular until the first half of the nineteenth century (Milizia, 1781). In his text Milizia combined the idealistic, normative and artistic tendencies that can be traced back to illuminist rationalism, inspired in turn by Vitruvius (80/70 BC–23 BC), a famous Roman architect of the period of Augustus. Vitruvius was the author of De architectura, whose basic principles for good building were solidity, utility, beauty.

As part of his treatise, Milizia dealt with the problem of relationship between houses and earthquakes. He suggested that a home would withstand an earthquake if it was made entirely of wood (in contrast, therefore, to the classical historical canons which he declared inspired much of his thinking). The wooden house should be built without foundations and leaning upon a stone floor or a wooden base. Such a house should have proportions such that its height is no greater than its width and its length. In practice, this rule led to a shape that was a straightforward parallelepiped or a cube, which according to Milizia could never collapse. Indeed, this house was just a solid box structure resting upon the ground. Milizia's project is an abstract exercise in building safety, because he completely neglected the weight of the house in his analysis and the value of the cultural characteristics of houses in which people would want to live. Indeed, very few people in those times, once the fear of an earthquake had subsided, would have accepted living in a simple box-shaped dwelling, no matter how safe from earthquakes it might have been. It may be surprising that Milizia, faced with the danger of the earthquake, had completely surrendered his very precise aesthetic canons from classical architecture.

A few decades before Milizia's treatise, in 1756 another Italian architect had dealt with the problem of antiseismic houses. This person was Eusebio Sguario, a physician and mathematician of Venice, an expert in civilian and military construction, and the author of a treatise that is far less famous than that of Milizia (Sguario, 1756; Barbisan and Laner, 1983, 1995). Sguario's basic concept was that reinforcing the walls and floors/ceilings of a building improved its seismic response. Sguario also suggested that top loading would raise the geometrical barycentre and hence reduce the movement of oscillations in the upper part of a building (Figure 11.20).

Going even further back in time, one finds another proposal of an antiseismic house that is dated 1571. It is inserted in a treatise by Pirro Ligorio, a famous architect of the Mannerist pontifical court in Rome, then at the court of Ferrara. As Ligorio was greatly experienced as a constructor, he had carefully observed in person how a house collapse occurred as the result of the great impact of an earthquake, and he wondered about the reasons for collapses under such circumstances. Ligorio investigated the reasons behind the weakness of the buildings

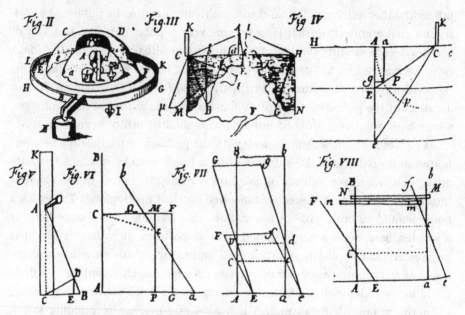

Figure 11.20 Starting from the assumption that both the walls and floors were completely rigid, Sguario (1756) suggests that an increase in top loading in a building would raise its geometrical barycentre and thereby reduce movement at the top of the building (from Barbisan and Laner, 1995).

(today this is called their *vulnerability*). He also designed a house with the specific aim of making it safe from diagonal shaking, instead of being adequate only to support vertical loads. This project, which was based on technical expertise, is the first known for the modern era in the Western world (see Figure 11.21 and below for more details). Unfortunately, Ligorio's handwritten codex was published only in the year 2005, and so his project did not spread to the constructors of his or later times.

Some scholars (e.g. Baratta, 1903; Di Teodoro and Barbi, 1983) observed that the problem of designing earthquake-resistant buildings had also been dealt with by Leonardo da Vinci in the years between 1492 and 1503. An analysis of some of his notes, present in various codices, appears to show that Leonardo too had posed himself this problem in an empirical and scientific way. Baratta (1903) observes that in the margin of the codex A of the Institut de France, Leonardo had drawn some arches and had written beside them '*Del riparo a' terremoti*' ('in defence from earthquakes'). However, Baratta noted that Leonardo's sketch could not be understood as the drawing was upside-down. In actual fact, Leonardo's drawing, as was later shown more carefully by Di Teodoro and Barbi (1983), really did portray some upside-down arches resting upon the ground. As can

*teste, di venticinque di ve teste. Tutti li muri quasi uogliono, che sono fatti senza
difesa, nelle cantonate tutti si senigano, et fanne separazioni o per solitudine o per
rovture. Tutto L'intento dene hauere l'artefice di fare i muri con legamenti, legare essi colle pietre, legare le cantonate colle grossede et colle chiui di ferro, o di gran
sassi, per che queste sono le recine, che mantengono le cantonate, per ciò che nulli li
cantoni che hanno li loro debii ripieni, o li suoi ferri ascesi deno possono chiamarsi
si curi: per questo gli antichi faceuano doppie le cantonate inquesta guisa che sono disopra
re queste stanze per li mostranzine:*

Figure 11.21 Model of an earthquake-resistant house (elevation and plan) as designed by Pirro
Ligorio (ms. 1571, ed. 2005) (State Archives of Turin, codex Ja. II. 15).

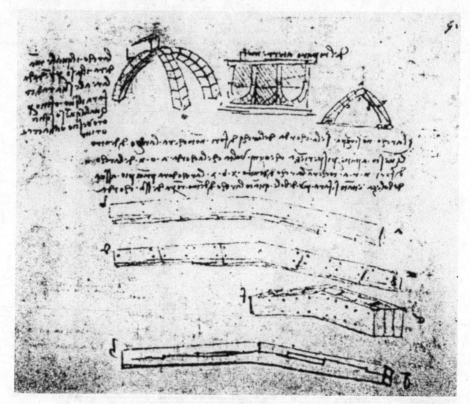

Figure 11.22 Drawing by Leonardo da Vinci of the upturned small arches (top right) in a buried foundation, made up of connected pillars; perhaps a foundation capable of limiting the risk of differential subsidence due to a differing resistance of the ground (*I manoscritti dell'Institut de France*, ed. 1990, codex A, fol. 51r).

be seen in Figure 11.22, Leonardo's drawing is inserted within a series of more general observations on arches, and the overturned arches support some pillars (or columns). Then Leonardo deals with the lengths of rafters (Figure 11.23) and writes:

> *Each rafter must go through its walls to be stopped beyond by those walls with enough chains, because often it can be seen from (owing to) earthquakes the rafters come out from the walls and cause the walls and the ceilings to collapse; instead, if they (the rafters) are chained, they will keep the walls still and the walls will uphold the ceilings.*

According to Di Teodoro and Barbi (1983), Leonardo thus paid specific attention to the resistance of buildings in earthquakes, outlining one of the basic

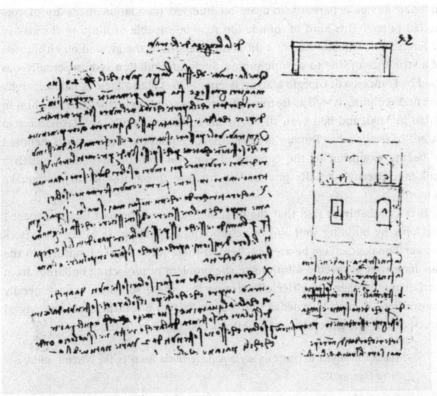

Figure 11.23 Drawing by Leonardo da Vinci of the correct construction of buildings, and in particular (top right), as a defence against earthquakes, the deep interlinking between wooden beams within the load-bearing walls (*I manoscritti dell'Institut de France*, ed. 1990, codex A, fol. 53r).

principles for anti-seismic masonry structures, that is the rigid connection between the ceiling and the perimeter walls.

Leonardo's idea about chaining rafters to supporting walls was not the first to have been proposed. A member of the generation prior to Leonardo, Leon Battista Alberti (1406–1472) was a Renaissance architect as well as philosopher, painter and musician. It was also Alberti's opinion that rafters and walls should be joined together in a strong manner with chains and hooks in order to increase a building's resistance to horizontal ground-shaking. This concept, later developed independently and in greater depth, was at the basis of the project of Pirro Ligorio, as has been seen above. Leonardo's contribution was his suggestion that the rafters should be rested upon some shelves in order not to tear the load-bearing walls during the seismic shaking.

Leonardo's sketch of the upturned arches is a particular development that was not subsequently revived. According to Di Teodoro and Barbi (1983) these

inverted arches apparently indicate an interred foundation made up of connected pillars. This kind of foundation may be capable of limiting the risk of
differential subsidence due to a different strength of the ground on either side
of a structure or due to seismic inputs. Such a foundation system actually was
used by Francesco di Giorgio Martini (Siena, 1439–1502), a famous painter, architect and sculptor, as well as fortress builder. The young Leonardo had met him in
Milan in 1490, and had been struck by his colleague's skills, and Leonardo had
carefully studied his *Trattato di architettura*. Leonardo's innovation is believed
to be the extension of the concept of consolidation of the soils beneath a
building, already partially practised in his day, to be used as an earthquake
defence.

It can probably be said that the great architects of the past had understood
that only by building well and by reinforcing a building could the most serious earthquake damage be averted. Beyond these ideas, most of cultures in the
past have sought specific solutions to the problem of protecting buildings from
earthquake damage. Whether effective or not, these solutions were generally
transmitted through knowledge that was unwritten but rather that was passed
directly from one generation of architects and builders to the next.

The oldest project to create an earthquake-resistant house in the Western world:
Pirro Ligorio, 1571

On 17 November 1570 three strong earthquakes, a few hours from one
another, struck the city of Ferrara (northern Italy), the seat of one of the most
prestigious European courts and one of the most renowned cities of Renaissance
town planning. These earthquakes were apparently experienced by Pirro Ligorio
(Naples 1513 – Ferrara 1583), a famous architect of the Roman pontifical court.
Ligorio had been a guest of the Duke Alfonso II d'Este for a few years, working
as the antiquarian and historian of the court, when the earthquakes occurred.
He turned his experiences into an opportunity to study earthquakes. He was
particularly interested in the causes of the earthquakes he experienced, the
occurrences in history of other earthquakes, and in finding ways to put right the
damage that buildings suffered in the earthquakes he experienced. In the month
of December 1570 he started writing a treatise, which he perhaps completed in
the spring of 1571. This treatise is a codex preserved today in the State Archives
of Turin (*Antichità Romane*, codex 28), with the verbose title *Libro o trattato de
diversi terremoti raccolti da diversi autori per Pyrro Ligorio cittadino romano mentre la
città di Ferrara è stata percossa et ha tremato per un simile accidente del moto de la terra*
(ed. 2005).

Ligorio had observed the collapses of the buildings in Ferrara due to the earthquake, and he had ascribed the reasons for the collapses to specific building

flaws. The part of the manuscript entitled *Essaminatione et conclusione nella fabricatione secondo il suo dovere* (Examination and conclusions in the construction according to his duty) (fol. 60v) seems to be a conclusion to his thorough analysis. In the text, which is dry and forthright, Ligorio quickly gets to the heart of the project, repeating the concept that to build properly one needs *rules*. He regards excessive cost-cutting, and thus the use of shoddy materials and the application of bad techniques, as the arch-enemies of safety in constructed housing. In Ligorio's opinion, the quality and quantity of the materials used in construction and the proportions of the building being erected should guide the architect as to whether he ought to reinforce the building with the iron fittings to strengthen the weakest points of the structure (a practice that can be applied to good effect today as well).

According to Ligorio, the thicknesses of walls should be precisely measured and have the same size throughout the building. Ligorio is deeply convinced of the crucial role of clamps (keys or iron tie-beams), which are used even today to reinforce structures. He writes:

> All the walls that are well tied and of equal thickness resist and lie to their advantage, and resist one upon the other; and being helped by the fittings, they do not move, and only tremble, and do not jerk one another in the fashion of rams. (fol. 60.)

According to this thinking, a strengthening of the building by tying its various components together is the objective to be achieved.

According to Ligorio there are typical parts and characteristics of the building that are cracked during earthquakes. These most easily damaged elements are the frames of doors and windows and the cornerstones. The disconnecting of one part of the structure from another is the reason for the damage, and to avoid it Ligorio indicates that reinforcements should be inserted into the frames and the cornerstones:

> And the thick walls are loosened, they move and knock against one another, with the striking of partitions become cracked and those cracks reach the frames of the windows. Thus it is a worthy idea to make some reinforcements over the frames and in the corners and make the walls stronger. (fol. 60v.)

Ligorio returns to this concept on another page, broadening the observations and specifying that the clamping could be done with keys made of stone or iron or with reinforcements in the masonry. He then provides the proportions for the height and the breadth of the walls in order to build safe houses, with those measurements in 'feet' and 'brick-heads'. If one considers that the width of a brick in the sixteenth century was about 12.5 cm, the proportions singled out for different walls appear approximately to be a thickness of 1 m for walls 20 m high, a thickness of 50 cm for walls 10 m high and a thickness of 38 cm for walls about 7.5 m high.

The uniform thicknesses of the walls, the correct proportions between height and breadth, and the clamps ('clamps of iron or stone') are indicated as the indispensable requisites for a residential building stock resistant to earthquakes. The main aim of the builder, according to Ligorio, must be to connect the building in a compact and coherent structure, with keys of iron and marble:

> *All the intentions must have the idea of making walls with clamps, binding the corners to the thicknesses, and with the keys of iron or great stones, that for these are the meshes, which maintain the corners, so that all the cornerstones that have their gaps filled in, or their concealed irons can call themselves safe. For this the people of old made double corners in this manner that are designed for these rooms by way of demonstration.* (fol. 61.)

Ligorio then explains his house project, which today we can define as *anti-seismic*, the first such specified design that is so far known for the Western world. Was this low-lying, solid house, with rectangular windows and doors, with solidly reinforced with arches and with reinforcements in its corners and internal walls, a model that was followed in the reconstruction of Ferrara? This hypothesis might allow modern researchers to understand better the transformation of the city after the earthquakes of 1570–1572, even if the new design method went unrecorded by the historians of Ferrarese architecture. But whether or not its ideas were employed in the reconstruction of Ferrara, the evidence remains of the original project. It is significant that the ideas proposed by Ligorio were based on a proper analysis of the causes of damage in earthquakes and those solutions that derived directly from that analysis. It is also significant that Ligorio's ideas were unaffected by the natural philosophy of the times. Ligorio's treatise was certainly far ahead of its time, and it could almost be considered as an act of volition that sought to challenge the ineluctability of seismic disasters.

11.2.5 Earthquake responses in traditional buildings

The hypothesis that societies living in earthquake-prone areas had developed over time specific anti-seismic construction methods is a research field that is more relevant to engineers and architects than it is to historians. However, since it seems clear that many anonymous local builders have dealt with the earthquake problem in times past, and this realization helps those interested in historical seismology to better identify the relationship between society and earthquakes. It is necessary to point out that by phrase *traditional architecture* we mean the common construction types that date from historical times and are characteristic of a given geographical area. Traditional architecture preserves a cultural character of an area, often with its own language or nomenclature, and it can characterize entire inhabited areas even today. The earthquake-resistant elements in traditional architecture are the result of the accumulation and passing down of centuries-old knowledge of the local builders regarding the unique challenges posed by earthquakes. Traditional buildings in different geographical areas obviously have very different characteristics. Their designs and specification are not found in treatises on architecture, but rather they belong to a vast body of unwritten knowledge and therefore can only be observed and documented on the buildings that still exist.

Through local architecture and the different techniques adopted to create it, over the centuries local builders have expressed their environmental experiences, their community values and their ways of living. In areas with a high level of seismicity construction methods, and the various adaptations intended to solve specific problems, were controlled by the functional needs and nearly always scarce economic resources of the local residents. It is also worthwhile mentioning that the widespread presence of traditional buildings in seismically active areas today gives rise to many problems in applying the seismic provisions in modern building codes. According to experts of traditional architecture, the application of modern building codes to historic building assets cannot be carried out by specifying quantitative design parameters to be applied to the buildings, but rather the codes should specify methods for earthquake-resistant reinforcement (Cardona, 2005).

Some traditional architectures in seismically active areas have been studied with particular attention. Amongst these the houses on the island of Santorini (Greece) and Leukas (Helly, 2005), the houses of the Kasbah of Algiers that were built using specific anti-seismic systems during the Ottoman period (Foufa, 2005), the houses of Garfagnana and Lunigiana, northern Italy (Pierotti and Ulivieri, 1998, 2001), and the traditional houses in the *bareque* in Colombia (Cardona, 2005) have been analysed in some detail. These studies have been carried out mainly by engineers and architects who deal with seismic vulnerability and seismic hazard or restoration. The use of this information in historical seismology may provide a positive contribution when it is important to know about the seismic history of a site in the long term or to better understand the persistence of some building characteristics in relation the earthquakes that have occurred in the past.

The earthquake-resistant traditional techniques used at the Kasbah of Algiers after the 1716 earthquake

A study by Amina Foufa (2005) was carried out to catalogue the urban, architectural and structural aspects of the traditional earthquake-resistant techniques used in the Kasbah of Algiers. These techniques were investigated by detailed historical research of documentary sources (written and graphical sources, files, etc.) together with an on-site architectural investigation and a comparison the traditional architecture to modern seismic design practices. The Kasbah of Algiers is located in an area with a relatively high seismotectonic hazard, and it still stands today despite the numerous seismic events that have hit the site in the past. This implies that in the past there was an awareness of the seismic risk that led to the development of earthquake-resistant techniques during the reconstruction of structures after past seismic disasters. For example, the 1716 earthquake, whose intensity was estimated at degree IX, seriously damaged the Kasbah of Algiers. Following that earthquake disaster, the authority of that time (the *Dey*, that is, the Governor) imposed construction measures designed to prevent damage from earthquakes upon the Algiers population. The work by Foufa (2005) outlines the construction techniques that were incorporated through time in the Kasbah.

Two types of earthquake-resistant preventative techniques are observed today in the Kasbah of Algiers:

(a) Techniques at the urban scale (i.e. for the Kasbah as a whole) that add strength to this entire urban structure, which is composed of a large number of individual buildings that have been connected together through various means, when exposed to dynamic earthquake shaking;

(b) Techniques on the scale of the individual building that are part of its construction, techniques that reduce the vulnerability of the building to earthquake loads to the point where the building complies with the current seismic codes established for masonry buildings.

Some specific traditional techniques that provide earthquake resistance on the urban scale are:

(1) The *sabats*. A large number of streets are covered by galleries over which the houses extend, thereby creating roofed passageways called the *sabats*. These act as elements of reinforcement that play a critical role in the bracing of the walls and structures between the *sabats* (Figure 11.24).

Figure 11.24 The Kasbah of Algiers: the *sabats*, discharging arches built out of stones or bricks (from Foufa, 2005).

(2) Discharging arches. The Kasbah of Algiers has, in its urban framework, a number of arches built out of stone or bricks called *discharging arches*. These arches have a flexibility and elasticity that is greater than that of the concrete at their points of support, and they allow the transfer of horizontal stresses to the ground. The individual constructed buildings are no longer considered as isolated elements but rather as a compact dynamic block of structures when exposed to earthquake shaking (Figures 11.24 and 11.25).

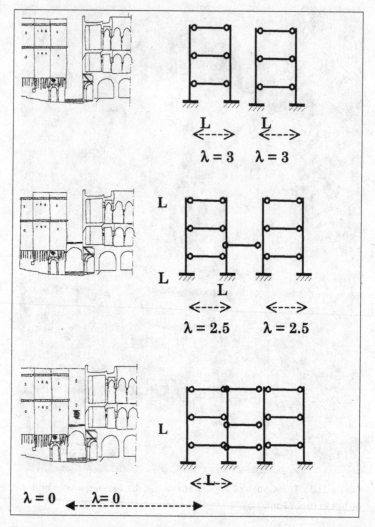

Figure 11.25 Role of the *sabats* and the discharging arches in an urban structure (from Foufa, 2005).

Some specific traditional techniques that provide earthquake resistance on the individual building scale are:

(1) Masonry made with two brick walls. The bricks are about $3 \times 10 \times 20$ cm in size and are bound by a sand and lime mortar. The space between the walls is filled with an ordinary material like broken bricks. The total wall thickness is 60 cm and is built with a stratification composed of layers between which are inserted, at regular intervals, *thuya* logs. These logs, about 10 cm in diameter, are

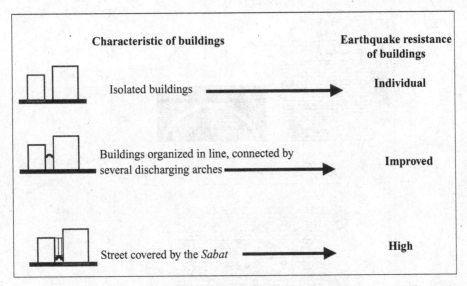

Figure 11.26 Morphology and earthquake resistance of buildings (from Foufa, 2005).

not squared, and they help withstand any tensional force on the structure (Figure 11.27a).

(2) Unique arches. The Algiers arches follow a rather particular constructive technique, which can be seen in the articulation of the arch with the column. On the top of the capital, on the level of the abacus and at the departure of the arch, there is the placement of three *thuya* logs on a base of two layers of earthenware brick (Figure 11.27b) and the perpendicular superposition of two lines of three *thuya* logs to the earthenware bricks.

This layered system of two materials, one rigid and the other flexible, guarantees that some minor slip or rolling movements can take place in the column when shaken laterally, which helps dissipate the shear stresses on the arch during earthquake. The system of arcades is thus less susceptible to damage in earthquakes, and this preventive construction technique contributes to earthquake load resistance in the building as a whole.

According to Foufa (2005), the Kasbah of Algiers contains an earthquake-resistant system on an urban scale as well as on the individual-building scale. It uses preventative technologies that have been adapted to the architectural styles developed during the eighteenth century. This system allowed the Kasbah to withstand the various earthquakes that have followed that of 1716.

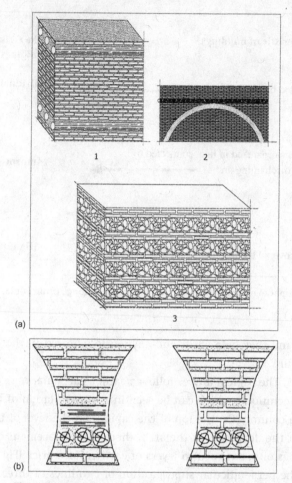

Figure 11.27 (a) The typology structure of walls in the Kasbah of Algiers (from Foufa, 2005). 1, regular superposition of bricks and logs of wood (*thuya*); 2, arch made of bricks inserted into the regular masonry; 3, *opus mixtum* bricks and stones. (b) Details of arch–column articulation, where one or two layers of logs (*thuya*) are inserted between the masonry to help relieve shear (lateral) stresses (from Foufa, 2005).

Ancient villages in a seismic area: the case of the Tuscan mountains

The villages of the Tuscan mountains in the historic region of the Garfagnana (northern Italy) are mostly built of stone, and they are representative of the building cultures in this part of the Apennines (Pierotti and Ulivieri, 1998, 2001). In some of these villages it is possible to detect ancient systems for building consolidation techniques that were deliberately made to withstand earthquakes. The analysis of buildings in this area has also shown some types of damage that was caused by the last strong earthquake in this area, which occurred on 7 September 1920 ($I_0 = X$; $M = 6.5$). Some of this damage can still be

Figure 11.28 Examples of vaulted passages, architraves and stone and wooden door jambs in Garfagnana, Tuscany, northern Italy (Pierotti and Ulivieri, 1998, 2001).

found today through an examination of the restoration work that was done. The visibility of this old earthquake damage is due to the fact that the repairs were not done following the traditional architecture but rather according to modern criteria that were demanded by a law that regulated the reconstruction. Such a law did not recognize any earthquake-resistant characteristics in the traditional building methods. Instead, the regulation imposed reinforced concrete as the method to consolidate the structures, thus substantially modifying the physical features of the historical buildings. In these villages some vaulted passages are used that create, thanks to their continuity, some local tunnels (Figure 11.28). The tunnel has a precise function in linking the internal and external space, and it serves the function of connecting the lower and upper floors of a building. It is also an element of reinforcement as it connects and helps support the opposing and nearby buildings. Generally speaking, the neighbourhoods (borghi) constructed in this way contain a compact, continuous structure with very few empty spaces. In the nineteenth century new urban planning regulations imposed by the state transformed the roads and the tunnels from being common areas into state property. This prevented the inhabitants from reinforcing the corners of their buildings with the construction of tunnels and staircases, as they had been accustomed to doing in the past. Instead, they were obliged to use iron chains and tie-rods for reinforcement. Although today many of these buildings are a hybrid of architectural styles and techniques inspired

Figure 11.29 Examples of vaults and small arches, architectural elements used for consolidation and reinforcement after earthquakes at Noli, Cervo and Sanremo in Liguria (northern Italy).

by different customs and regulations at different times, one can still note some interesting details in the buildings. External doors and windows are outlined with stone or wooden architraves (Figure 11.28), and the frames and architraves normally have a greater thickness in their central parts. Buttresses and thickened walls, either sloping or parallel, between two buildings were nearly always used to reinforce walls that suffered from localized lateral pressures and deformations. External arches between two buildings were made to repair cracks that stem from the pressure of the roofs or form wall rotations. Vaults and the arches are not just present in Garfagnana. Many other villages in western Liguria and eastern Liguria (Lunigiana) and along the Campania coast have these characteristics. Arches represent a practical and relatively inexpensive system for the consolidation of houses after an earthquake, if those houses had sustained offset and cracked walls rather than actual collapses (Figure 11.29).

Deriving earthquake source and shaking parameters and tsunami parameters from historical data

For a seismologist, the information uncovered from historical seismological research about past earthquakes can represent important data to be used to infer the source parameters, ground-shaking strength and other natural effects due to those earthquakes. For historical earthquakes, the *source parameters* that are typically sought by seismologists from historical seismological data are the *date*, *origin time* (time of the first seismic waves were generated by the earthquake), *epicentre*, *focal depth* (depth of the earthquake focus), *magnitude* and *seismic moment*. If an accurate description of ground deformation related to surface faulting was preserved in historical sources, then the fault upon which the earthquake took place along with the direction and amount of slip on that fault due to the earthquake might also be estimated, and confirmation for such an estimate can be sought with field geological investigations. Seismologists also are interested in using historical seismological data to compile maps estimating the strength of the ground-shaking from historic earthquakes. Such maps are necessary for engineers who wish to calculate the possible damage that might be inflicted on contemporary cities and towns if earthquakes similar to those historic events were to repeat in modern times. The natural effects induced by earthquakes that have been described in some historical documents include tsunamis, landslides, lateral ground spreading and ground liquefaction, and these descriptions are of great interest for modern seismic hazard studies.

In recent years, there has been a resurgence of research by earthquake seismologists using historical seismological data, and that resurgence has been accompanied by a number of new ideas and analysis techniques for eliciting earthquake source and ground-shaking parameters from historical data. The classical approach by research seismologists to the analysis of historical seismological data was to assign a macroseismic intensity to each locality where the earthquake under study caused damage or was felt, and then to determine the location, magnitude and strongest ground-shaking of the event based on the greatest intensity values (often denoted I_0 or $Imax$) and where those were reported. Even today, the locations and sizes of most historical earthquakes in earthquake catalogues for many areas throughout the world still are based on such an analysis. However, in the past few decades methods that use the full data set of felt and damage effects from historical earthquakes have been developed to make estimations of the epicentres, magnitudes, seismic moments and focal depths of the recorded seismic events. These methods use observations of modern earthquakes for which there are good instrumental data as well as many detailed descriptive accounts of damage, ground-shaking and other earthquake-induced effects as standards against which to compare historical seismological data. The new methods typically use the full spatial pattern of felt and damage reports, often combined with theoretical computations of the excitation and propagation of seismic waves from earthquakes, to make statistical estimates of the location, magnitude and seismic moment of an earthquake. Some of these new methods also provide quantitative estimates of other earthquake source or ground shaking parameters. Thus, the new seismological research methods that utilize historical seismological data often require the research seismologist to re-examine extant historical seismological reports as well as to seek previously undiscovered reports to augment and expand the extant data set.

One aspect of the new methods for finding seismological source and ground-motion parameters for historical earthquakes is that most are statistically based, meaning that they make use not only of the value of a seismological measurement (like seismic intensity or amount of building damage) but also of the uncertainty of that measurement. Because there can be ambiguity in the interpretations of the historical earthquake information, there is a corresponding uncertainty in any quantitative values assigned to those reports. As an example, a historical report may contain a description of ground-shaking that leaves seismologists uncertain about whether to assign one or another macroseismic intensity degree for the locality described in the report. This can have a critical effect on the interpretation of the location and size of the earthquake if only the site and value of the maximum intensity are used to estimate these seismological source parameters. On the other hand, many new analysis methods transform the uncertainties in all the intensity values available for a seismic event into an uncertainty in the earthquake source parameters that are determined from the historic records for that event, and so the estimated earthquake source parameters depend less on the exact intensity value assigned to any one particular site or observation.

This chapter summarizes how research seismologists use historical seismological information to infer earthquake source and ground-shaking parameters. While classical methods are discussed, the emphasis in this chapter is on modern methods, most of which make use of as much historical seismological information as possible. Each subsection in this chapter surveys some of the latest seismological methods for the determination particular types of source and ground-shaking parameters of historical earthquakes.

12.1   On the dates and times of earthquakes and tsunamis from historical records

12.1.1   Earthquake date

As described earlier in this book, for the historian there are many important reasons why the date and time of an earthquake or tsunami need to be ascertained. Just to name a few, these reasons include: to identify correlative sources, to evaluate the context of the reported earthquake, to estimate the historical authenticity of the seismic event, and to make literary and historical comparisons of different texts. For the research seismologist, the exact date and time of an earthquake or tsunami that was recorded historically also are vital parameters if the earthquake report is to be useful for scientific purposes. To a scientist, time is a metronomically ticking clock that began at the origin of the universe and continues steadily and inexorably into the future. Many theories of natural events require that those events be referenced to a common time standard that is agreed to by scientists in all parts of the world and at all times. Once an event is referenced to the common time standard, theories that look for patterns in the temporal occurrences of the events can be tested, and other natural observations (such as geological data, Earth and Moon orbits, or earthquakes from other regions) can be used for correlative studies.

Seismological science has adopted the Gregorian calendar for dates and coordinated universal time (UT) (which is derived from Greenwich Mean Time or GMT) as the timescale to which all seismic and tsunami events are to be referenced. This standard became accepted in October 1884 when an international scientific conference of 25 nations in Washington, DC established Greenwich as the prime meridian and defined a standardized time system. Following this conference, seismologists adopted GMT (or Greenwich Civil Time (GCT) as it was called in the first half of the twentieth century) as their common time standard for exchanging seismological information. Thus, all earthquake catalogues compiled prior to 1884, and some after that date, deviate from this later practice by using earlier or local dating systems, as has been extensively discussed earlier in this book. This is especially problematic for interpreting those early earthquake compilations in which there is no or inadequate information on the date and time system used for individual earthquake entries or reports. Confusion can result, leading to mistaken earthquake entries propagating into modern earthquake catalogues. Sometimes errors result from the earthquake catalog compiler either making an incorrect assumption or simply not understanding the date or time system implicit in the report of an historical earthquake. Sometimes mistakes arise when reports for the same earthquake from different localities each use a different date or time system. Earthquake listings and catalogues that mix different date or time standards may give inadequate information on the date and time systems for the entries of individual events, which can easily mislead subsequent users of those earthquake entries.

The 5 February 1663 earthquake in Quebec

Jesuit missionaries as well as others in Quebec in Canada reported a major earthquake followed by an active and energetic aftershock sequence in 1663. The Jesuit Relations (1663) states: 'It began half an hour after the close of Benediction on Monday, the 5th of February, the feast of our holy martyrs of Japan, namely at about 5 1/2 o'clock.' The French colony of Quebec used the Gregorian calendar, so this account gives unambiguous information concerning the date and time of this earthquake. However, the British colonies of New England still followed the Julian calendar at this time, and so they reported an earthquake on the evening of 26 January 1662 (sometimes written as 1662–1663 since the new year did not begin on the Julian calendar until March 25). At Boston, Massachusetts, three or four aftershocks of the main shock were felt during the next day or so. As pointed out by Ebel (1996), many earthquake catalogues contain mistaken entries for earthquakes on 26 January 1662 and 26 January 1663, errors arising from catalogue compilers who did not account for the different calendars used in Quebec and Massachusetts. Modern earthquake catalogues correctly list this earthquake as occurring on 5 February 1663 at about 22:00 GMT on the Gregorian calendar (see also Gouin, 2001).

As was detailed in Chapter 7, in Europe and the Middle East the calendar systems in historic times varied from region to region, even within the same linguistic area (Guidoboni et al., 1994; Gasperini and Ferrari, 2000). Gasperini and Ferrari (2000) highlight this problem for Italy by describing the six different dating systems used in the Italian region until the early eighteenth century, and they note that the dating style varied with the historical period, the geographical area and the type of historical source. As has been emphasized in Chapter 7, documentation of the date conversions to the Gregorian calendar that are used is essential to explain how the date of a historical earthquake in a modern earthquake catalogue or seismological study was determined from the original historical sources.

One area of seismological research that has grown greatly in importance during recent time is the discipline of *palaeoseismology*. Palaeoseismological research usually involves geological investigations of trenches dug across active faults or across other earthquake-induced features (sand blows, landslides, etc.) to investigate the history of past strong earthquakes that activated the fault or other feature. The most recent earthquakes invariably are the best resolved in this research, with the evidence for increasingly older earthquakes preserved in ever deeper soil layers in the investigation trench. At localities where a strong earthquake was well recorded in historic times, the evidence for that earthquake can be identified and then compared to the evidence for earlier strong earthquakes at the study site. Radiocarbon age dating of organic materials of the soil layers at different depths in the trench can give estimates of the dates to within about 20–50 years for earthquakes during the past 11 800 years. There are larger uncertainties for older ages. Organic samples younger than about 50 000 years before present cannot be analysed by radiocarbon age dating. Palaeoseismology research sometimes interfaces with *archaeoseismology* research, as described in Chapter 11.

The results of palaeoseismological investigations have provided new information about past strong earthquakes in areas where historical information about the earthquake may still exist. For example, southeast of Bree, Belgium, Meghraoui *et al.* (2000) found geological evidence that a strong earthquake with a magnitude probably somewhere between 6.2 and 6.6 took place sometime between the years AD 610 and 890. They note that historic records from Aachen, Germany indicate several strong earthquakes at the beginning of the ninth century AD, and that these earthquakes took place during a period of notable earthquake activity from AD 789 to 823 in the lower Rhine Valley. Although the historic earthquake descriptions do not unambiguously indicate which of the described events (if any) is that discovered by the palaeoseismological research, they do show how important historical seismological information can be for palaeoseismological investigations. In this case, further historical seismology discoveries may help identify the exact date when the large earthquake took place.

The 1812 southern California earthquake

It was long known from historical records at the Franciscan Mission at San Juan Capistrano in southern California that a strong earthquake took place on 8 December 1812. This earthquake collapsed a tower onto the mission church, killing 40 people who were inside at the time. The earthquake was also felt at other coastal communities in southern California, but its extent into the uninhabited inland regions is not reported in surviving documents. Other strong earthquakes were reported in southern California before and after this earthquake. Lacking other evidence, throughout most of the twentieth century seismologists had assigned an epicentre near San Juan Capistrano and a magnitude of about 7 to this earthquake. Later research by Jacoby *et al.* (1988) discovered that a major earthquake of about magnitude 8 took place on the San Andreas Fault at Wrightwood in 1812, damaging trees that could be precisely dated through tree-ring analysis. The obvious conclusion was that the 1812 earthquake that damaged the San Juan Capistrano mission as well as other missions in southern California took place on the San Andreas Fault and was not centred elsewhere. This led to the surprising result that strong earthquakes occurred on the San Andreas Fault at Wrightwood in 1812 and again in 1857, with the 45-year time difference being much smaller than the mean repeat time of 105 to 135 years estimated for major San Andreas earthquakes in this area (Biasi *et al.*, 2002). Palaeoseismological research (e.g. Fumal *et al.*, 2002) has shown that strong San Andreas Fault earthquakes in this part of southern California do not occur at regular intervals but rather have widely different time intervals from one event to the next.

The 1700 earthquake and tsunami on the Pacific coast of Oregon and Washington

Palaeoseismological evidence discovered in the 1970s and 1980s indicated that a strong earthquake and tsunami had affected the Pacific coastal region of the states of Oregon and Washington. Low-lying coastal areas with dead trees showed that a major earthquake had caused land subsidence, and marine sand layers atop the soils in which the trees had grown were strong evidence that a major tsunami had accompanied the earthquake. Analysis of the rings of the drowned trees (called *dendrochronology*) precisely dated the death of the trees to the year 1700. Historical research in Japan discovered that a tsunami had been observed on 27 January 1700 at about midnight. No record exists of a strong local earthquake in Japan or nearby areas on this date, and it is now accepted that the Japanese tsunami was from a moment-magnitude about 9.0 earthquake at Oregon and Washington. From the experience of major earthquakes and their resulting tsunamis around the Pacific Rim during the twentieth century, it is well established that a moment magnitude 9.0 earthquake at Oregon and Washington could cause an observable tsunami in Japan (see also Chapter 5.4).

12.1.2 *Earthquake origin time*

The proper interpretation of historical earthquakes, especially correlating the reports of an earthquake from one locality to another, is made easier if the times of the main earthquake and its aftershocks and foreshocks (if any) are recorded. Furthermore, as already stated it is the desire of seismologists to relate the dates and times of all earthquakes to the Gregorian calendar (although, strictly speaking, a single universal calendar cannot be made for all the known historical earthquakes) and UT so that an accurate time history of earthquake occurrences on a single time standard can be analysed for temporal patterns. Of course, where they exist in the days before mechanical chronometry, time estimates of earthquake occurrence are almost always quite crude and usually are related to the local solar time or other local events. Semantic interpretations sometimes can come into play, since words like 'day', 'night', 'morning' and 'evening' have had different meanings or connotations in different cultures or at different localities within a common linguistic or cultural population, a problem discussed extensively in Chapter 7. Unfortunately, for many historical earthquakes there is no record at all of the times of the events; for many other historical earthquakes, the reported times of an earthquake can appear to be different from one locality to another or one country to another, sometimes leading to questions about whether one or multiple earthquakes took place (Gasperini and Ferrari, 2000).

Before time was standardized in the nineteenth century, including periods for which mechanical clock time was available, there is invariably some uncertainty about the exact time that an earthquake took place. For example, Gasperini and Ferrari (2000) point out that in the late eighteenth and early nineteenth centuries, two different time systems were in use in Italy, each with a different beginning time for a new day. This can lead to confusion about which day and time of day that an earthquake occurred. Even in places where a single time system was used, drift of clock time from the common time standard (if there was such a standard) is inevitable. In the nineteenth century, there is often an uncertainty of half an hour or more in the time of an earthquake simply due to variations in the reported times from different sources. This usually reflects both differences in clock settings among the different sources as well as the habit of rounding the event time to some easily remembered time reference, such as the nearest half hour. Even today, many people in the general public will reference the time of an earthquake to the nearest full 10 or 5 minutes no matter how accurate their local clocks are.

Another insidious problem with determining the time of an earthquake is found in those records, such as personal diaries, where a time is given with no reference to a.m. or p.m. If there is no other clue to resolve this conundrum in the record itself, the researcher is left to rely on other extant records for resolution of this question. If no such information can be found, one is left with an inherently irresolvable 12-hour ambiguity in the earthquake origin time.

12.2   Macroseismic intensity and historical reports

12.2.1   *The macroseismic intensity scales in use today in the world*

For over a century, it has been the custom of seismological researchers to categorize descriptive reports of earthquake ground shaking using one or another *macroseismic intensity scale* (also called *seismic intensity*). Macroseismic intensity scales use descriptions of common felt and damage effects from earthquakes to classify the strength of the ground-shaking that was described in an earthquake report. Because macroseismic intensity is not an instrumental measure of earthquake shaking, seismologists use Roman numerals to indicate the *degree* or level of intensity assigned to each earthquake report. Different intensity scales have evolved in different parts of the world, in part to accommodate local building construction methods or local linguistic customs in describing earthquake shaking and in part due to the preferences of local researchers.

The history of the development of intensity scales to describe earthquake-shaking effects reflects the needs of early earthquake researchers to classify the earthquake observations with which they were working. As documented by Ferrari and Guidoboni (2000), at least 61 macroseismic intensity scales have been published, with the beginnings of the modern intensity scales arising during the second half of the nineteenth century. Many of the 61 intensity scales were later modifications or combinations of earlier intensity scales, and this is often reflected in the names given to those intensity scales. As time progressed during the twentieth century, seismologists came to accept a small subset of the available intensity scales for general use. Different parts of the world have tended to adopt their own, preferred intensity scale. As an illustration of an intensity scale, Table 12.3 lists the different levels of the Modified Mercalli intensity scale, one of the commonly used intensity scales.

In Europe, some of the most important early seismic intensity scales began with efforts in the 1870's by the Italian Central Geodynamic Observatory and Archive to use macroseismic questionnaires, commonly distributed in a postcard format, to collect information on the local seismic effects of earthquakes that took place in Italy (Guidoboni, 2000). Modifications and input from researchers in other countries led to the development of the MCS (Mercalli–Cancani–Sieberg)

Table 12.1 *Modified Mercalli intensity scale of 1931*

(The approximate earthquake magnitude at which this intensity is first experienced is given in parentheses after the description.)

I   Not felt except by a very few under especially favourable circumstances (1.5).

II Felt only by a few persons at rest, especially on upper floors of buildings. Delicately suspended objects may swing (2.0).

III Felt quite noticeably indoors, especially on upper floors of buildings, but many people do not recognize it as an earthquake. Standing motor cars may rock slightly. Vibration like passing of truck. Duration estimated (2.6).

IV During the day felt indoors by many, outdoors by few. At night some awakened. Dishes, windows, doors disturbed; walls make cracking sound. Sensation like heavy truck striking building. Standing motor cars rocked noticeably (3.2).

V   Felt by nearly everyone; many awakened. Some dishes windows, etc., broken; a few instances of cracked plaster; unstable objects overturned. Disturbance of trees, poles and other tall objects sometimes noticed. Pendulum clocks may stop (3.8).

VI Felt by all; many frightened and run outdoors. Some heavy furniture moved; a few instances of fallen plaster or damaged chimneys. Damage slight (4.4).

VII Everybody runs outdoors. Damage *negligible* in buildings of good design and construction; *slight* to moderate in well-built ordinary structures; *considerable* in poorly built or badly designed structures; some chimneys broken. Noticed by persons driving motor cars (5.0).

VIII Damage *slight* in specially designed structures; *considerable* in ordinary substantial buildings with partial collapse; *great* in poorly built structures. Panel walls thrown out of frame structures. Fall of chimneys, factory stacks, columns, monuments, walls. Heavy furniture overturned. Sand and mud ejected in small amounts. Changes in well water. Persons driving motor vehicles are disturbed (5.6).

IX Damage *considerable* in specially designed structures; well-designed frame structures thrown out of plumb; *great* in substantial buildings, with partial collapse. Buildings shifted off foundations. Ground cracked conspicuously. Underground pipes broken (6.2).

X   Some well-built wooden structures destroyed; most masonry and frame structures destroyed along with foundations; ground badly cracked. Rails bent. Landslides considerable from river banks and steep slopes. Sand and mud shifted. Water splashed (slopped) over banks (6.8).

XI Few, if any (masonry), structures remain standing. Bridges destroyed. Broad fissures in ground. Underground pipe lines completely out of service. Earth slumps and land slips in soft ground. Rails bent greatly (7.3).

XII Damage total. Waves seen on ground surfaces. Lines of sight and level distorted. Objects thrown upward into the air (7.9).

intensity scale (Sieberg, 1932a; Ferrari and Guidoboni, 2000), which seismologists in western and central Europe came to adopt for most macroseismic intensity observations. The MCS scale and its later modifications (such as the EMS scale: Grünthal, 1993, 1998) remain widely used in Europe today for intensity studies.

In the United States and Canada, the first macroseismic intensity scale that was widely used was the Rossi–Forel (RF) scale, which rated earthquake shaking on a 10-degree scale (Bath, 1979; Steinbrugge, 1982). This was used commonly for earthquakes in the late nineteenth century and early twentieth century. However, Wood and Neumann (1931) published the Modified Mercalli intensity scale (MMI), a version of the MCS scale and its earlier variants modified for application to North American construction and felt reports. This scale then became the principal seismic intensity scale used by scientists in the US government agencies that issued reports on earthquakes, and it was widely used by other scientists in North America as well. Richter (1958) further revised the MMI scale to better define the interpretation of good and poor brick and masonry construction when assigning intensity values. During much of the twentieth century, the United States Geological Survey (USGS) sent out questionnaires on postcards to postmasters in areas where earthquakes were felt or caused damage. The answers to these questionnaires were then interpreted by USGS or other government scientists to assign MMI values for each locality from which a postcard was returned. From 1932 to 1986 these report were published in an annual series called *US Earthquakes*.

Russian seismologists developed their own version of commonly used macroseismic intensity scales for their studies of earthquakes in eastern Europe and northern Asia. The Russian intensity scale is derived from the MCS scale and is denoted as the MSK scale after the seismologists Medvedev, Sponheuer and Kárník, who published it in 1964 (Medvedev *et al.*, 1964). Earthquake reports in the Russian literature invariably use the MSK scale for intensity reports. In Japan in 1949, a seven-degree seismic intensity scale was developed for application to Japanese earthquakes (Kawasumi, 1951). Called the JMA scale after the Japanese Meteorological Association, which is the government agency that conducts earthquake monitoring in Japan, this scale has been used for intensity reports of all earthquakes in Japan.

Most recently, efforts to gather descriptive macroseismic data and to assign intensity values to those reports have moved to the internet and the World Wide Web. Wald *et al.* (1999a) described a questionnaire that can be accessed and filled out on a Web page posted on the internet and how the answers to those questions can be interpreted in terms of MMI values. Community Internet Intensity Maps (CIIM) are then constructed by computer from the accumulated data and are posted as a public Web page. Unlike other methods of dealing with macroseismic descriptions that rely on the subjective judgement of the interpreter to assign an intensity value to each report, the CIIM method uses a mathematical scheme to convert from the intensity questionnaire answers to an MMI value for that site.

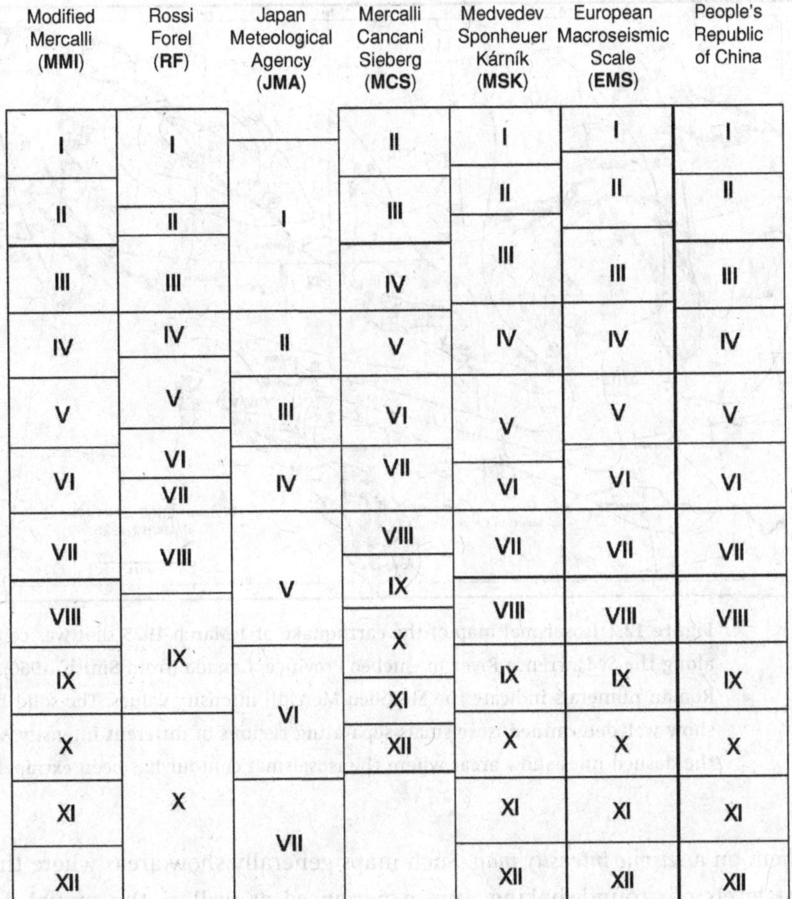

Modified Mercalli (MMI)	Rossi Forel (RF)	Japan Meteorological Agency (JMA)	Mercalli Cancani Sieberg (MCS)	Medvedev Sponheuer Kárník (MSK)	European Macroseismic Scale (EMS)	People's Republic of China
I	I		II	I	I	I
II	II	I	III	II	II	II
III	III		IV	III	III	III
IV	IV	II	V	IV	IV	IV
V	V	III	VI	V	V	V
VI	VI / VII	IV	VII	VI	VI	VI
VII	VIII	V	VIII / IX	VII	VII	VII
VIII	IX		X	VIII	VIII	VIII
IX		VI	XI	IX	IX	IX
X	X		XII	X	X	X
XI		VII		XI	XI	XI
XII				XII	XII	XII

Figure 12.1 The approximate correspondence of a number of different macroseismic intensity scales, modified from Krinitzsky and Chang (1988) and Reiter (1990). References for the macroseismic intensity scales are as follows: MMI and RF, Richter (1958); JMA, Kawasumi (1951); MCS, Sieberg (1932a); MSK, Medvedev *et al.* (1964); EMS, Grünthal (1993, 1998); and People's Republic of China, Hsieh (1957).

One important question faced by those who use macroseismic intensity values for scientific analysis is how the intensity values from one scale relate to those from another scale. Reiter (1990) showed graphically how some of the most commonly used macroseismic intensity scales approximately relate to each other (Figure 12.1).

12.2.2   Isoseismal maps

Once a number of macroseismic intensity values for an earthquake have been accumulated from different localities, it is the normal practice to display

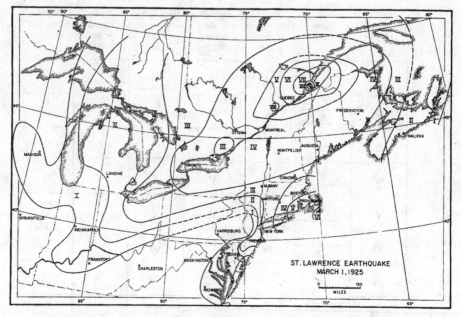

Figure 12.2 Isoseismal map of the earthquake of 1 March 1925 that was centred along the St Lawrence River in Quebec Province, Canada (from Smith, 1966). The Roman numerals indicate the Modified Mercalli intensity values. The solid lines show well-determined isoseismals separating regions of different intensity, while the dashed lines show areas where the isoseismal contour has been extrapolated.

them on a *seismic intensity map*. Such maps generally show areas where the highest levels of ground-shaking were experienced as well as the spatial decay of ground-shaking away from the area where the earthquake was centred. On seismic intensity maps, contours dividing regions of different macroseismic intensity values, called *isoseismals*, are usually drawn (Figure 12.2). Sometimes, maps are created that only show the isoseismals and not the intensity observations upon which they are based. Other maps may show only partial intensity contours where constrained by the observational data. On some such maps, the compiler may draw extrapolations of the contours through the areas with no observations in order to show the complete expected extent of the partial contour that was defined by the data.

Seismologists have no exact rules explaining how to draw the isoseismals on a seismic intensity map. Because the strongest earthquake ground-shaking is expected at or very near the earthquake epicentre with decreasing ground-shaking with distance from the epicentre, seismologists usually draw isoseismals as a set of concentric contours of decreasing value away from the epicentre. Sometimes isoseismals are drawn as rather smooth ovals or circles

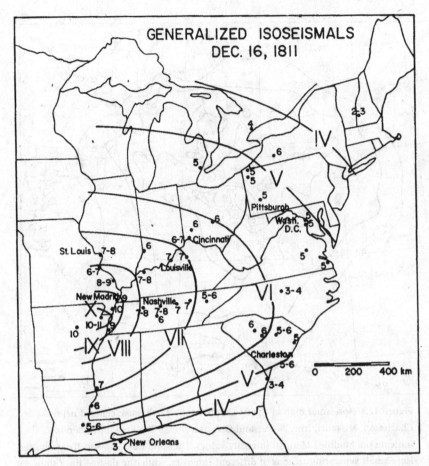

Figure 12.3 Isoseismal map of the 16 December 1811 earthquake centred near New Madrid, Missouri, from Nuttli (1973). Each labelled contour is the approximate outer limit where that Modified Mercalli intensity was probably experienced. The Arabic numbers at different localities on the map show the estimated Modified Mercalli values for those sites based on historic reports.

(Figure 12.3), reflecting an idealized view of how seismological theory expects the isoseismals to appear. Other interpreters have drawn very detailed isoseismals for some earthquakes (Figure 12.4), taking care to show the many areas where the earthquake effects were either enhanced or diminished compared to surrounding areas. For example, towns in river valleys often report stronger levels of ground-shaking than towns on upland or mountainous areas. This can lead to isoseismals following the courses of rivers or valleys in a rather finger-like fashion or to locally closed contours containing higher degrees of intensity (Figure 12.5).

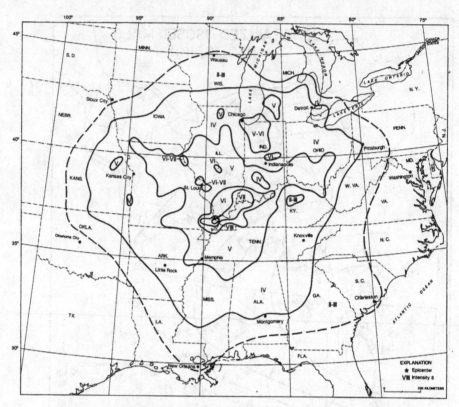

Figure 12.4 Isoseismal map of the 31 October 1895 earthquake centred near Charleston, Missouri, from Stover and Coffman (1993). The Roman numerals indicate the Modified Mercalli intensity values. The solid lines show well-determined isoseismals separating regions of different intensity, while the dashed lines show areas where the isoseismal contour has been extrapolated.

Multiple intensity reports from a local area such as a town or city often show differences of one or more intensity degrees. These differences arise in part from natural local variations in earthquake shaking at sites separated in distance by a few kilometres or less; they also exist because the sources of the reports are untrained observers who may not remember the earthquake effects exactly or who may distort their reports of the event. Thus, there can be an ambiguity in the mind of a seismologist concerning what intensity value should be assigned to a town or city where multiple intensity levels have been reported. If damage to some structures was reported, most seismologists will assign a macroseismic intensity value to that locality that reflects the reported damage, even if most structures sustained no damage from the earthquake. Exceptions to this tendency arise when the seismologist suspects that the reported damage may not have been caused by the earthquake but was a pre-existing condition. For towns

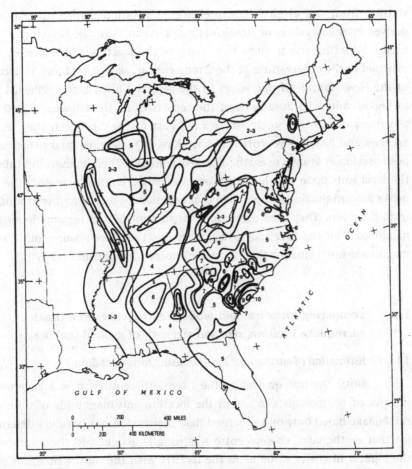

Figure 12.5 Isoseismal map of the 31 August 1886 earthquake centred near Charleston, Missouri, from Nuttli *et al.* (1986). The Arabic numbers on the map show regions where that particular Modified Mercalli intensity was experienced. Many of the elongated, closed contours follow river valleys, especially the Mississippi and Ohio rivers.

or cities where no damage was reported but where a diversity of intensity levels are found, some seismologists will assign an intensity value consistent with the most frequently reported intensity while others will assign a value reflecting some other aspect of the local reports (such as the higher value if two different intensity levels were commonly observed).

Research seismologists regard published earthquake intensity maps as the particular interpretations of earlier researchers who used subjective data and often ill-defined methods to interpret those data. For this reason, many modern researchers prefer to conduct new analyses of historical earthquakes by starting from the original macroseismic observations rather than from the derived intensity values or isoseismal maps. As an example, Hough *et al.* (2000) used original reports to show that many of the macroseismic intensity values assigned to the observations of the strong earthquakes in 1811 and 1812 centred in the New Madrid seismic zones in the central United States reflected a bias to higher values because most of the reports came from towns located along the rivers of eastern North America. As part of their research they searched archives and found new written descriptions of these earthquake that had not been previously known to earthquake seismologists. Their analysis indicates that the local soils upon which these towns were sited tended to amplify the earthquake ground-shaking compared to that experienced on the higher ground away from the rivers. They used corrections for this site bias to recompute estimated magnitudes for the 1811 and 1812 earthquakes that were about about 0.5 to 0.8 magnitude units smaller than those determined by Johnston (1996b).

12.3   Comparing historical and modern earthquakes to estimate earthquake location, size and strength of ground-shaking

12.3.1   *Estimation of earthquake location using historical data*

Until the last quarter of the twentieth century, it was generally the practice of seismologists to assign the location and magnitude of a historical earthquake based only on the largest macroseismic intensity values determined for that earthquake. The epicentre was usually assigned to the location with the highest intensity value or to the locality with the most damage. If several localities reported equal intensities or comparable damage, the judgement of the compiler of the earthquake catalogue was used to affix an epicentre to the event.

For many years, the shortcoming of this traditional method for assigning locations to historical earthquakes has been well recognized. Most modern earthquake epicentres do not lie directly under a city or town from which an intensity determination was derived. In fact, for many modern earthquakes the greatest macroseismic intensity may be found many kilometres from the fault upon which the earthquake takes place. This is especially true for towns where local soils significantly amplify the ground-shaking, causing stronger ground-shaking and damage than at towns and cities founded on hard rock even if they are closer to the earthquake epicentre. For example, in 1979 there was a magnitude 5.1 earthquake with an epicentre in the hills of central Kentucky near the town of Sharpsburg. There was a region of MMI VII reports within about 20 km of the epicentre, but the town of Maysville, located about 50 km north of the epicentre, also experienced localized MMI VII ground-shaking (Mauk *et al.*, 1982). The damage to structures at Maysville apparently resulted from local ground-shaking amplification due to the soft Ohio River bottom soils upon which this town was built. Focusing or defocusing of seismic energy by underground rock formations and the effect of surface hills and valleys can also cause significant local variations in macroseismic intensity reports, as can variations in the way the rock slip occurred on the fault during an earthquake. For all of these reasons, it has long been recognized that the locality that experienced the maximum macroseismic intensity may be at some distance from the instrumental earthquake epicentre and from the fault on which the earthquake rupture took place.

Even though the locality or localities where the maximum macroseismic intensity was experienced may not always be found at the epicentre of an earthquake, intensity and isoseismal maps for modern earthquakes where many intensity reports were collected at a large range of epicentral distances demonstrate that on average there is a general decrease of intensity value with distance from an earthquake epicentre (Figure 12.6). Furthermore, this is expected from the basic propagation laws of seismic waves, which are known to decrease in amplitude as they spread away from an earthquake focus. Thus, seismologists would expect to observe smaller average macroseismic intensity values at large distances from an earthquake epicentre and larger average intensity values at small distances. The epicentre of an earthquake should lie somewhere at or near the centre of the intensity observations.

Using this idea, the epicentres of some historic earthquakes have been estimated from macroseismic intensity maps, especially those for which isoseismals had been drawn. This idea has been employed especially for those earthquakes where an earthquake was felt along a coastal region and the epicentre was suspected to lie somewhere offshore. Figure 12.7 shows an example of such an analysis, for the case of the major catastrophic earthquake at Lisbon, Portugal on 1 November 1755. This earthquake caused damage on the southwestern Iberian peninsula, was felt throughout much of western Europe and northwestern Africa, and raised a damaging tsunami along the Portuguese coast. A generalized isoseismal map for this earthquake from Johnston (1996b) shows smooth isoseismal arcs, and the centres of the isoseismals converge in the Atlantic Ocean offshore of the southwestern tip of Portugal. This offshore epicentre can explain the sizeable tsunami that was generated by the earthquake, which Johnston (1996b) estimated had a moment magnitude of about 8.4. While not giving a very accurate estimate of the location of the earthquake epicentre, this method does permit seismologists to use onshore macroseismic observations to estimate offshore earthquake locations.

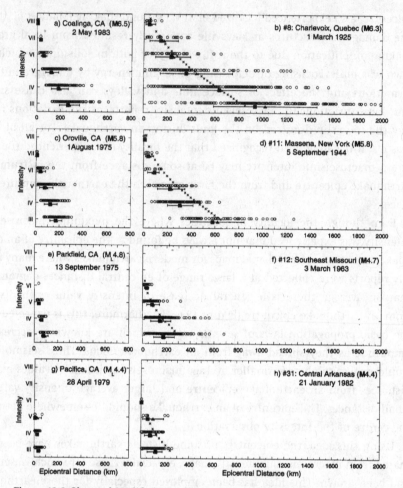

Figure 12.6 Observed Modified Mercalli intensity values plotted versus distance from the earthquake epicentre for several earthquakes in California (left column) and in eastern North America (right column), from the study of Bakun *et al.* (2003). The circles are the individual intensity values, while the average intensity values are shown as solid diamonds. The solid square shows the median distance for each intensity value (i.e. exactly half of the individual intensity values are on either side of the median value). The horizontal bars show the range of distances where 67% of the observations are found. Differences in the decrease of macroseismic intensity with distance for the two regions is quite obvious in these plots.

These earlier methods for earthquake location rely heavily on the interpretation of the person assigning the macroseismic intensities or drawing the isoseismal maps. Thus, their results are quite subjective and open to reinterpretation. In the 1960s and 1970s, better statistical models for analysing data sets with significant uncertainties combined with rapidly improving computing capabilities

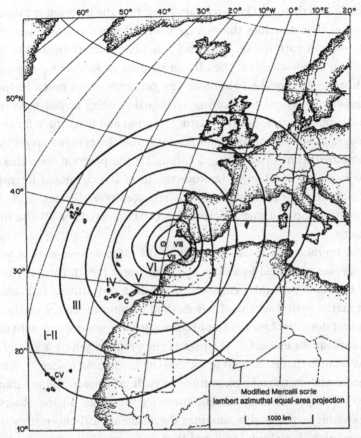

Figure 12.7 Isoseismal map of the 1 November 1755 earthquake, from Johnston (1996b). The epicentre of the earthquake is shown as a circle with a dot within the intensity VIII contour.

meant that the analyses of macroseismic observations for historical earthquake locations and magnitudes could be moved to a more quantitative basis. Beginning in the 1980s seismologists started developing methods that averaged many macroseismic observations at multiple distances from an earthquake epicentre to estimate the location of that epicentre. The greatest advantage of these new methods is that they make use of some or all of the available data for an earthquake, meaning that they are much less biased by the intensity value assigned to just one or a few localities or by the subjective way that isoseismals are drawn.

One statistically based method to locate an earthquake using all of the available macroseismic intensity data was developed by Seeber and Armbruster (1987). This method uses a computer algorithm called MACRO, in which the earthquake location is determined that gives the closest match to all of the intensity values for the earthquake. One necessary input into the program is a mathematical function that describes the average decay of macroseismic intensity with distance from an earthquake. Once this function is determined from modern earthquake data for a region, then it can be used in MACRO for the computation of historic earthquake epicenters. The program works much in the same way as modern methods for locating earthquakes using arrival time readings from seismographic stations: (1) a starting location and magnitude for the event is guessed; (2) the program computes how well the observations are fit by intensities computed from this starting location; (3) the program then determines where to move the location to improve the fit of the computed intensities to the observations. This process is typically repeated several times until the best match of the computed and observed values is achieved for all the intensity observations.

Another method that makes use of all of the macroseismic data to locate an earthquake was published by Bakun and Wentworth (1997). Their method uses many of the same principles employed in MACRO, but utilized in a somewhat different mathematical scheme. Their method requires as input a mathematical function that describes how the amplitudes of seismic waves decay with distance from an earthquake epicentre. The user of the program selects a grid of points that cover an area that is thought to include the earthquake epicentre. Assuming an epicentre and earthquake magnitude at each grid point, the program then computes the expected macroseismic intensity values for all the observations for the earthquake, and it also computes the mathematical uncertainty that the earthquake might have been located at that grid point. The grid point with the smallest uncertainty is taken as the most probable epicentre for the earthquake. Thus, rather than trying to find the best match of the theoretical and observed intensity values as in MACRO, the Bakun and Wentworth (1997) method looks for the smallest uncertainty in the computed event location and magnitude.

Gasperini et al. (1999) proposed a statistical method to find the geographic centre of the strongest macroseismic observations, which they called the earthquake barycentre (Gasperini and Ferrari, 2000). Using observations of the strongest macroseismic intensities for an earthquake, the Gasperini et al. (1999) method calls for a series of statistical tests on the data to find a set of macroseismic epicentre estimates and measures of how well each fits the observations. The macroseismic epicentre is selected from within the geographical area defined by the set of individual epicentre estimates.

A quite different approach for estimating the epicentre of a historical earthquake was taken by Ebel (2000) for an earthquake of magnitude 5.6 which took place in Massachusetts in 1727. That earthquake was followed by a very active and protracted series of aftershocks that were individually listed by local observers

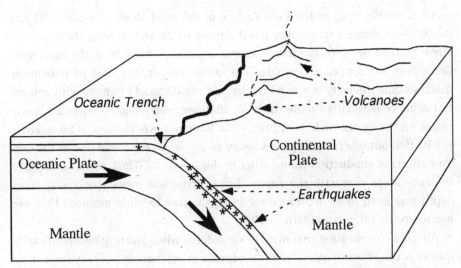

Figure 12.8 Schematic drawing of a subduction zone, where an oceanic plate on the left subducts beneath a continental plate to the right. A line of mountains with active volcanoes is found on the continent parallel to the shoreline. Earthquakes take place along the upper interface of the subducting oceanic plate.

in church records and diaries. Aftershocks are small earthquakes that take place along the fault that had movement during the main earthquake, and the smaller the earthquake the smaller the area over which it is felt. Thus, if many aftershocks are felt at only one or a few localities in close proximity to each other, then these localities must be very close to the main earthquake fault. Using this reasoning, Ebel (2000) estimated the location of the 1727 earthquake using the felt locations of the small aftershocks rather than macroseismic intensities for the main shock. For this earthquake the greatest macroseismic intensities in the main shock were reported from the same localities where the largest number of aftershocks were felt, thus providing another line of evidence to argue for the location of the mainshock epicentre.

Modern knowledge of plate tectonics and how plate tectonics affect the distribution of earthquakes throughout the world helps seismologists interpret some reports of historical earthquakes. For example, in a number of places on the Earth there are *subduction zones*, places where a section of oceanic crust is sliding back into the Earth beneath another tectonic plate (Figure 12.8). Some subduction zones occur beneath chains of islands (called *island arcs*), such as beneath the Aleutian, Kurile or Vanuatu islands. Other subduction zones are found beneath large islands, such as at Japan or the Philippines. Still other subduction zones are found beneath continents, such as along the west coasts of Central America and South America. At subduction zones, earthquake foci can occur at depths ranging from the surface of the earth down to almost 700 km depth. For shallow earthquakes (focal depths of 20 km or less), there is usually a marked decay of the macroseismic intensities away from the epicentre. Conversely, for deeper earthquakes (30 km or deeper) the area of maximum macroseismic intensity is much broader and the decay of intensity with epicentral distance is usually smaller than for shallower earthquakes. There are a few places on the Earth, such as at Vrancea in Romania, where there is no modern subduction but where earthquakes below 50 km depth are found due to the leftover effects of subduction from earlier geological times. Thus, since earthquake intensity maps vary with the depth of the earthquake focus, earthquake focal depth may need to be accounted for in earthquake location methods that use macroseismic intensities (Båth, 1979).

Another complication that must be considered when locating historical earthquakes is the effect of extended fault rupture in earthquakes of very large magnitude. A number of studies, such as that of Wells and Coppersmith (1994), document the lengths and widths of faults that slip in large earthquakes. For earthquakes above about magnitude 7.0, the fault rupture zone can be long enough that the area experiencing the maximum earthquake shaking, and hence maximum intensity, will be elongated along the fault zone. For earthquakes of magnitude 8.0 and greater, the zones of maximum shaking will be 100 km long or more. An example of a very long fault zone that ruptured in a single event is that of the 1906 San Francisco earthquake (Figure 12.9), where a fault rupture of over 300 km was documented. Many earthquakes with long fault ruptures take place at subduction zones, where the fault rupture might be confined to an offshore area but where the intensity reports are confined to onshore areas. In such cases, the isoseismals will outline elongated areas parallel to the offshore rupture, and the distance from the fault rupture is more important than the distance to the epicentre in controlling the strength of the earthquake ground-shaking at onshore sites. Consistency with plate tectonics has become an important constraint in the seismological interpretation of historical earthquake data.

12.3.2 Estimation of earthquake magnitude and seismic moment using historical data

In earlier earthquake compilations, the greatest macroseismic intensity found for an earthquake (I_0 or Imax) was frequently used as a primary measure of its size, and twentieth-century compilers of earthquake catalogues almost invariably listed I_0 for each historical event. After the first instrumental earthquake magnitude scale was introduced by Richter (1935), seismologists became interested in finding the magnitudes of historical earthquakes. Since there were no instrumental recordings for most historical earthquakes,

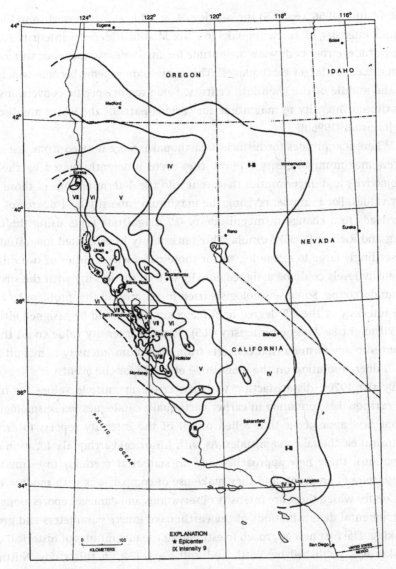

Figure 12.9 Isoseismal map of the 18 April 1906 earthquake that had an epicentre near San Francisco, California, from Stover and Coffman (1993). The Roman numerals indicate the Modified Mercalli intensity values. The solid lines show well-determined isoseismals separating regions of different intensity, while the dashed lines show areas where the isoseismal contour has been extrapolated.

seismologists were forced to estimate earthquake magnitudes from the macro-seismic intensity effects. Before the late 1970s, the only method in use to esti-mate the magnitudes of historic earthquakes was based on a direct conversion from maximum macroseismic intensity to magnitude using a relation that had been derived from a number of empirical observations. For example, Gutenberg and Richter (1956) reported the relation $M = 2/3\,I_0 + 1$ to convert from the maximum intensity of an earthquake (on the Modified Mercalli intensity scale) to magnitude (either body-wave magnitude for smaller earthquakes or surface-wave magnitude for larger earthquakes). This conversion scheme became widely used in the middle of the twentieth century. Some region-specific conversions from maximum intensity to magnitude for various parts of the world are discussed by Johnston (1996b).

Where magnitudes for historical earthquakes were derived from just one or a few maximum intensity reports, they were inherently biased by the same subjectivity and uncertainties that went into the determinations of those intensity values. For example, revising the maximum intensity by 1 degree of intensity leads to a change in magnitude of 0.7 magnitude units using the Gutenberg and Richter (1956) formula. This uncertainty in estimated magnitude was unsettlingly large to seismologists for those cases where either of two different intensity levels could be assigned equally well to the locality with the strongest ground-shaking. Some seismologists tried to get around this problem by assigning intensity at the 0.5 degree level (i.e. a site that could be assigned either VI or VII might be assigned intensity VI.5). Such an intensity value could then be input into any formula that converts from maximum intensity to magnitude to get a finer resolution on the magnitude estimate for the event.

By the 1970s, dissatisfaction with the crude magnitude values for historical earthquakes contained in earlier earthquake catalogues led seismologists to adopt new approaches that relied on all of the intensity reports to estimate historical earthquake magnitudes. As with historical earthquake location determinations, these new approaches rely on statistical methods to estimate the magnitude of an event, and they make use of comparisons with modern earthquakes for which there are intensity observations and damage reports along with instrumental determinations of the earthquake source parameters and ground-shaking. The first new approach to estimating the magnitudes of historical earthquakes was published by Nuttli and Zollweg (1974). In this study, Nuttli and Zollweg (1974) showed that the total felt areas of modern earthquakes increased systematically with the magnitudes of those earthquakes, and they developed a relationship between felt area and magnitude for the central and eastern United States. This idea was used by such investigators as Street and Lacroix (1979) to determine the magnitudes of historical earthquakes throughout the eastern half of North America.

A comprehensive global study of felt effects and earthquake magnitude in stable continental regions was conducted by Johnston (1996a, 1996b). In particular, Johnston (1996b) examined how the felt area as well as the area within each isoseismal of increasing value correlated with magnitude for a large number of instrumentally recorded events. He presented a hierarchy of methods for analysing earthquake intensities and isoseismal maps to determine the best estimate of the magnitudes of historical events. He also analysed the amount of uncertainty in magnitude estimates from intensity data. Since for many historical earthquakes it is much easier to determine the outer limits of the macroseismic intensity area at some higher intensity level than it is to find the total felt area (especially if the earthquake happened at night, offshore or near a coastal area, or in a region with a large unpopulated area), methods such as those of Johnston (1996b) eliminate the necessity of finding the total felt area of an earthquake when a magnitude estimate must be found from macroseismic intensity data.

Another approach to estimating the magnitude of an historic earthquake was used by Ebel (2000) for the 1727 earthquake in northeastern Massachusetts. In this study, Ebel (2000) found the earthquake that best matched the observed decrease of observed macroseismic intensities with distance from the estimated earthquake epicenter to predictions of those intensities from a mathematical relation that gives MMI as a function of event magnitude and distance from the epicentre. He preferred his magnitude of 5.6 for this earthquake over an earlier magnitude of 5.0 published by Street and Lacroix (1979), since the latter study were forced to compute felt and MMI IV areas by guessing about three-quarters of those isoseismal contours due to lack of observations in some onshore areas and in the offshore areas to the east of the epicenter. Both the Ebel (2000) and the Street and Lacroix (1979) studies used all of the intensity observations to estimate the magnitude of this earthquake, but each used those observations in a different way. Both studies would have benefited from the inclusion of more macroseismic intensity observations at all distances around the earthquake epicentre.

Because one must know the epicentre of an earthquake or the location of the fault upon which an earthquake took place before the magnitude of the earthquake can be properly determined, some of the new statistical methods combine the determination of epicenter and magnitude into a single calculation. This is true for both the MACRO method of Seeber and Armbruster (1987) and the Bakun and Wentworth (1997) method. The MACRO method effectively makes a determination of the sizes of different isoseismal contours as part of its computation, and these are then used for magnitude estimations. The Bakun and Wentworth (1997) method includes the best estimate of the event magnitude as well as earthquake epicentral location by finding the location/magnitude combination with the smallest uncertainty.

Just as for earthquake location, the depth of an earthquake also has a direct influence on the estimate of event magnitude from intensity data. As earthquake hypocentres get deeper, the epicentral intensity tends to decrease more slowly with distance, while the felt area tends to increase (Blake, 1941). Furthermore, the deeper the earthquake faulting, the smaller the maximum intensity experienced at the Earth's surface for a given magnitude. These effects are not accounted for in the methods described in the previous two paragraphs, since those studies looked at regions where the deepest earthquakes extend down only to about 25 km, and for such shallow-focus earthquakes the focal-depth effects on the surface intensity patterns are not significant.

As described for earthquake locations earlier, very large earthquakes can have a fault rupture that extends many tens of kilometres, and for the largest known earthquakes many hundreds of kilometres. For such very large earthquakes, evidence of the extent of the fault rupture can be used to estimate the size of the earthquake, as was done in such studies as that of Kelleher *et al.* (1973) for large earthquakes at the Pacific and Caribbean plate boundaries. In fact, for earthquakes above about magnitude 7.5, a determination of the extent of fault rupture may be the best way to estimate the size of the earthquake. One important way to delimit the fault size associated with an earthquake rupture is to find the spatial extent of the smaller aftershocks that follow the main shock. The wider the area over which the aftershocks extend, the greater the magnitude of the main shock. The fault rupture lengths, and hence estimated magnitudes, of many of the earthquakes in Kelleher *et al.* (1973) were based on reports of aftershocks along with determinations of the areas where the strongest ground-shaking was experienced.

Tsunami observations can also aid in assessing the magnitudes of large earthquakes that took place offshore. Several magnitude scales have been proposed based on tsunami observations (Papadopoulos, 2003), with the Abe (1979, 1989) magnitude scale probably being the one most commonly used today. At a fixed distance from an earthquake fault, the height of a tsunami on tide gauges approximately scales with the tsunami magnitude of the earthquake (Papadopoulos, 2003), although *earthquake directivity* effects (the focusing of seismic energy in one particular direction as rupture spreads along a fault) can enhance the tsunami in some directions and diminish it in other directions. During the twentieth century, earthquakes that approached or exceeded moment magnitude 8.5 at some subduction zones caused tsunamis that caused damage at large distances across major ocean basins. For example, Hawaii was inundated by tsunamis from major earthquakes off southern Alaska and in the Aleutian Islands in 1946, 1957 and 1964. Thus, reports of tsunamis at large distances across oceans from the earthquake that generated the tsunamis give an indication that the earthquake may have equalled or exceeded moment magnitude 8.5.

While magnitude is the most frequently cited measure of earthquake size by the general public, a much more fundamental measure of an earthquake is the seismic moment. Seismic moment is a better indication of the true size of an earthquake than are the classical earthquake magnitude scales such as the Richter (or local) magnitude, the body-wave magnitude and the surface-wave magnitude. Seismic moment is also a parameter that is directly related to the total rock deformation in an area caused by local plate tectonic movements. Today, GPS observations of regional tectonic deformations of the Earth's surface are being compared to the spatial and temporal rates in the seismic moment release in the earthquakes of an area.

A number of studies have looked into the question of how to estimate the seismic moment of an earthquake from macroseismic intensity data. For example, Johnston (1996b) argued for continental seismic events that the felt area and the areas within the different isoseismals of an earthquake can be used to estimate the seismic moment of that earthquake. Gasperini and Ferrari (2000) surveyed several different methods to estimate seismic moment from macroseismic intensity observations. Using the results of such analyses combined with assessing the fault size and orientation from intensity data, Gasperini and Valensise (2000) argued that fault segmentation and earthquake recurrence models can be derived, improving estimates of the seismic hazard of a region.

Analysis of the geological effects of historical earthquakes has developed into another tool that can be used to make determinations of the sizes of past seismic events. Two different lines of geological research have been successful at constraining the magnitudes of some past historical seismicity.

The first line of research involves estimating the magnitude of earthquakes from the ground deformations observed in trenches dug across the fault that generated large earthquakes that were experienced in historical time (McCalpin, 1996). Obviously, this idea requires that the fault that generated the earthquake be known, be observable at the ground surface, and can be studied at localities where the local geological conditions across the fault are favourable such that the fault movement can be dated and the amount of fault movement in each documented earthquake can be estimated. In some cases, it was through geological work that the fault that generated a documented historical earthquake was verified and the size of the historical earthquake was determined. Such was the case of the 1812 earthquake in southern California, which had long been known only from historical reports at the San Juan Capistrano mission and other coastal missions and settlements in California.

A second geological method for estimating the magnitude of past historical earthquakes involves studies of secondary ground deformation features, such as sand blows and lateral spreads, that are caused by large earthquakes. For example, some historical reports contain descriptions of earthquake-induced sand blows, and sometimes these sand blows can be located and studied in detail. Sand blows are generated only when earthquake ground-shaking exceeds some lower limit that depends on the size of the earthquake, the frequency content of the earthquake shaking and the distance of the site from the earthquake fault (Obermeier and Pond, 1999). Thus, the existence of a sand blow provides some minimum constraint on the magnitude of its causative earthquake, especially if the location of the earthquake faulting is known. The sizes and spatial extent of sand blows also provide important constraints on the size of the earthquake that generated the features (Obermeier and Pond, 1999). Thus, verifiable historical reports of sand blows, lateral spreads, landslides and other ground deformations that are induced by strong earthquake ground-shaking are of great interest to seismologists and geologists who are looking to determine the sizes of the earthquakes that generated these features.

A unique historical record: turning data on past Italian earthquakes into virtual tectonic information

Italy represents unusually fertile ground for the application of the methods of historical seismology. This is due both to the availability of an exceptional amount of data that exists on past earthquakes and to the development over more than 20 years of highly specialized research methods to mine those data for information on historical earthquakes. These circumstances have resulted in exceptionally rich and homogeneous catalogues of earthquake data, the *Catalogue of Strong Earthquakes in Italy* (Boschi et al., 1995 and subsequent updates published in 1997 and 2000; Guidoboni et al., 2007) being the most significant example. The quality of this large data set has favoured a new trend in the use of historical data; in addition to bringing information on the location and severity of an earthquake, intensity data are now being used to derive quantitative tectonic information.

How has this been accomplished? Active faults and intensity data have indeed very little in common. In fact, they represent such different phenomena that until just a few years ago the only commonality between them was the derivation of possible fault orientations from the elongation of subjectively drawn isoseismals. Gasperini et al. (1999) and Gasperini and Valensise (2000) proposed a strategy for using historical information to obtain equivalent seismogenic sources, which are described by the same set of physical parameters normally used to describe instrumentally documented sources, although with a lesser degree of confidence.

Historically derived seismogenic sources have been extensively used by the compilers of Italy's *Database of Individual Seismogenic Sources* (Valensise and Pantosti, 2001, and its current update, DISS Working Group, 2007; Basili *et al.*, 2008) to complement and strengthen the usually limited instrumental or surface faulting evidence upon which the current earthquake distributions and recurrence models are based. Although the procedure was initially conceived to exploit the rich Italian historical record, it was eventually extended to several European countries in the framework of the project Faust (FAUlts as a Seismologists' Tool) funded by the European Community (Mucciarelli and FAUST Working Group, 2000). The information gathered by this project was later organized into the Database of Potential Sources for Earthquakes Larger than M 5.5 in Europe (Valensise *et al.*, 2002).

The proposed analytical strategy of the DISS involves three fundamental steps:

(1) estimating the location of the seismogenic source, expressed as the centre of the distribution of damage, and its size, to be derived from the overall extent of the damage pattern;

(2) assessing the physical dimensions of the seismogenic source (length and width) using empirical relationships (Wells and Coppersmith, 1994);

(3) calculating the orientation of the seismogenic source using an original algorithm based on the highest intensity data points.

The estimated tectonic parameters are calibrated using earthquakes for which both intensity and modern instrumental data are available. The procedure is most accurate for earthquakes of M 5.5 and greater, but the earthquake location and size can also be confidently obtained for smaller events. Each historical earthquake source is then conceptually and graphically delineated as an oriented rectangle representative of the fault at depth. This rectangle is meant to represent either the actual surface projection of the seismogenic fault or, at least, the projection of the portion of the Earth's crust where a given seismic source is more likely to be located. Gasperini *et al.* (1999) summarized all these capabilities in a user-friendly software package named 'Boxer' that can be easily downloaded from the internet and freely used (currently at http://ibogfs.df.unibo.it/user2/paolo/www/boxer/boxer.html).

In the framework of the DISS database the method was applied systematically to all Italian earthquakes having $M > 5.5$ and for which the historical record is sufficiently rich. Figure 12.10 shows the virtual seismogenic sources obtained

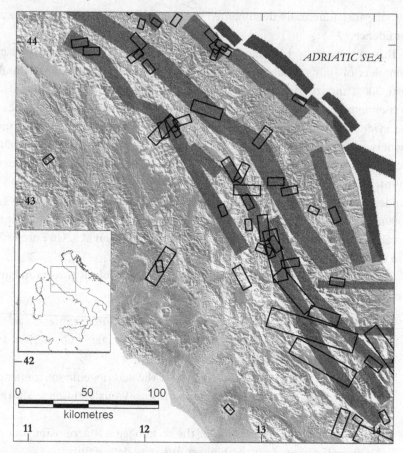

Figure 12.10 A snapshot of the DISS 3.0.4 database for a portion of central Italy. Virtual seismogenic sources obtained using Boxer for earthquakes of M 5.5 and greater (white empty boxes) are plotted above Seismogenic Areas (elongated grey patterned areas) from the latest version of the database. Intensity-based sources help to delineate the main tectonic trends in areas where the evidence is elusive or ambiguous. They also carry useful information on which parts of the main fault systems have ruptured historically and which are currently quiescent.

from Boxer plotted above Seismogenic Areas from the latest version of DISS (3.0.4).

The procedure described allows identified seismogenic zones to be located more precisely, especially in areas of diffuse blind and hidden faulting, and their seismogenic potential and behaviour to be assessed more confidently. The existence of additional and often perfectly unknown seismogenic faults can be inferred from the analysis of the spatial relations between adjacent intensity-based sources, or between historical and instrumental earthquakes, or by recourse to more focused geological observations. The process allows the extent of overlap between adjacent sources, that is, the regularity of the seismic release in space, to be tested (if the sources tend to a *characteristic* behaviour they should not overlap nor leave unruptured gaps between them). Hence, the systematic quantification of historical earthquakes through this procedure may have obvious and important implications for the assessment of regional seismic hazard.

(This section was written by Gianluca Valensise, Istituto Nazionale di Geofisica e Vulcanologia, Rome.)

12.3.3 *Estimating the level of ground-shaking in historical earthquakes*

Modern seismic hazard studies for a region compute the spatial patterns of ground-shaking that can be expected from future earthquakes in the region. Many seismic hazard studies employ a probabilistic seismic hazard analysis (PSHA) methodology in what the ground-shaking levels at a site from all future earthquakes in the surrounding region are computed, and then the probabilities of different levels of ground-shaking in some specified time period (such as 50 years) are determined. Maps of the ground-shaking levels that have only some small chance of being exceeded (such as only a 2% chance of being exceeded in 50 years) are a typical result of PHSA projects, and these maps can form the basis for building codes that specify seismic resistance in the construction of buildings and other structures. While not an earthquake source parameter, earthquake ground-shaking estimates in historical earthquakes are important for seismic hazard studies.

Just as for earthquake magnitude determinations, the classical seismological approach to estimating the quantitative levels of ground-shaking at a site that experienced an earthquake was to make a direct conversion from macroseismic intensity to some measure of ground-shaking. The common measure that was used was the *peak ground acceleration* (PGA), which is the highest ground acceleration value that would be measured by a seismic instrument called an *accelerograph* that was designed to sense and record ground accelerations. Another simple ground-motion measure that better correlates with earthquake damage is *peak ground velocity* (PGV). Conversions between an intensity scale such as MMI and the ground-motion measures pga and pgv were developed from modern earthquakes for which both determinations were available (e.g. Trifunac and Brady, 1975; Murphy and O'Brien, 1977; Wald *et al.*, 1999b), and these could then be applied to the macroseismic intensity determination from historical earthquakes. Such conversion schemes from macroseismic intensity to peak ground acceleration or velocity suffer from all of the same drawbacks as the direct relations of maximum intensity to earthquake magnitude.

Recent research by seismologists has led to more sophisticated methods to estimate earthquake ground-shaking levels from historical reports of ground-shaking. Atkinson and Sonley (2000) Ebel and Wald (2003) and Kaka and Atkinson (2004) have developed mathematical relationships that use the macroseismic intensity level, the distance from a site to the earthquake epicentre and the magnitude of the earthquake to convert from Modified Mercalli intensity to various ground-shaking measures for earthquakes in the eastern and western United States. These methodologies operate under the assumption that two sites with the same intensity value but at different epicentral distances experienced different levels of ground-shaking in the earthquake.

Conversions from the amounts of damage in historical earthquakes to estimates of the levels of ground-shaking experienced at those damaged sites have also undergone new analyses in recent times. Formerly, this was carried out by finding the macroseismic intensity for a locality based on the reported damage, and then converting from intensity to ground motion level. New studies attempt to determine the ground motion level directly from the kind and amount of reported damage at the locality. For example, Whitman (2002) and Ebel (2006) used the percentage of damaged chimneys in Boston in the 1755 Cape Ann earthquake along with the types and amounts of reported damage to masonry structures to estimate the PGA in Boston due to that earthquake. The assumption behind this work is that a community with a majority of its chimneys damaged in an earthquake must have experienced stronger ground-shaking than a community where only a very small percentage of its chimneys were damaged. As such analysis methods are developed further, there will undoubtedly arise increased interest in the details of damage to cities and towns that experienced historic earthquakes.

12.4   Estimating tsunami parameters from historical data

Tsunamis have been reported from all of the world's major ocean basins (Pacific, Atlantic and Indian) as well as from large inland seas such as the Mediterranean. Most tsunamis are generated by undersea earthquakes, and those coastal areas at subduction zones that experience strong earthquakes are the most prone to damaging tsunamis. Thus, the subduction zones along the Pacific coasts of the Americas and the island-arc regions of the western Pacific Ocean all are susceptible to strong tsunamis, as is the Indonesian island-arc system. Numerous damaging tsunamis have been reported from islands in the Caribbean, from Italy, Greece and the eastern Mediterranean coasts, and from the Atlantic coasts of Portugal and Spain. In addition to the tsunamis generated by undersea earthquakes, some tsunamis have been caused by major eruptions of volcanic islands, and a few have been caused landslides of large masses of rock off of hillsides into a body of water at the base of the mountain. Volcanic eruptions that are followed by a collapse of the volcano below sea level can cause significant tsunamis, as happened during the 1883 eruption of Krakatoa or the great Minoan eruption of the island of Santorini. In a few instances, major landslides have caused locally strong tsunamis, even in relatively small water bodies like inland coves and seas.

The magnitude 9.0 earthquake in December 2004 at Sumatra and the resulting tsunami that spread across the Indian Ocean renewed worldwide interest in the problem of damaging tsunamis that strike at large distances from the earthquakes that generated the tsunamis. While the tsunamis from most strong earthquakes under the ocean floor are damaging only along coastlines near the fault on which the earthquakes take place, the 2004 tsunami highlighted once again the fact that some major tsunamis can travel across ocean basins and cause damage along shorelines at great distances, even thousands of miles, from the location of the earthquake that generated the tsunami. For example, the Hawaiian Islands have experienced several catastrophic tsunamis that arose from earthquakes that were centred thousands of kilometres from the islands. This same phenomenon of transoceanic tsunamis certainly must have happened in the historical past, and it represents a special challenge for research into historical tsunamis and their causative earthquakes. Not only must research be carried out to determine the extent of the coastal regions affected by a tsunami and to estimate the height of the tsunami at different coastal locations, so also evidence needs to be sought that points to the location of the earthquake or other cause that generated the tsunami. The researcher must not lose sight of the possibility that the earthquake that was the source of a tsunami was not local but rather may have been centred on some submarine fault that is at a great distance from the localities where the tsunami was reported. Thus, matching up tsunami reports to the location of the earthquake that caused the tsunami and estimating the magnitude of an earthquake that generated a historical tsunami can be quite difficult challenges when the historical reports about the tsunami and its causative earthquake are fragmentary or spatially limited. This is especially a problem when the historical record contains indications of a tsunami but no reports of contemporaneous seismic shaking, especially in cases where it might not be possible to rule out a volcanic eruption or landslide as the cause of the tsunami.

12.4.1 Evaluating tsunami inundation and run-up

For all tsunami reports, regardless of whether the source of the tsunami was local or was thousands of kilometres away, the most important parameter that scientists want to understand is amount of sea-level change associated with the tsunami. Sea-level change in the open ocean can be measured directly today by specially instrumented offshore buoys and underwater sensors, but such equipment did not exist in historical times. Remote sensing measurements from satellites can also give a snapshot of sea-level height changes in the open ocean. The most important way that sea-level rise can be estimated along an affected shoreline is to determine the tsunami *inundation*, or the most inland point affected by the tsunami, and then to observe the elevation relative to sea level of this inland point, called the tsunami *run-up*.

Estimations of tsunami inundation, run-up and sea-level rise at many places along a coast are critical to scientists who wish to discover the true strength and extent of the tsunami. Local nearshore conditions have a major effect in controlling the run-up of a tsunami. In the deep ocean the sea-level rise associated with a tsunami is rarely more than 1 m spread over a wavelength of 100 km or more. Such a wave travels at about 800 km/hr. As the tsunami approaches land, shallowing of the ocean causes the tsunami to slow down and the amplitude of the sea-level rise to increase. The local offshore sea-bottom bathymetry is one major factor affecting tsunami run-up. A long, gentle slope of the sea bottom to the shore can cause a large breaking wave to form. A moderate seafloor slope onshore will cause a run-up without a breaking wave, while for a steep slope the run-up is reduced and there is no breaking wave. Offshore islands and shallow seafloor regions can serve to focus or defocus incoming tsunami waves, causing strong variations in tsunami run-up along stretches of coastline a few tens of kilometres long. For an incoming tsunami the side of a large island opposite that from which the tsunami is coming will experience smaller run-ups than side of the island in the direct tsunami path. It is for these reasons that one historical report of a tsunami is much less useful to scientists than many detailed reports from long stretches of coastline, and it is vital to the scientist to know the exact coastal point from which each historical report comes.

Tsunami reports show some variety, both in the way the tsunami begins and how long it lasts. Some tsunamis begin with an abnormal rise of the sea, while others begin with an unusual drawdown of the sea to abnormally low levels. Whether a rise or fall, the initial phase of a tsunami usually lasts about 5–10 minutes, followed by the sea moving in the opposite direction during the subsequent 10–20 minutes. Several subsequent oscillations of the sea may be observed, each lasting typically about 10–20 minutes. The greatest run-up may not be observed during the initial phase of the tsunami, but it may occur during the second or third sea oscillation. The oscillations of the sea associated with a tsunami may last hours, although the heights of the later oscillations are usually smaller than the first few.

An initial withdrawal of the sea as a tsunami approaches land can lead to a large loss of life if people go down to the water to investigate what is happening. The sudden withdrawal of water can leave fish stranded on the beach, inducing people to come collect the free catch. However, this unexpected boon is deceptive, since the sea returns at a speed faster than people can run. Thus, many people get caught by the rapid return of the sea during the run-up phase of a tsunami, and this is often the cause of large numbers of human deaths. Many historical records report tsunamis that began with an initial withdrawal of the ocean.

Tsunamis can be quite turbulent as they run up onto a shore, stirring up sand and silt from the ocean bottom and carrying it onshore. Strong tsunamis will also sweep up local vegetation and manmade structures in their paths and deposit this material randomly as the water recedes. Thus, there is usually a jumble of mud, sand, plants and manmade objects left in the inundated areas after a major tsunami. Geologically, unusual layers of sand and mud deposited in tsunami-prone coastal areas may be taken as evidence of prehistoric tsunamis. Historical reports of such deposits may be an indicator that a tsunami was experienced along a coastal region.

12.4.2 Matching tsunamis with their physical sources

Most tsunamis are caused by strong offshore earthquakes centred near the coastal areas affected by the tsunami. Earthquakes that cause significant vertical displacements of the ocean floor will cause sizeable tsunamis, as will major submarine landslides perhaps triggered by strong earthquake shaking. In these cases, the coastal areas affected by the tsunami are likely to have felt strong, and possibly damaging, earthquake ground motions. For these localities, differentiating between the damage to structures caused by the earthquake shaking and the damage caused by the tsunami may be difficult from the historical reports alone. Structures along a coast may be susceptible to many effects caused by strong earthquake shaking, including: ground-shaking amplification due to soft soils; liquefaction of water-saturated soils under a structure that cause it to sink, tilt or collapse; and lateral spreading of soils coastal soils toward the water that also cause structures to sink, tilt or collapse. Lateral spreading can be a problem especially for historical manmade landfill along a water's edge. Since all of these ground-shaking effects might be most profound for structures located on the low ground next to an ocean or sea, it is possible that historical observers or modern interpreters of historical records might mistake earthquake shaking damage as the effects of a tsunami (or vice versa).

Interpreting historical tsunami reports at localities where no earthquake shaking was reported presents a set of challenges that should be considered. The first question that must be addressed is whether the event was a tsunami, a storm surge or some other local sea-level aberration. If the evidence points to a tsunami, then the possible source of the tsunami needs to be assessed. A tsunami might be generated by a local earthquake, a distant earthquake, a volcanic eruption or a local landslide into a water body. Some earthquakes, such as the Sanriku, Japan earthquake of 1968, are called *tsunamigenic* or *slow earthquakes* because they generate a significant tsunami even though they generate relatively little earthquake ground-shaking and have much smaller magnitudes than most earthquakes that generate sizeable tsunamis. It is possible that many tsunamigenic earthquakes are earthquakes of moderate magnitude (magnitude 6 to 7.5), but they generate major submarine slumps that are the real cause of the tsunami. Modern tsunamigenic earthquakes all have taken place at oceanic subduction zones, and so the same probably should be expected for historical times as well. Thus, if a coastal area at a subduction zone experienced a tsunami but no earthquake is known from historical records, a tsunamigenic earthquake might be suspected. If the coastal area is not at or near a subduction zone, then a local earthquake source probably would be regarded as unlikely.

If a tsunami report cannot be attributed to a local earthquake, then a distant earthquake source might have caused the tsunami. In some cases, it may be possible to correlate earthquake reports from hundreds or even thousands of kilometres away to a local tsunami. However, not all distant earthquakes can be candidates as the source of a reported tsunami. For example, the earthquake source should be in the same ocean or sea basin as the location of the tsunami report. An earthquake in the Atlantic Ocean might cause a significant tsunami that traverses the Atlantic, but it would not cause a damaging tsunami in the Pacific Ocean, although a slight, transient sea-level rise and fall might be observed. Since good seismological observations began at the end of the nineteenth century, only earthquakes with magnitudes well above 8.0 have caused tsunamis that traverse a large ocean basin. Thus, the ground-shaking and tsunami reports from near the epicentre of the suspected earthquake must be consistent with a very large magnitude. Finally, the observation of the tsunami at a distant site must be within hours of the time that the distant earthquake was reported. It takes less than about 24 hours for a tsunami to traverse the Pacific Ocean, the largest of Earth's ocean basins. Thus, a tsunami report that is more than a day after a strong earthquake likely cannot be attributed to that earthquake. Similarly, an earthquake that occurred too close in time to a tsunami that was experienced on the other side of a large ocean probably did not cause the tsunami.

The 1755 tsunami at St Martin's in the West Indies

On 18 November 1755 there was a strong earthquake that was felt along much of the east coast of North America. This earthquake is now called the Cape Ann earthquake because its epicentre is thought to have been just east of Cape Ann, Massachusetts. Winthrop (1757–58) wrote a detailed article about this earthquake, which was felt strongly in eastern Massachusetts and caused damage at Boston. One observation reported by Winthrop (1757–58) was of a tsunami at the harbour of St Martin, where the sea withdrew and then returned about six feet (2 metres) higher than normal. Winthrop (1757–58) claims this tsunami took place on 18 November at 2 p.m., about 9 hours after the earthquake at Boston. An investigation by Rothman (1968) used a number of contemporaneous accounts to conclude that the tsunami at St Martin was experienced not on 18 November as reported by Winthrop (1757–58) but rather on 1 November 1755 at 2 p.m. This latter date corresponds to that of the major earthquake and tsunami at Lisbon, Portugal, and a tsunami on the afternoon of 1 November 1755 in the West Indies would be consistent with the time of the Lisbon earthquake. The magnitude of the Lisbon earthquake is listed by the US Geological Survey as 8.7, whereas for the Cape Ann earthquake the magnitude is estimated to have been about 6.3 (Ebel, 2006). No tsunami was reported along coastal Massachusetts on 18 November, while a large tsunami was experienced along the Portuguese coast on 1 November. Thus, the sizes of these two earthquakes, the tsunami reports from near their epicentres (or lack thereof) and the contemporary reports from the West Indies are all consistent with a trans-Atlantic tsunami from the Lisbon earthquake and not from the Cape Ann earthquake.

A major volcanic eruption of an oceanic volcanic island is another possible tsunami source. While much rarer than tsunamis caused by earthquakes, volcano-generated tsunamis can be very large and devastating. For example, the 26 August 1883 eruption of Krakatoa caused a major tsunami in the Indian Ocean, reaching heights exceeding 35 m on the island of Java (Wharton and Evans, 1888; Pararas-Carayannis, 2003; Pelinovsky et al., 2006). The celebrated eruption of the island of Santorini in the Mediterranean about 1627–1600 BC (Friedrich et al., 2006; Manning et al., 2006) is thought to have caused a devastating tsunami on the island of Crete (Yokoyama, 1978; McCoy and Heiken, 2000; Minoura et al., 2000; Dominey-Howes, 2004; Pareschi et al., 2006). Volcanic eruptions cause tsunamis when they cause major rock movements on the islands where the eruptions take place. In the cases of both Krakatoa and Santorini, the volcanic eruptions lead to a collapse below sea level of the centre of the island. In both cases, the sea was suddenly able to rush into the depression (called a *caldera*) left by the collapse of the volcano centre. In each case, the tsunami was probably associated with the rapid inundation of the central caldera during its sudden collapse. Major landslides of rock down the sides of a volcano into the sea can also be a source of major local tsunamis associated with volcanoes.

Some tsunamis can be caused by major landslides into local water bodies. The best known tsunami generated by a local landslide took place in Lituya Bay, Alaska after an earthquake on 9 July 1958. The ground-shaking caused a large mass of rock on the side of a mountain to suddenly let go and fall into a local bay. The tsunami apparently exceeded 100 m in height as the water in the bay was pushed outward away from the landslide. A ship in the bay reported being carried by the tsunami over the land (with treetops visible below the bottom of the ship) and into the open ocean (Tocher and Miller, 1959; Miller, 1960; Mader, 1999, 2002). Sudden large landslides into lakes and reservoirs can also be associated with tsunamis in such confined water bodies. Such landslides can be caused by the strong ground-shaking associated with an earthquake.

12.4.3 Tsunami intensity and magnitude scales

Similar to their counterparts for describing earthquake damage and shaking effects, there have been several *tsunami intensity scales* that have been proposed. The first tsunami intensity scale comprising of six different degrees was proposed by Sieberg (1923, 1927), while an improved six-degree scale was published by Ambraseys (1962). Papadopoulos and Imamura (2001) proposed a 12-degree intensity scale which is the most comprehensive put forth to date (Table 12.2). The degrees on the scale are dependent on the local tsunami wave height and describe the different possible effects of the tsunami, taking into account the vulnerability of the manmade structures to tsunami inundation. In Papadopoulos and Imamura (2001) the damage grade levels are not defined. In a study of damage from the 2004 Indian Ocean tsunami, Maheshwari *et al.* (2006), which uses the Papadopoulos and Imamura (2001) intensity scale, also define the damage grades (see Table 12.3). Unlike macroseismic intensity due to earthquake shaking, tsunami intensity has not yet become widely used by scientists who investigate tsunamis.

In addition to tsunami intensity scales, seismologists have developed magnitude scales based on tsunami measurements. Iida *et al.* (1967) introduced a tsunami magnitude formula that was a modification of an earlier formula that had been introduced by Imamura (1949). The Imamura/Iida scale used measurements of the maximum run-up height from localities within 250 km of the tsunami source to produce a tsunami magnitude designated as *m*. The Imamaura/Iida magnitude scale usually gives values between −1 and 5, and so its values are not equivalent to the earthquake magnitude for those tsunamis that are generated by undersea seismic events. Hatori (1986) introduced a further modification to the Imamura/Iida scale by adding a small correction factor that

Table 12.2 *Tsunami intensity scale of Papadopoulos and Imamura (2001)*

(The shoreline tsunami wave height in meters is listed for the larger intensity values. The damage grades are those of the EMS-98 intensity scale.)

I Not felt.

II Scarcely felt.

    (a) Felt by few people on board small vessels. Not observed on the coast.

    (b) No effect on objects like boats.

    (c) No damage to coastal structures.

III Weak.

    (a) Felt by most people on board small vessels. Observed by a few people on the coast.

    (b) No effect on objects like boats.

    (c) No damage to coastal structures.

IV Largely observed.

    (a) Felt by all on board small vessels and by few people on board large vessels. Observed by most people on the coast.

    (b) Few small vessels move slightly onshore.

    (c) No damage to coastal structures.

V Strong. (Wave height 1 m)

    (a) Felt by all on board large vessels and observed by all on the coast. Few people are frightened and run to higher ground.

    (b) Many small vessels move strongly onshore, few of them crash into each other or overturn. Traces of sand layer are left behind on ground with favourable circumstances. Limited flooding of cultivated land.

    (c) Limited flooding of outdoor facilities (such as gardens) of nearshore structures.

VI Slightly damaging. (2 m)

    (a) Many people are frightened and run to higher ground.

    (b) Most small vessels move violently onshore, crash strongly into each other or overturn.

    (c) Damage and flooding in a few wooden structures. Most masonry buildings withstand.

VII Damaging. (4 m)

    (a) Many people are frightened and try to run to higher ground.

    (b) Many small vessels damaged. Few large vessels oscillate violently. Objects of variable size and stability overturn and drift. Sand layer and accumulations of pebbles are left behind. Few aquaculture rafts washed away.

    (c) Many wooden structures damaged, few are demolished or washed away. Damage of grade 1 and flooding in a few masonry buildings.

VIII Heavily damaging. (4 m)

    (a) All people escape to higher ground, a few are washed away.

    (b) Most of the small vessels are damaged, many are washed away. Few large vessels are moved ashore or crash into each other. Big objects are drifted away. Erosion and littering of the beach. Extensive flooding. Slight damage in tsunami-control forests and stop drifts. Many aquaculture rafts washed away, few partially damaged.

    (c) Most wooden structures are washed away or demolished. Damage of grade 2 in a few masonry buildings. Most reinforced-concrete buildings sustain damage, in a few damage of grade 1 and flooding is observed.

(cont.)

Table 12.2 *(cont.)*

IX Destructive. (8 m)

(a) Many people are washed away.

(b) Most small vessels are destroyed or washed away. Many large vessels are moved violently ashore, few are destroyed. Extensive erosion and littering of the beach. Local ground subsidence. Partial destruction in tsunami-control forests and stop drifts. Most aquaculture rafts washed away, many partially damaged.

(c) Damage of grade 3 in many masonry buildings, few reinforced-concrete buildings suffer from damage grade 2.

X Very destructive. (8 m)

(a) General panic. Most people are washed away.

(b) Most large vessels are moved violently ashore, many are destroyed or collide with buildings. Small boulders from the sea bottom are moved inland. Cars overturned and drifted. Oil spills, fires start. Extensive ground subsidence.

(c) Damage of grade 4 in many masonry buildings, few reinforced-concrete buildings suffer from damage grade 3. Artificial embankments collapse, port breakwaters damaged.

XI Devastating. (16 m)

(b) Lifelines interrupted. Extensive fires. Water backwash drifts cars and other objects into the sea. Big boulders from sea bottom are moved inland.

(c) Damage of grade 5 in many masonry buildings. Few reinforced-concrete buildings suffer from damage grade 4, many suffer from damage grade 3.

XII Completely devastating. (32 m)

(c) Practically all masonry buildings demolished. Most reinforced-concrete buildings suffer from at least damage grade 3.

Table 12.3 *Grades of structural damage used in the tsunami intensity scale of Papadopoulos and Imamura (2001) as defined by Maheshwari et al. (2006)*

Damage category	General extent of damage
Grade 1: slight nonstructural damage	Thin cracks in plaster, falling of small pieces of plaster
Grade 2: slight structural damage	Small cracks in wall, falling of fairly large pieces of plaster
Grade 3: moderate structural damage	Large and deep cracks in walls; widespread cracking of walls, columns, and piers; tilting or falling of chimneys. The load-carrying capacity of the structure is partially reduced. Roads are eroded
Grade 4: severe structural damage	Gaps occur in walls, inner or outer walls collapse, and separate parts of the building lose their cohesion. Severe damage to roads, tilting of structures, and exposure of foundations by erosion
Grade 5: collapse	Partial or total collapse of the building, complete/partial washout of roads

depends on the distance from the tsunami source to the locations where the tsunami run-up height is measured.

Another tsunami magnitude scale denoted M_t was introduced by Abe (1979). The M_t scale is based on the maximum amplitude of the tsunami waves as measured at localities beyond about 1000 km from the earthquake source that generated the tsunami, and it is designed to produce values that are approximately comparable to the moment magnitudes of the earthquake sources that generated the tsunamis. For example, the moment magnitudes of the 1960 Chile earthquake and the 1964 Alaska earthquake are 9.5 and 9.2, and their tsunami magnitudes M_t are 9.4 and 9.1, respectively. On average, for most earthquakes the Abe (1979) tsunami magnitudes are within about 0.3 magnitude units of the moment magnitudes for those earthquakes.

13

Cooperation in historical seismology research

The previous chapters in this volume have presented practical discussions of the issues that both historians (including archaeologists) and seismologists must face and understand when their research involves earthquakes and tsunamis that happened during historic, or even prehistoric, times. Some of these issues involve a basic understanding of the research tools and interpretation methods that are important in each discipline, while other issues arise from the diversity of linguistic, political, cultural and social realities that existed in the areas affected by earthquakes and tsunamis in the past. The complexity of these issues is large, and for this reason we advocate in this chapter a high level of cooperation between knowledgeable seismologists, engineers, architects, archaeologists and historians when their research involves past earthquakes and tsunamis. Such cooperation can benefit the research of all groups by improving the interpretation of earthquake and tsunami data used in both the historical and seismological fields.

For those historians and archaeologists whose research concerns geographical areas that are known or suspected to be susceptible to earthquakes, a thorough knowledge of the possibility of earthquakes and their impact on society is important for their work. For times before the advent of seismic instrumentation, indications of past earthquakes and tsunamis in historical records or at archaeological sites are the primary and usually the only sources of information about these events. Thus, if an earthquake is suspected from evidence in the historical record, seismologists cannot provide independent evidence to confirm or deny the existence of the event. However, seismologists can play a vital role in evaluating the historical evidence to see if it conforms to the modern understanding of earthquakes and their effects. For example, if a seismologist was confronted with historical evidence that a suspected earthquake apparently caused major damage in one city but no damage at all at another city only a few kilometres away, he or she would be highly suspicious that an earthquake was indeed the cause of the reported damage. As noted earlier in this book, perhaps a landslide not associated with an earthquake was the real cause of the reported disaster.

For research seismologists, accurate interpretations of the full range of information on historical earthquakes are vital if they are to apply quantitative analysis methods to historic earthquake data. No longer are seismologists satisfied to know merely when an earthquake occurred and where it caused damage. Now seismologists want detailed information on exactly how much damage was experienced by different kinds of structures, where those structures were located, how strongly the earthquake was felt at all localities where no damage took place and where the earthquake was not felt at all. Seismologists also need to know if a lack of earthquake reports at localities means that no earthquakes were experienced there or that historical information on possible earthquake reports is lacking. These are questions that are best addressed by cooperative work involving both scientists and historians.

13.1   The accuracy of historical earthquake and tsunami data

Both seismologists and historians require an accurate picture of the information on an historic earthquake or tsunami for their own research needs. For example, the date and time of the event need to be determined as well as possible, the exact effects of the event on both manmade structures and the physical environment must be discerned, and the geographic extent of the event should be determined. This is best accomplished through cooperative work involving both seismologists and historians. Historians may discover and extract earthquake reports from the archives in which they work, but seismologists are in the best position to judge how consistent those report are with the known patterns of earthquake effects.

One topic in which cooperative work between seismologists and historians is especially important is in the evaluation of the severity and spatial distribution of both damage and felt effects from an earthquake or tsunami known from historical sources. By studying the damage patterns and distributions of felt reports of modern earthquakes at many different magnitudes and from many different parts of the world, seismologists now have a good understanding of the expected damage and felt effects as a function of earthquake magnitude for most areas of the Earth. Furthermore, seismologists can make educated guesses about the largest possible earthquakes that might be expected in different areas of the globe. Thus, seismologists can now make good judgements about whether historical earthquake or tsunami reports from different parts of a region appear consistent with a single seismic event or whether some other explanation is needed. If some other explanation is required to interpret the data, that explanation might be a seismological one, such as two independent earthquakes at remote localities, or it might require historians to reinterpret their sources if the historical data are clearly at odds with a single seismic event. Clearly, a mutual interpretation involving both seismologists and historians of the historical data, with give and take between the two groups, can lead to the most robust results.

Seismological interpretations can have a major influence on the direction and emphasis of historical seismology research. The interpretation of damage reports of a strong earthquake that should have been widely felt can lead seismologists to encourage earthquake historians to seek reports of felt earthquake shaking at large distances from the centre of damage. While this may seem to the earthquake historian to be a search for minor information, to the seismologist the documentation of far-flung felt effects may be the evidence necessary to clinch the argument for the suspected magnitude of the earthquake. In the same way, seismologists' knowledge of the severity and spatial extent of the ground-shaking generated by earthquakes of different magnitudes can help historians limit the geographical extent of their search for new sources of information on a past seismic event.

The Port Royal (Jamaica) earthquake and 'tsunami' of 1692

There is a well-documented earthquake which destroyed the settlement of Port Royal on the island of Jamaica on 7 June 1692 (Julian date). The historical reports indicate that a strong earthquake was experienced at Port Royal, followed by an incursion by the sea that inundated part of the town. When the event was over, a large part of the lowest-lying area of the town had dropped permanently below sea level. On the face of it, a simple interpretation of the historical reports is that the earthquake caused a tsunami that flooded the town and washed part of it into the sea. However, this explanation is problematic from a seismological point of view because a tsunami by itself does not cause the land to subside. Another explanation of the earthquake reports is that the faulting associated with the earthquake occurred very near or under Port Royal and therefore it was the faulting and not a tsunami that cause the inundation and subsidence. With this explanation, a tsunami probably would have been caused by the earthquake, although its height would be uncertain. If the subsidence was associated with major offshore faulting, then a larger tsunami might have been generated. However, if the subsidence was associated with faulting beneath the seashore itself, then only a very small tsunami would have arisen. This explanation presents some problems from a modern plate-tectonic point of view, since Jamaica is located along a transform plate boundary, which means that it is most probable that a major earthquake would have been associated with strike-slip faulting that very likely would not generate much or any tsunami at all. A third explanation is that the ground-shaking from the earthquake caused a lateral spread along the low-lying areas of the town (Kozák and Ebel, 1996). In this scenario, the lateral spread would cause part of the town to break apart and slide down into the water. This would lead to a permanent subsidence and inundation of part of the town. With this explanation, no tsunami at all is needed, nor would any permanent vertical movement of the land due to earthquake faulting need have taken place.

These three explanations of the Port Royal reports of the 1692 event affect the historical seismological research that could still be carried out concerning this earthquake. If a major tsunami was the cause of the devastation, then it might have been reported from other coastal settlements around the Caribbean. This would entail a search of documents from many different areas, from settlements on islands as well on the mainlands surrounding the Caribbean. Palaeoseismological evidence of the tsunami should still be preserved in places along the shore of Jamaica and perhaps as well on other nearby islands. On the other hand, if the devastation represented local lateral spreading at Port Royal, then a broader search for historical reports or geological evidence of a tsunami would be fruitless. Obviously, a lack of tsunami reports from other localities does not prove that lateral spreading was the cause of the destruction at Port Royal; it only means that a large tsunami cannot be proven. Confirmation of a tsunami or of lateral spreading at Port Royal from the 1692 event could come from detailed geological and geotechnical engineering work at Port Royal. This work would require the expertise of several different types of experts, who would need to be guided by the historical data to those localities where the evidence of the 1692 event would be found.

13.2   Improving earthquake catalogues

As has been discussed in detail earlier in this book, an accurate and complete earthquake catalogue is one of the important goals of historical seismology research. Most existing earthquake catalogues, especially older catalogues, contain many kinds of errors, including those arising from misinterpretations of historical sources, mistakes in the determination of dates or times, and transcription or printing errors. Errors in existing catalogues can result in missing earthquakes or earthquakes with significantly underestimated or overestimated event magnitudes. Users of such catalogues get a distorted picture of the earthquake history of a region, whether they are interested in the effects of particular earthquakes or in the temporal and spatial patterns of earthquakes across the region.

In their own individual fields, the research of both earthquake seismologists and historians demands an earthquake catalogue that is a reliable source of information. Thus cooperation between these two diverse disciplines is important for research needs in their own fields individually as well as for projects that involve a joint research effort. There are several ways in which existing earthquake catalogues can be improved by cooperative efforts involving seismologists and earthquake historians.

(1) *Evaluating the accuracy of existing earthquake catalogues*   It can be a major research task in its own right simply to evaluate the accuracy and completeness of existing catalogues of historical earthquakes. Historians inherently play an important role here, since they are the ones best able to sort out the many knotty problems, described throughout this book, that must be faced when conducting research with historical sources. These problems include determining the proper interpretation of the reports used to show the existence, location and size of an earthquake, evaluating whether the sources used were indeed primary sources or not, and verifying the correct date, time and location from which the earthquake reports came. There are likely many mistakes that still remain in modern catalogues of historical earthquakes due to these problems. For example, it was pointed out earlier that a common mistake found in modern earthquake catalogues is multiple versions of a single earthquake with different dates, times and/or locations. Typically such a mistake was made in an earlier catalogue and then was propagated to later catalogues by compilers who used the earlier catalogue as a source of their information. However, a more insidious problem is where a mistaken earthquake is propagated forward in time through the historical sources themselves. In this case, one must trace back through the historical evidence to discover whether or not the earthquake actually did take place. Since later earthquake catalogue compilers are almost invariably considerate enough to cite their sources, tracing back such mistakes in earthquake catalogues may be relatively straightforward. On the other hand, many historical sources do not state their sources of information on earthquakes that occurred earlier in time, and so it can take a major research effort to learn whether or not the cited earthquake actually did take place.

(2) *Finding new information about historical earthquakes already in existing
     catalogues* The new methods that seismologists have devised to
     estimate the epicentres, magnitudes and ground-shaking of historical
     earthquakes rely on detailed reports concerning each earthquake from
     many different localities. The details these methods look to use are the
     specific amounts and kinds of damage that were reported due to the
     earthquake shaking, the specific locations where damage occurred or
     where the earthquake was felt, and the full spatial extent over which
     the earthquake shaking was felt. Reports on those localities where
     aftershocks were felt, especially smaller aftershocks that were felt over
     only very limited areas, are important pieces of information for
     constraining the location of the earthquake and the fault upon which
     the earthquake may have taken place. For many known historical
     earthquakes there is still much information that would be important
     for modern seismological analyses that must be accumulated. Since
     some of this information must be acquired at large distances from the
     location where the earthquake was presumably centred, and getting
     this information may well require research into sources from different
     countries, linguistic traditions, types of records, etc. Accumulating this
     additional information on known earthquakes demands close
     interaction between historians, who need to be aware of what
     information is most important to the research seismologists, and
     seismologists, who must understand the difficulties and problems that
     the historians must confront.

(3) *Continuing the search for historical earthquakes that were not previously
     known* While this is an effort that can seismologists can carry out
     separately from historians, interaction between experts from each area
     is highly desirable, especially when information concerning previously
     unknown earthquakes is found. In many cases, the search for new
     earthquakes in the historical record will be focused on a particular
     area for a particular reason, such as looking for past earthquake
     activity on a known or suspected active fault, seeking information on
     the past tsunamis in a coastal area, or trying to discover reports of
     aftershocks from a known strong earthquake. However, in some cases
     there will be serendipitous discoveries of previously unknown
     earthquakes by researchers who are investigating archival material for
     other purposes. It is important that these researchers be sufficiently
     aware of the field of historical seismology that they bring this
     information to the attention of those who are interested in historical

earthquakes. The same can be said of new information concerning historical earthquakes that are already known.

13.3   Improving seismic hazard estimations

The analysis of the seismic hazard of a site or of a region is one of the most important kinds of studies where the fruits of historical seismology research have practical application. Seismic hazard analyses rely on the complete and accurate knowledge of all of the strong earthquakes (typically magnitude 5 and greater) that have taken place in the area encompassing and surrounding the site or region where the analyses are being carried out. The magnitudes and locations of the past earthquakes must be known as well as possible in order for a seismic hazard analysis to be as accurate as possible. Furthermore, many modern seismic hazard analyses make use of the recurrence times of strong earthquakes on active faults. For some parts of the world, useful information on the repeat times of strong earthquakes on those faults can sometimes be achieved by combining information from geological, historical and archaeological research for the fault and its surrounding area. The North Anatolian Fault in Turkey is one such feature where the geological, historical and archaeological data together are producing an increasingly detailed history of fault movements throughout most of the past two millennia.

For seismic hazard analyses, it is the seismologists who specify the data that are required as input into their analyses. Because such analyses may require new assessments of the locations and magnitudes of the past strong earthquakes that are known for their analysis region, seismologists may seek new information from the historical records concerning these past earthquakes. For this reason, historians may be pressed into service to seek out the additional information that is needed. One example of this might be the search for the full spatial extent of the felt shaking from an earthquake in order to better estimate the magnitude of that event. In such cases, it is generally the seismologist who will direct the research. However, it is important that the seismologist appreciates what knowledge about the earthquake history of the region that the historian can produce and what uncertainties the historian must face.

Documentation of silent earthquake areas is one aspect of historical seismology research that is especially important for seismic hazard analyses. Just as the computation of the seismic hazard of a region demands an accurate picture of the earthquakes in the seismically active areas, it is also important for these studies to know which areas are undoubtedly seismically inactive. Areas of lower seismic activity are often areas of lower seismic hazard. However, seismologists must know whether an area is quiet because no earthquakes have occurred in that area or whether it is quiet because of inadequate research into the historical seismological record of that area. This can be especially problematic for those areas where few or no records exist that might document the past earthquake history. If seismologists suspect that an area has experienced past earthquakes but those earthquakes are somehow missing from the historical earthquake catalogues, then they may extrapolate their current estimations of the seismic activity of that region backward in time to estimate how many earthquakes should have taken place in historic and prehistoric time. This can happen for those areas where modern instrumental earthquake monitoring is detecting regular earthquake activity for an area that is at a higher rate than that suggested from the historical earthquake catalogues.

Modern seismic hazard analyses typically need to know the complete earthquake record for their region of analysis going several thousand years in time backward from the present. For example, many building codes that contain provisions for earthquake resistance base their analyses on earthquake ground motions that are experienced no more than once every 1000 years or even 2500 years. For seismic hazard studies it is vital that as much information as possible be sought for even the earliest earthquakes in the historical records. Thus, the historian and the archaeologist both play vital roles in uncovering data on the occurrences of the earliest known earthquakes for a region.

13.4   Bringing seismologists, historians and archaeologists together

A major purpose of this book is to provide a common knowledge base to improve the communication between seismologists, who come from a scientific background where literal facts are the important data, and historians, for whom the contexts and motivations of the creators of historical documents may be of primary interest. The book also seeks to foster constructive contacts between these two groups and archaeologists, who focus on deciphering the past from fragmentary and very incomplete data. The better the cooperation among these groups when addressing issues related to historical earthquakes, the better the knowledge of past earthquake history and its effects on society in the historic past.

It is also hoped that the information presented in this book will encourage the funding of more interdisciplinary projects concerning historical earthquakes. Ideally these projects should involve some combination of seismologists, geologists, earthquake engineers, historians and archaeologists, depending on the project and its goals. As much recent research has shown, there is still much to be learned about earthquakes in the historic past, and much of this new understanding has come from interdisciplinary research.

Historical seismology is, in one sense, a very old field. In another sense historical seismology is a new and emerging field that is bringing new research tools and perspectives to the study of historical earthquakes. It has benefited from new ideas and analysis methods that have arisen in both seismology and historical research. We hope that this book will help promote the emergence and maturity of historical seismology and encourage even more developments in this field of research.

Glossary

Accelerograph An instrument that detects and records the strong shaking from large earthquakes.

Accelerometer A scientific instrument that senses ground acceleration; the signals from a accelerometer are transmitted to an accelerograph for recording.

Active fault A fault that has had earthquake slip in geologically recent time (within the past 500 000 years) and is capable of generating a future strong earthquake.

Aftershock A smaller earthquake that follows the occurrence of a larger earthquake and occurs on or in the vicinity of the same fault that had slip in the larger earthquake. Most aftershocks occur within a few months of the larger event, but some can occur years later.

Annal A chronological record of events, kept on a year-by-year basis.

Anomaly (in local construction) A building construction practice that deviates from the normal practice followed in a local area.

Archaeoseismology Analysis and interpretation of the archaeological evidence of seismic effects, elaborated within the framework of the stratigraphic method and concerning specific seismological issues.

Archives Orderly and systematic collections of deeds and documents, produced by public, private, ecclesiastic or monastic institutions, offices or administrations, whose conservation is believed to have an historical interest.

Attenuation (of seismic waves) The decrease in the amplitude or strength of seismic waves as they travel away from the seismic source that generated the waves.

Barycentre (of an earthquake) The geographical centre of the strongest macroseismic intensity observations of an earthquake.

Canonical hours Ancient hours ended up giving their name to the phases of monastic prayer and being defined as canonical.

Capitular archive The archive maintained by a cathedral, usually containing documents pertaining to the administration of a Catholic diocese.

Caldera (of a volcano) The collapsed centre of a volcano, where the collapse occurred following a strong volcanic eruption.

Catalogue (of earthquakes) see *Earthquake catalogue*.

Chancellery The department, staff or office of a high official or secretary that helps administer a government, church or university.

Chronicle A record of historic events that cover some specific time period.

Classification (of references) The assignment of a value of importance to each text in a reference list according to some specified historiographic criteria.

Codex Ancient manuscript (paper or parchment) of several papers put together in book form; diplomatic codex: ancient manuscript that collects together the documents relating to the history of a place, a community, a monastery.

Colophon Final part of a book or manuscript, with information on the author or the person commissioning the work, the place, the date. In the medieval Armenian manuscripts this part is often highly developed, at times in the form of a short chronicles of important events.

Community Internet Intensity Map A map of earthquake intensities that is created using observations of earthquake shaking and damage effects that were collected from persons who filled out a questionnaire that is posted on the internet.

Completeness (of an earthquake catalogue) The degree to which all of the earthquakes at or above some magnitude are thought to be contained in the earthquake catalogue for a region.

Degree (of macroseismic intensity) The value, normally specified as a Roman numeral, of the macroseismic intensity that is assigned to a report of felt or damage effects due to an earthquake.

Depth (of an earthquake) The depth below the surface of the earthquake where the initial slip on the fault takes place; also called the focal depth of the earthquake.

Deterministic seismic hazard analysis A computation of the estimated level of earthquake ground motion at a site due to a postulated earthquake at a specific location or on a specific fault with a specified magnitude.

Diary A daily record of events, observations, ideas, etc. maintained by an individual.

Duration (of an earthquake) The total amount of time that ground-shaking is felt by an individual during the occurrence of an earthquake.

Earthquake A natural occurrence in which sudden movement or slip along a fault releases seismic waves that are strong enough to be felt or be detected by seismic instrumentation.

Earthquake catalogue A chronological listing of the of occurrences of past earthquakes for an area, where the list normally includes such parameters as the date, time, location and strength of each earthquake, as well as other parameters deemed important by the compiler of the catalogue.

Earthquake shaking The shaking of the ground during the passage of the seismic waves that are generated by the fault slip in an earthquake.

Elastic rebound theory The theory that fault slip during an earthquake takes place to relieve the elastic pressures that build up along a fault due to rock movements that are in opposite directions on either side of the fault; the rock movements are driven by the movements of the tectonic plates on the Earth.

Epicentre (of an earthquake) The location on the surface of the Earth above the earthquake focus; the location in the Earth where the earthquake fault slip initiated.

Epigraphy The study of inscriptions on monuments, tombstones, buildings, etc.

Epigraphic source (of an earthquake) An inscription on a monument, tombstone, building, etc. that refers either directly or indirectly to an earthquake.

Ethnographic source (of an earthquake) An unwritten description of an earthquake or of a seismic effect that is contained in the oral tradition or in the myths and legends of a people.

False earthquake An entry in an earthquake catalogue of a seismic event that did not take place.

Fault An approximately planar crack or fracture in the Earth along which there has been displacement of rock in one direction on one side of the fault and in the opposite direction on the other side of the fault.

Fault length (of an earthquake) The horizontal dimension of the fault that has movement in an earthquake.

Fault slip (of an earthquake) The distance that by which two points that were adjacent to each other on either side of a fault before an earthquake were displaced by the fault movement during the earthquake.

Filter (of a historical source) The act of recording only some aspects of a historical event or of recording facts and conjectures that support a particular viewpoint or purpose.

Focal depth (of an earthquake) The depth below the surface of the Earth where the fault slip initiated; also called the earthquake depth.

Focus (of an earthquake) The point within the Earth where the fault slip initiated; also called the earthquake hypocentre.

Forgotten earthquake An earthquake that was known at one time in history but was subsequently forgotten by historians or eliminated from earthquake catalogues; also called a lost earthquake.

Foreshock A smaller earthquake that occurs at or near the focus of a subsequent larger earthquake and that occurs within a few weeks prior to the subsequent larger earthquake.

Greenwich Mean Time (GMT) The standard time at Greenwich Observatory in England; also called coordinated universal time (UTC).

Historical context (in historical seismology) Synthetic concept introduced in order to evaluate the data on the effects of past earthquakes in relation to specific historical, cultural and economic situations.

Historical seismology (not to be confused with history of seismology) Historical research method applied to the questions pertaining to seismology in order to know about the effects of the past earthquakes and to evaluate with appropriate methods the parameters of those earthquakes.

Historical source (in historical seismology) Original and authoritative historical text, produced by single individuals or institutions. The research of historical seismology based on the sources qualify the data subsequently used in the data sets of the earthquake catalogues.

Hypocentre (of an earthquake) The point within the Earth where the fault slip initiated; also called the earthquake focus.

Iconography Paintings, drawings, prints, relief carvings, or other artistic depictions.

Imax The largest of all of the macroseismic intensity values of an earthquake.

Indiction A Roman chronological system based on a 15-year cycle, originally arising due to a schedule of Roman tax assessments on land.

Inscription A written text that is engraved or painted onto a durable material such as a stone wall or a clay tablet.

Intensity (of an earthquake) see *Macroseismic intensity*.

Intensity barycentre The geographical centre of the area that contains the greatest macroseismic intensity reports for an earthquake.

Intensity (degree) The value assigned to a macroseismic intensity report, normally reported as a Roman numeral.

Intensity scale see *Macroseismic intensity scale*.

Interpolation The act of inserting words or phrases into descriptions of events in order to interpret those events.

Inundation see *Tsunami inundation*.

I_0 The epicentral intensity of an earthquake.

Island arc An arcuate chain of volcanic islands that overlie an oceanic subduction zone.

Isoseismal A line that divides regions of different macroseismic intensity values for an earthquake.

Isoseismal map A map showing the isoseismals of an earthquake.

Lateral spread (due to an earthquake) Deformation of the flat-lying surface soils along the coast of a water body where the land surface partially spreads toward the water body due to strong earthquake ground-shaking.

Left-lateral strike-slip fault A vertical or near-vertical fault where an observer on one side of the fault sees the opposite side of the fault moved horizontally to the left during earthquake movement.

Length (of a fault) see *Fault length*.

Lexicon The words used in the past to describe items or events and that have their own specific origin and meaning.

Liquefaction The loss of strength of a soil due to strong earthquake ground-shaking in which the soil has little resistance to vertical loads placed on the soil and items therefore sink into the soil.

Lost earthquake An earthquake that was known at one time in history but was subsequently forgotten; also called a forgotten earthquake.

Love wave A type of seismic wave that travels along the surface of the Earth and shakes the ground in a horizontal manner as the wave passes.

Macroseismic intensity A non-instrumental estimate of the strength of earthquake shaking at a location based descriptions of the observed shaking effects and damage (if any) due to the earthquake shaking; also called intensity.

Macroseismic intensity scale A set of descriptions of different levels of ground-shaking strength based on reports of felt and damage effects, where each level is assigned a value that is usually specified by a Roman numeral.

Macroseismic intensity questionnaire A questionnaire that contains a series of questions concerning descriptions of the observed shaking effects and damage (if any) due to the earthquake shaking at a location; the answers to the questions are used to assign a macroseismic intensity value to the ground-shaking at the location.

Mainshock The largest earthquake in a sequence of earthquakes that take place on the same fault and within a few years of each other.

Magnitude (of an earthquake) The strength of an earthquake source determined directly from instrumental observations or inferred indirectly from macroseismic intensity observations; as first published by Charles Richter, earthquake magnitude scales are logarithmic in that an increase of one unit of magnitude corresponds to a factor of 10 increase in the amplitude of the seismic waves at the earthquake source.

Maximum intensity see *Imax*.

Moment magnitude (of an earthquake) An earthquake magnitude scale that is based on the seismic moment of an earthquake; moment magnitude is the most accurate measure of the size of the largest earthquakes that take place on the Earth.

Normal fault A fault that dips into the Earth where the rock that overlies the fault slips downward relative to the rock that underlies the fault.

Notary A person who is authorized to draw up or certify contracts, deeds and other legal paperwork.

Notula A notation that was written in the blank spaces along the margins or on the cover page of a manuscript text that is already complete.

Origin time (of an earthquake) The time the initiation of the radiation of seismic waves from a fault that undergoes earthquake slip.

Palaeoseismology The study of historic and prehistoric earthquakes through investigations of the rocks and sediments that contain evidence of fault slip or of secondary earthquake effects such as liquefaction, landslides or lateral spreading.

Peak ground acceleration The strongest ground acceleration value measured by an accelerograph due to the ground shaking of an earthquake.

Peak ground velocity The strongest ground velocity value measured by an accelerograph due to the ground shaking of an earthquake.

Periodization (of historical sources) The grouping of historical information or events into a time period in order to interpret or evaluate the information or events.

Plate boundary A fault or zone of faults where two tectonic plates meet and move relative to each other.

Probabilistic seismic hazard analysis A computation of the ground motion value that has a specified probability of being exceeded during some time period (for example, the peak ground acceleration that has a 2% chance of being exceeded in a 50-year time period) due to all earthquakes that can possibly be expected to take place around the site or area for which the analysis is carried out.

Production (of historical sources) Those processes and activities that lead to the creation of sources of historical information.

Rayleigh wave A seismic wave that moves along the ground surface and shakes the ground both vertically and horizontally in an elliptical motion.

Right-lateral strike-slip fault A vertical or near-vertical fault where an observer on one side of the fault sees the opposite side of the fault moved horizontally to the right during earthquake movement.

Run-up (of a tsunami) The elevation relative to sea level of the most inland point where a tsunami was observed along a coastline.

Sand blow A liquefaction feature where a subsurface sand layer spews sand and water onto the surface of the Earth during strong earthquake ground-shaking; sometimes called a sand or mud volcano, the erupted sand from an individual sand blow may spread anywhere from a few meters to over 100 m from the crack in the ground from which the sand and water was vented.

Seismic activity see *Seismicity*.

'Seismic local culture' The techniques, materials and formal solutions followed in local construction practice that are an effort to cope with the threat of earthquakes.

Seismicity The earthquakes of a region as observed over some time period; also called seismic activity.

Seismic hazard The probabilities that different levels of strong earthquake ground shaking will affect a site or region.

Seismic indicators (in archaeoseismology) Evidence or traces of seismic effects in the stratigraphy of an excavation.

Seismic moment (of an earthquake) A measure of the strength of the forces in the Earth that generated an earthquake, where seismic moment scales with the magnitude, fault length and fault slip in an earthquake.

Seismic scenario A postulated set of descriptions of the effects of a past earthquake as reconstructed from surviving evidence or of a future earthquake based on a proposed location and magnitude.

Seismic scenario of effects (in historical seismology) The set of past earthquake effects inferred from historical analyses; it can be on an urban or regional scale.

Seismic waves The vibrational waves that are released by slip on a fault during an earthquake.

Seismogenic structure A geological feature such as a fault, plate boundary, fold, volcano, etc., where earthquakes have occurred in the past and where an earthquake can be expected to occur again at some time in the future.

Seismograph A scientific instrument that records the motion of the ground.

Seismometer A scientific instrument that is designed to detect the motion of the ground; the signals from a seismometer are transmitted to a seismograph for recording.

Seismotectonics The evolution of the geology of a region based on the movements and interactions of the tectonic plates and on the occurrences of earthquakes that are caused by the plate movements.

Selection (of historical sources) The processes by which deliberate decisions or inadvertent events lead to the present survival of historic sources of information, in contrast with those sources that were destroyed at some point in the historic past.

SGA (initials standing for Storia Geofisica Ambiente) A private research institute operating in Bologna (Italy) from 1983, specializing in the multidisciplinary study of the extreme natural events occurring in historical times, particularly the earthquakes in Italy and the Mediterranean area.

Shakemap A map showing the spatial distribution of a ground-shaking parameter like peak ground acceleration or peak ground velocity that was generated by the occurrence of an earthquake.

Silent area (regarding earthquakes) An area for which there are no known historic earthquakes, either because there have been no earthquakes in the area or because researchers have not discovered the evidence of past earthquakes.

Slip See *Fault slip*.

Slow earthquake An earthquake in which the fault slip takes place slowly enough that earthquake generates only weak seismic waves; slow earthquakes can generate large tsunamis even when they have relatively small magnitudes.

Soil liquefaction See *Liquefaction*.

Source parameters (of an earthquake) Values determined by seismologists, usually from instrumental measurements, that quantify the location, magnitude, focal depth, seismic moment, fault length, fault slip and other information about the earthquake source.

Spectral acceleration A parameter that gives the ground acceleration at a specific frequency of vibration.

Stratigraphic continuum (in archaeology) Uninterrupted stratigraphic sequence, i.e. the whole complex of depositional events concerning a site.

Stratigraphic method (in archaeology) The current method of archaeological excavation, developed by W. Harris in the 1970s. It is based on the removal of deposits (stratigraphic units) in reverse order to the depositional sequence.

Stratigraphic sequence (in archaeology) Sequence of stratigraphic units/deposits.

Stratigraphy A set of descriptions of the different accumulated layers of soils and rocks of a site or area.

Strike-slip fault A vertical or near-vertical fault where the slip on either side of the fault is purely horizontal.

Subduction zone A boundary zone between two tectonic plates that are coming together where an oceanic plate is sliding into the Earth below the other tectonic plate, which can be either a continent or another ocean basin.

Surface faulting (in an earthquake) The circumstance where the slip on a fault during an earthquake comes to the ground surface and therefore the fault offset can be observed on the Earth's surface.

Technical report (of an earthquake) A report written by a scientist or engineer describing observations and analyses of the effects of an earthquake.

Tectonic plate A part of the outermost part of the Earth, approximately 60 to 100 km in thickness, that moves as an approximately rigid block over the Earth; a tectonic plate may be comprised of a continent, an ocean basin, or some combination of both continent and ocean.

Thrust fault A fault that dips into the Earth where the rock that overlies the fault slips upward relative to the rock that underlies the fault.

Traditional architecture The common construction types that date from historical times and are characteristic of a given geographical area.

Tsunami A series of sea waves that are generated by an undersea earthquake, by the eruption of a volcanic island or an undersea volcano, or by a large landslide into a water body; in some cases a tsunami can be experienced at large distances (thousands of kilometres) from the source that generated the tsunami.

Tsunamigenic earthquake An earthquake under the sea that generates a tsunami.

Tsunami height The height above the average sea level of the highest amplitude of a tsunami.

Tsunami intensity scale A set of descriptions of different levels of run-up and inundation based on reports of wave observations and damage effects, where each level is assigned a value that is usually specified by a Roman numeral.

Tsunami inundation The area along a coast that is flooded by a tsunami.

Tsunami magnitude scale An estimate of the strength of the source of a tsunami make using measurements of maximum run-up heights.

Universal Time The common standard of time used worldwide for scientific observations; also called Coordinated Universal Time (UTC) or Greenwich Mean Time (GMT).

Bibliographical summaries

Chapter 1 What is historical seismology?

Albini *et al.*, 2004; Vannucci *et al.*, 1999b.

Chapter 2 The importance of historical earthquake and tsunami data

Albini *et al.*, 2004; Ambraseys, 1962; Baratta, 1901; Bertrand, 1757; Bonito, 1691; Caputo and Faita, 1984; Filippo da Secinara, 1652; Galanopoulos, 1960; Guéneau de Montbeillard, 1761; Guidoboni and Stucchi, 1993; Kárník, 1968–71; Kondorskaya and Shebalin, 1982; Ligorio, ed. 2005; Lycosthenes, 1557; Mallet, 1852–54; Manetti, *De terraemotu*, ms. 1457; Mercalli, 1883; Montessus de Ballore, 1884, 1906, 1923; Moreira de Mendonça, 1758; Morelli, 1942; Musha, 1941–43, 1951; Papadopoulos and Chalkis, 1984; Perrey, 1845–46, 1846, 1850; Roux, 1932; Schmidt, 1881; Seyfart, 1756; Sieberg, 1932b; Soloviev, 1990; Soloviev *et al.*, 2000; Tinti *et al.*, 2004; von Hoff, 1840–41; Wilson, 1930.

Chapter 3 Written historical sources and their use

Acosta, 1590; Ahrweiler, 1973; Albini, 1993; Albini *et al.*, 1991; Alexandre, 1989, 1990; Alishan, 1881, 1899; Ambraseys, 1988, 2005; Ambraseys and Melville, 1982; Ambraseys *et al.*, 2002; Arslan, 1939; Autino, 1987; Bacchielli, 1995; Barbano *et al.*, 1994; Bartolomeo da Rossano, ed. 1972; Batlló, 2002, 2003; Benjamin of Tudela, ed. 1907; Ben-Menahem, 1979; Bentor, 1989; Beschaouch, 1975; *Biblia Hebraica Stuttgartensia*, ed. 1997; Biblioteca, 1348; Bischoff and Koehler, 1939; Bonito, 1691; Boschi *et al.*, 1995, 2000; Burkert, 1985; Burkhardt, and Smith, 1985–; Burnand, 1984; Busi, 1994; Busi, 1995; Camodeca, 1971; Capelle, 1908, 1924; Cartledge, 1976; Cergol and Slejko, 1991; Chatelain, 1909; Ciccarello, 1996; Cignitti and Caronti, 1956; *Chronicon Sublacense*, ed. 1927; Comparetti, 1914; Constantinides and Browning, 1993; *Copia de una lettra*, 1542; Croke, 1981; Dagron, 1981; Daltri and Albini, 1991; Darrouzès, 1950, 1951, 1953, 1956; De Felice, 1994; De la Torre, 1995; Delehaye, 1907; De Noronha Wagner, 1993; Dever, 1992; Di Carlo, 1955–56; *Die Fragmente der griechischen Historiker*, ed. 1923–58; Dio Cassius, ed. 1895–1931; Di Vita, 1964, 1980, 1982; Downey, 1955; Dubois de Montpéreux, 1839–43; Ducat, 1984; Dumézil, 1966; Egidi, 1904; Evagrius, ed. 2000; Evangelatou-Notara, 1982, 1993; Faura, 1880; Favreau, 1989; Ferrari, 1988; Filangieri, 1958;

Finkelstein and Silberman, 2002; Gellius, ed. 1967–89; Gerola, 1932; Giessberger, 1922; Goethe, 1816; Gonzalez de Clavijo, ed. 1971; Gouin, 1979, 2001; Guidoboni, 1989; Guidoboni and Comastri, 2005; Guidoboni and Ferrari, 1986; Guidoboni and Marmo, 1989; Guidoboni and Poirier, 2004; Guidoboni *et al.*, 1994, 2007; Gutdeutsch *et al.*, 1987; Gutenberg and Richter, 1949; Haigh and Madabhushi, 2002; Hakobyan, 1951–56; Hamilton, 1783; Harland, 1942; Harris and Beardow, 1995; Henry, 1985; Herman, 1946; Hermann, 1962; Herzog, 1999, 2002; Hine, 2002; *Historicorum Romanorum reliquiae*, ed. 1906–14; Homer, ed. 1991; Ibn al-Qalanisi, ed. 1908; Jacques and Bousquet, 1984; Jerome, *Vita sancti Hilarionis*, ed. 1983; Karakhanian and Abgaryan, 2004; Kircher, 1664–65; Knauf, 2002; Kovach, 2004; Lampros, 1910, 1922; Lanciani, 1918; Lattuada, 2002; Lemaire, 1997; Lepelley, 1981, 1984; Lycosthenes, 1557; MacBain, 1982; Macelwane, 1936; Malalas, ed. 2000; Marco and Agnon, 2002; Milne, 1911; Molin *et al.*, 2000; Momigliano, 1930; Montessus de Ballore, 1915; Morandi, 1964; Musson, 1986; Nissen, 1883; Nur and Ron 1997; Olshausen and Sonnabend, 1998; Ovid, ed. 1977; Pacho, 1827–29; Pallas, 1802–03; Papazian, 1991; Pausanias, ed. 1973–81; Perrey, 1845; Piccardi, 2005; Piccardi and Masse, 2007; Placanica, 1985b; Plutarch, ed. 1927–2004; Porter, 1915–17; Prawer, 1988; Procopius, ed. 1962–64; Propp, 1971; Pseudo-Apollodorus, ed. 1963–67; Puzzolo Sigillo, 1949; Quenet, 2005; Rebuffat, 1980; Richard, 1955; Riera Melis *et al.*, 1993; Robert, 1978; Saderra Masó, 1895; Sanudo, ed. 1879–1903; Sarconi, 1784; Schreiner, 1975–79; Seeber and Armbruster, 1987; Ševčenko, 1973; *Sibylline Oracles*, ed. 1998; Smidt, 1970; Smirnov, 1931; Soggin, 1970; Spyridon of the Laura and Eustratiades, 1925; Stauder, 1962; Stelluti, 1997; Stolz, 1957; Strauss, 1975; Synesius of Cyrene, ed. 2003; Theophanes, ed. 1883; Thompson, 1937; Thucydides, ed. 1964–72; Traina, 1989, 1994; Turyn, 1980; Udías, 1999, 2003; Udías and Stauder, 1996; Villani, ed. 1990–91; Vitaliano, 1973; Vogt, 2005; von Hoff, 1840–41; Waldherr, 1997; Xac'ikyan, 1955–67; Xenophon, ed. 1918–22; Yadin, 1961; [Yovsevp'ean], 1951.

Chapter 4 Types of scientific sources

Acosta, 1590; Agathias, ed. 1967; Agricola, 1546, 1556; Aki, 1966; Albertus Magnus, *De mineralibus*, ed. 1569; Amontons, 1703; Anaxagoras, ed. 1952; Anaximenes, ed. 1951; Anonymous, 1749; Aquila, 1784; Archelaus, ed. 1952; Aristotle, *De mundo*, ed. 1955; *Metaphysica*, ed. 1831; *Meteorologica*, ed. 1952; Averroes, *Commentaries*, ed. 1983, *The incoherence*, ed. 1978; Avicenna, ed. 1927; Baldini, 1981; Baratta, 1896; Beccaria, 1753, 1776; Bede, ed. 1862; Beroaldo, 1505; Bertelli, 1887; Bertholon, 1787; Bertrand, 1757; Bina, 1751a, 1751b; Boschi and Guidoboni, 2003; Buffon, 1749; Buoni, 1571; Burkhardt and Smith, 1985, 1994; Burnet, 1681–89; Cavallo, 1777; Cecchi, 1876; *Ciclopedia*, 1754; Cirillo, 1733–34, 1747; Comastri, 1986; Chrysippus, ed. 1903; Davison, 1927; Dean, 1989; Dell'Aquila, ed. 1990; Democritus of Abdera, ed. 1952; Descartes, 1644; Diogenes Laertius, 1999; Diogenes of Apollonia, ed. 1952; Di Teodoro and Barbi, 1983; Epicurus, ed. 1887; Ewing, 1880, 1881; Feijóo y Montenegro, 1756; Ferrari, 1992, 2002; Ferrari and Guidoboni, 2000; Ferrari *et al.*, 2000; Franklin, 1751; Gallo, 1784; Galvani, 1791; Gray, 1883; al-Ghazali, *The incoherence*, ed. 2000; Gutenberg and Richter, 1954; Hales, 1749–50; Hamilton, 1783, 1783–85; Heilbron, 1979; Holbach, 1765; Isidore of Seville, ed. 2002; Kant, 1756a, 1756b, 1756c; Khalidi, 2005; Kircher, 1664–65; Leibniz, 1749; Leonardo da Vinci, *Atlantic Codex*, ed. 2000, *Codex Leicester*, ed. 1980; Ligorio, ed. 2005; Lister, 1684; Lloyd, 1970, 1973, 1991; Lucretius, ed. 1947; Maggio, 1571; Mallet, 1862; Manetti, *De terraemotu*, ms.

1457, *De dignitate*, ed. 1532; Marmo, 1989a; Melville, 1985; Mezzavacca, 1672; Michell, 1761; Milne, 1897, 1898, 1901; Needham, 1959; Oldham, 1899; Oldroyd *et al.*, 2007; Philastrius, ed. 1898; Phillips, 1581; Pirona and Taramelli, 1873; Placanica, 1984, 1985a; Pliny the Elder, ed. 1958–67; Poirier, 2005; Posidonius, ed. 1989–99; Priestley, 1775; Ray, 1691; Reale and Bos, 1995; Reid, 1911; Romei, 1587; Russo, 2004; Sarconi, 1784; Sardi, 1586; Sarti, 1783; Schiantavelli and Stile, 1784; Scopelliti, 1983; Seneca, ed. 1971–72; Serpieri, 1873, 1876; Stoetzer, 1986; Strato of Lampsacus, ed. 1969; Stukeley, 1750; Taramelli and Mercalli, 1886, 1888; Telesio, 1586; Terrenzi, 1887; Thales of Miletus, ed. 1951; Theophrastus, ed. 1866; Traina, 1989; Travagini, 1669; Turner, 1922; Udías, 1986, 1999; Vannucci, 1787; Vivenzio, 1783, 1788; Wadati, 1928; Whiston, 1696; Winkler, 1745; Woodward, 1695; Yeomans, 1991; Zuccolo, 1571; Zupo, 1784.

Chapter 5 Other types of sources

Almagià, 1914; Atwater *et al.*, 2005; Baratta, 1901, 1910; Becchetti and Ferrari, 2004; Blumetti *et al.*, 1988; Boschi *et al.*, 1995; Cole *et al.*, 1996; Davison, 1927; De Poardi, 1627; Ferrari, 1991; Frižot, 1998; Guidoboni *et al.*, 2007; Hutchinson and McMillan, 1997; Kozák and Ebel, 1996; Losey, 2002; Ludwin *et al.*, 2005; Mallet, 1862; Margottini and Kozák, 1992; Minor and Grant, 1996; Molin and Margottini, 1982; Mongitore, 1727; Montessus de Ballore, 1923; Moroni and Stucchi, 1993; Nelson *et al.*, 1995; Satake *et al.*, 1996; Schiantarelli and Stile, 1784; Taramelli and Mercalli, 1886, 1888; Volger, 1856.

Chapter 6 Potential problems in historical records

Abrahamyan, 1976; Alexandre, 1990; Allen, 1890; Ambraseys and Finkel, 1991; Ambraseys and Vogt, 1988; Ambraseys *et al.*, 1994; *Annales Corbeienses*, ed. 1864; *Annales Ratisponenses*, ed. 1861; *Annales sancti Emmerammi*, ed. 1861; Augustine, *Sermons*, ed. 1961; Babayan, 2006; Bächtold, 1993; Baratta, 1901; Benouar, 1994, 2004; Bilham, 1994; Boschi *et al.*, 1995; Camassi and Stucchi, 1997; Chesneau, 1892; *Chronicon Parmense*, ed. 1902–04; Conant, 1939; CRAAG, 1994; D'Addezio *et al.*, 1995; Downey, 1955; Dunbar *et al.*, 1992; Ergin *et al.*, 1967; Evangelatou-Notara, 1993; Fabretti and Vayra, 1891; Ferrari, 2004; Follieri, 1977, 1979; Gregoras, ed. 1829–55; Grumel, 1958; Gruppo del Lavoro CPTI, 2004; Guidoboni, 1985; Guidoboni and Comastri, 2002, 2005; Guidoboni and Traina, 1995; Guidoboni *et al.*, 1994, 2003a, 2007; Harbi *et al.*, 2003; Lampros, 1910; Lepelley, 1984; Leydecker and Brüning, 1988; Lilla, 1970; Mazza, 1998; Mercalli, 1883; Montessus de Ballore, 1892, 1906; Muratori, 1726; Musson, 1998; Omori, 1909; Pantosti *et al.*, 1996; Perrey, 1848, 1850; Roller, 1861; Saint Francis of Assisi, ed. 2007; Salimbene de Adam, ed. 1905–13; Shebalin *et al.*, 1974; Sigonio, 1591; Turyn, 1980; Valensise and Guidoboni, 1995, 2000b; Vallerani, 1995; Vogt, 1987; Vogt and Ambraseys, 1991, 1992; Wirth, 1966; *Zibaldone da Canal*, ed. 1967.

Chapter 7 Determination of historical earthquakes: dates and times

Acosta, 1590; *Biblia Hebraica Stuttgartensia*, ed. 1997; Bilfinger, 1892; Boschi *et al.*, 1995, 2000; Cassiodorus, ed. 1894; *Chronicon Paschale*, ed. 1832; *Codex Theodosianus*, ed. 1905; Diodorus Siculus, ed. 1888–1906; Dohrn-van Rossum, 1996; Ferrari and Marmo, 1985;

Freeman-Grenville, 1995; Gouin, 1979; Guido Pisanus, ed. 1963; Guidoboni *et al.*, 2005, 2007; Justinian, ed. 1912; Landes, 1983; Lane, 1867; Livy, ed. 1900–08; Nejjar, 1974; Pliny the Elder, ed. 1958–67; Quatremère, 1845; Renou and Filliozat, 1953; Stein, 1968; Theophanes, ed. 1883.

Chapter 8 Planning the goals of the analysis of historical earthquake data

Albarello and Muccarelli, 2002; Alexandre and Vogt, 1994; Boschi *et al.*, 1995, 2000; Carvalhao Buescu and Cordeiro, 2005; DISS Working Group, 2005; Figliuolo, 1988–89; Fischer *et al.*, 2001; Fracassi and Valensise, 2004, 2007; Gasperini *et al.*, 1999; Grünthal, 2004; Grünthal and Fischer, 1998, 2001; Guidoboni and Comastri, 2005; Guidoboni and Mariotti, 2005; Guidoboni *et al.*, 1994, 2007; Kendrick, 1956; Levret, 1991; Magri and Molin, 1984, 1985; Martínez Solares, 2001; Martínez Solares and Lopez Arroyo, 2004; Meletti *et al.*, 1988–89; Nur, 2002; Pereira de Sousa, 1919–28; Poirier, 2005; Roux, 1932; Tagliapietra, 2004; Vogt and Grünthal, 1994; Voltaire, *Poème*, ed. 1876, *Candide*, ed. 1980.

Chapter 9 Processing historical records

Guidoboni *et al.*, 2004, 2007.

Chapter 10 From interpretation of historical records to historical seismic scenarios

Albini *et al.*, 1995; Ambraseys, 1971; Ambraseys and Melville, 1995; Ambraseys *et al.*, 1994; Ammianus Marcellinus, ed. 1999; Athanasius, ed. 1985; Augustine, *Confessiones*, ed. 1992; Bacchielli, 1995; Baratta, 1910; Beschaouch, 1975; Bidez, 1930; Blanchard-Lemée, 1984; Boccaletti *et al.*, 2001; Boschi and Guidoboni, 2003; Boschi *et al.*, 1995, 2000; Buonocore, 1992; Cagiano de Azevedo, 1974; Camodeca, 1971; Cassian, ed. 1958; *Chronicon Salernitanum*, ed. 1956; *Chronicon Vulturnense*, ed. 1925–40; *Codex Iustinianus*, ed. 1877; *Codex Theodosianus*, ed. 1905; Comninakis and Papazachos, 1982; Constantine, ed. 1974; *Consularia Constantinopolitana*, ed. 1993; Crespellani *et al.*, 1992; Croke, 2001; Demetrius of Callatis, ed. 1923–58; Di Somma, 1641; Di Vita, 1964, 1982, 1990; Doglioni *et al.*, 1994; Dufour and Raymond, 1994; El-Sayed *et al.*, 2000; *Fasti Vindobonenses posteriores*, ed. 1892; *Fasti Vindobonenses priores*, ed. 1892; Fatouros *et al.*, 2002; Felgentreu, 2004; Fokaefs and Papadopoulos, 2007; Galadini and Galli, 2004; Galetti, 1985; Galli and Bosi, 2004; Ganas *et al.*, 2001; Gizzi and Masini, 2004; Goodchild, 1966–67; Grünthal, 1993; Guerrieri and Vittori, 2007; Guidoboni and Comastri, 1997, 2005; Guidoboni and Ferrari, 1995; Guidoboni *et al.*, 1994, 2003a, 2003b, 2007; Henry, 1985; Hough, 2004; Issel, 1888; Jackson, 2006; Jacques and Bousquet, 1984; Jerome, *Chronicon*, ed. 1984, *Commentariorum*, ed. 1963, *Vita sancti Hilarionis*, ed. 1983; Jones *et al.*, 1971; Kelly, 2004; Kenrick, 1986; Lawson, 1908; Lepelley, 1984, 1994; Libanius, ed. 1904; *Liber Pontificalis*, ed. 1886–1957; Lorito *et al.*, 2008; Mallet, 1862; Marcellinus Comes, ed. 1894; McMahon, 1897; Molin and Paciello, 1999; MunichRe, 1991; Mutafian and Van Lauve, 2001; Nicolaus Damascenus, ed. 1923–58; Nuffelen, 2006; Obsequens, ed. 1910; Optatus, ed. 1893; Papathanassiou and Pavlides, 2007; Papazachos and Papazachos, 1997; Pavlides *et al.*, 1999, 2002; Petrarca, ed. 1934; Pilla, 1846; Pirazzoli and

Thommeret, 1977; Pirazzoli et al., 1982, 1992, 1996; Placanica, 1970, 1979, 1985a; Price et al., 2002; Rebuffat, 1980; Restifo, 1995; Riccò, 1907; Schiantarelli and Stile, 1784; Serva, 1994; Serva et al., 2007; Shaw et al., 2008; Shebalin et al., 1974; Sievers, 1868; Sigonio, 1580; Skuphos, 1894; Socrates Scholasticus, ed. 1995; Sozomen, ed. 1995; Spratt, 1865; Stiros, 2001; Stiros and Papageorgiou, 2001; Taramelli and Mercalli, 1888; Tatevossian, 2007; Teti, 2004; Thucydides, ed. 1964–72; Tobriner, 1982; Utsu, 2002; Valensise and Guidoboni, 2000a; Vannucci et al., 1999a; Vogt, 1977a, 1977b, 1984a, 1984b, 1993; Ward-Perkins, 1984; Westerbergh, 1956; William of Rubruck, ed. 1929; Zosimus, ed. 1971–89.

Chapter 11 Traces of earthquakes in archaeological sites and in monuments

Adam, 1989; Agnello, 1930; Annales Cremonenses, ed. 1903; Baratta, 1903; Barbisan and Laner, 1983, 1995; Bassani, 1895–1902; Butzer, 1982; Cairoli Giuliani et al., 2007; Cardona, 2005; Ciuccarelli, 2003; Cuneo, 1968, 1988; D'Agostino, 2002, 2003, 2005; D'Agostino and Bellomo, 2007; De Bernardi Ferrero, 1993; Del Moro, 1895; De Rossi, 1874; Di Teodoro and Barbi, 1983; Fernández, 1990; Ferrigni et al., 2005; Foufa, 2005; Galli and Galadini, 2003; Gebhard, 1996; Guidoboni, 2002; Guidoboni et al., 1994, 2000; Helly, 2005; Karcz and Kafri, 1978, 1981; Kasangian, 1981; Lamboley, 1998; Lanciani, 1918; Leonardo, Il manoscritto A, ed. 1990; Ligorio, ed. 2005; MacGuire et al., 2000; Marchini, 1978; Martini, ed. 1979; Milizia, 1781; Molin, 2001; Nur, 2002; Nur and Reches, 1979; Nur et al., 1989; Papageorgiou and Stiros, 1996; Papastamatiou and Psycharis, 1996; Pierotti and Ulivieri, 1998, 2001; Porter, 1915–17; Rocchi, 1978; Seneca, ed. 1971–72; Sguario, 1756; Sieberg, 1932b; Spondilis, 1996; Stiros, 1988; Stiros and Jones, 1996; Supplementum Annalium Cremonensium, ed. 1903; Tacitus, ed. 1970; Tobriner, 1983; Vannucci, 2000; Willis, 1928; Wood and Johnson, 1978; Zarian, 1992.

Chapter 12 Deriving earthquake source and shaking parameters and tsunami parameters from historical data

Abe, 1979, 1989; Ambraseys, 1962; Atkinson and Sonley, 2000; Bakun and Wentworth, 1997; Bakun et al., 2003; Basili et al., 2008; Båth, 1979; Biasi et al., 2002; Blake, 1941; Boschi et al., 1995, 1997, 2000; DISS Working Group, 2007; Dominey-Howes, 2004; Ebel, 1996, 2000, 2006; Ebel and Wald, 2003; Ferrari and Guidoboni, 2000; Friedrich et al., 2006; Fumal et al., 2002; Gasperini and Ferrari, 2000; Gasperini and Valensise, 2000; Gasperini et al., 1999; Gouin, 2001; Grünthal, 1993, 1998; Guidoboni, 2000; Guidoboni et al., 1994, 2007; Gutenberg and Richter, 1956; Hatori, 1986; Hough et al., 2000; Hsieh, 1957; Iida et al., 1967; Imamura, 1949; Jacoby et al., 1988; Jesuit Relations, 1663; Johnston, 1996a, 1996b; Kaka and Atkinson, 2004; Kawasumi, 1951; Kelleher et al., 1973; Krinitzsky and Chang, 1988; Mader, 1999, 2002; Maheshwari et al., 2006; Manning et al., 2006; Mauk et al., 1982; McCalpin, 1996; McCoy and Heiken, 2000; Medvedev et al., 1964; Meghraoui et al., 2000; Miller, 1960; Minoura et al., 2000; Mucciarelli and FAUST Working Group, 2000; Murphy and O'Brien 1977; Nuttli, 1973; Nuttli and Zollweg, 1974; Nuttli et al., 1986; Obermeier and Pond, 1999; Papadopoulos, 2003; Papadopoulos and Imamura, 2001; Pararas-Carayannis, 2003; Pareschi et al., 2006; Pelinovsky et al., 2006; Reiter, 1990; Richter, 1935, 1958; Rothman, 1968; Seeber and

Armbruster, 1987; Sieberg, 1923, 1927, 1932a; Smith, 1966; Steinbrugge, 1982; Stover and Coffman, 1993; Street and Lacroix, 1979; Tocher and Miller, 1959; Trifunac and Brady, 1975; Valensise and Pantosti, 2001; Valensise *et al.*, 2002; Wald *et al.*, 1999a, 1999b; Wells and Coppersmith, 1994; Wharton and Evans, 1888; Whitman, 2002; Winthrop, 1757–58; Wood and Neumann, 1931; Yokoyama, 1978.

Chapter 13 Cooperation in historical seismology research

Kozák and Ebel, 1996.

References

Ancient and medieval works

Agathias. *Historiarum libri quinque*, ed. R. Keydell. (Corpus fontium historiae Byzantinae.) Berlin: W. de Gruyter, 1967. (English translation: *The Histories*, transl. with an introduction and short explanatory notes by D. J. Frendo. Berlin: W. de Gruyter, 1975.)

Albertus Magnus. *Phisicorum sive auditus phisici Alberti magni tractatus primus*. Venice: Gregorio de' Gregori, 1494.

Albertus Magnus. *De mineralibus et rebus metallicis libri quinque*. Cologne: J. Birckmann & Th. Baum, 1569. (English translation: *Book of Minerals*, transl. D. Wyckoff. Oxford: Clarendon Press, 1967.)

Ammianus Marcellinus. *Rerum gestarum libri qui supersunt*, ed. W. Seyfarth. 2 vols. (Bibliotheca scriptorum graecorum et romanorum teubneriana.) Stuttgart: B. G. Teubner, 1999. (English translation by J. C. Rolfe. 3 vols. (The Loeb Classical Library.) Cambridge, Mass.: Harvard University Press; London: Heinemann, 1971–72.)

Anaxagoras. *Die Fragmente der Vorsokratiker: vol. 2, Die Fragmente der Philosophen des sechesten und fünften Jahrhunderts (und unmittelbarer Nachfolgen)*, eds. H. Diels and W. Kranz, 6th edn, no. 59, 5–44. Berlin: Weidmann, 1952.

Anaximenes. *Die Fragmente der Vorsokratiker: vol. 1, Die Fragmente der Philosophen des sechesten und fünften Jahrhunderts (und unmittelbarer Nachfolgen)*, eds. H. Diels and W. Kranz, 6th edn, no. 13, 90–96. Berlin: Weidmann, 1951.

Annales Corbeienses, ed. Ph. Jaffé (Bibliotheca Rerum Germanicarum 1: Monumenta Corbeiensia.) Berlin: Weidmann, 1864, 33–43.

Annales Cremonenses, ed. O. Holder-Egger (Monumenta Germaniae Historica, Scriptores 31.) Hanover: Hahnsche Buchhandlung, 1903, 3–21.

Annales Ratisponenses, ed. W. Wattenbach (Monumenta Germaniae Historica, Scriptores 17.) Hanover: Hahnsche Buchhandlung, 1861, 579–788.

Annales sancti Emmerammi saeculi XI, ed. W. Wattenbach (Monumenta Germaniae Historica, Scriptores 17.) Hanover: Hahnsche Buchhandlung, 1861, 571.

Archelaus. *Die Fragmente der Vorsokratiker: vol. 2, Die Fragmente der Philosophen des sechesten und fünften Jahrhunderts (und unmittelbarer Nachfolgen)*, eds. H. Diels and W. Kranz, 6th edn, no. 60, 44–49. Berlin: Weidmann, 1952.

[Aristotle]. *De mundo. On sophistical refutations. On coming-to-be and passing-away*, ed. and transl. E. S. Forster. *On the cosmos*, ed. and transl. D. J. Furley. (The Loeb Classical Library.) Cambridge, Mass.: Harvard University Press; London: Heinemann, 1955.

Aristotle. *Metaphysica*, in *Aristoteles Graece*, ed. I. Bekker. Berlin: Reimer, 1831. (Repr. ed. O. Gigon, Berlin: de Gruyter, 1960–61.) (English translation: *Aristotle's Metaphysics. A Revised Text with Introduction and Commentary*, ed. W. D. Ross. Oxford: Clarendon Press, 1924, 3rd edn 1953.)

Aristotle. *Meteorologica*, ed. and transl. H. D. P. Lee. (The Loeb Classical Library.) Cambridge, Mass.: Harvard University Press; London: Heinemann, 1952.

Athanasius of Alexandria. *Index to the Festal Letters*, ed. with French transl. by A. Martin and M. Albert, *Histoire "acéphale" et Index syriaque des Lettres festales d'Athanase d'Alexandrie*. (Sources chrétiennes 317.) Paris: Cerf, 1985, pp. 224–277. (English translation: *The festal epistles of St Athanasius, Bishop of Alexandria*, transl. from the Syriac with notes and indices by H. Burgess, ed. H. G. Williams. Oxford: John Henry Parker, 1854.)

Augustine. *Confessiones*, ed. M. Simonetti. (Scrittori greci e latini.) Rome: Fondazione Lorenzo Valla; Milan: A. Mondadori, 1992 (English translation: *Confessions*, transl. F. J. Sheed, with notes by M. P. Foley, 2nd edn. Indianapolis, Ind.; Cambridge: Hackett, 2006.)

Augustine. *Sermones de Vetere Testamento, id est Sermones 1–50*, ed. C. Lambot (Corpus Christianorum, Series Latina 41.) Turnholt: Brepols, 1961. (English translation: *Sermons (1–19) on the Old Testament*, ed. J. E. Rotelle, translation and notes E. Hill. Brooklyn, NY: New City Press, 1990.)

Averroes. *Middle commentaries on Aristotle's Categories and De interpretatione*, ed. and transl. Ch. E. Butterworth. Princeton, N.J.: Princeton University Press, 1983.

Averroes. *Tahafut al-tahafut (The incoherence of the incoherence)*. Translated from the Arabic with introduction and notes by S. Van den Bergh. London: the Trustees of the E. J. W. Gibb Memorial, 1978. 2 vols. in 1. (Originally published in 1954.)

Avicenna. *De mineralibus: Avicennae De congelatione et conglutinatione lapidum*. Being sections of the *Kitâb al-shifâ'*. The Latin and Arabic texts, edited with an English translation of the latter and with critical notes by E. J. Holmyard and D. C. Mandeville. Paris: P. Guethner, 1927.

Bartolomeo da Rossano, *Vitae sancti Nili iunioris*, ed. G. Giovanelli. Grottaferrata: Badia di Grottaferrata, 1972.

Bede, *De natura rerum liber*, in *Opera omnia*, ed. J.-P. Migne. (Patrologia Latina 90.) Paris: Garnier, 1862. (Repr. Turnhout: Brepols, 1968.)

Benjamin of Tudela. *The Itinerary of Benjamin of Tudela*, ed. and transl. M. N. Adler. London: Henry Frowde, 1907. (Repr. Frankfurt am Main: Institute for the History of Arabic-Islamic Science, 1995.)

Biblia Hebraica Stuttgartensia, ed. R. Kittel, K. Ellinger, W. Rudolph; textum Masoreticum ed. H. P. Rüger, 5th edn ed. A. Schenker. Stuttgart: Deutsche Bibelgesellschaft, 1997.

Biblioteca del Monumento Nazionale di Santa Scolastica, Subiaco, arca XXIII, parchment 7, Deed by the notary Paolo di Cervara, Subiaco, 13 September 1348.

Cassian, John. *Conlationes*, ed. E. Pichery, vol. 2 (Sources chrétiennes.) Paris: Les Editions du Cerf, 1958. (English translation: *The Conferences*, transl. B. Ramsey. New York: Paulist Press, 1997.)

Cassiodorus. *Chronica ad a. 519*, ed. Th. Mommsen. (Monumenta Germaniae Historica, Auctores Antiquissimi 11.) Berlin: Wiedmann, 1894, 120–161.

Chronicon Paschale, ed. L. Dindorf. (Corpus scriptorum historiae Byzantinae.) Bonn: Weber, 1832. (English translation: *Chronicon Paschale 284–628 AD*, translated with notes and introduction by M. Whitby and M. Whitby (Translated texts for historians 7.) Liverpool: Liverpool University Press, 1989.)

Chronicon Salernitanum. A critical edition with studies on literary and historical sources and on language, ed. U. Westerbergh. (Acta Universitatis Stockholmiensis. Studia Latina Stockholmiensia.) Stockholm: Almqvist & Wiksell, 1956.

Chronicon Vulturnense, ed. V. Federici. (Fonti per la storia d'Italia 58–60.) Rome: Tipografia del Senato, 1925–40.

Chronicon Sublacense, aa. 593–1369, ed. R. Morghen (Rerum Italicarum Scriptores 24/6.) Bologna: Zanichelli, 1927.

Chronicon Parmense, ab anno 1038 usque ad annum 1338, ed. G. Bonazzi (Rerum Italicarum Scriptores 9/9.) Città di Castello: S. Lapi, 1902–04.

Chrysippus. *Stoicorum veterum fragmenta*, ed. H. F. A. von Arnim. Vol. 2, *Chrysippi fragmenta logica et physica*. Vol. 3, *Chrysippi fragmenta moralia; fragmenta successorum Chrysippi*. Stuttgart: B. G. Teubner, 1903.

Codex Iustinianus, ed. P. Krüger. Berlin: Weidmann, 1877. (Repr. Goldbach: Keip, 1998.)

Codex Theodosianus, eds. Th. Mommsen and P. M. Meyer. Berlin: Weidmann, 1905. (Repr. Hildesheim: Weidmannsche Verlagsbuchhandlung, 2002.) (English translation: *The Theodosian code and novels* and *The Sirmondian constitutions*, a translation, with commentary, glossary and bibliography by C. Pharr; in collaboration with T. S. Davidson and M. B. Pharr; with an introduction by C. D. Williams. Union, N.J.: Lawbook Exchange, 2001.) (Originally published: Princeton, N.J.: Princeton University Press, 1952.)

Constantine, Bishop of Assiut. *Encomia in Athanasium duo*, ed. T. Orlandi. (Corpus Scriptorum Christianorum Orientalium 349–350, Scriptores Coptici 37–38.) Vol. 1, Coptic text; vol. 2, Latin translation. Louvain: Peeters, 1974.

Consularia Constantinopolitana, in *The Chronicle of Hydatius and the Consularia Constantinopolitana: Two contemporary accounts of the final years of the Roman Empire*, ed. with an English translation by R. W. Burgess. Oxford: Clarendon Press, 1993.

Corpus Inscriptionum Graecarum, ed. A. Boeckh, vol. 2 (nos. 1793–3809). Berlin: ex Officina Academica, 1843.

Dell'Aquila, Matteo. *Tractatus de cometa atque terraemotu: Cod. Vat. Barb. Lat. 268*, ed. B. Figliuolo. Salerno: P. Laveglia, 1990.

Demetrius of Callatis. *Die Fragmente der griechischen Historiker*, no. 85, ed. F. Jacoby, 15 vols. Berlin: Weidmann; Leiden: Brill, 1923–58.

Democritus of Abdera. *Die Fragmente der Vorsokratiker: vol. 2, Die Fragmente der Philosophen des sechesten und fünften Jahrhunderts (und unmittelbarer Nachfolgen)*, eds. H. Diels and W. Kranz, 6th edn, no. 68, 81–230. Berlin: Weidmann, 1952. (Other edn., *Democritea*, ed. S. Lur'e [Luria]. Leningrad: Izd-vo "Nauka", 1970 (in Russian, Original texts in Greek or Latin).)

Die Fragmente der griechischen Historiker, ed. F. Jacoby, 15 vols. Berlin: Weidmann; Leiden: E. J. Brill, 1923–58.

Dio Cassius Cocceianus. *Historiarum Romanarum quae supersunt*, ed. U. P. Boissevain, 5 vols. Berlin: Weidmann, 1895–1931.

Diodorus Siculus. *Bibliotheca historica*, ed. F. Vogel and C. Th. Fischer. (Bibliotheca scriptorum graecorum et romanorum teubneriana.) Leipzig: Teubner, 1888–1906. (English translation: *Diodorus of Sicily in twelve volumes*, transl. C. H. Oldfather *et al.* (The Loeb Classical Library.) Cambridge, Mass.: Harvard University Press; London: W. Heinemann, 1976–2006.)

Diogenes Laertius. *Vitae philosophorum*, ed. M. Marcovich, vols. 1–2. (Bibliotheca scriptorum graecorum et romanorum teubneriana.) Stuttgart: B. G. Teubner, 1999. *Indices*, vol. 3, ed. H. Gärtner. Munich: Saur, 2002. (English translation by R. D. Hicks: *Lives of Eminent Philosophers*, 2 vols. (The Loeb Classical Library.) Cambridge: Harvard University Press; London: Heinemann, 1980.

Diogenes of Apollonia. *Die Fragmente der Vorsokratiker: vol. 2, Die Fragmente der Philosophen des sechesten und fünften Jahrhunderts (und unmittelbarer Nachfolgen)*, eds. H. Diels and W. Kranz, 6th edn, no. 64, 51–69. Berlin: Weidmann, 1952.

Epicurus. *Epicurea*, ed. H. Usener. Leipzig: B. G. Teubner, 1887.

Eusebius–Jerome. *Chronicon*, ed. R. Helm, 2nd edn. (Die griechischen christlichen Schriftsteller der ersten Jahrhunderte; Eusebius Werke 7.) Berlin: Akademie-Verlag, 1984.

Evagrius Scholasticus. *Historia Ecclesiatica*, eds. J. Bidez and L. Parmentier. London: Methuen, 1898. (English translation: *The ecclesiastical history of Evagrius Scholasticus*, transl. M. Whitby. Liverpool: Liverpool University Press, 2000.)

Fasti Vindobonenses posteriores, ed. Th. Mommsen. (Monumenta Germaniae Historica, Auctores Anquissimi 9.) Berlin: Weidmann, 1892, 274–334.

Fasti Vindobonenses priores, ed. Th. Mommsen. (Monumenta Germaniae Historica, Auctores Anquissimi 9.) Berlin: Weidmann, 1892, 274–336.

Gellius, Aulus. *Noctes Atticae*, ed. R. Marache. Paris: Les Belles Lettres, 1967–89. (English translation: *The Attic Nights*, transl. J. C. Rolfe. (The Loeb Classical Library.) Cambridge, Mass.: Harvard University Press, 2002.)

al-Ghazali. *The incoherence of the philosophers*. A parallel English–Arabic text, ed. and trans. M. E. Marmura. Provo, Ut.: Brigham Young University Press, 1997, 2000.

Gonzalez de Clavijo, Ruy. *The Spanish embassy to Samarkand 1403–1406*, original Spanish text with Russian translation and notes by I. Sreznevskj. London: Variorum Reprints, 1971. (English translation: *Embassy to Tamerlane 1403–1406*, transl. G. Le Strange, London: G. Routledge & Sons, 1928.) (Repr. Frankfurt am Main: Institute for the History of Arabic-Islamic Science, 1994).

Gregoras, Nicephorus. *Byzantina Historia*, eds. L. Schopen and I. Bekker, 3 vols. (Corpus Scriptorum Historiae Byzantinae.) Bonn: Weber, 1829–55.

Guido Pisanus. Notula (annotation on the earthquakes of 3 January 1117) in the codex *Vaticanus Latinus* 11564, fol.184, of the Biblioteca Apostolica Vaticana, in G. Scalia, "Epigraphica pisana. Testi latini sulla spedizione contro le Baleari del 1113–1115 e su altre imprese antisaracene del secolo XI," *Miscellanea di Studi Ispanici* [1963]. Pisa: Università di Pisa, 285–286.

Historicorum Romanorum reliquiae, ed. H. Peter. Leipzig: B. G. Teubner, 1906–14.

Homer. *Odyssea*, ed. H. van Thiel. Hildesheim: G. Olms, 1991. (English translation: *The Odyssey*, transl. A. T. Murray, revised by G. E. Dimock. (The Loeb Classical Library.) Cambridge, Mass.: Harvard University Press, 1995.)

Ibn al-Qalanisi. *Dhayl tarikh Dimashq (History of Damascus, 363–555 A.H.)*, ed. H. F. Amedroz. Leiden: E. J. Brill, 1908. (English translation: *The Damascus chronicle of the Crusades, extracted and translated from the Chronicle of Ibn al-Qalanisi* [by] H. A. R. Gibb. Mineola, New York: Dover Publications, 2002.)

Infimae Aetatis: A Textual Data Bank of Late Antigue and Medieval Inscriptions, ed. J. M. Mansfield. Available at http://132.236.125.30/JMM/ICE_ICK_top.html

Isidore of Seville. *De natura rerum liber*, ed. J. Fontaine. (Collection des études augustiniennes 39.) Paris: Institut d'études augustiniennes, 2002 (French and Latin).

Jerome. *Chronicon*. See Eusebius–Jerome.

Jerome. *Commentariorum in Esaiam libri*, ed. M. Adriaen and G. Morin, 2 vols. (Corpus Christianorum, Series Latina, 73, 73A.) Turnholt: Brepols, 1963.

Jerome. *Vita sancti Hilarionis*, in *Vita di Martino, Vita di Ilarione, In memoria di Paola*, ed. A. A. R. Bastianensen and J. W. Smit. (Scrittori greci e latini.) Milan: Fondazione Lorenzo Valla, A. Mondadori, 1983. (English translation: *Certaine selected epistles of S. Hierome, as also the lives of Saint Pavl the first hermite, of Saint Hilarion the first monke of Syria, and of S. Malchvs*, transl. H. Hawkins. Ilkley: Scolar Press, 1975. (Reprint of the 1630 edn published by English College Press, St Omer.)

Justinian. *Corpus iuris civilis*. Vol. 3, *Novellae*, eds. R. Schoell and W. Kroll, 4th edn. Berlin: Weidmann, 1912.

Libanius. *Orationes XII–XXV*, vol. 2, in *Libanii Opera*, ed. R. Foerster. (Bibliotheca scriptorum graecorum et romanorum teubneriana.) Leipzig: Teubner, 1904 (English translation: *Selected works*, introduction and notes by A. F. Norman, vol. 1, *The Julianic orations*. (The Loeb Classical Library.) London: Heinemann, 1969.)

Liber Pontificalis, vols. 1–2: ed. L. Duchesne; vol. 3: ed. C. Vogel. Paris: Thorin, 1886–92; E. de Boccard, 1957.

Livy. *Ab urbe condita libri*, ed. W. Weissenborn, M. Müller and G. Heraeus. (Bibliotheca scriptorum graecorum et romanorum teubneriana.) Leipzig: G. B. Teubner, 1900–08. (English translation: *Livy in fourteen volumes*, transl. B. O. Foster *et al.*, 14 vols. London: Heinemann; New York: Putnam's Sons; Cambridge, Mass.: Harvard University Press, 1919–59.)

Lucretius. *De rerum natura libri sex*, ed. C. Bailey, with prolegomena, critical apparatus, translation, and commentary. 3 vols. Oxford: Clarendon Press, 1947.

Malalas. *Chronographia*, ed. I. Thurn. (Corpus fontium historiae Byzantinae.) Berlin: W. de Gruyter, 2000. (English translation: *The chronicle of John Malalas*, transl. E. Jeffreys, M. Jeffreys and R. Scott. Melbourne: Australian Association for Byzantine Studies, 1986.)

Manetti, Giannozzo. *De dignitate & excellentia hominis libri 4*. Basel: A. Cratander, 1532. (Rep. Frankfurt am Main: Minerva, 1975.) (Other edn.: ed. E. R. Leonard. Padua: Antenore, 1975. Introduction in English.)

Manetti, Giannozzo. *De terraemotu libri tres*, 1457. Biblioteca Apostolica Vaticana, *Urbinati Latini*, 1077.

Marcellinus Comes. *Chronicon ad annum 518, continuatum ad A. 534, cum addimento ad A. 548*, ed. Th. Mommsen (Monumenta Germaniae Historica, Auctores Antiquissimi 11). Berlin: Weidmann, 1894. (English translation: *The Chronicle of Marcellinus: A Translation and commentary with a reproduction of Mommsen's edition of the text*, transl. B. Croke (Byzantina Australiensia 7). Sydney: Australian Association for Byzantine Studies, 1995.)

Nicolaus Damascenus. *Die Fragmente der griechischen Historiker*, no. 90, ed. F. Jacoby, 15 vols. Berlin: Weidmann; Leiden: Brill, 1923–58.

Obsequens, Julius. *Prodigiorum liber*, in *T. Livi periocae omnium librorum fragmenta Oxyrhynchi reperta*, vol. 1, *Obsequentis Prodigiorum liber*, ed. O. Rossbach. Leipzig: Teubner, 1910. (English translation: *Livy, History of Rome*, vol. 14, transl. A. C. Schlesinger. (The Loeb Classical Library.) Cambridge, Mass.: Harvard University Press, 2004.)

Optatus, Bishop of Milevis. *Optati Milevitani libri VII*, ed. C. Ziwza. (Corpus Scriptorum Ecclesiasticorum Latinorum 26.) Vienna: Tempsky, 1893. (English translation: *Against the donatists*, ed. and transl. by M. Edwards. (Translated texts for historians 27.) Liverpool: Liverpool University Press, 1997.)

Oracula Sibyllina, ed. A. Kurfess, 2nd edn ed. J.-D. Gauger. Dusseldorf: Artemis & Winkler, 1998. (English translation: *The Sibylline oracles*, transl. H. N. Bate. London: S.P.C.K., 1918.)

Ovid. *Metamorphoses*, ed. W. S. Anderson. (Bibliotheca scriptorum graecorum et romanorum teubneriana.) Stuttgart: B. G. Teubner, 1977.

Pausanias. *Graeciae descriptio*, ed. M. H. Rocha-Pereira. (Bibliotheca scriptorum graecorum et romanorum teubneriana.) Leipzig: B. G. Teubner, 1973–81. (English translation: *Guide to Greece*, transl. P. Levi. Harmondsworth: Penguin, 1971.)

Petrarca, Francesco. *Le familiari [Familiarum rerum libri]*, vol. 2, Libri V–XI, ed. V. Rossi (Edizione nazionale delle opere di Francesco Petrarca 11). Florence: Sansoni, 1934. (English translation: *Letters on familiar matters*, transl. A. S. Bernardo, vol. 2. Boltimore, Md.: Johns Hopkins University Press, 1982.)

Philastrius, *Liber de haeresibus*, ed. F. Marx. (Corpus Scriptorum Ecclesiasticorum Latinorum 38.) Prague: F. Tempsky; Leipzig: G. Freytag, 1898.

Pliny the Elder. *Naturalis historia*, ed. and transl. H. Rackham, W. H. S. Jones, and D. E. Eichholz, 10 vols. (The Loeb Classical Library.) Cambridge, Mass.: Harvard University Press; London: Heinemann, 1958–1967.

Plutarch. *Moralia*, transl. F. C. Babbit *et al.*, 16 vols. (The Loeb Classical Library.) Cambridge, Mass.: Harvard University Press; London: Heinemann, 1927–2004.

Posidonius. *The Fragments*. Vol. 1, *The fragments*, ed. L. Edelstein and I. G. Kidd. 2nd edn, 1989. Vol. 2, *The Commentary*. Part I: *Testimonia and fragments 1–149*, Part II: *Fragments 150–293*, ed. I. G. Kidd, 1988. Vol. 3, *The translation of the fragments*, ed. I. G. Kidd, 1999. Cambridge: Cambridge University Press, 1989–99.

Procopius. *Bella*, in *Opera omnia*, eds. J. Haury and G. Wirth. (Bibliotheca scriptorum graecorum et romanorum teubneriana.) Leipzig: B. G. Teubner, 1962–64. (English translation: *History of the wars*, transl. H. B. Dewing. (The Loeb Classical Library.) Cambridge, Mass.: Harvard University Press; London: Heinemann, 1961.)

Pseudo-Apollodorus. *Bibliotheca*, transl. J. G. Frazer, 2 vols. (The Loeb Classical Library.) Cambridge, Mass.: Harvard University Press London: Heinemann, 1963–67.

Saint Francis of Assisi. *I Fioretti di San Francesco*, versione inedita di p. B. Bughetti, 3rd edn. Rome: Città Nuova, 2007. (English translation: *The little flowers of Saint Francis*, transl. Th. Okey. Mineola, N.Y.: Dover, 2003.)

Salimbene de Adam, *Cronica*, ed. O. Holder-Egger. (Monumenta Germaniae Historica, Scriptores 32.) Hanover: Hahnsche Buchhandlung, 1905–13. (English translation: *The chronicle of Salimbene de Adam*, eds. J. L. Baird, G. Baglivi and J. R. Kane. Binghamton, N.Y.: Medieval and Renaissance Texts & Studies, 1986.)

Seneca. *Naturales quaestiones*, ed. and transl. Th. H. Corcoran, 2 vols. (The Loeb Classical Library.) Cambridge, Mass.: Harvard University Press; London: Heinemann, 1971–72.

Sibylline Oracles see *Oracula Sybillina*.

Socrates Scholasticus. *Historia ecclesiastica*, ed. G. C. Hansen. (Die griechischen christlichen Schriftsteller der ersten Jahrhunderte Neue Folge 1.) Berlin: Akademie Verlag, 1995. (English translation: *A select library of Nicene and post-Nicene Fathers of the Christian Church*, vol. 2, 2nd series, *Socrates, Sozomenus: Church histories*, translated ... under the editorial supervision of P. Schaff and H. Wace. Edinburgh: T & T Clark; Grand Rapids, Mich.: Eerdmans, 1997.)

Sozomen. *Historia ecclesiastica*, ed. J. Bidez and G. C. Hansen, 2nd edn. (Die griechischen christlichen Schriftsteller der ersten Jahrhunderte Neue Folge 4.) Berlin: Akademie Verlag, 1995 (English translation: *A select library of Nicene and post-Nicene Fathers of the Christian Church*, vol. 2, 2nd series, *Socrates, Sozomenus: Church histories*, translated ... under the editorial supervision of P. Schaff and H. Wace. Edinburgh: T & T Clark; Grand Rapids, Mich.: Eerdmans, 1997.)

Strato of Lampsacus. *Die Schule des Aristoteles. Texte und Kommentar*, vol. 5, ed. F. Wehrli. Basel–Stuttgart: Benno Schwabe, 1950, 2nd edn 1969.

Supplementum Annalium Cremonensium, ed. O. Holder-Egger. (Monumenta Germaniae Historica, Scriptores 31.) Hanover: Hahnsche Buchhandlung, 1903, 185–188.

Synesius of Cyrene. *Epistulae*, ed. A. Garzya, French translation D. Roques, 2 vols. Paris: Les Belles Lettres, 2003. (English translation: *The letters of Synesius of Cyrene*, transl. A. FitzGerald. London: Humphrey Milford, 1926.)

Tacitus. *Annales ab excessu divi Augusti*, ed. E. Koestermann. (Bibliotheca scriptorum graecorum et romanorum teubneriana.) Leipzig: Teubner, 1965. (English translation: *The Annals*, transl. J. Jackson (The Loeb Classical Library.) London: Heinemann; Cambridge, Mass.: Harvard University Press, 1970.)

Thales of Miletus. *Die Fragmente der Vorsokratiker*, vol. 1, *Die Fragmente der Philosophen des sechesten und fünften Jahrhunderts (und unmittelbarer Nachfolgen)*, eds. H. Diels and W. Kranz, 6th edn, no. 11, 67–81. Berlin: Weidmann, 1951.

Theophanes. *Chronographia*, ed. C. de Boor. Leipzig: G. B. Teubner, 1883. (Repr. Hildesheim, New York: G. Olms, 1980). (English translation: *The chronicle of Theophanes Confessor. Byzantine and Near Eastern history, AD 284–813*, transl. with introduction and commentary by C. Mango and R. Scott, with the assistance of G. Greatrex. Oxford: Clarendon Press, 1997.)

Theophrastus. *Opera quae supersunt omnia*, ed. F. Wimmer. Paris: F. Didot, 1866. (Repr. Frankfurt am Main: Minerva, 1964.)

Thucydides. ed. J. De Romilly, 5 vols. Paris: Les Belles Lettres, 1964–72. (English translation: *History of the Peloponnesian war*, transl. R. Warner with an introduction and notes by M. I. Finley (revised, with a new introduction and appendices). Harmondsworth: Penguin Books, 1972.)

Villani, Giovanni. *Nuova cronica*, ed. G. Porta, 3 vols. Parma: U. Guanda, 1990–91.

William of Rubruck. *Itinerarium*, in *Sinica Franciscana*, vol. 1, ed. A. Van den Wyngaert. Quaracchi (Florence): Collegium s. Bonaventurae, 1929. (English translation: *The mission of Friar William of Rubruck: his journey to the court of the Great Khan Möngke 1253–1255*, transl. P. Jackson. London: Hakluyt Society, 1990.)

Xenophon. *Hellenica*, with an English translation by Ch. L. Brownson, 3 vols. New York: G. P. Putnam; London: Heinemann, 1918–22.

Zibaldone da Canal. Manoscritto mercantile del sec. XIV, ed. A. Stussi. Venice: Comitato per la publicazione delle fonti relative alla storia di Venezia, 1967. (English translation: *Merchant culture in fourteenth-century Venice: the Zibaldone da Canal*, transl. J. E. Dotson. Binghampton, N.Y.: Medieval and Renaissance Texts and Studies, 1994.)

Zosimus. *Historia nova*, ed. F. Paschoud, 5 vols. (Collection des universités de France.) Paris: Société d'édition les Belles Lettres, 1971–1989 (English translation: *New history*, a translation with commentary by R. T. Ridley. Canberra: Australian Association for Byzantine Studies, 1982.)

Modern and contemporary works and studies

Abe, K. 1979. "Size of great earthquakes of 1837–1974 inferred from tsunami data," *Journal of Geophysical Research*, 84: 1561–1568.

Abe, K. 1989. "Quantification of tsunamigenic earthquakes by the M_t scale," *Tectonophysics*, **166**: 27–34.

Abrahamyan, A. G. 1976. "Chronological corrections," *Annual Review*, **90**: 139–168 (in Armenian).

Acosta, José de. 1590. *Historia Natural y Moral de las Indias*. Seville: J. de Leon; another edition, Madrid: A. Martin, 1608. (English translation: *The naturall and morall historie of the East and West Indies*. London: E. Blount and W. Aspley, 1604. *Natural and moral history of the Indies*, ed. J. E. Mangan; with an introduction and commentary by W. D. Mignolo; transl. F. M. López-Morillas. Durham, N.C.: Duke University Press, 2002.)

Adam, J.-P. 1989. "Osservazioni tecniche sugli effetti del terremoto di Pompei del 62 d.C.," in *I terremoti prima del Mille in Italia e nell'area mediterranea. Storia archeologia sismologia*, ed. E. Guidoboni. Bologna: ING–SGA, 460–474.

Agnello, G. 1930. *Guida del Duomo di Siracusa*. Milan: Officina d'arte grafica A. Lucini & C.

Agricola, Giorgius. 1546. *De natura fossilium*, in *De ortu & causis subterraneorum libri V; De natura eorum quae effluunt ex terra libri III; De natura fossilium libri X; De ueteribus & nouis metallis libri II; Bermannus, siue De re metallica Dialogus*. Basel: H. Froben & N. Episcopius. (English translation: *De natura fossilium* (*Textbook of mineralogy*) translated from the first Latin edn of 1546 by M. C. Bandy and J. A. Bandy for the Mineralogical Society of America. New York: Geological Society of America, 1955.)

Agricola, Giorgius. 1556. *De re metallica libri XII*. Basel: H. Froben & N. Episcopius. (English translation: Giorgius Agricola, *De re metallica*, translated from the first Latin edn of 1556 by H. C. Hoover and L. H. Hoover. New York: Dover, 1950; reprint of the translation published by the *Mining Magazine*, London, in 1912.)

Ahrweiler, H. 1973. "Les inscriptions historiques de Byzance," in *Akten des VI Internationalen Kongresses für griechische und lateinische Epigraphik: München 1972*. Munich: Beck.

Aki, K. 1966. "Generation and propagation of G waves from the Niigata earthquake of June 16, 1964: Part 2. Estimation of earthquake moment, released energy and stress drop from the G wave spectra," *Bulletin of Earthquake Research Institute, University of Tokyo*, **44**: 73–88.

Albarello, D. and M. Mucciarelli. 2002. "Seismic hazard estimates using ill-defined macroseismic data at site," *Pure and Applied Geophysics*, **159**: 1289–1304.

Albini, P. 1993. "Investigation of seventeenth and eighteenth centuries earthquakes in the documents of governors and representatives of the Republic of Venice," in *Materials of the CEC project. Review of Historical Seismicity in Europe*, vol. 1. Milan: CNR, 55–74.

Albini, P., A. Moroni and A. Bellani. 1991. "The 1846 Orciano (Pisa) earthquake in published sources and government survey documents," *Tectonophysics*, **193**: 117–130.

Albini, P., N. Ambraseys and J. Vogt. 1995. "Some aspects of historical research on 'over borders' earthquakes in eighteenth century Europe", in *Earthquakes in the*

Past: Multidisciplinary Approaches, eds. E. Boschi, R. Funiciello, E. Guidoboni, and A. Rovelli, special issue of *Annali di Geofisica*, **38**: 541–550.

Albini, P., V. García Acosta, R. M. W. Musson and M. Stucchi, eds. 2004. *Investigating the Records of Past Earthquakes*, 21st Course of the International School of Geophysics, Erice Sicily, 2002, special issue of *Annals of Geophysics*, **47**: 335–911.

Alexandre, P. 1989. "Le prétendu séisme de Tongres vers 600: une invention hagiographiques," *Ciel et Terre*, **105**: 11–12.

Alexandre, P. 1990. *Les séismes en Europe occidentale de 394 à 1259. Nouveau catalogue critique*. Brussels: Observatoire royal de Belgique.

Alexandre, P. and J. Vogt. 1994. "La crise séismique de 1755–1762 en Europe du Nord-Ouest. Les secousses des 26 et 27.12.1755: recensement des matériaux," in *Materials of the CEC Project: Review of Historical Seismicity in Europe*, eds. P. Albini and A. Moroni. Milan: CNR, 37–75.

Alishan, L. 1881. *Sissouan ou l'Arméno-Cilicie: Description géographique et historique*. Venice: S. Lazare.

Alishan, L. 1899. *Sissouan ou l'Arméno-Cilicie: Description géographique et historique avec carte et illustrations, traduit du texte arménien*. Venice: S. Lazare.

Allen, T. W. 1890. *Notes on Greek Manuscript in Italian Library*. London: D. Nutt. (Repr., with additions, from the *Classical Review*, 1889–1890.)

Almagià, R. 1914. "Intorno ai primi saggi di carte sismiche," *Rivista Geografica Italiana*, **21**: 463–470.

Ambraseys, N. N. 1962. "Data for the investigation of seismic sea waves in the Eastern Mediterranean," *Bulletin of the Seismological Society of America*, **52**: 895–913.

Ambraseys, N. N. 1971. "A note on an early earthquake in Macedonia," in *Proceedings of the 3rd European Symposium on Earthquake Engineering*, Sofia, Bulgaria, September 14–17, 1970. Sofia: Bulgarian Academy of Sciences Press, 73–78.

Ambraseys, N. N. 1988. "Engineering seismology," *Earthquake Engineering and Structural Dynamics*, **17**: 1–105.

Ambraseys, N. N. 2005. "Historical earthquakes in Jerusalem: a methodological discussion," *Journal of Seismology*, **9**: 329–340.

Ambraseys, N. N. and C. F. Finkel. 1991. "Long-term seismicity of Istanbul and of the Marmara Sea region," *Terra Nova*, **3**: 527–539.

Ambraseys, N. N. and C. P. Melville 1982. *A History of Persian Earthquakes*. Cambridge: Cambridge University Press.

Ambraseys, N. N. and C. P. Melville. 1995. "Historical evidence of faulting in Eastern Anatolia and Northern Syria," *Annali di Geofisica*, **38**: 337–343.

Ambraseys, N. N. and J. Vogt. 1988. "Material for the investigation of the seismicity of the region of Algiers," *European Earthquake Engineering*, **3**: 16–29.

Ambraseys, N. N., C. P. Melville, and R. D. Adams. 1994. *The Seismicity of Egypt, Arabia and the Red Sea: A Historical Review*. Cambridge: Cambridge University Press.

Ambraseys, N. N., J. A. Jackson and C. P. Melville. 2002. "Historical seismicity and tectonics: the case of the Eastern Mediterranean and the Midde East," in *International Handbook of Earthquake and Engineering Seismology*, eds. W. H. K. Lee,

H. Kanamori, P. C. Jennings, and C. Kisslinger. Amsterdam: Academic Press, 747–763.

Amontons, G. 1703. "Que les nouvelles expériences que nous avons du poids et du ressort de l'air, nous font connaître qu'un degré de chaleur médiocre, peut réduire l'air dans un état assez violent pour causer seul de très grands tremblements et bouleversements sur le globe terrestre," *Mémoires de l'Académie des Sciences*, 101–108 (1720, 2nd edn). Available at http://ads.ccsd.cnrs.fr/docs/00/10/47/42/PDF/p101_108_vol3483m.pdf

Anonymous. 1749. *A true and particular relation of the dreadful earthquake, which happen'd at Lima, the capital of Peru, and the neighboring port of Callao, on the 28th of October, 1746*. Philadelphia, Penn.: B. Franklin and D. Hall.

Aquila, B. 1784. *Dissertazione critica-filosofica su le riflessioni pubblicate in rapporto alla cagione fisica dei tremuoti delle Calabrie nell'anno 1783*.

Arslan, W. 1939. *L'architettura romanica veronese*. Verona: La Tipografica veronese.

Atkinson, G. M. and E. Sonley. 2000. "Empirical relationships between modified Mercalli Intensity and response spectra," *Bulletin of the Seismological Society of America*, **90**: 537–544.

Atwater, B. F., M.-R. Satoko, S. Kenji, *et al.* 2005. *The Orphan Tsunami of 1700: Japanese Clues to a Parent Earthquake in North America*. Reston, Va.: US Geological Survey; Seattle, Wash.: University of Washington Press.

Autino, P. 1987. "I terremoti nella Grecia classica," *Memorie dell'Istituto Lombardo – Accademia di Scienze e Lettere, Classe di Lettere, Scienze morali e storiche*, **38**: 355–446.

Babayan, T. H. 2006. *Atlas of Strong Earthquakes of the Republic of Armenia, Artsakh and Adjacent Territories from Ancient Times through 2003*. Yerevan: National Academy of Sciences of the Republic of Armenia, Institute of the Geophysics and Engineering Seismology.

Bacchielli, L. 1995. "A Cyrenaica earthquake *post* 364 A.D.: written sources and archaeological evidences," in *Earthquakes in the Past: Multidisciplinary Approaches*, eds. E. Boschi, R. Funiciello, E. Guidoboni and A. Rovelli, special issue of *Annali di Geofisica*, **38**: 977–982.

Bächtold, H. U. 1993. "Das Erdbeben von Ferrara 1570. Fundgrube Simmlersche Sammlung," in *Zentralbibliothek Zürich: Alte und neue Schätze*, eds. A. Cattani, M. Kotrba and A. Rutz. Zurich: Verlag NZZ, 78–81, 202–204.

Bakun, W. H. and C. M. Wentworth. 1997. "Estimating earthquake location and magnitude from seismic intensity data," *Bulletin of the Seismological Society of America*, **87**: 1502–1521.

Bakun, W. H., A. C. Johnston and M. G. Hopper. 2003. "Estimating locations and magnitudes of earthquakes in eastern North America from modified Mercalli intensities" *Bulletin of the Seismological Society of America*, **93**: 190–202.

Baldini, U. 1981. "Cirillo, Nicola," in *Dizionario Biografico degli Italiani*, vol. 25. Rome: Istituto della Enciclopedia Italiana, 801–805.

Baratta, M. 1896. "Ricerche storiche sugli apparecchi sismici," *Annali dell'Ufficio Centrale Meteorologico e Geodinamico Italiano*, ser. II, **17**: 1–37.

Baratta, M. 1901. *I terremoti d'Italia. Saggio di storia geografia e bibliografia sismica italiana*. Torino: Fratelli Bocca. (Repr. Sala Bolognese: Forni, 1979.)

Baratta, M. 1903. *Leonardo da Vinci e i problemi della Terra*. Turin: Fratelli Bocca.

Baratta, M. 1910. *La catastrofe sismica calabro-messinese 28 Dicembre 1908. Relazione alla Società Geografica Italiana*, 2 vols. Rome: Società Geografica Italiana. (Repr. Sala Bolognese: Forni, 1985.)

Barbano, M. S., D. Bellettati and D. Slejko. 1994. "Sources for the study of the Eastern Alps earthquakes at the turn of the 17th century," in *Materials of the CEC Project: Review of Historical Seismicity in Europe*, vol. 2, eds. P. Albini P. and A. Moroni. Milan: CNR, 115–132.

Barbisan, U. and F. Laner. 1983. *Terremoto ed architettura: Il trattato di Eusebio Sguario e la sismologia nel settecento*. Venice: Cluva università.

Barbisan, U. and F. Laner. 1995. "Wooden floors: part of historical antiseismic building systems," in *Earthquakes in the Past: Multidisciplinary Approaches*, eds. E. Boschi, R. Funiciello, E. Guidoboni and A. Rovelli, special issue of *Annali di Geofisica*, **38**: 775–784.

Basili, R., G. Valensise, P. Vannoli *et al.* 2008. "The Database of Individual Seismogenic Sources (DISS), version 3: summarizing 20 years of research on Italy's earthquake geology," *Tectonophysics*, **452**, doi:10.1016/j.tecto.2007.04.014.

Bassani, C. 1895–1902. "Prime ricerche sulla provenienza del terremoto di Firenze nella sera 18 maggio 1895," *Bollettino Mensuale dell'Osservatorio Centrale del Real Collegio Carlo Alberto in Moncalieri*, ser. II, **15**: 4–7, 25–32, 41–52, 63–70; **16**: 12–22, 35–52, 72–88, 108–124; **17**: 1–7, 17–24, 33–42, 49–60, 65–69, 81–88, 97–102; **18**: 1–9, 17–24, 33–37, 49–59, 65–75; **19**: 13–19, 32–35; **20**: 14–20, 29–35, 45–50, 61–65; **21**: 1–7, 22–28, 40–45, 55–63; **22**: 24–28.

Båth, M. 1979. *Introduction to Seismology*. Basel: Birkhäuser.

Batlló, J. 2002. "Sismologia colonial: la introducció de la sismologia instrumental a les illes Filipines 1865–1901," in *Actes de la VI Trobada d'Història de la Ciència i de la Tècnica*, eds. J. Batlló, P. Bernat and R. Puig. Barcelona: Institut d'Estudis Catalans, 215–224.

Batlló, J. 2003. "Los sismógrafos del observatorio de Cartuja," in *Historia del Observatorio de Cartuja 1902–2002: Nuevas investigaciones* (CD-ROM). Gránada: Instituto Andaluz de Geofísica, Universidad de Granada.

Beccaria, G. 1753. *Dell'elettricismo artificiale e naturale libri due*. Turin: Filippo Antonio Campana.

Beccaria, G. 1776. *A treatise upon artificial electricity, in which are given solutions of a number of interesting electric phoenomena, hitherto unexplained. Translated from the original Italian*. London: J. Nourse.

Becchetti, P. and G. Ferrari. 2004. "Fotografia e osservazione scientifica: Robert Mallet e il reportage fotografico nelle aree del terremoto del 16 dicembre 1857," in *Viaggio nelle aree del terremoto del 16 dicembre 1857. L'opera di Robert Mallet nel contesto scientifico e ambientale attuale del Vallo di Diano e della Val d'Agri*, ed. G. Ferrari. Bologna: SGA, 63–92.

Ben-Menahem, A. 1979. "Earthquake catalogue for the Middle East (92 BC – 1980 AD)," *Bollettino di Geofisica Teorica ed Applicata*, **21** (84): 245–310.

Benouar, D. 1994. "Materials for the investigation of the seismicity of Algeria and adjacent regions during the twentieth century," *Annali di Geofisica*, **37**: 459–860.

Benouar, D. 2004. "Materials for the investigation of historical seismicity in Algeria from the records of past earthquakes," in *Investigating the Records of Past Earthquakes*, eds. P. Albini, V. G. Acosta, R. M. W. Musson and M. Stucchi, special issue of *Annals of Geophysics*, **47**: 555–560.

Bentor, Y. K. 1989. "Geological events in the Bible," *Terra Nova*, **1**: 326–338.

Beroaldo, Filippo. 1505. *Opusculum de terraemotu et pestilentia, cum annotamentis Galeni.* Bologna: Benedetto Faelli.

Bertelli, T. 1887. "Di alcune teorie e ricerche elettro-sismiche antiche e moderne," *Bullettino di bibliografia e storia delle scienze matematiche e fisiche*, **20**: 481–542.

Bertholon, P. 1787. *De l'éléctricité des météores*, 2 vols. Lyon: Bernuset.

Bertrand E. 1757. *Mémoires historiques et physiques sur les tremblements de terre*. The Hague: Pierre Gosse. Available at http://gallica2.bnf.fr/ark:/12148/bpt6k109068r; http://fr.wikisource.org/wiki/Mémoires_historiques_et_physiques_sur_les_tremblemens_de_terre

Beschaouch, A. 1975. "A propos de recentes découvertes épigraphiques dans le pays de Carthage," *Comptes Rendus de l'Académie des Inscriptions et Belles-Lettres*, **1975**: 101–111.

Biasi, G. P., R. J. Weldon II, T. E. Fumal and G. G. Seitz. 2002. "Paleoseismic event dating and the conditional probability of large earthquakes on the southern San Andreas fault, California," *Bulletin of the Seismological Society of America*, **92**: 2761–2781.

Bidez, J. 1930. *La vie de l'empereur Julien*. Paris: Les Belles Lettres. (Repr. Paris: Les Belles Lettres, 1965).

Bilfinger, G. 1892. *Die mittelalterlichen Horen und die modernen Stunden: Ein Beitrag zur Kulturgeschichte*. W. Kohlhammer. (Repr. Wiesbaden: M. Sändig, 1969.)

Bilham, R. 1994. "The 1737 Calcutta earthquake and cyclone evaluated," *Bulletin of the Seismological Society of America*, **84**: 1650–1657.

Bina, A. 1751a. *Electricorum effectuum explicatio, quam ex principis newtonianis deduxit, novisque experimentis ornavit*. Padua: Giovanni Battista Conzatti.

Bina, A. 1751b. *Ragionamento sopra la cagione de' terremoti ed in particolare di quello della Terra di Gualdo di Nocera nell'Umbria seguito l'A. 1751*. Perugia: Costantini & Maurizi. (2nd edn Carpi: Francesco Torri, 1756.)

Bischoff, B. and W. Koehler. 1939. "Eine illustrierte Ausgabe der spätantiken Ravennater Annalen," in *Medieval Studies in Memory of A. Kingsley Porter*, ed. W. R. W. Koehler. Cambridge, Mass.: Harvard University Press, vol. 1, 125–138.

Blake, A. 1941. "On the estimation of focal depth from macroseismic data," *Bulletin of the Seismological Society of America*, **31**: 225–231.

Blanchard-Lemée, M. 1984. "Cuicul, le 21 juillet 365: critiques archéologique et historique de l'idée de séisme," in *Tremblements de terre histoire et archéologie*,

IVèmes Rencontres internationales d'archéologie et d'histoire d'Antibes, eds. B. Helly and A. Pollino. Valbonne: APDCA, 207–219.

Blumetti, A. M., A. M. Michetti and L. Serva. 1988. "The ground effects of the Fucino earthquake of Jan. 13th, 1915: an attempt for the understanding of recent geological evolution of some tectonic structures," in *Workshop on Historical Seismicity of Central-Eastern Mediterranean Region. Proceedings, Rome, ENEA CRE Casaccia*, eds. C. Margottini and L. Serva. Rome: ENEA-IAEA, 297–320.

Boccaletti M., G. Corti, P. Gasperini, *et al.* 2001. "Seismic zonation and active tectonics of the urban area of Florence," *Pure and Applied Geophysics*, **158**: 2313–2332.

Bonito, M. 1691. *Terra tremante, o vero continuatione de' terremoti dalla Creatione del Mondo sino al tempo presente*. Naples: Parrino. (Repr. Sala Bolognese: Forni, 1980.)

Boschi, E. and E. Guidoboni. 2003. *I terremoti a Bologna e nel suo territorio dal XII al XX secolo*. Rome, Bologna: INGV, Editrice Compositori, SGA.

Boschi, E., G. Ferrari, P. Gasperini, *et al.* 1995. *Catalogo dei forti terremoti in Italia dal 461 a.C. al 1980* (with database on CD-ROM). Bologna: ING-SGA.

Boschi, E., E. Guidoboni, G. Ferrari, P. Gasperini and G. Valensise. 1997. *Catalogo dei forti terremoti in Italia dal 461 a.C. al 1990*, vol. 2 (with database on CD-ROM). Bologna: ING-SGA.

Boschi, E., E. Guidoboni, G. Ferrari, *et al.*, eds. 2000. *Catalogue of Strong Italian Earthquakes from 461 BC to 1997*, Introductory texts and CD-ROM, Version 3 of the *Catalogo dei forti terremoti in Italia*, special issue of *Annali di Geofisica*, **43**: 609–868.

Buffon, G.-L. L., comte de 1749. *Histoire naturelle générale et particulière*, vol. 1, *Histoire et théorie de la Terre*, Paris: Imprimerie royale.

Buoni, Iacomo Antonio. 1571. *Del terremoto dialogo di Iacomo Antonio Buoni medico Ferrarese distinto in quattro giornate*. Modena: Paolo Gadaldini & fratelli.

Buonocore, M. 1992. "Una nuova testimonianza del rector provinciae Autonius Iustinianus e il macellum di Saepinum," *Athenaeum*, **80**: 484–486.

Burkert, W. 1985. *Greek Religion: Archaic and Classical*, transl. J. Raffan. Oxford: Blackwell. (Originally published in German as *Griechische Religion der archaischen und klassischen Epoche*. Stuttgart: Kohlhammer, 1977.)

Burkhardt, F. and S. Smith, eds. 1985–. *The Correspondence of Charles Darwin*. Cambridge: Cambridge University Press. Available at www.lib.cam.ac.uk/ Departments/Darwin/

Burkhardt, F. and S. Smith, eds. 1994. *A Calendar of the Correspondence of Charles Darwin, 1821–1882, with supplement*, 2nd edn. Cambridge: Cambridge University Press. (1st edn New York: Garland, 1985.)

Burnand, Y. 1984. "Terrae Motvs. La documentation épigraphique sur les tremblements de terre dans l'Occident romain," in *Tremblements de terre histoire et archéologie, IVèmes Rencontres internationales d'archéologie et d'histoire d'Antibes*, eds. B. Helly and A. Pollino. Valbonne: APDCA, 173–182.

Burnet, Th. 1681–89. *Telluris theoria sacra*. London: G. Kettilby.

Busi, G. 1994. "Earthquakes in the Hebrew Bible," in *Catalogue of Ancient Earthquakes in the Mediterranean Area up to the Tenth Century*, eds. E. Guidoboni, A. Comastri and G. Traina. Bologna: ING–SGA, 90–94.

Busi, G. 1995. "The seismic history of Italy in the Hebrew sources," in *Earthquakes in the Past: Multidisciplinary Approaches*, eds. E. Boschi, R. Funiciello, E. Guidoboni and A. Rovelli, special issue of *Annali di Geofisica*, 38: 473–478.

Butzer, K. W. 1982. *Archaeology as Human Ecology*. Cambridge: Cambridge University Press.

Cagiano de Azevedo, M. 1974. "Aspetti urbanistici delle città altomedievali," in *Topografia urbana e vita cittadina nell'alto Medioevo in Occidente*, Settimane di Studio del Centro Italiano di Studi sull'Alto Medioevo, Spoleto: Centro italiano di studi sull'alto Medioevo, vol. 2, 641–677.

Cairoli Giuliani, F., M. L. Conforto, S. D'Agostino, *et al.* 2007. "The conception of building in ancient architecture and the 'rules of art'," in *Structural Studies, Repairs and Maintenance of Heritage Architecture X*, ed. C. A. Trebbia. Southampton and Boston, Mass.: WIT Press, 599–610.

Camassi, R. and M. Stucchi, eds. 1997. *NT4.1 un catalogo parametrico di terremoti di area italiana al di sopra della soglia del danno. A parametric catalogue of damaging earthquakes in the Italian area*, con la collaborazione di D. Molin, G. Monachesi, A. Rebez and A. Zerga. Milan: GNDT. Available at http://emidius.mi.ingv.it/NT/

Camodeca, G. 1971. "Fabius Maximus e la creazione della provincia del Samnium," *Atti dell'Accademia di scienze morali e politiche della Società nazionale di scienze, lettere ed arti in Napoli*, 82: 249–264.

Capelle, W. 1908. "Erdbeben im Altertum," *Neue Jahrbücher für das klassische Altertum, Geschichte und deutsche Literatur und für Pädagogik*, 21: 603–633.

Capelle, W. 1924. s.v. "Erdbebenforschung," in *Paulys Realencyclopädie der classischen Altertumswissenschaft*, Suppl. IV: 344–374. Stuttgart: J. B. Metzler.

Caputo, M. and G. Faita. 1984. "Primo catalogo dei maremoti delle coste italiane," *Accademia Nazionale dei Lincei, Classe Scienze fisiche, matematiche e naturali*, ser. VIII, 17(1): 213–356.

Cardona, O. D. 2005. "The seismic proof efficiency of traditional building construction," in *Ancient Buildings and Earthquakes: The Local Seismic Culture Approach – Principles, Methods, Potentialities*, eds. F. Ferrigni, B. Helly, L. Mendes Victor, P. Pierotti, A. Rideaud, P. Teves Costa *et al.* Centro Culturale Europeo per i Beni Culturali, Ravello, Council of Europe, Strasbourg. Bari: Edipuglia, 103–109.

Cartledge, P. A. 1976. "Seismicity and Spartan society," *Liverpool Classical Monthly*, 1: 25–28.

Carvalhao Buescu, H. and G. Cordeiro. 2005. *O grande terramoto de Lisboa: ficar diferente*. Lisbon: Gradiva.

Cavallo, T. 1777. *A Complete Treatise on Electricity in Theory and Practice, with Original Experiments*. London: E. & C. Dilly.

Cecchi, F. 1876. "Sismografo elettrico a carte affumicate scorrevoli," *Atti della Accademia Pontificia de' Nuovi Lincei*, 29: 421–428.

Cergol, M. and D. Slejko. 1991. "I terremoti del 1511 e del 1690 nelle Alpi Orientali," in *Atti del Convegno, Pisa, 25–27 giugno 1990, Macrosismica*, 2, eds. P. Albini and M. S. Barbano. Milan: GNDT, 69–91.

Chatelain, L. 1909. "Théories d'auteurs anciens sur les tremblements de terre," *Mélanges d'archéologie et d'histoire de l'École française de Rome*, **29**: 87–101.

Chesneau, M. 1892. "Note sur les tremblements de terre en Algérie," *Annales des Mines*, ser. IX, **1**: 5–46.

Ciccarello, S. 1996. "A short note on some Arabic inscriptions recording seismic effects in the Mediterranean area 472 H./1079 A.D. – 703 H./1303–1304 A.D.," *Annali di Geofisica*, **39**: 487–491.

Ciclopedia ovvero Dizionario universale delle arti e delle scienze . . . Tradotto dall'inglese, e di molti articoli accresciuto da Giuseppe Maria Secondo, 9 vols. Naples: Giuseppe de Bonis, 1747–1754.

Cignitti, B. and L. Caronti. 1956. *L'abbazia nullius Sublacense: le origini, la commenda*. Rome: Lozzi.

Cirillo, N. 1733–34. "Historia terraemotus Apuliam & totum fere Neapolitanum regnum, anno 1731, vexantis," *Philosophical Transactions*, **38**: 79–84.

Cirillo, N. 1747. "The history of an earthquake, which shook Apulia, and almost the whole Kingdom of Naples, in 1731," in *The Philosophical Transactions (from the year 1732, to the year 1744) abridged, and disposed under general heads*. London: W. Innys, vol. 8, pt II, 682–684.

Ciuccarelli, C. 2003. "Il santuario della Madonna di S. Luca: restauri sismici dimenticati," in E. Boschi and E. Guidoboni, *I terremoti a Bologna e nel suo territorio dal XII al XX secolo*. Rome, Bologna: INGV-SGA, 391–400.

Cole, S. C., B. F. Atwater, P. T. McCutcheon, J. K. Stein and E. Hemphill-Haley. 1996. "Earthquake-induced burial of archaeological sites along the southern Washington coast," *Geoarchaeology*, **11**: 165–177.

Comastri, A. 1986. "Un terremoto in cerca di spiegazione: la teoria elettricista di Giuseppe Vannucci," in *Il terremoto di Rimini e della costa romagnola, 25 dicembre 1786: analisi e interpretazione*, eds. E. Guidoboni and G. Ferrari. Bologna: SGA, 157–165.

Comastri, A. in preparation. "A hitherto unknown letter by Charles Darwin to the French seismologist Alexis Perrey."

Comninakis, P. E. and B. C. Papazachos. 1982. *A Catalogue of Historical Earthquakes in Greece and Surrounding Area: 479 BC – 1900 AD*. Thessaloniki: University of Thessaloniki Geophysical Laboratory.

Comparetti, D. 1914. "Iscrizione cristiana di Cirene," *Annuario della Scuola archeologica Italiana di Atene e delle Missioni in Oriente*, **1**: 161–167.

Conant, K. J. 1939. "The first dome of the Hagia Sophia and its rebuilding," *American Journal of Archeology*, **43**: 589–591.

Constantinides, C. and R. Browning. 1993. *Dated Greek Manuscripts from Cyprus to the Year 1570*. Washington, D.C.: Dumbarton Oaks Research Library and Collection; Nicosia: Cyprus Research Centre.

Copia de una lettra che contiene li spaventosi & horrendi Terremoti venuti ne la Isola di Sicilia & la gran rovina & strage che han fatto in quella con danno grandissimo de paesani. Questo terremoto fu adì XVII Decembrio MDXLII. Venice, 1542.

CRAAG (Centre de Recherche en Astronomie Astrophysique et Géophysique). 1994. *Les séismes de l'Algérie de 1365.* Algiers: CRAAG.

Crespellani, T., G. Vannucchi and X. Zeng. 1992. "Seismic hazard analysis in the Florence area," *European Earthquake Engineering*, 3: 33–42.

Croke, B. 1981. "Two early Byzantine earthquakes and their liturgical commemoration," *Byzantion*, 51: 122–147.

Croke, B. 2001. *Count Marcellinus and his Chronicle.* Oxford: Oxford University Press.

Cuneo, P. 1968. "Introduzione all'architettura medievale armena," in *Architettura medievale armena*, Roma, Palazzo Venezia, June 10–30, 1968. Rome: De Luca, 47–54.

Cuneo, P. 1988. *Architettura armena dal quarto al diciannovesimo secolo*, 2 vols. Rome: De Luca.

D'Addezio, G., F. R. Cinti and D. Pantosti. 1995. "A large unknown historical earthquake in the Abruzzi region Central Italy: combination of geological and historical data," *Annali di Geofisica*, 38: 491–501.

D'Agostino, S. 2002. "Historical buildings as an archive of the material history of construction," in *Towards a History of Construction: Dedicated to Edoardo Benvenuto*, eds. A. Becchi, M. Corradi, F. Foce and O. Pedemonte. Berlin: Birkhäuser, 369–376.

D'Agostino, S. 2003. "The ancient approach to construction and the modern project," in *Proceedings of the 1st International Congress on Construction History*, ed. S. Huerta. Madrid: Instituto Juan De Herrera, vol. 1, 669–676.

D'Agostino, S. 2005. "Il contributo dell'ingegneria per i beni culturali alla conservazione del patrimonio costruito," in *Teoria e pratica del costruire, Saperi, strumenti, modelli: Esperienze didattiche e di ricerca a confronto*, ed. G. Mochi. Ravenna: Moderna, vol. 3, 1073–1077.

D'Agostino, S. and M. Bellomo. 2007. "Seismic risk and conservation of architectural heritage in the Mediterranean basin," in *Structural Studies, Repairs and Maintenance of Heritage Architecture X*, eds. C. A. Trebbia. Southampton and Boston, Mass.: WIT Press, 611–620.

Dagron, G. 1981. "Quand la terre tremble . . . ," *Travaux et Mémoires*, 8: 87–103.

Daltri, A. and P. Albini. 1991. *L'Archivio di Stato di Venezia: studio sulla potenzialità della documentazione ivi depositata ai fini delle ricerche sui terremoti europei di interesse per il Progetto CEE "RHISE"*, unpublished progress report, Istituto di Ricerca sul Rischio Sismico, Milan: CNR.

Darrouzès, J. 1950. "Les manuscrits originaires de Chypre à la Bibliothèque Nationale de Paris," *Revue des Etudes Byzantines*, 8: 162–197.

Darrouzès, J. 1951. "Un obituaire chypriote: le Parisinus Graecus 1588," *Kupriakai spoudai [Cypriot Studies]*, 11: 25–62. (Repr. in *Littérature et histoire des textes byzantins*. London: Variorum Reprints, 1972.)

Darrouzès, J. 1953. "Notes pour servir à l'histoire de Chypre," *Kupriakai spoudai [Cypriot Studies]*, **17**: 83–102. (Repr. in *Littérature et histoire des textes byzantins*. London: Variorum Reprints, 1972.)

Darrouzès, J. 1956. "Notes pour servir à l'histoire de Chypre (deuxième article)," *Kupriakai spoudai [Cypriot Studies]*, **20**: 33–63. (Repr. in *Littérature et histoire des textes byzantins*. London: Variorum Reprints, 1972.)

Davison, Ch. 1927. *The Founders of Seismology*. Cambridge: Cambridge University Press. (Repr. New York: Arno Press, 1978.)

Dean, D. R. 1989. "Benjamin Franklin and earthquakes," *Annals of Science*, **46**: 481–495.

De Bernardi Ferrero, D. 1993. "Hierapolis," in *Arslantepe, Hierapolis, Iasos, Kyme. Scavi archeologici italiani in Turchia*. Venice: Marsilio, 104–187.

De Felice, E. 1994. *Forma Italiae 36: Larinum*. Florence: Olschki.

De la Torre, F. R. 1995. "Spanish sources concerning the 1693 earthquake in Sicily," in *Earthquakes in the Past: Multidisciplinary Approaches*, eds. E. Boschi, R. Funiciello, E. Guidoboni and A. Rovelli, special issue of *Annali di Geofisica*, **38**: 523–539.

Delehaye, H. 1907. "Saints de Chypre," *Analecta Bollandiana*, **26**: 161–301.

Del Moro, L. 1895. "Relazione," in *Deputazione secolare di Santa Maria del Fiore, Relazione sui danni arrecati ai monumenti insigni dal terremoto del 18 maggio 1895*. Firenze: G. Carnesecchi e Figli, 5–12.

De Noronha Wagner, M. 1993. "As frontes para o estudo da sismicidade histórica do Algarve na 1ª metade do século XVIII: a investigação no Arquivo Nacional da Torre do Tombo, Lisboa," in *Review of Historical Seismicity in Europe*, ed. M. Stucchi. Milan: CNR, 207–224.

De Poardi, G. V. 1627. *Nuova relatione del grande e spaventoso terremoto successo nel Regno di Napoli, nella Provincia di Puglia, in Venerdì li 30 luglio 1627*. Rome.

de Rossi, M. S. 1874. "La antica basilica di S. Petronilla presso Roma testè discoperta crollata per terremoto," *Bullettino del Vulcanismo Italiano*, **1**: 62–65.

Descartes, R. 1644. *Principia philosophia*, in *Oeuvres de Descartes*, eds. Ch. Adam and P. Tannery, vol. 8. Paris: Léopold Cerf, 1905. (English translation: René Descartes, *A Discourse on Method: Meditations on the First Philosophy and Principles of Philosophy*. London: Everyman, 1994.)

Dever, W. G. 1992. "A case-study in biblical archaeology: the earthquake of *ca*. 760 BCE," *Eretz Israel*, **23**: 27*–35*.

Di Carlo, E. 1955–56. "Precisazioni sul soggiorno di Goethe a Messina," *Archivio storico messinese*, ser. VII, **3**: 9–23.

Di Somma, A. 1641. *Historico racconto de i terremoti di Calabria dall'anno 1638 fin'anno 41*. Naples: Camillo Cavallo.

DISS Working Group. 2005. *Database of Individual Seismogenic Sources*, v. 3.0. Available at http://www.ingv.it/banchedati/banche.html

DISS Working Group. 2007. *Database of Individual Seismogenic Sources (DISS)*, v. 3.0.4: *A Compilation of Potential Sources for Earthquakes Larger than M 5.5 in Italy and Surrounding Areas*, © INGV 2005, 2007. Available at http://www.ingv.it/DISS/.

Di Teodoro, F. and L. Barbi. 1983. "Leonardo da Vinci: 'Del riparo a' terremoti'," *Physis*, **25**: 5-39.

Di Vita, A. 1964. "Archaeological News: Tripolitania," *Libya Antiqua*, **1**: 133-142.

Di Vita, A. 1980. "Evidenza dei terremoti del 306-310 e del 365 d.C. in Tunisia," *Antiquités africaines*, **15**: 303-307.

Di Vita, A. 1982. "Evidenza dei terremoti del 306-310 e del 365 in monumenti e scavi di Tunisia, Sicilia, Roma e Cirenaica," *Africa*, **7-8**: 127-139.

Di Vita, A. 1990. "Sismi, urbanistica e cronologia assoluta. Terremoti e urbanistica nelle città di Tripolitania fra il I secolo a.C. ed il IV d.C.," in *L'Afrique dans l'Occident romain (Ier siècle av. J.-C. - IVe siècle ap. J.-C.)*. Rome: Ecole française de Rome.

Doglioni, F., A. Moretti and V. Petrini, eds. 1994. *Le chiese e il terremoto: Dalla vulnerabilità constatata nel terremoto del Friuli al miglioramento antisismico nel restauro, verso una politica di prevenzione.* Trieste: LINT.

Dohrn-van Rossum, G. 1996. *History of the Hour: Clocks and Modern Temporal Orders*, transl. Th. Dunlap. Chicago, Ill.: University of Chicago Press.

Dominey-Howes, D. 2004. "Re-analysis of the Late Bronze Age eruption and tsunami of Santorini, Greece, and the implications for the volcano-tsunami hazard," *Journal of Volcanology and Geothermal Research*, **130**: 107-132.

Downey, G. 1955. "Earthquakes at Constantinople and Vicinity, AD 342-1454," *Speculum*, **30**: 596-600.

Dubois de Montpéreux, F. 1839-43. *Voyage autour du Caucase, chez les Tcherkesses et les Abkhases, en Colchide, en Géorgie, en Arménie et en Crimée*, 6 vols. Paris: Gide.

Ducat, J. 1984. "Le tremblement de terre de 464 et l'histoire de Sparte," in *Tremblements de terre histoire et archéologie, IVèmes Rencontres internationales d'archéologie et d'histoire d'Antibes*, eds. B. Helly and A. Pollino. Valbonne: APDCA, 73-85.

Dufour, L. and H. Raymond. 1994. *1693 Val di Noto: La rinascita dopo il disastro.* Catania: Sanfilippo.

Dumézil, G. 1966. *Archaic Roman Religion.* Chicago, Ill.: University of Chicago Press.

Dunbar, P. K., P. A. Lockridge and L. S. Whiteside. 1992. *Catalog of Significant Earthquakes 2150 BC - 1991 AD including Quantitative Casualties and Damage.* NOAA, National Geophysical Data Center Report SE-49. Boulder, Colo.: US Department of Commerce.

Ebel, J. E. 1996. "The seventeenth century seismicity of Northeastern North America," *Seismological Research Letters*, **67**: 51-68.

Ebel, J. E. 2000. "A reanalysis of the 1727 earthquake at Newbury, Massachusetts," *Seismological Research Letters*, **71**: 364-374.

Ebel, J. E. 2006. "The Cape Ann, Massachusetts earthquake of 1755: a 250th anniversary perspective," *Seismological Research Letters*, **77**: 74-86.

Ebel, J. E. and D. J. Wald. 2003. "Bayesian estimations of peak ground acceleration and 5% damped spectral acceleration from modified Mercalli intensity data," *Earthquake Spectra*, **19**: 511-529.

Egidi, P. 1904. "Notizie storiche," in *I monasteri di Subiaco*, eds. P. Egidi, G. Giovannoni and F. Hermann. Rome: Ministero della pubblica istruzione, vol. 1, 45–247.

El-Sayed, A., F. Romanelli and G. Panza. 2000. "Recent seismicity and realistic waveforms modeling to reduce the ambiguities about the 1303 seismic activity in Egypt," *Tectonophysics*, **328**: 341–57.

Ergin, K., U. Güçlü, and Z. Uz. 1967. *Türkie ve civarinin deprem katalogu Milattan Sonra 11 yilindan 1964 sonuna kadar [A Catalogue of Earthquakes for Turkey and Surrounding Area 11 AD to 1964 AD]*. Istanbul: Istanbul Teknik Üniversitesi, Maden Fakültesi, Arz Fizigi Enstitüsü.

Evangelatou-Notara, F. 1982. *Notes in Greek Manuscripts as a Source for the Study of the Social and Economic life of Byzantium from the Ninth Century to the Year 1204*. Athens: Ethnikon kai Kapokistriakon Philosophike Skole (in Greek).

Evangelatou-Notara, F. 1993. *Earthquakes in Byzantium from the Thirteenth to the Fifteenth century: Historical Research*. Athens: Parousia (in Greek).

Ewing, J. A. 1880. "On a new seismograph for horizontal motion," *Transactions of the Seismological Society of Japan*, **2**: 45–49.

Ewing, J. A. 1881. "On a new seismograph," *Proceedings of the Royal Society of London*, **31**: 440–446.

Fabretti, A. and P. Vayra. 1891. *Il processo del diavolo ad Issime nella valle di Gressoney*. Turin: A. Fabretti.

Fatouros, G., T. Krischer and W. Portmann, eds. 2002. *Libanios. Kaiserreden. Eingeleitet, Übersetzt und Kommentiert*. Stuttgart: Hiersemann.

Faura, F. 1880. *Observaciones seismométricas de los terremotos del mes de Julio de 1880*. Manila: La Oceanía.

Favreau, R. 1989. "L'Épigraphie médiévale: naissance et développement d'une discipline," *Comptes Rendus de l'Académie des Inscriptions et Belles-Lettres*, **1989**: 328–363.

Feijóo y Montenegro, B. J. 1756. *Nuevo systhema sobre la causa physica de los terremotos explicado por los phenomenos electricos, y adaptado al que padeciò Espana en primero de noviembre del ano antecedente de 1755*. Puerto de Santa Maria: Casa Real de las Cadenas; Lisbon: Joseph da Costa Coimbra.

Felgentreu, F. 2004. "Zur Datierung der 18. Rede des Libanius," *Klio* **86**: 206–217.

Fernández, M. 1990. *Artificios del Barroco: México y Puebla en el siglo XVII*. Mexico City: Universidad Nacional Autónoma de México.

Ferrari, G. 1988. "Some aspects of the seismological interpretation of information on historical earthquakes," in *Workshop on Historical Seismicity of Central–Eastern Mediterranean Region*, eds. C. Margottini and L. Serva. Rome: ENEA-IAEA, 45–63.

Ferrari, G. 1991. "The 1887 Ligurian earthquake: a highly detailed study from contemporary scientific observations," in *Investigation of Historical Earthquakes in Europe*, eds. M. Stucchi, D. Postpischl and D. Slejko, special issue of *Tectonophysics*, **193**: 131–139.

Ferrari, G., ed. 1992. *Two hundred years of seismic instruments in Italy 1731–1940*. Bologna: ING-SGA.

Ferrari, G. 2002. "Census, filing and elaboration of scientific letters in the Earth sciences," *Nuncius*, **17**(1): 307–320.

Ferrari, G., ed. 2004. *Viaggio nelle aree del terremoto del 16 dicembre 1857: L'opera di Robert Mallet nel contesto scientifico e ambientale attuale del Vallo di Diano e della Val d'Agri*, 2 vols. Bologna: SGA.

Ferrari, G. and E. Guidoboni. 2000. "Seismic scenarios and assessment of intensity: some criteria for the use of the MCS scale," in *Catalogue of Strong Italian Earthquakes from 461 BC to 1997*, eds. E. Boschi, E. Guidoboni, G. Ferrari, D. Mariotti, G. Valensise and P. Gasperini, special issue of *Annali di Geofisica*, **43**: 707–720.

Ferrari, G. and C. Marmo. 1985. "Il 'quando' del terremoto," in *Terremoti e storia*, ed. E. Guidoboni, *Quaderni storici*, n.s., **60**: 691–715.

Ferrari, G., D. Albarello and G. Martinelli. 2000. "Tromometric measurements as a tool for crustal deformation interpretation," *Seismological Research Letters*, **71**: 562–569.

Ferrigni, F., B. Helly, A. Mauro, *et al.*, eds. 2005. *Ancient Buildings and Earthquakes: The Local Seismic Culture Approach – Principles, Methods, Potentialities*. Centro Culturale Europeo di Ravello, Council of Europe, Strasburg. Bari: Edipuglia.

Figliuolo, B. 1988–89. *Il terremoto del 1456*, 2 vols. Altavilla Silentina: Studi storici meridionali.

Filangieri, R. 1958. *I registri della Cancelleria angioina ricostruiti da Riccardo Filangieri con la collaborazione degli archivisti napoletani*, vol. 11, doc. no. 151. Naples: Presso l'Accademia.

Filippo da Secinara. 1652. *Trattato universale di tutti li terremoti occorsi, e noti nel mondo, con li casi infausti, ed infelici pressagiti da tali terremoti*. Aquila: Gregorio Gobbi.

Finkelstein, I. and N. A. Silberman. 2002. *The Bible Unearthed: Archeology's New Vision of Ancient Israel and the Origin of its Sacred Texts*. New York: Simon and Schuster.

Fischer, J., G. Grünthal and J. Schwarz. 2001. "Das Erdbeben vom 7. Februar 1839 in der Gegend von Unterriexingen, Thesis," *Wissenschaftliche Zeitschrift der Bauhaus-Universität Weimar*, 1/2: 8–30.

Fokaefs, A. and G. Papadopoulos. 2007. "Testing the new INQUA intensity scale in Greek earthquakes," *Quaternary International*, **173–174**: 15–22.

Follieri, E. 1977. "La minuscola libraria dei secoli IX e X," in *La Paléographie grecque et byzantine*. Paris: CNRS, 139–165. (Repr. in E. Follieri, *Byzantina et Italograeca: Studi di Filologia e di Paleografia*, ed. A. Acconcia Longo, L. Perria, and A. Luzzi. Rome: Edizioni di Storia e Letteratura, 1997, 205–225.)

Follieri, E. 1979. "Due codici greci già cassinesi oggi alla Biblioteca Vaticana: gli *Ottob. gr.* 250 e 251," in *Paleographica, diplomatica et archivistica: Studi in onore di Giulio Battelli*, ed. Scuola speciale per archivisti e bibliotecari dell'Università di Roma. Rome: Edizioni di Storia e Letteratura, vol. 1, 159–221. (Repr. in *Byzantina et Italograeca: Studi di Filologia e di Paleografia*, ed. A. Acconcia Longo, L. Perria and A. Luzzi. Rome: Edizioni di Storia e Letteratura, 1997, 271–336.)

Foufa, A. 2005. "Contribution for a catalogue of earthquake resistant traditional techniques in northern Africa: the case of the Casbah of Algiers, Algeria," *European Earthquake Engineering*, **2**: 23–39.

Fracassi, U. and G. Valensise. 2004. "The 'layered' seismicity of Irpinia: important but incomplete lessons learned from the 23 November 1980 earthquake," in *The Many Facets of Seismic Risk, Proceedings of the Workshop on Multidisciplinary Approach to Seismic Risk Problems*, eds. M. Pecce, G. Manfredi and A. Zollo. Naples: Centro Regionale di Competenza Analisi e Monitoraggio del Rischio Ambientale, 46–52.

Fracassi, U. and G. Valensise. 2007. "Unveiling the sources of the catastrophic 1456 multiple earthquake: hints to an unexplored tectonic mechanism in southern Italy," *Bulletin of the Seismological Society of America*, **97**: 725–748.

Franklin, B. 1751. *Experiments and Observations on Electricity Made at Philadelphia in America*. London: E. Cave. (Repr. *Benjamin Franklin's Experiments: A New Edition of Franklin's Experiments and Observations on Electricity*, edited, with a critical and historical introduction, by I. B. Cohen. Cambridge, Mass.: Harvard University Press, 1941.)

Freeman-Grenville, G. S. P. 1995. *The Islamic and Christian Calendars AD 622–222 AH 1–1650: A Complete Guide for Converting Christian and Islamic Dates and Dates of Festivals*. Reading, UK; Concord, Mass.: Garnet.

Friedrich, W. L., B. Kromer, M. Friedrich, *et al.* 2006. "Santorini eruption radiocarbon dated to 1627–1600 B.C.," *Science*, **312**: 548, doi: 10.1126/science.1125087.

Frizot, M. 1998. *A New History of Photography*. Koel: Koeneman. (1st edn 1994, in French.)

Fumal, T. E., R. J. Weldon II, G. P. Biasi, *et al.* 2002. "Evidence for large earthquakes on the San Andreas fault at the Wrightwood, California, paleoseismic site: AD 500 to present," *Bulletin of the Seismological Society of America*, **92**: 2726–2760.

Galadini, F. and P. Galli. 2004. "The 346 AD earthquake (Central–Southern Italy): an archaeoseismological approach," *Annals of Geophysics*, **47**: 885–905.

Galanopoulos, A. G. 1960. "Tsunamis observed on the coasts of Greece from antiquity to present time," *Annali di Geofisica*, **13**: 369–386.

Galetti, P. 1985. "Struttura materiale e funzioni negli insediamenti urbani e rurali della Pentapoli," in *Ricerche e studi sul "Breviarium Ecclesiae Ravennatis"* (Studi Storici, 148–149). Rome: Istituto storico italiano per il Medio Evo, 109–124.

Galli, P. and V. Bosi. 2004. "Catastrophic 1638 earthquakes in Calabria southern Italy): new insights from paleosismological investigation," *Journal of Geophysical Research*, **108**, B1, doi:10.1029/2001.

Galli, P. and F. Galadini. 2003. "Disruptive earthquakes revealed by faulted archaeological relics in Samnium Molise, southern Italy," *Geophysical Research Letters*, **30**, 1266, doi:10.1029/2002GL016456.

Gallo, A. 1784. *Lettere . . . dirizzate al sig. cavaliere N.N. delle Reali Accademie di Londra, Bordò, ed Upsal pelli terremoti del 1783. Con un giornale meteorologico de' medemi. Aggiuntavi anche la relazione di quei di Calabria con li paesi distrutti, ed il numero de' morti*. Messina: Giuseppe di Stefano.

Galvani, L. 1791. "De viribus electricitatis in motu musculari commentarius," *De Bononiensi Scientiarum et Artium Instituto atque Academia Commentarii*, 7: 363–431 (reprinted, with Italian translation, Bologna: Forni, 1998).

Ganas, A., G. Papadopoulos and S. B. Pavlides. 2001. "The 7th September 1999 Athens 5.9 M_s earthquake: remote sensing and digital elevation model inputs towards identifying the seismic fault," *International Journal of Remote Sensing*, 22: 191–196.

Gasperini, P. and G. Ferrari. 2000. "Deriving numerical estimates from descriptive information: the computation of earthquake parameters," in *Catalogue of Strong Italian Earthquakes from 461 BC to 1997*, eds. E. Boschi, E. Guidoboni, G. Ferrari, D. Mariotti, G. Valensise and P. Gasperini, special issue of *Annali di Geofisica*, 43: 729–746.

Gasperini, P. and G. Valensise. 2000. "From earthquake intensities to earthquake sources: extending the contribution of historical seismology to seismotectonic studies," in *Catalogue of Strong Italian Earthquakes from 461 BC to 1997*, eds. E. Boschi, E. Guidoboni, G. Ferrari, D. Mariotti, G. Valensise and P. Gasperini, special issue of *Annali di Geofisica*, 43: 765–785.

Gasperini, P., F. Bernardini, G. Valensise and E. Boschi. 1999. "Defining seismogenic sources from historical felt reports," *Bulletin of the Seismological Society of America*, 89: 94–110.

Gebhard, E. R. 1996. "Evidence for an earthquake in the theatre at Stobi, c. AD 300," in *Archaeoseismology*, ed. S. C. Stiros and R. E. Jones. Fitch Laboratory Occasional Paper 7. Athens: Institute of Geology and Mineral Exploration, The British School at Athens, 55–61.

Gerola, G. 1932. *Monumenti veneti nell'isola di Creta*, vol. 4. Venice: Istituto Italiano Arti Grafiche.

Giessberger, H. 1922. "Die Erbeben Bayerns," *Abhandlungen der der königlich-bayerischen Akademie der Wissenschaften*, Mathematisch-physikalischen Klasse, 29: 1–72.

Gizzi, F. T. and N. Masini. 2004. "Damage scenario of the earthquake on 23 July 1930 in Melfi: the contribution of technical documentation," *Annals of Geophysics*, 47: 1641–1663.

Goethe, J. W. 1816. *Italiänische Reise*. Munich. (English translation: *Italian Journey, 1786–1788*, transl. W. H. Auden and E. Mayer. London: Collins, 1962, reissued 2004.)

Goodchild, R. G. 1966–67. "A coin-hoard from Balagrae (El-Beida), and the earthquake of A. D. 365," *Libya Antiqua*, 3–4: 203–211.

Gouin, P. 1979. *Earthquake History of Ethiopia and the Horn of Africa*. Ottawa: International Development Research Centre.

Gouin, P. 2001. *Tremblements de terre "historiques" au Québec (de 1534 à mars 1925) identifiés et interprétés à partir de textes originaux contemporains*. Montreal: Guérin.

Gray, T. 1883. "On the Gray and Milne seismographic apparatus," *Quarterly Journal of the Geological Society of London*, 39: 218–223.

Grumel, V. 1958. *Traité d'études byzantines*, vol. 1, *La chronologie*. Paris: Presses Universitaires de France.

Grünthal, G., ed. 1993. *European Macroseismic Scale 1992 (updated MSK-scale)*, Cahiers du Centre Européen de Géodynamique et de Séismologie, 7. Luxembourg: Conseil de l'Europe.

Grünthal, G., ed. 1998. *European Macroseismic Scale 1998*, Cahiers du Centre Européen de Géodynamique et de Séismologie, 15. Luxembourg: Conseil de l'Europe.

Grünthal, G. 2004. "The history of historical earthquake research in Germany," in *Investigating the Records of Past Earthquakes*, eds. P. Albini, V. G. Acosta, R. M. W. Musson and M. Stucchi, special issue of *Annals of Geophysics*, **47**: 631–643.

Grünthal, G. and J. Fischer. 1998. "Die Rekonstruktion des 'Torgau'-Erdbebens vom 17. August 1553," *Brandenburgische geowissenschaftliche Beiträge*, **5**(2): 43–60.

Grünthal, G. and J. Fischer. 2001. "Eine Serie irrtümlicher Schadenbeben im Gebiet zwischen Nördlingen und Neuburg an der Donau vom 15. bis zum 18. Jahrhundert," *Mainzer naturwissenschaftliches Archiv*, **39**: 15–32.

Gruppo di lavoro CPTI. 2004. *Catalogo Parametrico dei Terremoti Italiani*, versione 2004 (CPTI04). Bologna: INGV. Available at http://emidius.mi.ingv.it/CPTI04/

Guéneau de Montbeillard, P. 1761. "Liste chronologique des éruptions de volcans, des tremblements de terre, . . . jusqu'en 1760," in *Collection académique composée des mémoires, actes ou Journaux des plus célèbres Académies et sociétés littéraires de l'Europe, partie étrangère*. Paris, vol. 6, 488–676.

Guerrieri, L. and E. Vittori, eds. 2007. *Environmental Seismic Intensity Scale ESI 2007*, Memorie Descrittive della Carta Geologica d'Italia 74. Rome: Servizio Geologico d'Italia – Dipartimento Difesa del Suolo, APAT. Available online at www.apat.gov.it/site/_Files/Inqua/ESI07_definition_intensity_degrees.pdf.

Guidoboni, E. 1985. "Immagini e interpretazioni di fenomeni naturali: il "terremoto" di Issime del 1600–1601," in *Terremoti e storia*, ed. E. Guidoboni, *Quaderni Storici*, **60**: 811–838.

Guidoboni, E. 1989. "Catalogo delle epigrafi latine riguardanti terremoti," in *I terremoti prima del Mille in Italia e nell'area mediterranea: Storia archeologia sismologia*, ed. E. Guidoboni. Bologna: ING-SGA, 135–168.

Guidoboni, E. 2000. "Method of investigation, typology and taxonomy of the basic data: navigating between seismic effects and historical contexts," in *Catalogue of Strong Italian Earthquakes from 461 BC to 1997*, eds. E. Boschi, E. Guidoboni, G. Ferrari, D. Mariotti, G. Valensise and P. Gasperini, special issue of *Annali di Geofisica*, **43**: 621–666.

Guidoboni, E. 2002. "L'archeologia sismica," in *Il mondo dell'archeologia*. Rome: Istituto della Enciclopedia Italiana, vol. 1, 220–222.

Guidoboni, E. and A. Comastri. 1997. "The large earthquake of 8 August 1303 in Crete: seismic scenario and tsunami in the Mediterranean area," **1**: 55–72.

Guidoboni, E. and A. Comastri. 2002. "A 'belated' collapse and a false earthquake in Constantinople: 19 May 1346," *European Earthquake Engineering*, **3**: 22–26.

Guidoboni, E. and A. Comastri. 2005. *Catalogue of Earthquakes and Tsunamis in the Mediterranean Area from the Eleventh to the Fifteenth Century*. Rome, Bologna: INGV–SGA.

Guidoboni, E. and G. Ferrari, eds. 1986. *Il terremoto di Rimini e della costa romagnola: 25 dicembre 1786: Analisi e interpretazione.* Bologna: SGA.

Guidoboni, E. and G. Ferrari. 1995. "Historical cities and earthquakes: Florence during the last nine centuries and evaluations of seismic hazard," in *Earthquakes in the Past: Multidisciplinary Approaches,* eds. E. Boschi, R. Funiciello, E. Guidoboni and A. Rovelli, special issue of *Annali di Geofisica,* **38**: 617–647.

Guidoboni, E. and D. Mariotti. 2005. *Revisione e integrazione di ricerca riguardante il massimo terremoto storico della Puglia: 20 febbraio 1743,* unpublished technical report RPT 272A/05, October 2005. Bologna: INGV-SGA. (CD-ROM available in the libraries of the Istituto Nazionale di Geofisica e Vulcanologia in Rome and Milan.)

Guidoboni, E. and C. Marmo. 1989. "Impossibili o probabili? i terremoti nelle fonti agiografiche," in *I terremoti prima del Mille in Italia e nell'area mediterranea: Storia archeologia sismologia,* ed. E. Guidoboni. Rome, Bologna: ING–SGA, 330–334.

Guidoboni, E. and J.-P. Poirier. 2004. *Quand la terre tremblait.* Paris: Odile Jacob.

Guidoboni, E. and M. Stucchi. 1993. "The contribution of historical records of earthquakes to the evaluation of seismic hazard," in *Global Seismic Hazard Assessment Program,* eds. D. Giardini and P. Basham, special issue of *Annali di Geofisica,* **36**: 201–215.

Guidoboni, E. and G. Traina. 1995. "A new catalogue of earthquakes in the historical Armenian area from antiquity to the twelfth century," *Annali di Geofisica,* **38**: 85–147.

Guidoboni, E., A. Comastri and G. Traina. 1994. *Catalogue of Ancient Earthquakes in the Mediterranean Area up to the Tenth Century.* Bologna: ING–SGA.

Guidoboni, E., A. Muggia and G. Valensise. 2000. "Aims and methods in territorial archaeology: possible clues to a strong fourth-century AD earthquake in the Straits of Messina (southern Italy)," in *Archaeology of Geological Catastrophes,* ed. W. J. McGuire, D. R. Griffiths, P. L. Hancock and I. S. Stewart. Geological Society Special Publication 171. London: The Geological Society, 45–70.

Guidoboni, E., D. Mariotti, M. S. Giammarinaro and A. Rovelli. 2003a. "Identification of amplified damage zones in Palermo, Sicily (Italy), during the earthquakes of the last three centuries," *Bulletin of the Seismological Society of America,* **93**: 1649–1669.

Guidoboni, E., R. Haratounian and A. Karakanian. 2003b. "The Garnì (Armenia) large earthquake on 14 June 1679: a new analysis," *Journal of Seismology,* **7**: 301–328.

Guidoboni, E., F. Bernardini, A. Comastri and E. Boschi. 2004. "The large earthquake on 29 June 1170 (Syria, Lebanon and central southern Turkey)," *Journal of Geophysical Research,* **109**, B7, B07304, doi:10.1029/2003JB002523.

Guidoboni, E., A. Comastri and E. Boschi. 2005. "The 'exceptional' earthquake of 3 January 1117 in the Verona area (northern Italy): a critical time review and detection of two lost earthquakes (lower Germany and Tuscany)," *Journal of Geophysical Research,* **110**, B12, B12309, doi:10.1029/2005JB003683.

Guidoboni, E., G. Ferrari, D. Mariotti, *et al.* 2007. *CFTI4Med, Catalogue of Strong Earthquakes in Italy (461 BC – 1997) and Mediterranean Area (760 BC – 1500)*. Bologna: INGV-SGA. Available at http://storing.ingv.it/cfti4med/

Gutdeutsch, R., Chr. Hammerl, I. Mayer and K. Vocelka. 1987. *Erdbeben als historisches Ereignis: Die Rekonstruktion des Bebens von 1590 in Niederösterreich*. Berlin: Springer-Verlag.

Gutenberg, B. and C. F. Richter. 1949. *Seismicity of the Earth and Associated Phenomena.* Princeton, N.J.: Princeton University Press.

Gutenberg, B. and C. F. Richter. 1954. *Seismicity of the Earth and Associated Phenomena,* 2nd edn. Princeton, N.J.: Princeton University Press.

Gutenberg, B. and C. F. Richter. 1956. "Earthquake magnitude, intensity, energy and acceleration," *Bulletin of the Seismological Society of America,* **46**: 105–145.

Haigh, S. K. and S. P. G. Madabhushi. 2002. "Dynamic centrifuge modelling of the destruction of Sodom and Gomorrah," in *Proceedings of the 1st International Conference on Physical Modelling in Geotechnics*, St. John's, Newfoundland, Canada, eds. R. Phillips, P. Guo and R. Popescu. Leiden: A. A. Balkema, 507–512.

Hakobyan, V. A. 1951–56. *13th–18th Century Minor Chronicles*, 2 vols. Erevan: Publishing House of the Academy of Sciences of Armenian SSR (in Armenian).

Hales, S. 1749–50. "Some considerations on the causes of earthquakes," *Philosophical Transactions,* **46**: 669–681.

Hamilton, W. 1783. "An account of the earthquakes which happened in Italy, from February to May 1783," *Philosophical Transactions,* **73**: 169–208.

Hamilton, W. 1783–85. Letters written between 18 February 1783 and 29 March 1785. The National Archives, *General Correspondence, Sicily and Naples*, F.O. 70/2, 70/3.

Harbi, A., D. Benouar and H. Benhallou. 2003. "Re-appraisal of seismicity and seismotectonics in northeastern Algeria. Part I: Review of historical seismicity," *Journal of Seismology,* **7**: 115–136.

Harland, J. P. 1942. "Sodom and Gomorrah: the location of the Cities of the Plain," *The Biblical Archaeologist,* **5**: 17–32.

Harris, G. and A. Beardow. 1995. "The destruction of Sodom and Gomorrah: a geotechnical perspective," *Quarterly Journal of Engineering Geology,* **28**: 349–362.

Hatori, T. 1986. "Classification of tsunami magnitude scale," *Bulletin of Earthquake Research Institute, University of Tokyo,* **61**: 503–515 (in Japanese with English abstract).

Heilbron, J. L. 1979. *Electricity in the Seventeenth and Eighteenth Centuries: A Study of Early Modern Physics*. Berkeley, Calif.: University of California Press.

Helly, B. 2005. "Case studies," in *Ancient Buildings and Earthquakes: The Local Seismic Culture Approach – Principles, Methods, Potentialities*, eds. F. Ferrigni, B. Helly, A. Mauro *et al.* Centro Culturale Europeo di Ravello, Council of Europe, Strasburg. Bari: Edipuglia.

Henry, M. 1985. "Le témoignage de Libanius et les phénomènes séismiques du IVe siècle de notre ère: Essai d'interprétation," *Phoenix,* **39**: 36–61.

Herman, E. 1946. "Chiese private e diritto di fondazione negli ultimi secoli dell'impero bizantino," *Orientalia Christiana Periodica*, **12**: 302–321.

Hermann, A. 1962. *s.v.* "Erdbeben," *Reallexikon für Antike und Christentum*, **5**: 1070–1113.

Herzog, Z. 1999. "Deconstructing the walls of Jericho," *Ha'aretz*, 29 October 1999.

Herzog, Z. 2002. "The fortress mound at Tel Arad: an interim report," *Tel Aviv*, **29**(1): 3–109.

Hine, H. M. 2002. "Seismology and vulcanology in Antiquity?," in *Science and Mathematics in Ancient Greek Culture*, eds. C. J. Tuplin and T. E. Rihll. Oxford: Oxford University Press, 56–75.

Holbach, P. H. T., baron d'. 1765. *s.v.* "Tremblements de terre," in *Encyclopédie ou dictionnaire raisonné des sciences, des arts et des métiers*, eds. D. Diderot and J.-B. D'Alembert. Neufchastel [i.e. Paris]: Samuel Faulche & compagnie, vol. 16, 580–583.

Hough, S. E. 2004. "Scientific overview and historical context of the 1811–1812 New Madrid earthquake sequence," *Annals of Geophysics*, **47**: 523–537.

Hough, S. E., J. G. Armbruster, L. Seeber and J. F. Hough. 2000. "On the modified Mercalli intensities and magnitudes of the 1811–1812 New Madrid earthquakes," *Journal of Geophysical Research*, **105**: B10: 23 839–23 864.

Hsieh, Yushow. 1957. "A new scale of seismic intensity adapted to the condition in China," *Acta Geophysica Sinica*, **6**: 35–47.

Hutchinson, I. and A. D. McMillan. 1997. "Archaeological evidence for village abandonment associated with late Holocene earthquakes at the northern Cascadia subduction zone," *Quaternary Research*, **48**: 79–87.

Iida, K., D. C. Cox and G. Pararas-Carayannis. 1967. *Preliminary Catalog of Tsunamis Occurring in the Pacific Ocean*. Data Report 5, HIG–67–10. Honolulu, Ha.: Hawaii Institute of Geophysics, University of Hawaii.

Imamura, A. 1949. "List of tsunamis in Japan," *Journal of the Seismological Society of Japan*, **2**: 23–28 (in Japanese).

Issel, A. 1888. *Il terremoto del 1887 in Liguria*. Rome: Tipografia Nazionale.

Jackson, J. 2006. "Fatal attraction: living with earthquakes, the growth of villages into megacities, and earthquake vulnerability in the modern world," *Philosophical Transactions of the Royal Society of London*, **364A**: 1911–1925.

Jacoby, G. C., P. R. Sheppard and K. E. Sieh. 1988. "Irregular recurrence of large earthquakes along the San Andreas fault: evidence from trees," *Science*, **241**: 196–199.

Jacques, F. and B. Bousquet. 1984. "Le raz de marée du 21 juillet 365: Du cataclysme local à la catastrophe cosmique," *Mélanges de l'École française de Rome – Antiquité*, **96**(1): 423–461.

Jesuit Relations. 1663. *The Jesuit Relations and Allied Documents: Travels and Explorations of the Jesuit Missionaries in New France 1610–1791*. The original French, Latin, and Italian texts, with English translation and notes, ed. R. G. Thwaites. Clark, N.J.: Lawbook Exchange, 2005.

Johnston, A. C. 1996a. "Seismic moment assessment of earthquakes in stable continental regions. Part I: Instrumental seismicity," *Geophysical Journal International*, **124**: 381–414.

Johnston, A. C. 1996b. "Seismic moment assessment of earthquakes in stable continental regions. Part III: New Madrid 1811–1812, Charleston 1886 and Lisbon 1755," *Geophysical Journal International*, **126**: 314–344.

Jones, A. H. M., J. R. Martindale and J. Morris, eds. 1971. *The Prosopography of the Later Roman Empire*, vol. 1, *AD 260–395*. Cambridge: Cambridge University Press.

Kaka, S. I. and G. M. Atkinson. 2004. "Relationships between instrumental ground-motion parameters and modified Mercalli intensity in eastern North America," *Bulletin of the Seismological Society of America*, **94**: 1728–1736.

Kant, I. 1756a. "Von der Ursachen der Erderschütterungen bei Gelegenheit des Unglücks, welches die westliche Länder von Europa gegen das Ende des vorigen Jahres betroffen hat," *Königsbergischen Wöchentlichen Frag- und Anzeigungs-Nachrichten*, **4** (24 January), **5** (31 January). (Repr. in *Kant's gesammelte Schriften*. Berlin: Georg Reimer, 1900, vol. 1, 417–427.)

Kant, I. 1756b. *Geschichte und Naturbeschreibung der merkwürdigsten Vorfälle des Erdbebens, welches an dem Ende des 1755sten Jahres einen großen Theil der Erde erschüttert hat*. Königsberg: J. H. Hartung. (Repr. in *Kant's gesammelte Schriften*. Berlin: Georg Reimer, 1900, vol. 1, 429–461.)

Kant, I. 1756c. "Fortgesetzte Betrachtung der seit einiger Zeit wahrgenommenen Erderschütterungen," *Königsbergischen Wöchentlichen Frag- und Anzeigungs-Nachrichten*, **15** (10 April), **16** (17 April). (Repr. in *Kant's gesammelte Schriften*. Berlin: Georg Reimer, 1900, vol. 1, 463–472.)

Karakhanian, A. and Y. Abgaryan. 2004. "Evidence of historical seismicity and volcanism in the Armenian Highland (from Armenian and other sources)," in *Investigating the Records of Past Earthquakes*, eds. P. Albini, V. G. Acosta, R. M. W. Musson and M. Stucchi, special issue of *Annals of Geophysics*, **47**: 793–810.

Karcz, I. and U. Kafri. 1978. "Evaluation of supposed archaeoseismic damage in Israel," *Journal of Archaeological Science*, **5**: 237–253.

Karcz, I. and U. Kafri. 1981. "Studies in archaeoseismicity of Israel: Hisham's Palace, Jericho," *Israel Journal of Earth Sciences*, **30**: 12–23.

Kárník, V. 1968–71. *Seismicity of the European Area*. Dordrecht: D. Reidel.

Kasangian, H. 1981. "Comportement statistique de quelques coupoles arméniennes," in *The 2nd International Symposium on Armenian Art, Erevan 1978, Collection of Reports*. Erevan: Publishing House of the Academy of Sciences of Armenian SSR, vol. 2, 32–42.

Kawasumi, H. 1951. "Measures of earthquake danger and expectancy of maximum intensity throughout Japan as inferred from the seismic activity in historical times," *Bulletin of Earthquake Research Institute, University of Tokyo*, **29**: 469–482.

Kelleher, J., L. Sykes and J. Oliver. 1973. "Possible criteria for predicting earthquake locations and their application to major plate boundaries of the Pacific and Caribbean," *Journal of Geophysical Research*, **78**: 2547–2585.

Kelly, G. 2004. "Ammianus and the Great Tsunami," *Journal of Roman Studies*, **94**: 141–165.

Kendrick, T. C. 1956. *The Earthquake of Lisbon*. Philadelphia, Penn.: J. B. Lippincott.

Kenrick, Ph. M. 1986. *Excavations at Sabratha 1948–1951* (Journal of Roman Studies Monographs, 2). London: Society for the Promotion of Roman Studies.

Khalidi, M. A. 2005. *Medieval Islamic Philosophy*. Cambridge: Cambridge University Press.

Kircher, A. 1664–65. *Mundus subterraneus in XII libros digestus*. 2 vols. Amsterdam: Johannes van Waesberge and Elizaeus Weyerstraten (reprint of 1678 edn, ed. G. B. Vai, Sala Bolognese: Forni, 2004).

Knauf, E. A. 2002. "Excavating Biblical history," *Revelations from Megiddo: The Newsletter of the Megiddo Expedition*, **6**. Available at http://megiddo.tau.ac.il/publications_newsletter6.html4

Kondorskaya, N. V. and N. V. Shebalin. 1982. *New Catalog of Strong Earthquakes in the USSR from Ancient Times through 1977*. World Data Center A for Solid Earth Geophysics Report SE-31. Boulder, Colo.: US Department of Commerce.

Kovach, R. L. 2004. *Early Earthquakes of the Americas*. Cambridge: Cambridge University Press.

Kozák, J. and J. E. Ebel. 1996. "Macroseismic information from historic pictorial sources," *Pure and Applied Geophysics*, **146**: 103–111.

Krinitzsky, E. L. and F. K. Chang. 1988. "Intensity-related earthquake ground motions," *Bulletin of the Association of Engineering Geologists*, **25**: 425–435.

Lamboley, J. L. 1998. "Vaste Lecce: La porte Est et la tombe des Pizzinaghe," *Studi di Antichità*, **9**(1996): 361–430.

Lampros, S. 1910. "First collection of prodigies found in chronicles or other records," *Neos Hellenomnemon*, **7**: 113–313 (in Greek).

Lampros, S. 1922. "Second collection of prodigies found in chronicles or other records," *Neos Hellenomnemon*, **16**: 407–420 (in Greek).

Lanciani, R. 1918. "Segni di terremoti negli edifizi di Roma antica," *Bullettino della Commissione Archeologica Comunale di Roma*, **45**: 1–28.

Landes, D. S. 1983. *Revolution in Time: Clocks and the Making of the Modern World*. Cambridge, Mass.: Belknap Press of Harvard University Press.

Lane, E. W. 1867. *Arabic-English Lexicon*. London: Williams and Norgate, 1863–1893. (Lithographic reproduction, Cambridge: Islamic Texts Society, 2 vols., 2003.)

Lattuada, R. 2002. "La ricostruzione a Napoli dopo il terremoto del 1688: architetti, committenti e cultura del ripristino," in *Contributi per la storia dei terremoti nel bacino del Mediterraneo*, ed. A. Marturano. Salerno: Laveglia, 203–231.

Lawson, A. C. 1908. *The California Earthquake of April 18, 1906: Report of the State Earthquake Investigation Commission*. Carnegie Institution of Washington Publication 87, 2 vols. Washington, D.C.: Carnegie Institution. (Repr. 1969.)

Leibniz, G. W. 1749. *Protogaea sive de prima face telluris*. Göttingen: C. L. Scheidt.

Lemaire, A. 1997. "Deir 'Alla Inscriptions," in *The Oxford Encyclopedia of Archaeology in the Near East*, ed. E. M. Meyers. Oxford: Oxford University Press, vol. 2, 138–140.

Leonardo da Vinci. *Atlantic Codex: Il codice atlantico della Biblioteca Ambrosiana di Milano*, trascrizione critica di Augusto Marinoni; 3 vols. Florence: Giunti Barbèra, 2000.

Leonardo da Vinci. *Codex Leicester, formerly Codex Hammer: Il codice di Leonardo da Vinci della biblioteca di Lord Leicester in Holkham Hall*, ed. G. Calvi. Florence: Giunti Barbèra, 1980. (Repr. of ed. published: Milan: Cogliati, 1909.) (English translation: *The Codex Hammer of Leonardo da Vinci*, ed. C. Pedretti. Florence: Giunti Barbèra, 1987.) (Facsimile in Italian, written in reverse script.)

Leonardo da Vinci. *Il manoscritto A.*, in *I manoscritti dell'Institut de France*, facsimile edition under the patronage of the Commissione Nazionale Vinciana and the Institut de France, critical and diplomatic transcription by A. Marinoni. Florence: Giunti-Barbèra, 1990. (English translation: *The manuscripts of Leonardo da Vinci in the Institut de France*, transl. and annot. John Venerella, Vol. 1, *Manuscript A*. Milan: Ente Raccolta Vinciana, 1999.)

Lepelley, C. 1981. *Les Cités de l'Afrique romaine au Bas-Empire*, vol. 2. Paris: Etudes Augustiniennes.

Lepelley, C. 1984. "L'Afrique du Nord et le prétendu séisme universel du 21 juillet 365," *Mélanges de l'École française de Rome – Antiquité*, **96**(1): 463–491.

Lepelley, C. 1994. "Le présage du nouveau désastre de Cannes: la signification du raz de marée du 21 juillet 365 dans l'imaginaire d'Ammien Marcellin," *Kokalos*, **36-37** (1990–91): 259–274.

Levret, A. 1991. "The effects of November 1, 1755 'Lisbon' earthquake in Morocco," *Tectonophysics*, **193**: 83–94.

Leydecker, G. and H. J. Brüning. 1988. "Ein vermeintliches Schadenbeben im Jahre 1046 im Raum Höxter und Holzminden in Norddeutschland: Über die Notwendigkeit des Studiums der Quellen historischer Erdbeben," *Geologisches Jahrbuch Reihe E: Geophysik*, **E42**: 119–125.

Ligorio, Pirro. *Libro di diversi terremoti*, ed. E. Guidoboni (Edizione Nazionale delle Opere di Pirro Ligorio, Libri delle antichità, Torino, vol. 28, Codice Ja.II.15). Rome: De Luca editori d'arte, 2005.

Lilla, S. 1970. *Il testo tachigrafico del "De divinis nominibus" Vat. Gr. 1809*. Studi e Testi 263. Vatican City: Biblioteca Apostolica Vaticana.

Lister, M. 1684. "Three papers of dr. Martin Lyster, the first of the nature of earth-quakes; more particularly of the origine of the matter of them, from the pyrites alone," *Philosophical Transactions*, **14**: 512–515.

Lloyd, G. E. R. 1970. *Early Greek Science: Thales to Aristotle*. New York: W. W. Norton.

Lloyd, G. E. R. 1973. *Greek Science after Aristotle*. New York: W. W. Norton.

Lloyd, G. E. R. 1991. *Methods and Problems in Greek Science*. Cambridge: Cambridge University Press

Lorito, S., M. M. Tiberti, R. Basili, A. Piatanesi and G. Valensise. 2008. "Earthquake-generated tsunamis in the Mediterranean Sea: scenarios of potential threats to Southern Italy," *Journal of Geophysical Research*, **113**, B01301, doi:10.1029/2007JB004943.

Losey, R. J. 2002. "Communities and catastrophe: tillamook response to the AD 1700 earthquake and tsunami, northern Oregon coast", Ph.D. dissertation, University of Oregon, Eugene, Oreg.

Ludwin, R. S., R. Dennis, D. Carver, *et al.* 2005. "Dating the 1700 Cascadia earthquake: great coastal earthquakes in native stories," *Seismological Research Letters*, **76**: 140–148.

Lycosthenes, Conradus [Conrad Wolffhart]. 1557. *Prodigiorum ac ostentorum chronicon.* Basel: Heinrich Petri.

MacBain, B. 1982. *Prodigy and Expiation: A Study in Religion and Politics in Republican Rome.* Brussels: Latomus.

Macelwane, J. B. 1936. *Introduction to Theoretical Seismology. Part I: Geodynamics.* New York: John Wiley.

MacGuire, B., D. R. Griffiths, P. L. Hancock and I. S. Stewart, eds. 2000. *The Archaeology of Geological Catastrophes.* Geological Society Special Publication 171. London: The Geological Society.

Mader, C. L. 1999. "Modeling the 1958 Lituya Bay mega-tsunami," *Science of Tsunami Hazards*, **17**: 57–67.

Mader, C. L. 2002. "Modeling the 1958 Lituya Bay mega-tsunami. Part II," *Science of Tsunami Hazards*, **20**: 241–250.

Maggio, Lucio. 1571. *Del terremoto dialogo del signor Lucio Maggio gentil'huomo bolognese.* Bologna: Alessandro Benacci.

Magri, G. and D. Molin. 1984. *Il terremoto del dicembre 1456 nell'Appennino centro-meridionale*, RT/AMB(83)8. Rome: ENEA.

Magri, G. and D. Molin. 1985. "The earthquake of December 1456 in Central-Southern Italy," in *Atlas of Isoseismal Maps of Italian Earthquakes*, ed. D. Postpischl. Quaderni de "La Ricerca Scientifica", 114, 2A. Rome: CNR-PFG, 20–23.

Maheshwari, B. K., M. L. Sharma and J. P. Narayan. 2006. "Geotechnical and structural damage in Tamil Nadu, India, from the December 2004 Indian Ocean tsunami," *Earthquake Spectra*, **22**: S475–S493.

Mallet, R. 1852–54. "Catalogue of recorded earthquakes from 1606 BC to AD 1850: Third Report on the facts of earthquake phaenomena," *Report of the 22nd Meeting of the British Association for the Advancement of Science*, **1852**: 1–176; *Report of the 23rd Meeting of the British Association for the Advancement of Science*, **1853**: 118–212; *Report of the 24th Meeting of the British Association for the Advancement of Science*, **1854**: 1–326.

Mallet, R. 1862. *Great Neapolitan Earthquake of 1857: The First Principles of Observational Seismology.* London: Chapman and Hall, 2 vols. (Repr. in *Mallet's Macroseismic Survey on the Neapolitan Earthquake of 16th December, 1857*, ed. E. Guidoboni and G. Ferrari. Bologna: SGA, 1987.)

Manning, S. W., C. B. Ramsey, W. Kutschera, *et al.* 2006. "Chronology for the Aegean Late Bronze Age 1700–1400 B.C.," *Science* 312, 565–569, doi: 10.1126/science. 1125682.

Marchini, G. P. 1978. "Verona romana e paleocristiana," in *Ritratto di Verona: lineamenti di una storia urbanistica*, ed. L. Puppi. Verona: Banca Popolare di Verona, 23–134.

Marco, S. and A. Agnon. 2002. "Armageddon quakes," *Revelations from Megiddo: The Newsletter of the Megiddo Expedition*, **6**. Available at http://megiddo.tau.ac.il/publications_newsletter6.html4

Margottini, C. and J. Kozák. 1992. *Terremoti in Italia dal 62 AD al 1908*. Rome: ENEA.

Marmo, C. 1989. "Le teorie del terremoto da Aristotele a Seneca," in *I terremoti prima del Mille in Italia e nell'area mediterranea: Storia archeologia sismologia*, ed. E. Guidoboni. Rome, Bologna: ING-SGA, 170–177.

Martínez Solares, J. M. 2001. *Los efectos en España del terremoto de Lisboa (1 de noviembre de 1755)*. Madrid: Ministerio de Fomento, Dirección General del Instituto Geográfico Nacional.

Martínez Solares, J. M. and A. Lopez Arroyo. 2004. "The great historical 1755 earthquake: effects and damage in Spain," *Journal of Seismology*, 8: 275–294.

Martini, Francesco di Giorgio. *Trattato di architettura. Il codice Ashburnham 361 della Biblioteca Medicea Laurenziana di Firenze*, ed. P. C. Marani, 2 vols. Florence: Giunti Barbèra, 1979.

Mauk, F. J., D. Christensen and S. Henry. 1982. "The Sharpsburg, Kentucky, earthquake 27 July 1980: main shock parameters and isoseismal maps," *Bulletin of the Seismological Society of America*, 72: 221–236.

Mazza, V. 1998. "The supposed Egyptian earthquakes of 184 and 95 BC: critical review and some lines of research in historical seismology using Greek papyri from Egypt," *Annali di Geofisica*, 41: 121–125.

McCalpin, J. P. 1996. *Paleoseismology*. San Diego, Calif.: Academic Press.

McCoy, F. W. and G. Heiken. 2000. "Tsunami generated by the Late Bronze Age eruption of Thera (Santorini), Greece," *Pure and Applied Geophysics*, 157: 1227–1256.

McMahon, A. H. 1897. "The southern borderlands of Afghanistan," *Geographical Journal*, 9: 393–415.

Medvedev, S. V., W. Sponheuer and V. Kárník. 1964. "Seismic Intensity Scale, MSK 1964," *Academy of Sciences of the USSR, Soviet Geophysical Committee*, 13.

Meghraoui, M., T. Camelbeeck, K. Vanneste, M. Brondeel and D. Jongmans. 2000. "Active faulting and paleoseismology along the Bree fault, lower Rhine graben, Belgium," *Journal of Geophysical Research*, 105: 13 809–13 841.

Meletti, C., E. Patacca, P. Scandone and B. Figliuolo. 1988-89. "Il terremoto del 1456 e la sua interpretazione nel quadro sismotettonico dell'Appennino meridionale," in *Il terremoto del 1456*, ed. B. Figliuolo. Altavilla Silentina: Edizioni Studi Storici Meridionali, vol. 1, 71–108; vol. 2, 35–163.

Melville, C. P. 1985. "Terremoti britannici anteriori al 1800: alcuni problemi di localizzazione irrisolti," in *Terremoti e storia*, ed. E. Guidoboni, *Quaderni storici*, 60: 717–741.

Mercalli, G. 1883. *Vulcani e fenomeni vulcanici in Italia*. Milan: Vallardi. (Repr. Sala Bolognese: Forni, 1981.)

Mezzavacca, F. 1672. *De terraemotu libellus in quo curiosa aperitur terraemotus doctrina, & agitur de terraemotu huius anni 1672*. Bologna: G. B. Ferroni.

Michell, J. 1761. "Conjectures concerning the cause and observations upon the phenomena of earthquakes," *Philosophical Transactions*, 51: 566–634.

Milizia, F. 1781. *Principi di architettura civile*. Finale: Jacopo de' Rossi, 3 vols.

Miller, D. J. 1960. *Giant Waves in Lituya Bay, Alaska*. US Geological Survey Professional Paper 354-C. Washington, D.C.: US Government Printing Office.

Milne, J. 1897. "Seismological investigations: first report," *Report of the British Association for the Advancement of Science*, **1896**: 180–230.

Milne, J. 1898. "Seismological investigations: second report," *Report of the British Association for the Advancement of Science*, **1897**: 129–206.

Milne, J. 1901. "Seismological investigations: sixth report," *Report of the British Association for the Advancement of Science*, **1901**: 40–54.

Milne, J. 1911. "A catalogue of destructive earthquakes AD 7 to AD 1899," *Report of the British Association for the Advancement of Science*, **1911**: 649–740.

Minor, R. and W. C. Grant. 1996. "Earthquake-induced subsidence and burial of late Holocene archaeological sites, northern Oregon coast," *American Antiquity*, **61**: 772–781.

Minoura, K., F. Imamura, U. Kuran, *et al.* 2000. "Discovery of Minoan tsunami deposits," *Geology*, **28**: 59–62.

Molin, D. and C. Margottini. 1982. "Il terremoto del 1627 nella Capitanata settentrionale," in *Contributo alla caratterizzazione della sismicità del territorio italiano*, Memorie presentate al Convegno annuale del Progetto Finalizzato "Geodinamica" del CNR sul tema: "Sismicità dell'Italia: stato delle conoscenze scientifiche e qualità della normativa sismica", Udine. Rome: Tipografia CSR, 251–279.

Molin, D. and A. Paciello. 1999. "Seismic hazard assessment in Florence city, Italy," *Journal of Earthquake Engineering*, **3**: 475–494.

Molin, D., A. Rossi and A. Tertulliani. 2000. "Researches of historical seismology for a low seismicity area: Aniene upper–valley Central Italy," in *Papers and Memoranda from the 1st Workshop of the ESC Working Group "Historical Seismology,"* ed. V. Castelli with the collaboration of G. Monachesi and H. Coppari. Macerata: Osservatorio Geofisico Sperimentale di Macerata, 27–32.

Molin, K. 2001. *Unknown Crusader Castles*. London: Hambledon.

Momigliano, A. 1930. "Erodoto e Tucidide sul terremoto di Delo," *Studi italiani di filologia classica*, n.s., **8**: 87–89.

Mongitore, A. 1727. *Palermo ammonito, penitente e grato, nel formidabil terremoto del primo settembre 1726*. Palermo: engraver Antonino Bova.

Montessus de Ballore, F. de. 1884. *Temblores y erupciones volcánicas en Centro-América*. San Salvador: F. Sagrini.

Montessus de Ballore, F. de. 1892. "La France et l'Algérie sismiques," *Annales des Mines*, ser. IX, **2**: 317–328.

Montessus de Ballore, F. de. 1906. *Les Tremblements de terre: géographie séismologique*. Paris: Armand Collin.

Montessus de Ballore, F. de. 1915. "La sismologia en la Biblia," *Boletín del Servicio Sismológico de Chile*, **11**: 27–158.

Montessus de Ballore, F. de. 1923. *Ethnographie sismique et volcanique ou les tremblements de terre et les volcans dans la religion, la morale, la mythologie et le folklore de tous les peuples*. Paris: E. Champion.

Morandi, U. 1964. *Le Biccherne senesi: Le tavolette della Biccherna, della Gabella e di altre magistrature dell'antico stato senese conservate presso l'Archivio di Stato di Siena*. Siena: Monte dei Paschi di Siena.

Moreira de Mendonça, J. J. 1758. *Historia universal dos terremotos que tem havido no mundo, de que ha noticia, desde a sua creaçaõ até o seculo presente: com huma narraçam individual do terremoto do primeiro de Novembro de 1755*. Lisbon: A.V. da Silva.

Morelli, C. 1942. *Carta sismica dell'Albania*. Reale Accademia d'Italia, Commissione italiana di studio per i problemi del soccorso alle popolazioni 10. Florence: Le Monnier.

Moroni, A. and M. Stucchi. 1993. "Materials for the investigation of the 1564, Maritime Alps earthquake," in *Historical Investigation of European Earthquakes, Materials of the CEC Project Review of Historical Seismicity in Europe*, ed. M. Stucchi. Milan: CNR, vol. 1, 101–125.

Mucciarelli, M. and FAUST Working Group. 2000. "Faults as a seismologists' tool: current advancement of the EC project 'FAU.S.T'," in *Proceedings of the 27th General Assembly of the European Seismological Commission (ESC)*, Lisbon, 319–324.

MunichRe, 1991. *World Map of Natural Hazards*, DIN A4, Order 17272-V-e. Munich: Münchener Rückversicherung-Gesellschaft.

Muratori, L. A., ed. 1726. "Memoriale potestatum Regiensium," in *Rerum italicarum scriptores*. Milan: Societatis Palatinae, vol. 8, 1073–1174.

Murphy, J. R. and L. J. O'Brien. 1977. "The correlation of peak ground acceleration amplitude with seismic intensity and other physical parameters," *Bulletin of the Seismological Society of America*, 67: 877–915.

Musha, K., ed. 1941–43. *Zotei Dai-nihon Jishin Shiryo [Collection of Historical Documents on Earthquakes in Great Japan]*, 3 vols. Earthquake Prevention Council, Ministry of Education. Tokyo: Japan Society for the Promotion of Science (in Japanese).

Musha, K., ed. 1951. *Nihon Jishin Shiryo [Collection of Historical Documents on Earthquakes in Japan]*. Tokyo: Mainichi Newspapers (in Japanese).

Musson, R. M. W. 1986. "The use of newspaper data in historical earthquake studies," *Disasters*, 10: 217–223.

Musson, R. M. W. 1998. "Inference and assumption in historical seismology," *Surveys in Geophysics*, 19: 189–203.

Mutafian, C. and E. Van Lauve. 2001. *Atlas historique de l'Arménie: Proche-Orient et Sud-Caucase du VIIIe siècle av. J.-C. au XXI siècle*. Paris: Edition Autrement.

Needham, J. 1959. *Science and Civilisation in China*, vol. 3, *Mathematics and the Sciences of the Heavens and the Earth*, with the collaboration of Wang Ling. Cambridge: Cambridge University Press.

Nejjar, S., transl. 1974. *Jalal ad-Din as-Suyut'i, Kashf aç-Calçala 'an waçf az-Zalzala*. Rabat: Centre universitaire de la recherche scientifique.

Nelson, A. R., B. F. Atwater, P. T. Bobrowsky, *et al*. 1995. "Radiocarbon evidence for extensive plate-boundary rupture about 300 years ago at the Cascadia subduction zone," *Nature*, 378: 371–374.

Nissen, H. 1883. *Italische Landeskunde*, vol. 1, *Land und Leute*. Berlin: Weidmann.

Nuffelen, P. van 2006. "Earthquakes in AD 363–368 and the date of Libanius, *oratio* 18," *The Classical Quarterly*, 56(2): 657–661.

Nur, A. 2002. "Earthquakes and archaeology," in *International Handbook of Earthquake and Engineering Seismology*, ed. W. H. K. Lee, H. Kanamori, P. C. Jennings and C. Kisslinger. Amsterdam: Academic Press, 765–774.

Nur, A. and Z. Reches. 1979. "The Dead Sea rift: geophysical, historical and archaeological evidence for strike slip motion," *Eos (Transactions of the American Geophysical Union)*, 60(18), 322.

Nur, A. and H. Ron. 1997. "Armageddon's earthquakes," *International Geology Review*, 39: 532–541.

Nur, A., H. Ron and D. Tal. 1989. "Earthquake parameters inferred from archaeological evidence," in *Israel Geological Society, Annual Meeting*, Ramot, Israel, 56.

Nuttli, O. W. 1973. "Seismic wave attenuation and magnitude relations for eastern North America," *Journal of Geophysical Research*, 78: 876–885.

Nuttli, O. W. and J. E. Zollweg. 1974. "The relation between felt area and magnitude for central United States earthquakes," *Bulletin of the Seismological Society of America*, 64: 73–85.

Nuttli, O. W., G. A. Bollinger and R. B. Herrmann. 1986. *The 1886 Charleston, South Carolina, Earthquake: A 1986 Perspective*. US Geological Survey Circular 985. Washington, D.C.: US Government Printing Office.

Obermeier, S. F. and E. C. Pond. 1999. "Issues in using liquefaction features for paleoseismic analysis," *Seismological Research Letters*, 70: 34–58.

Oldham, R. D. 1899. "Report on the great earthquake of 12 June 1897," *Memoirs of the Geological Survey of India*, 29: 138–147.

Oldroyd, D. R., F. Amador, J. Kozák, A. Carneiro and M. Pinto. 2007. "The study of earthquakes in the hundred years following the Lisbon earthquake of 1755," *Earth Sciences History*, 26: 321–370.

Olshausen, E. and H. Sonnabend, eds. 1998. *Naturkatastrophen in der antiken Welt*. Stuttgarter Kolloquim zur historischen Geographie des Altertums 6. Wiesbaden: F. Steiner.

Omori, F. 1909. "Preliminary report on the Messina–Reggio earthquake of Dec. 28, 1908," *Bulletin of the Imperial Earthquake Investigation Committee*, 3(2): 37–46.

Pacho, J.-R. 1827–29. *Relation d'un voyage dans la Marmarique, la Cyrénaïque et les oasis d'Audjelah et de Maradeh*. Paris: Firmin-Didot, 2 vols. (Repr. Marseille: J. Laffitte, 1979.)

Pallas, P. S. 1802–03. *Travels through the Southern Provinces of the Russian Empire in the years 1793 and 1794*. Translated from the German by F. W. Blagdon, 2 vols. London: Longman & Rees.

Pantosti, D., G. D'Addezio and F. R. Cinti. 1996. "Paleoseismicity in the Ovindoli–Pezza fault, Central Appennines, Italy: a history including a large previously unrecorded earthquake in the Middle Ages 860–1300 AD," *Journal of Geophysical Research*, 101, B3: 5937–5959.

Papadopoulos, G. A. 2003. "Quantification of tsunamis; a review," in *Submarine Landslides and Tsunamis*, NATO Science Series, Series IV, Earth and Environmental Series 21, eds. A. C. Yalçiner, E. N. Pelinovsky, E. Okal and C. E. Synolakis. Dordrecht: Kluwer Academic Publishers, 285–291.

Papadopoulos, G. A. and B. J. Chalkis. 1984. "Tsunamis observed in Greece and the surrounding area from antiquity up to the present times," *Marine Geology*, **56**: 309–17.

Papadopoulos, G. A. and F. Imamura. 2001. "A proposal for a new tsunami intensity scale," in *International Tsunami Symposium 2001 Proceedings*, Seattle, Washington, 569–577.

Papageorgiou, S. and S. Stiros. 1996. "The harbour of Aigeira (north Peloponnese, Greece): an uplifted ancient harbour," in *Archaeoseismology*, ed. S. C. Stiros and R. E. Jones. Fitch Laboratory Occasional Paper 7. Athens: Institute of Geology and Mineral Exploration, The British School at Athens, 211–214.

Papastamatiou, D. and I. Psycharis. 1996, "Numerical simulation of the seismic response of megalithic monuments: preliminary investigations related to the Apollo temple at Vassai," in *Archaeoseismology*, ed. S. C. Stiros and R. E. Jones. Fitch Laboratory Occasional Paper 7. Athens: Institute of Geology and Mineral Exploration, The British School at Athens, 225–236.

Papathanassiou, G. and S. Pavlides. 2007. "Using the INQUA scale for the assessment of intensity: case study of the 2003 Lefkada (Ionian Islands), Greece earthquake," *Quaternary International*, **173–174**, 4–14, doi:10.1016/j.quaint.2006.10.038.

Papazachos, B. and C. Papazachos. 1997. *The Earthquakes of Greece*. Thessaloniki: Ziti. (1st edn Thessaloniki: Ziti, 1989 (in Greek).)

Papazian, G. 1991. "Richesse symbolique des khatchkars antérieurs au Xe siècle," in *Atti del Quinto simposio internazionale di arte armena*. San Lazzaro, Venice: Tipolitografia armena, 709–726.

Pararas-Carayannis, G. 2003. "Near and far-field effects of tsunamis generated by the paroxysmal eruptions, explosions, caldera collapses and massive slope failures of the Krakatau Volcano in Indonesia on August 26–27, 1883," *Science of Tsunami Hazards*, **21**: 191–211.

Pareschi, M. T., M. Favalli and E. Boschi. 2006. "Impact of the Minoan tsunami of Santorini: simulated scenarios in the eastern Mediterranean," *Geophysical Research Letters*, **33**, L18607, doi:10.1029/2006GL027205.

Pavlides, S. B., G. A. Papadopoulos and A. Ganas. 1999. "The 7th September, 1999, unexpected earthquake of Athens: preliminary results on the seismotectonic environment," in *Proceedings of 1st Conference on Advances in Natural Hazard Mitigation: Experience from Europe and Japan*, Reports, ed. G. A. Papadopoulos, 80–85.

Pavlides, S. B., A. Ganas and G. A. Papadopoulos. 2002. "The fault that caused the Athens September 1999 $M_s = 5.9$ earthquake: field observations," *Natural Hazards*, **27**: 61–84.

Pelinovsky, E., B. H. Choi, A. Stromkov, I. Didenkulova and H.-S. Kim. 2006. "Analysis of tide-gauge records of the 1883 Krakatau tsunami," in *Tsunamis: Case Studies and Recent Developments*, ed. K. Satake. Berlin: Springer-Verlag, 57–77.

Pereira de Sousa, F. L. 1919–28. *O Terremoto do 1° de Novembro de 1755 em Portugal e um estudo demográfico*, 3 vols. Lisbon: Serviços Geologicos.

Perrey, A. 1845. "Mémoire sur les tremblements de terre ressentis en France, en Belgique et en Hollande depuis le IVᵉ siècle jusqu'à nos jours," *Mémoires couronnés et mémoires des savants étrangers publiés par l'Académie royale des sciences, des lettres et des beaux-arts de Belgique*, **18**: 1–110.

Perrey, A. 1845–46. "Mémoire sur les tremblements de terre dans le bassin du Rhin," *Mémoires couronnés et mémoires des savants étrangers publiés par l'Académie royale des sciences, des lettres et des beaux-arts de Belgique*, **19**: 1–113.

Perrey, A. 1846. "Mémoire sur les tremblements de terre dans le bassin du Danube," *Annales des sciences physiques et naturelles d'agriculture et de industrie, Société royale d'Agriculture, histoire naturelle et arts utiles*, **9**: 333–414.

Perrey, A. 1848. "Note sur les tremblements de terre en Algérie et dans l'Afrique septentrionale," *Mémoires de l'Académie des sciences, arts et belles-lettres de Dijon*, 1845–1846: 299–323.

Perrey, A. 1850. "Mémoire sur les tremblements de terre ressentis dans la péninsule Turco-Hellénique et en Syrie," *Mémoires couronnés et mémoires des savants étrangers publiés par l'Académie royale des sciences, des lettres et des beaux-arts de Belgique*, **23**: 1–73.

Phillips, John. 1581. *The wonderfull worke of God shewed upon a chylde*. London: Robert Waldegrave.

Piccardi, L. 2005. "Paleoseismic evidence of legendary earthquakes: the apparition of Archangel Michael at Monte Sant'Angelo (Italy)," *Tectonophysics*, **408**: 113–128.

Piccardi, L. and W. B. Masse, eds. 2007. *Myth and Geology*. Geological Society Special Publication 273. London: The Geological Society.

Pierotti, P. and D. Ulivieri. 1998. *Culture sismiche locali: Garfagnana e Lunigiana* (CD-ROM). Pisa-Ravello: Conseil de l'Europe, Accord Partial Ouvert, Centro Europeo per i Beni Culturali di Ravello.

Pierotti, P. and D. Ulivieri. 2001. *Culture sismiche locali: Garfagnana e Lunigiana*. Pisa: Plus (in Italian and English).

Pilla, L. 1846. *Istoria del tremuoto che ha devastato i paesi della costa toscana il dì 14 agosto 1846*. Pisa: Vannucchi. (Repr. Sala Bolognese: A. Forni, 1985).

Pirazzoli, P. A. and J. Thommeret. 1977. "Datation radiométrique d'une ligne de rivage à + 2,5 m près de Aghia Roumeli, Crète, Grèce," *Comptes Rendus de l'Académie des Sciences de Paris*, **284**(14): 1255–1258.

Pirazzoli, P. A., J. Thommeret, Y. Thommeret et al. 1982. "Crustal block movements from Holocene shoreline: Crete and Antikythira (Greece)", *Tectonophysics*, **86**: 27–43.

Pirazzoli, P. A., J. Ausseil-Badie, P. Giresse et al. 1992. "Historical environmental changes at Phalasarna Harbor, West Crete," *Geoarchaeology*, **7**: 371–392.

Pirazzoli, P. A., J. Laborel and S. C. Stiros. 1996. "Earthquake clustering in the Eastern Mediterranean during historical times," *Journal of Geophysical Research*, **101**(B3): 6083–6097.

Pirona, G. A. and T. Taramelli. 1873. "Sul terremoto del Bellunese del 29 giugno 1873." *Atti del R. Istituto Veneto di Scienze, Lettere ed Arti*, ser. IV, **2**: 1523–1574.

Placanica, A. 1970. *Cassa sacra e beni della Chiesa nella Calabria del Settecento*. Naples: Università degli studi di Napoli.

Placanica, A. 1979. *All'origine dell'egemonia borghese in Calabria: La privatizzazione delle terre ecclesiastiche 1784–1815*. Salerno–Catanzaro: Società editrice meridionale.

Placanica, A. 1984. "Presentazione," in I. Kant, *Scritti sui terremoti*, ed. P. Manganaro. Salerno: 10/17, 5–9. (Repr. in A. Placanica, *Scritti*, ed. M. Mafrici and S. Martelli. Soveria Mannelli: Rubbettino, 2004, vol. 2, 237–240.)

Placanica, A. 1985a. *Il filosofo e la catastrofe: Un terremoto del Settecento*. Turin: Einaudi.

Placanica, A. 1985b. "Goethe davanti alle rovine di Messina: poesia e verità," *Intersezioni*, **5**(1): 63–87. (Repr. in A. Placanica, *Scritti*, ed. M. Mafrici and S. Martelli. Soveria Mannelli: Rubbettino, 2004, vol. 2, 241–268.)

Poirier, J.-P. 2005. *Le tremblement de terre de Lisbonne, 1755*. Paris: Odile Jacob.

Porter, A. K. 1915–17. *Lombard Architecture*, 4 vols. New Haven, Conn.: Yale University Press. (Repr. New York: Hacker Art Books, 1967.)

Prawer, J. 1988. *The History of the Jews in the Latin Kingdom of Jerusalem*. Oxford: Clarendon Press.

Price, S., T. Higham, L. Nixon and J. Moody. 2002. "Relative sea-level changes in Crete: reassessment of radiocarbon dates from Sphakia and West Crete," *Annual of the British School at Athens*, **97**: 171–200.

Priestley, J. 1775. *The History and Present State of Electricity, with Original Experiments*, 3rd edn corrected and enlarged. London: C. Bathurst *et al.*

Propp, V. I. 1971. *Morphology of the Folktale*, transl. L. Scott, 2nd edn revised. Austin, Tex.: University of Texas Press.

Puzzolo Sigillo, D. 1949. "Poesia e verità riguardanti Messina nel 'Viaggio in Italia' di W. Goethe, accertate con critica delle fonti e notizie e documenti inediti," *Archivio Storico Messinese*, ser. III, **1**: 33–163.

Quatremère, M. 1845. *Histoire des sultans Mamlouks de l'Égypte*, II/2. Paris: Oriental Translation Fund of Great Britain and Ireland.

Quenet, G. 2005. *Les Tremblements de terre aux XVIIe et XVIIIe siècle: La naissance d'un risque*. Champ Vallon: Seyssel.

Ray, J. 1691. *The wisdom of God manifested in the works of the creation*. London: S. Smith. (Repr. New York: Garland, 1979.)

Reale, G. and A. P. Bos. 1995. *Il trattato Sul cosmo per Alessandro attribuito ad Aristotele*, 2nd edn revised and enlarged. Milan: Vita e pensiero.

Rebuffat, R. 1980. "Cuicul, le 21 juillet 365," *Antiquités africaines*, **15**: 309–328.

Reid, H. F. 1911. "The elastic-rebound theory of earthquakes," *University of California Publications, Bulletin of the Department of Geology*, **6**: 413–444.

Reiter, L. 1990. *Earthquake Hazard Analysis*. New York: Columbia University Press.

Renou, L. and J. Filliozat. 1953. *L'Inde classique: manuel des études indiennes*, vol. 2, Appendice 3, *Notions de chronologie*. Paris: Imprimerie nationale.

Restifo, G. 1995. "Local administrative sources on population movements after the Messina earthquake of 1908," in *Earthquakes in the Past: Multidisciplinary Approaches*, eds. E. Boschi, R. Funiciello, E. Guidoboni, and A. Rovelli, special issue of *Annali di Geofisica*, **38**: 559–566.

Riccò, A. 1907. "Il terremoto del 16 novembre 1894 in Calabria e Sicilia. Parte I: Relazione sismologica," *Annali dell'Ufficio centrale meteorologico e geodinamico italiano*, ser. II, **19**(1897), pt. I: 7–261.

Richard, M. 1955. "Les Chapitres à Epiphane sur les hérésies de Georges Hiéromoine (VIIe siècle)," *Epeteris Etaireias Byzantinon Spoudon [Annual of the Society for Byzantine Studies]*, **25**: 331–362.

Richter, C. F. 1935. "An instrumental earthquake magnitude scale," *Bulletin of the Seismological Society of America*, **25**: 1–32.

Richter, C. F. 1958. *Elementary Seismology*. San Francisco, Calif.: W. H. Freeman.

Riera Melis, A, A. Roca and C. Olivera (with the collaboration of S. Plana and B. Martínez). 1993. "Analysis of the Pastoral Visit of 1432 to the Diocese of Girona for the study of the seismic series 1427–1428 in Catalonia," in *Historical Investigation of European Earthquakes: Materials of the CEC Project "Review of Historical Seismicity in Europe"*, ed. M. Stucchi with the collaboration of J. Vogt. Milan: CNR, 161–172.

Robert, L. 1978. "Documents d'Asie Mineure. V: Stèle funéraire de Nicomédie et séismes dans les inscriptions," *Bulletin de Correspondance Hellénique*, **102**: 395–408.

Rocchi, G. 1978. "Elementi genetici dell'architettura altomedievale armena: confronto con l'architettura medievale lombarda," in *Atti del Primo Simposio Internazionale di Arte Armena*, Bergamo, 1975. Venice: Tipolitografia armena, 555–588.

Roller, Th. 1861. *Il governo borbonico innanzi alla coscienza dell'umanità ossia i provvedimenti del governo nella tremenda catastrofe del terremoto del 16 dic. 1857*. Naples: Giuseppe Marghieri.

Romei, Annibale. 1587. *Dialogo del conte Annibale Romei gentil'huomo ferrarese. Diuiso in due giornate. Nella prima delle quali si tratta delle cause uniuersali del terremoto, e di tutte le impressioni, & apparenze, che, con stupor del volgo, nell'aria si generano. Nella seconda, del terremoto, della salsedine del mare, della via Lattea, e del flusso, e reflusso del mare s'assegnano cause particolari, diverse d'Aristotele, e da qualunque filosofo sin'ad hora ne habbi scritto*. Ferrara: Vittorio Baldini stampator ducale.

Rothman, R. L. 1968. "A note on the New England earthquake of November 18, 1755," *Bulletin of the Seismological Society of America*, **58**: 1501–1502.

Roux, G. 1932. "Notes sur les tremblements de terre ressentis au Maroc avant 1933," *Mémoires de la Société des sciences naturelles du Maroc*, **39**: 42–71.

Russo, L. 2004. *The Forgotten Revolution: How Science was Born in 300 BC and Why It Had to be Reborn*, with the collaboration of the translator, Silvio Levy. Berlin: Springer-Verlag.

Saderra Masó, M. 1895. *La seismología en Filipinas: Datos para el estudio de terremotos del archipiélago Filipino*. Manila: Observatorio de Manila.

Sanudo, Marin. *I Diarii (1496–1533)*, ed. F. Stefani *et al.*, 58 vols. Venice: Fratelli Visentini, 1879–1903.

Sarconi, M. 1784. *Istoria de' fenomeni del tremoto avvenuto nelle Calabrie, e nel Valdemone nell'anno 1783 posta in luce dalla Reale Accademia delle Scienze e delle Belle Lettere di Napoli*. Naples: Giuseppe Campo. (Repr. Catanzaro: Giuditta, 1987 and 1989.)

Sardi, Alessandro. 1586. *Discorsi. Della bellezza. Della nobiltà. Della poesia di Dante. De i precetti historici. Delle qualità del generale. Del terremoto. Di novo posti in luce*. Venice: Giolito de Ferrari.

Sarti, C. 1783. *Saggio di congetture su i terremoti*. Lucca: Francesco Bonsignori.

Satake, K., K. Shimazaki, Y. Tsuji and U. Ueda. 1996. "Time and size of a giant earthquake in Cascadia inferred from Japanese tsunami records of January 1700," *Nature*, **379**, 246–249.

Schiantarelli, P. and I. Stile. 1784. *Istoria de' fenomeni del tremoto avvenuto nelle Calabrie, e nel Valdemone nell'anno 1783 posta in luce dalla Reale Accademia delle Scienze e delle Belle Lettere di Napoli. Atlante*. Naples: Giuseppe Campo. (Repr. ed. E. Zinzi. Catanzaro: Giuditta, 1990.)

Schmidt, J. F. J. 1881. *Studien über Vulkane und Erdbeben*. Leipzig: Alwin Georgi.

Schreiner, P., ed. 1975–79. *Die byzantinischen Kleinchroniken*. (Corpus fontium historiae byzantinae, 12, 3 vols.) Vienna: Verlag der österreichischen Akademie der Wissenchaften.

Scopelliti, C., transl. 1983. *Giannozzo Manetti, De terraemotu*. Rome: ENEA, Romana Editrice.

Seeber, L. and J. G. Armbruster. 1987. "The 1886–1889 aftershocks of the Charleston, South Carolina, earthquake; a widespread burst of seismicity," *Journal of Geophysical Research*, **92**: 2663–2696.

Serpieri, A. 1873. "Rapporto delle osservazioni fatte sul terremoto avvenuto in Italia la sera del 12 marzo 1873," *Supplemento alla Meteorologia Italiana*, **1872**: 45–83.

Serpieri, A. 1876. "Documenti, nuove note e riflessioni sul terremoto della notte 17–18 marzo 1875," *Supplemento alla Meteorologia Italiana*, **1875**: 3–51.

Serva, L. 1994. "Ground effects in intensity scales," *Terra Nova*, **6**: 414–416.

Serva, L., E. Esposito, L. Guerrieri, *et al.* 2007. "Environmental effects from five historical earthquakes in southern Apennines (Italy) and macroseismic intensity assessment: Contribution to INQUA EEE Scale Project," *Quaternary International*, **173–174**: 30–44.

Ševčenko, I. 1973. "The corpus of dated Byzantine inscriptions," in *Akten des VI. Internationalen Kongresses für griechische und lateinische Epigraphik: München 1972*. Munich: Beck, 5.

Seyfart, J. F. 1756. *Algemeine Geschichte der Erdbeben*. Frankfurt: A. J. Felssecker.

Sguario, E. 1756. *Specimen physico-geometricum de terraemotu ad architecturae utilitatem concinnatum*. Venice: Giovanni Battista Recurti. (Repr. in U. Barbisan and F. Laner, *Terremoto ed architettura: il trattato di Eusebio Sguario e la sismologia nel Settecento*. Venice: Cluva università, 1983.)

Shaw, B., N. N. Ambraseys, P. C. England *et al.* 2008. "Eastern Mediterranean tectonics and tsunami hazard inferred from the AD 365 earthquake," *Nature Geoscience*, **1**: 268–276.

Shebalin, N. V., V. Kárník and D. Hadzievski. 1974. *UNDP–UNESCO Survey of the Seismicity of the Balkan Region: Catalogue of Earthquakes, Part I, 1901–1970, Part II, prior to 1901.* Skopje: UNESCO.

Sieberg, A. 1923. *Geologische, physicalische und angewandte Erdbebenkunde.* Jena: G. Fischer.

Sieberg, A. 1927. *Geologische Einführung in die Geophysik.* Jena: G. Fischer.

Sieberg, A. 1932a. "Die Erdbeben," in *Handbuch der Geophysik*, ed. B. Gutenberg. Berlin: Gebruder Borntraeger, vol. 4, 527–686.

Sieberg, A. 1932b. "Untersuchungen über Erdbeben und Bruchschollenbau im östlichen Mittelmeergebiet," *Denkschriften der medizinsch–naturwissenschaftlichen Gesellschaft zu Jena*, **18**: 161–273.

Sievers, G. R. 1868. *Das Leben des Libanius aus dem Nachlasse des Vaters hrsg. von Gottfried Sievers.* Berlin: Weidmann. (Repr. Amsterdam: Rodopi, 1969.)

Sigonio, C. 1580. *Historiarum de regno Italiae libri quindecim.* Bologna: Società tipografica bolognese.

Sigonio, C. 1591. *Historiarum de Regno Italiae libri vigenti.* Frankfurt: Andreas Erben Wechel, Claude Marne and Johann Aubry.

Skuphos, T. 1894. "Die zwei grossen Erdbeben in Lokris am 8/20 und 15/27 April 1894," *Zeitschrift der Gesellschaft für Erdkunde zu Berlin*, **24**: 409–474.

Smidt, T. C. 1970. "Tsunamis in Greek literature," *Greece and Rome*, **17**: 100–104.

Smirnov, M. V. 1931. "Katalog zemletrjasenij v Krymu," in *Izhutsenio Krima*. Pravlen: O-Va (in Russian).

Smith, W. E. T. 1966. "Earthquake of eastern Canada and adjacent areas, 1928–1959," *Publication of the Dominion Observatory, Ottawa*, **32**: 87–121.

Soggin, J. A. 1970. "Das Erdbeben von Amos 1 1 und die Chronologie der Könige Ussia und Jotham von Juda," *Zeitschrift für die alttestamentliche Wissenschaft*, **82**: 117–121.

Soloviev, S. L. 1990. "Tsunamigenic zones in the Mediterranean Sea," *Natural Hazards*, **3**: 183–202.

Soloviev, S. L., O. N. Solovieva, C. N. Go, K. S. Kim and N. A. Shchetnikov. 2000. *Tsunamis in the Mediterranean Sea 2000 BC–2000 AD.* Dordrecht: Kluwer.

Spondilis, I. 1996. "Contribution to a study of the configuration of the coast of Pylia, based on the location of new archaeological sites," in *Archaeoseismology*, ed. S. C. Stiros and R. E. Jones. Fitch Laboratory Occasional Paper 7. Athens: Institute of Geology and Mineral Exploration, The British School at Athens, 119–128.

Spratt, T. A. B. 1865. *Travels and Researches in Crete*, vol. 2. London: J. van Voorst.

Spyridon of the Laura and S. Eustratiades. 1925. *Catalogue of the Greek Manuscripts in the Library of the Laura on Mount Athos.* Cambridge, Mass.: Harvard University Press.

Stauder, W. 1962. "The focal mechanism of earthquakes," *Advances in Geophysics*, **9**: 1–76.

Stein, E. 1968. *Histoire du Bas-Empire*, vol. 1, *De l'état romain a l'état byzantin (284–476)*. Amsterdam: A. M. Hakkert.

Steinbrugge, K. V. 1982. *Earthquakes, Volcanoes and Tsunamis: An Anatomy of Hazards*. New York: Skandia America Group.

Stelluti, N. 1997. *Epigrafi di Larino e della bassa Frentania*, vol. 1, *Il repertorio*. Campobasso: Lampo.

Stiros, S. C. 1988. "Archaeology: a tool to study active tectonics – the Aegean as a case study," *Eos (Transactions of the American Geophysical Union)*, **69**: 1633, 1639.

Stiros, S. C. 2001. "The AD 365 Crete earthquake and possible seismic clustering during the fourth to sixth centuries AD in the Eastern Mediterranean: a review of historical and archaeological data," *Journal of Structural Geology*, **23**: 545–562.

Stiros, S. C. and R. E. Jones, eds. 1996. *Archaeoseismology*. Fitch Laboratory Occasional Paper 7. Athens: Institute of Geology and Mineral Exploration, The British School at Athens.

Stiros, S. C. and S. Papageorgiou. 2001. "Seismicity of Western Crete and the destruction of the town of Kisamos at AD 365: archaeological evidence," *Journal of Seismology*, **5**: 381–397.

Stoetzer, C. 1986. "El mundo ideal del Padre José de Acosta S.I. (1540–1600) 'el Plinio del Nuevo Mundo'," *Cuadernos salmantinos de filosofía*, **13**: 205–218.

Stolz, O. 1957. *Der geschichtliche Inhalt der Rechnungsbücher der Tiroler Landesfürsten von 1288–1350*. Innsbruck: Wagner.

Stover, C. W. and J. L. Coffman. 1993. *Seismicity of the United States, 1568–1989 (Revised)*. US Geological Survey Professional Paper 1527. Washington, D.C.: US Government Printing Office.

Strauss, W. 1975. *The German Single-Leaf Woodcut 1550–1600*, vol. 2. New York: Abaris Books.

Street, R. L. and A. Lacroix. 1979. "An empirical study of New England seismicity: 1727–1977," *Bulletin of the Seismological Society of America*, **69**: 159–175.

Stukeley, W. 1750. *The philosophy of earthquakes, natural and religious*. London: Charles Corbet.

Tagliapietra, A., ed. 2004. *Voltaire, Rousseau, Kant. Sulla catastrofe. L'illuminismo e la filosofia del disastro*. Milan: Bruno Mondadori.

Taramelli, T. and G. Mercalli. 1886. "I terremoti andalusi cominciati il 25 dicembre 1884," *Memorie della Reale Accademia dei Lincei*, ser. IV, **3**: 116–222.

Taramelli, T. and G. Mercalli. 1888. "Il terremoto ligure del 23 febbraio 1887," *Annali dell'Ufficio centrale meteorologico e geodinamico italiano*, ser. II, **8**, pt. 4 (1886): 331–626.

Tatevossian, R. E. 2007. "The Verny, 1887, earthquake in Central Asia: application of the INQUA scale, based on coseismic environmental effects," *Quaternary International*, **173–174**: 23–29.

Telesio, Bernardino. 1586. *De rerum natura iuxta propria principia. Libri IX*. Naples: Orazio Salviani.

Terrenzi, G. 1887. "L'inventore del sismografo a pendolo," *Rivista Scientifico-Industriale*, **19**: 52–55.

Teti, V. 2004. *Il senso dei luoghi: Memoria e storia dei paesi abbandonati*. Rome: Donzelli.

Thompson, R. C. 1937. "A new record of an Assyrian earthquake," *Iraq*, **4**: 186–189.

Tinti, S., A. Maramai and L. Graziani. 2004. "The new catalogue of the Italian tsunamis," *Natural Hazards*, **33**: 439–465.

Tobriner, S. 1982. *The Genesis of Noto: An Eighteenth Century Sicilian City*. London: A. Zwemmer.

Tobriner, S. 1983. "*La casa baraccata*: earthquake-resistant construction in 18th century Calabria," *Journal of the Society of Architectural Historians*, **42**: 131–138.

Tocher, D. and D. J. Miller. 1959. "Field observations on effects of Alaska earthquake of 10 July, 1958," *Science*, **129**: 394–395.

Traina, G. 1989. "Tracce di un'immagine: il terremoto fra prodigio e fenomeno," in *I terremoti prima del Mille in Italia e nell'area mediterranea: Storia archeologia sismologia*, ed. E. Guidoboni. Rome, Bologna, ING-SGA, 104–115.

Traina, G. 1994. "Sismicità storica delle Marche nell'antichità: Esame critico delle fonti letterarie," *Le Marche, Archeologia storia territorio* **1991/92/93**: 75–81.

Travagini, F. 1669. *Super observationibus a se factis tempore ultimorum Terraemotuum, ac potissimum Ragusiani physica disquisitio seu gyri terrae diurni indicium*. Leiden. (2nd edn Frankfurt, 1673.)

Trifunac, M. D. and A. G. Brady. 1975. "On the correlation of seismic intensity scales with the peaks of recorded strong ground motion," *Bulletin of the Seismological Society of America*, **65**: 139–162.

Turner, H. H. 1922. "On the arrival of earthquake waves at the antipodes, and on the measurement of the focal depth of an earthquake," *Monthly Notices of the Royal Astronomical Society, Geophysical Supplement*, **1**: 1–13.

Turyn, A. 1980. *Dated Greek Manuscripts of the Thirteenth and Fourteenth Centuries in the Libraries of Great Britain*. Dumbarton Oaks Studies 17. Washington, D.C.: Dumbarton Oaks Center for Byzantine Studies.

Udías, A. 1986. "José de Acosta (1539–1600) un pionero de la geofísica," *Jesuitas*, **9**: 22–24.

Udías, A. 1999. "Origenes de la sismologia en España," in *100 años de observaciones sismológicas en San Fernando 1898–1998*, Conferencias y trabajos presentados, eds. J. Martín and A. Pazos, *Boletín del Real Instituto y Observatorio de la Armada*, **5**: 11–17.

Udías, A. 2003. *Searching the Heavens and the Earth: The History of Jesuit Observatories*. Dordrecht: Kluwer.

Udías, A. and W. Stauder. 1996. "Jesuit contribution to seismology," *Seismological Research Letters*, **67** (3): 10–19. (Repr. in *International Handbook of Earthquake and Engineering Seismology*, eds. W. H. K. Lee, H. Kanamori, P. C. Jennings and C. Kisslinger. Amsterdam: Academic Press, 19–27.)

Utsu, T. 2002. "A list of deadly earthquakes in the world: 1500–2000," in *IASPEI Handbook of Earthquake Engineering and Seismology*, Part A, eds. W. H. Lee, H. Kanamori, P. Jennings and C. Kisslinger. Amsterdam; London: Academic Press, 691–717.

Valensise, G. and E. Guidoboni. 1995. "Verso nuove strategie di ricerca: zone sismogenetiche silenti o silenzio delle fonti?," in E. Boschi, G. Ferrari, P. Gasperini, E. Guidoboni, G. Smriglio and G. Valensise, *Catalogo dei forti terremoti in Italia dal 461 a.C. al 1980*, Rome, Bologna: ING–SGA, 112–127.

Valensise, G. and E. Guidoboni. 2000a. "Earthquake effects on the environment: from historical descriptions to thematic cartography," in *Catalogue of Strong Italian Earthquakes from 461 BC to 1997*, eds. E. Boschi, E. Guidoboni, G. Ferrari, D. Mariotti, G. Valensise and P. Gasperini, special issue of *Annali di Geofisica*, **43**: 747–763.

Valensise, G. and E. Guidoboni. 2000b. "Towards new research strategies: silent seismogenic areas or silent sources?," in *Catalogue of Strong Italian Earthquakes from 461 BC to 1997*, eds. E. Boschi, E. Guidoboni, G. Ferrari, D. Mariotti, G. Valensise and P. Gasperini, special issue of *Annali di Geofisica*, **43**: 797–812.

Valensise, G. and D. Pantosti, eds. 2001. "Database of potential sources for earthquakes larger than M 5.5 in Italy (DISS version 2.0), *Annali di Geofisica*, **44**(4), Suppl. 1: 797–964 (with database on CD-ROM).

Valensise, G., R. Basili, M. Mucciarelli and D. Pantosti, eds. 2002. *Database of Potential Sources for Earthquakes Larger than M 5.5 in Europe, a compilation of data collected by partners of the EU project FAUST*. Available at http://legacy.ingv.it/~roma/banche/catalogo_europeo/

Vallerani, M. 1995. "Urban decay and abandonment in western medieval Sicily: a problem for the study of strong earthquakes," in *Earthquakes in the Past: Multidisciplinary Approaches*, eds. E. Boschi, R. Funiciello, E. Guidoboni and A. Rovelli, special issue of *Annali di Geofisica*, **38**: 691–703.

Vannucci, G. 1787. *Discorso istorico-filosofico sopra il tremuoto che nella notte del di 24 venendo il 25 dicembre dell'anno 1786 dopo le ore 9 d'Italia scosse orribilmente la Città di Rimini, e varj Paesi vicini*, 3rd edn. Cesena: Gregorio Biasini.

Vannucci, G. 2000. "Individuazione di strutture attive nell'Appennino centro settentrionale sulla base di dati macrosismici storici" Ph.D. thesis, University of Camerino, Italy.

Vannucci, G., P. Gasperini and M. Boccaletti. 1999a. "Microzonation of the City of Florence by the analysis of the macroseismic data," in *EGS 24th General Assembly*, 19–23 April 1999, Hague (The Netherlands), *Geophysical Research Abstract*, 1, part IV, 841.

Vannucci, G., P. Gasperini, G. Ferrari and E. Guidoboni. 1999b. "Encoding and computer analysis of macroseismic effects," *Physics and Chemistry of the Earth*, **24**: 505–510.

Vitaliano, D. B. 1973. *Legends of the Earth: Their Geologic Origins*. Bloomington, Ind.: Indiana University Press.

Vivenzio, G. 1783. *Istoria e teoria de' tremuoti in generale ed in particolare di quelli della Calabria, e di Messina del 1783*. Naples: Stamperia regale.

Vivenzio, G. 1788. *Istoria de' tremuoti avvenuti nella provincia della Calabria ulteriore, e nella città di Messina nell'anno 1783 e di quanto nella Calabria fu fatto per lo suo risorgimento fino al 1787. Preceduta da una teoria, ed istoria generale de tremuoti*, 2 vols. Naples: Stamperia regale.

Vogt, J. 1977a. "Archives et géologie appliquée: séismes, glissements, éboulements, érosion anthropique," *La Gazette des archives*, n.s., **98**(3): 131–136.

Vogt, J. 1977b. "L'apport des archives à la connaissance géologique de Vaucluse: glissements, éboulements, séismes," *Etudes vauclusiennes*, **18**: 1–7.

Vogt, J. 1984a. "Mouvements de terrain associés aux séismes dans les Pyrénées," *Revue géographique des Pyrénées et du Sud-Ouest*, **55**(1): 49–56.

Vogt, J. 1984b. "Mouvements de terrain associés aux séismes en Afrique du Nord," *Méditerranée*, **1–2**: 43–48.

Vogt, J. 1987. "Un tremblement de terre énigmatique: le séisme de 29 novembre 1784, vers 22 h," *Bulletin de la Société belfortaine d'émulation*, 21–26.

Vogt, J. 1993. "Effets naturels des séismes: l'exemple du séisme antillais de 1690," *Revue de géomorphologie dynamique*, **42**(1): 11–14.

Vogt, J. 2005. "Review of Pierre Gouin, *Tremblements de terre 'historiques' au Québec*," *Annals of Geophysics*, **47**: 907–908.

Vogt, J. and N. N. Ambraseys. 1991. "Matériaux relatifs à la sismicité de l'Algérie occidentale au cours de la deuxième moitié du XIXe et au début du XXe siècle," *Méditerranée*, **74**(4): 39–45.

Vogt, J. and N. N. Ambraseys. 1992. "The seismicity of Algeria during the first half of the eighteenth century," in *Proceedings of the Workshop on Archaeoseismicity in the Mediterranean Region*, Damascus, 90–95.

Vogt, J. and G. Grünthal. 1994. "Die Erdbebenfolge vom Herbst 1612 im Raum Bielefeld: Ein bisher unberücksichtigtes Schadenbeben," *Geowissenschaften*, **12**(8): 236–240.

Volger, G. H. O. 1856. "Untersuchungen über das letzjährige Erdbeben in Central-Europe," *Mitteilungen aus Justus Perthes' geographischer Anstalt (Petermann's geographische Mitteilungen*, 1856), **2**(3): 85–102.

Voltaire. *Candide, ou l'optimisme*, in *The complete works of Voltaire (Les oeuvres complètes de Voltaire)*, vol. 48 (ed. by W. Barber), édition critique par R. Pomeau. Oxford: Voltaire Foundation at the Taylor Institution, 1980.

Voltaire. *Poème sur le désastre de Lisbonne*, in F.-M. Arouet, *Voltaire, Œuvres complètes*, vol. 12. Paris: Institut de France, 1876, 509–523.

von Hoff, K. E. A. 1840–41. *Chronik der Erbeben und Vulcan-Ausbrüche*, 2 vols. Gotha: Justus Perthes.

Wadati, K. 1928. "Shallow and deep earthquakes," *Geophysical Magazine*, **1**: 162–202.

Wald, D. J., V. Quitoriano, L. A. Dengler and J. W. Dewey. 1999a. "Utilization of the Internet for rapid community intensity maps," *Seismological Research Letters*, **70**: 680–693.

Wald, D. J., V. Quitoriano, T. H. Heaton and H. Kanamori. 1999b. "Relationships between peak ground acceleration, peak ground velocity, and modified Mercalli intensity in California," *Earthquake Spectra*, **15**: 557–564.

Waldherr, G. H. 1997. *Erdbeben, das aussergewöhnliche Normale: zur Rezeption seismischer Aktivitäten in literarischen Quellen vom 4. Jahrhundert v. Chr. bis zum 4. Jahrhundert n. Chr.* Geographica historica 9. Stuttgart: F. Steiner.

Ward-Perkins, B. 1984. *From Classical Antiquity to the Middle Ages: Public Building in Northern and Central Italy AD 300–850*. Oxford: Clarendon Press.

Wells, D. L. and K. J. Coppersmith. 1994. "New empirical relationships among magnitude, rupture length, rupture width, rupture area, and surface displacement," *Bulletin of the Seismological Society of America*, **84**: 974–1002.

Westerbergh, U. 1956. *Chronicon Salernitanum*, A critical edition with Studies on Literary and Historical Sources and on Language. Stockholm: Almqvist and Wiksell.

Wharton, W. J. L. and F. J. Evans. 1888. "On the seismic sea waves caused by the eruption of Krakatoa, August 26th and 27th, 1883," Part III of *The Eruption of Krakatoa and Subsequent Phenomena: Report of the Krakatoa Committee of the Royal Society*, ed. G. L. Symons. London: Trübner & Co., 89–151.

Whiston, W. 1696. *A New Theory of the Earth, from its Original, to the Consummation of all Things*. London: R. Roberts.

Whitman, R. V. 2002. "Ground motions during the 1755 Cape Ann earthquake," in *Proceedings of the 7th US National Conference on Earthquake Engineering, Urban Earthquake Risk (CD-ROM)*. Oakland, Calif.: Earthquake Engineering Research Institute.

Willis, B. 1928. "Earthquakes in the Holy Land," *Bulletin of the Seismological Society of America*, **18**: 72–103.

Wilson, A. T. 1930. "Earthquakes in Persia," *Bulletin of the School of Oriental Studies*, **6**(1): 103–131.

Winkler, J. H. 1745. *Die Eigenschaften der electrischen Materie und des electrischen Feuers aus verschiedenen neuen Versuchen erkläret, und, nebst etlichen neuen Maschinen zum Electrisiren*. Leipzig: Breitkopf. (Repr. Leipzig: Zentralantiquariat der DDR, 1983.)

Winthrop, J. II. 1757–58. "An account of the earthquake felt in New England, and the neighbouring parts of America, on the 18th of November 1755," *Philosophical Transactions*, **50**: 1–18.

Wirth, P. 1966. "Zur 'byzantinischen' Erdbebenliste," *Byzantinische Forschungen*, **1**: 393–399.

Wood, H. O. and F. Neumann. 1931. "Modified Mercalli intensity scale," *Bulletin of the Seismological Society of America*, **21**: 277–283.

Wood, W. R. and D. L. Johnson. 1978. "A survey of disturbance processes in archaeological site formation," in *Advances in Archaeological Method and Theory I*, ed. M. B. Schiffer. New York: Academic Press, 315–381.

Woodward, J. 1695. *An Essay toward a Natural History of the Earth*. London: Richard Wilkin.

Xac'ikyan, L. A. 1955–67. *Records of 15th Century Armenian Manuscripts* I [1955]; II [1958]; III [1967]. Erevan: Publishing House of the Academy of Sciences of Armenian SSR (in Armenian).

Yadin, Y. 1961. "Ancient Judean weights and the date of Samaria ostraca," *Scripta Hierosolymitana*, **8**: 9–25.

Yeomans, D. K. 1991. *Comets: A Chronological History of Observation, Science, Myth, and Folklore*. New York: John Wiley.

Yokoyama, I. 1978. "The tsunami caused by the prehistoric eruption of Thera," in *Thera and the Aegean World II*, vol. 2, ed. C. Doumas. London: The Thera Foundation, 277–283.

[Yovsevp'ean] G.(aregin I kat'olikos) 1951. *Records of the Manuscripts*, vol. 1, *From the 5th Century to 1250*. Ant'ilias (in Armenian).

Zarian, A. 1992. "L'effetto statico-dinamico in alcuni edifici paleocristiani armeni," in *Atti del Quinto Simposio Internazionale di Arte Armena, Venezia 1988*. Venice: Tipolitografia armena, 727–738.

Zuccolo, Gregorio. 1571. *Del terremoto. Trattato di M. Gregorio Zuccolo nobil fauentino, nel qual si vede intorno a questo movimento opinion diversa dall'altre publicate fin qui.* Bologna: Alessandro Benaccio.

Zupo, N. 1784. *Riflessioni su le cagioni fisiche dei tremuoti accaduti nelle Calabrie nell'anno 1783*. Naples: Giuseppe–Maria Porcelli.

Index